Wide Band Gap Semiconductor Nanowires 1

Series Editor
Robert Baptist

Wide Band Gap Semiconductor Nanowires 1

Low-Dimensionality Effects and Growth

Edited by

Vincent Consonni
Guy Feuillet

iSTE

WILEY

First published 2014 in Great Britain and the United States by ISTE Ltd and John Wiley & Sons, Inc.

ISTE Ltd
27-37 St George's Road
London SW19 4EU
UK

www.iste.co.uk

John Wiley & Sons, Inc.
111 River Street
Hoboken, NJ 07030
USA

www.wiley.com

Library of Congress Control Number: 2014941789

British Library Cataloguing-in-Publication Data
A CIP record for this book is available from the British Library
ISBN 978-1-84821-597-9

MIX
Paper from responsible sources
FSC www.fsc.org FSC® C013604

Printed and bound in Great Britain by CPI Group (UK) Ltd., Croydon, Surrey CR0 4YY

Contents

Preface

This book is devoted to the specific case of wires obtained from a given kind of semiconductors, namely the semiconducting materials with a direct and wide band gap (WBG). In short, semiconductors are considered as WBG semiconducting materials if their band gap energy is typically above 1.5/1.6 eV. The interest of these materials for optoelectronic devices lies in the fact that they are well-adapted for emission, detection or absorption processes in most of the visible range, and part of the UV range as well. From the more basic point of view, the large refractive index and high exciton binding energy as well as the strong photon/exciton interactions give rise to long sought effects such as polariton lasing at room temperature for instance. The two main materials composing the family of WBG semiconducting materials are GaN and ZnO. They have close band gap energy in the near UV region (i.e., around 3.3/3.4 eV), and have in common that their cationic alloys span the visible as well as the UV range (and also part of the near IR region for In-rich GaInN alloys). More importantly, they both crystallize, in standard conditions, in the strongly anisotropic wurtzite crystalline phase, leading to a large number of similar physical quantities such as lattice parameters and piezoelectric constants and of similar physical processes related for instance to polarity.

GaN and its alloys are now well-mastered and used in a flurry of industrial applications as optoelectronic devices. On the other hand, ZnO is less advanced in terms of industrial applications and its development is mainly hampered by the difficulty for controlling p-type doping. However, ZnO has a stronger exciton binding energy than GaN (60 meV vs. 25 meV) and also a stronger oscillator strength. GaN and related alloys are generally heteroepitaxially grown on foreign substrates since low-cost nitride substrates with large dimensions are still not available. In contrast, ZnO and related alloys can homoepitaxially be grown onto ZnO substrates with excellent structural properties but still with limited availability and sizes. Therefore, epitaxial growth is mostly carried out heteroepitaxially for both kinds of materials, typically yielding epitaxial planar layers with a high density of structural defects. If such WBG semiconducting materials with a rather poor structural quality are actually used for some optoelectronic devices such as commercial LEDs for the moment, the improvement of their overall structure would certainly be beneficial for additional potential optoelectronic devices but also for the understanding of the physical processes at stake in these devices.

The need for WBG semiconducting materials with better structural quality is one of the main reasons that propelled (nano)wires to their present day status in the field of semiconductor research: when grown onto foreign substrates, and as for the case of planar layers, wires can relax the elastic strain energy originating from large lattice mismatch by forming misfit dislocations. But these lie in the basal plane or bend towards the nearby lateral surfaces of wires, thus leaving defect-free materials in their core. This process whereby dislocations can bend towards the lateral growth front had been demonstrated beforehand in epitaxial lateral overgrowth (ELO).

The second reason behind the development of WBG semiconductor wires – considered for a long time as the unwanted result of wrong growth conditions when trying to synthesize 2-dimensional (2D) epitaxial layers– is related to the increasing interest for low-dimensionality objects, typically of sub-micron or nanometer size. The specific structural, optical, and electronic properties of these low-dimensionality objects open new opportunities for nanoscale optoelectronic devices, especially to fully exploit the strong photon/exciton interactions. As an example, wires allow for a full confinement of light in their section with free propagation along their axis. Such physics and the related optoelectronic applications are nonetheless limited by the large developed surfaces of the wires, for which surface passivation is for instance required in order to prevent light diffusion. Because of the presence of surface states, Fermi level pinning also leads to band bending affecting the carrier mobility along the wires and resulting in possible carrier trapping. In return, this specific property makes wires very invaluable objects to investigate surface effects in WBG semiconductors and can also be beneficial in photodetection applications.

Looking back in time, the first demonstration of semiconductor wire growth was achieved by the pioneering work of Wagner and Ellis in 1964 according to the vapor-liquid-solid (VLS) mechanism [WAG 64]. In the field of WBG semiconducting materials for optoelectronic devices, which are the materials that we are interested in in this book, one of the first "nano"objects that were looked into were dots, named quantum dots when the typical dimensions are smaller than the De Broglie's wavelength, inserted as they were in 2D epitaxial layers. For instance, the dots can be grown according to the so-called Stransky-Krastanov mode owing to the elastic stress relaxation processes at play in lattice mismatched heteroepitaxial systems. This is nevertheless limited somehow to heteroepitaxial layers in a state of compressive strain, and of medium lattice mismatch range (typically a few percent). For one heteroepitaxial system, such dots have once and for all a fixed size given by the nature of the involved materials. Thus, one had to think of other possibilities for making sub-micron or nano objects with an easier control over their sizes and shapes. Instead of playing for instance with strain to form dots, the easier way to grow low-dimensionality structures is to try and depart from the 2D growth conditions, thereby changing the atomic diffusion and incorporation processes, hence using growth modes different from the usual 2D mode. This time, this leads to the controlled formation of 1D objects, now referred to as nanowires, microwires or more generally wires, depending on their lateral dimensions, or also as nanocolumns, nanorods or microrods.

Interestingly, in terms of growth conditions, while most of the semiconductor (i.e., Si, Ge, arsenides, phosphides, ...) wires can exclusively be grown by VLS or vapor-solid-solid mechanisms in the bottom-up approach, one of the most amazing properties of GaN and ZnO is their ability to grow in the form of wires following catalyst-free approaches (i.e., self-induced growth, spontaneous growth, ...). These catalyst-free approaches are expected to reduce potential contamination into the wires and, more importantly, offer new valuable growth modes with great potentiality for optoelectronic devices. The first demonstrations of GaN and ZnO wire growth were shown in 1998 by molecular beam epitaxy [YOS 97, SAN 98] and in 2001 by vapor phase transport [HUA 01, PAN 01] and in solution [VAY 01], respectively. Basically, GaN wires can mainly be grown by molecular beam epitaxy and metal-organic chemical vapor deposition. In contrast, ZnO wires can additionally be deposited by vapor phase transport, pulsed-laser deposition or more specifically in solution *via* the low-cost and low-temperature chemical bath deposition technique for instance.

As discussed above, growing wires with dedicated properties in a reproducible way requires a good control of the growth conditions. When it comes to radial as well as axial heterostructures grown around or on top of the wires, things are somehow more complicated, since growth conditions very often have to be moved from the initial 1D case in order to stack the layers on top of each other. As in the case of any kind of heterostructures, managing the lattice mismatch issue may also be essential. This does depend upon the sizes involved and may potentially lead to the generation of misfit dislocations at the interfaces between the constituting layers. Moreover, owing to the specific geometry of the wires, other types of defects may also be introduced, such as stacking faults or inversion domain boundaries for instance, the origin of which has to be identified in order to better limit their occurrence. In return, identifying the right conditions for growing heterostructures with a good structural quality opens up a flurry of applications in the field of optoelectronics. These will benefit not only from the wave guiding properties of the wires (i.e., specific optical modes) but also from the control over the density of defects into the wires, leading to a decrease in the number of non-radiative recombination centers. These applications also take advantage of the larger surface to volume ratio at low-scale dimensions, leading for instance to much larger emitting or absorbing surfaces than in 2D layers or to efficient photodetectors.

The book has been organized along the lines of these introductory remarks.

Accordingly, it is the aim of the first part of volume 1 to focus on the specific properties of WBG semiconductor wires, in order to point out what differentiates these objects from their 2D counterparts. This appears as a necessary step in order to point out what these specificities could bring for the physics and applications of WBG semiconductors in the field of optoelectronics. It is nonetheless also the aim of this first part to try and pin-point the present day limitations associated with the use of WBG semiconductor wires, in order to draw possible solutions for a thorough use of these 1D objects. As for the second part of volume 1, it is dedicated to the different growth methods for the deposition of GaN and ZnO wires, stressing the mechanisms at play for the nucleation and growth of these 1D objects. The most interesting growth methods are discussed in detail with a special emphasis on the necessary ingredients to spontaneously grow GaN and ZnO wires. In volume 2, the first part aims at reviewing the different axial or radial heterostructures that can be integrated into GaN and ZnO wires. This is done to address relevant potential optoelectronic applications including LEDs, lasers, UV photodetectors and solar cells, which are presented and discussed in the second part of volume 2.

As revealed by the very numerous publications, the subject is far from being closed and new results emerge at a quick pace. With this in mind, this book is intended to give the reader a detailed overview of the current status of research in the field of WBG semiconductor wires for optoelectronic devices. As announced in the very title of this book, the choice was deliberately made to intermix chapters devoted to GaN and ZnO wires: the two materials have a lot in common, and the two communities will gain from mutual exchanges.

We hope that the reviews presented here by pioneering and world-leading scientists in the field, the discussion on the chemistry, physics, and applications of WBG semiconductor wires, together with the comparison between the two kinds of materials and between the different growth methods will be a useful source of information not only for the new comers in the field, but also for the already involved engineers and scientists who seek a detailed overview of the subject to give their work a new impulse.

Finally, we would like to warmly thank all our friends and colleagues who took part in this book project to create a lively, fruitful and high level place on the hot topic of WBG semiconductor wires.

Vincent CONSONNI
Guy FEUILLET
June 2014

Bibliography

[WAG 64] WAGNER R.S., ELLIS W.C., *Appl. Phys. Lett.*, 4, 89 (1964).

[YOS 97] YOSHIZAWA M., KIKUCHI A., MORI M., *et al.*, *Japanese J. Appl. Phys.*, 36, L459 (1997).

[SAN 98] SANCHEZ-GARCIA M.A., CALLEJA E., MONROY E., *et al.*, *J. Cryst. Growth,* 183, 23 (1998).

[HUA 01] HUANG M.H., MAO S., FEICK H., *et al.*, *Science*, 292, 1897 (2001).

[PAN 01] PAN Z.W., DAI Z.R., WANG Z.L., *Science*, 291, 1947 (2001).

[VAY 01] VAYSSIERES L., KEIS K., LINDQUIST S.E., *et al.*, *J. Phys. Chem.,* B 105, 3350 (2001).

Part 1

GaN and ZnO Nanowires: Low-dimensionality Effects

1

Quantum and Optical Confinement

1.1. Introduction

GaN and ZnO nanowires (NWs) are a fascinating photonics platform with the combination of one-dimensional (1D) structural geometry and the remarkable electronic and optical properties of wide band gap semiconductors [TAK 07, MOR 08]. GaN and ZnO have a direct gap ~3.4 eV at room temperature and an exciton oscillator strength, roughly two orders of magnitude larger than in GaAs [GIL 97, KLI 10], one of the most popular semiconductors for optoelectronics. Their surface recombination velocity is comparable to most of the other semiconductors, ~ 10^4–10^5 cm/s [ZHA 08, LIN 06, ALE 03], therefore, some care should be taken to manage unavoidable surface effects in NW structures. For further discussion on this topic, please refer to Chapter 3 of this book for GaN NWs and to Chapter 4 for ZnO NWs.

GaN and ZnO NWs are usually grown along the *c*-axis of the wurtzite crystal structure, with six hexagonal facets, tens to hundreds of nm in diameter and around 1 μm in length. These dimensions are orders of magnitude larger than the exciton Bohr radius ~2.8 nm in GaN and ~1.8 nm in ZnO; thus, electronic quantum confinement can only be obtained through material modulation, either along the growth axis with axial heterostructures or along the radial direction with core-shell heterostructures. This will be the subject of the different chapters in Volume 2 [CON 14]. However, NW diameters are of the same order or smaller than optical wavelengths in the near band edge region (~150 nm, assuming 2.5 refractive index); as a result, light is naturally confined in the NW cross-section plane and freely propagative along the NW length. This 1D wave guiding effect turns out to be a highly efficient way to extract or absorb light in a medium of high refractive index such as GaN and ZnO. In this chapter, quantum and optical confinement in NWs

---

Chapter written by Le Si DANG.

will be examined with three photonics topics of great promise: (1) all-optical integrated circuits with Bose exciton polaritons, involving 1D photon modes interacting with bulk excitons in the so-called strong coupling regime of the light-matter interaction; (2) high efficiency single photon sources (SPSs) for quantum information processing, based on single quantum dots (QDs) axially embedded in 1D photonic wires; (3) high efficiency photovoltaics with core–shell NW arrays.

1.2. All-optical integrated circuits with Bose exciton polaritons

All-optical networks were developed more than 20 years ago to overcome the electronic bottleneck of a few gigabit/s, with extra benefits of lower energy consumption and lower loss. For higher speeds, up to terabit/s, the signal must remain photonic all along its path using optical switching and routing. Recently, various all-optical devices based on exciton polaritons have been demonstrated at low temperature ~10 K, with very promising performance in terms of speed and control power [CER 13, BAL 13, NGU 13, STU 14].

Exciton polaritons (or polaritons) in semiconductors are bosonic quasi-particles resulting from the strong interaction between photon and exciton modes [HOP 65, KAV 03]. Due to their composite half-light half-matter nature, polaritons possess a unique combination of physical properties. From the half-light part, they can be easily manipulated by standard optical spectroscopy techniques. They can travel at high speed in semiconductors (~1% speed of light, [FRE 00]), and feature an ultralight mass (~10^{-4}–10^{-5} the electron mass), favoring transition to quantum-condensed phases at elevated temperatures, e.g. Bose–Einstein condensation [KAS 06]. On the other hand, the half-exciton part brings in Coulombic interaction, spin polarization, and optical nonlinearities that could initiate many physical situations of interest for photonics, e.g. parametric scattering, quantum correlation and entanglement [POR 10].

For practical control and manipulation, polaritons are preferably generated in heterostructures combining both photonic and electronic confinements, the most popular one being a (two-dimensional) 2D photonic microcavity embedding a few 2D quantum wells [KAV 03, WEI 92]. Such a device has been used to generate the first low-dimensional polaritons in 1992 [WEI 92] and to achieve their Bose–Einstein condensation in 2006 [POR 10]. Shortly after, a new "polaritronics" field emerged prospecting for innovative photonic applications based on exciton polaritons [DEV 08, ESP 13]. Indeed, polariton potentialities for all-optical operations have been convincingly demonstrated for ultrafast (on the picosecond time scale) and low power (on the fJ energy scale) spin switch [CER 13], transistor [BAL 13], resonant tunneling diode [NGU 13] and giant phase shift in Mach–Zehnder interferometer [STU 14]. The proof of principle of the polariton transistor by Ballarini *et al.* is depicted in Figure 1.1

[BAL 13]. The sample used is a GaAs-based planar microcavity embedding three InGaN quantum wells. A monomode laser is tuned to inject (at low level) two types of polaritons in the microcavity: control polaritons with wave vector K_C by in-resonance pumping and address polaritons with wave vector K_A by slightly off-resonance pumping, as shown in Figure 1.1(b). The low population of address polaritons corresponds to the OFF state of the transistor. By increasing the laser power, one can induce a global polariton blue shift through the polariton-polariton interaction [KAV 03], and bring address polaritons into resonance with the laser. This triggers a sharp increase in their population (Figure 1.1(c)), switching the transistor to the ON state. In this demonstration device, the gain I_A/I_C, defined as the population ratio of address and control polaritons, is around 15. Furthermore, Ballarini *et al.* show that their polariton all-optical transistor is fully cascadable, and can be used as a building block for AND/OR logic gates [BAL 13]. With a fast switching time ~10 ps (polariton lifetime) and a control energy as low as ~1 fJ, this work is a very important step toward the development of all-optical networks based on exciton polaritons.

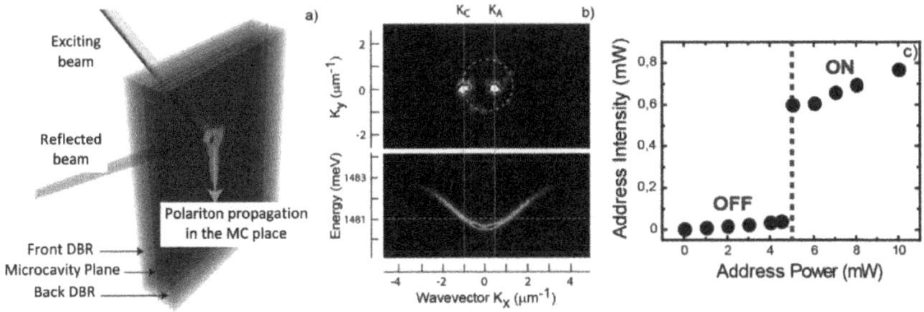

Figure 1.1. *Demonstration of all-optical polariton transistor at 10 K. a) Experimental configuration: polaritons are optically injected into the planar Fabry–Perot microcavity using a cw-laser. During their lifetime ~10 ps, they propagate in the microcavity plane over 50–100 µm. The (relative) polariton population can be extracted by measuring its photoluminescence intensity. b) Bottom panel: polariton dispersion (E, K_x) visualized by photoluminescence imaging at low pumping laser power. The horizontal dotted line indicates the pumping laser energy used in this experiment, and the two vertical solid lines the wave vectors K_C and K_A of control and address polaritons injected into the microcavity, respectively. The laser is in resonance with control polaritons and slightly detuned with address polaritons. Top panel: same data as in bottom panel, but displayed in wave vector space. c) Photoluminescence intensity of address polaritons as a function of the pumping laser power, featuring the ON and OFF states of the polariton transistor. (Reprinted with permission from Ballarini et al. [BAL 13]. Copyright 2013 Nature Publishing Group). For a color version of this figure, see www.iste.co.uk/consonni/nanowires1.zip*

Interests of wide band gap semiconductors

All the above "polaritronics" works have been carried out at low temperatures (~10 K), using GaAs-based 2D microcavities and 1D wave guiding structures etched out of planar microcavities. Extension to higher temperatures calls for similar structures made of materials with larger exciton binding energies, such as wide band gap semiconductors. In fact, GaN- and ZnO-based 2D microcavities have been successfully used to realize polariton "lasing" at room temperature [CHR 08, LU 12, LI 13]. However, their spatial homogeneities are still too limited for polariton propagation, a desirable feature for practical applications. In this context, readily available photonic alternatives for all-optical polariton networks are as grown NWs of wide band gap semiconductors.

NWs with wavelength size diameters can sustain ID optical modes, with full confinement of light in the NW cross-section plane and free propagation along the NW length axis. Owing to the large refractive index ~2.5 in GaN and ZnO, standing optical waves called whispering gallery modes (WGMs) can be formed by total internal reflection at the semiconductor–air interface, as sketched in Figure 1.2.

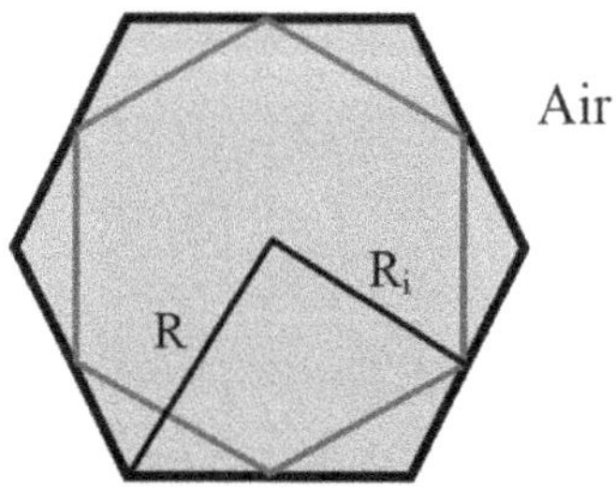

Figure 1.2. *Cross-section of a hexagonal nanowire of radius R. The gray line represents the path of a whispering gallery mode confined by total internal reflection at the semiconductor–air interface. For a color version of this figure, see www.iste.co.uk/consonni/nanowires1.zip*

The gray line represents the path of such WGM. In a simple plane wave model, energies E of WGMs in a hexagonal cavity of radius R (>> optical wavelength) are given by [NOB 04]:

$$6R_i = \frac{hc}{nE}\left[N + \frac{6}{\pi}\arctan(\beta\sqrt{3n^2 - 4}\right]$$

where N is the mode number, R_i = sqrt(3) R/2, n is the refractive index, $\beta = 1/n$ for transverse magnetic (TM) polarization (electric field E is parallel to c-axis), $= n$ for transverse electric (TE) polarization (E perpendicular to c-axis), h is the Planck's constant and c is the speed of light in vacuum. WGMs with large mode numbers N are well confined in NWs. For example, the spatial intensity distribution of modes with N ~15 is more than 95% inside NWs of radius ~ 500 nm, which explains the strong interaction with bulk excitons, yielding Rabi splittings Ω_{Rabi} nearly as large as in the ideal case of light matter interaction in bulk materials [KAL 07].

WGM polaritons resulting from the strong coupling between WGMs and bulk excitons have been evidenced at room temperature in ZnO [SUN 08, TRI 11] and GaN microwires [TRI 12]. Their 1D nature has been assessed by angle-resolved measurements of their far-field photoluminescence (Figure 1.3, [TRI 11]). In Figure 1.3(a), the emission direction is defined by two angles θ and Φ, that are one-to-one related to wave vectors k_z along the *c*-axis and k_x in the cross-section plane by $k_z = E\sin(\theta)/\hbar c$ and $k_x = E\sin(\Phi)/\hbar c$, respectively. As shown in Figures 1.3(b) and (c), several polariton modes are observed in the near band edge region 3.1 – 3.3 eV of ZnO microwires, with $E(k_z = 0) \neq 0$ corresponding to WGMs of different mode numbers N and polarizations TE/TM. The strong polariton dispersion around $k_z \sim 0$ in Figure 1.3(b) can be approximated by a parabola and an ultralight effective mass ($\sim 10^{-4}$ the electron mass), related to a very low density of states ($\sim 10^{-4}$ the exciton density of states) as well as a very fast group velocity $1/h\ \delta E(k_z)/\delta k_z$ ($\sim$ 1% the speed of light [FRE 00]). A remarkable consequence of this strong dispersion is the quenching of acoustic phonon scatterings as long as $k_BT < \Omega_{Rabi}/2$ [SAV 97, BOR 00]. This dispersion gradually flattens out when it gets closer to the excitonic resonances X_A, X_B and X_C at larger k_z, reflecting the anticrossing behavior of the strong coupling regime [KAV 03, WEI 92]. However, polariton dispersions along Φ (or k_x) in Figure 1.3(c) are rigorously flat, as expected for a total confinement in the cross-section plane.

The combination of large exciton oscillator strength and strong confinement of WGMs contributes to Rabi splittings up to 115 meV in GaN [TRI 12], and 200 meV in ZnO NWs [TRI 11], generating a kind of potential trap in *k*-space, as shown in Figure 1.3(b).

Then, in addition to the quenching of the acoustic phonon interaction mentioned above [SAV 97, BOR 00], longitudinal optical (LO) phonon interaction could also be inhibited if the trap is deep enough. Specifically, when the energy separation between the trap bottom and the exciton states at large k_z is larger than the LO phonon energy, scatterings of $k_z \sim 0$ polaritons by LO phonon would only involve final polariton states in the trap. The strength of such scattering events is several orders of magnitude weaker than the usual exciton–LO phonon scatterings, because of the very low polariton density of states ($\sim 10^{-4}$ exciton density of states). Quenching of LO phonon scatterings is clearly evidenced in ZnO NWs, in which the trap depth is $\sim \Omega_{Rabi}/2 \sim 100$ meV > LO phonon energy ~ 72 meV [TRI 11].

In Figures 1.3(b) and (c), the polariton full-width-at-half maximum is ~ 4 meV at room temperature, as compared to ~ 40 meV broadening of exciton resonance in bulk ZnO [KLI 07]. This complete phonon quenching is shown to persist up to 550 K [ZHA 12]. For GaN NWs, the situation is less favorable with $\Omega_{\text{Rabi}}/2$ ~ 57 meV < LO phonon energy ~ 92 meV. Nevertheless, WGMs with negative detunings ~ –20 meV with respect to exciton resonances can yield polariton traps that are deep enough to satisfy the LO phonon quenching condition. The full-width-at-half-maximum of these polaritons slightly increases from 6.5meV at 5 K to 7.5 meV at room temperature, attesting the severe reduction of phonon scatterings [TRI 12]. With GaN and ZnO WGM polaritons, we have a unique situation in the solid state of complete decoupling from the surrounding lattice vibrations as a result of ultrastrong coupling to photons.

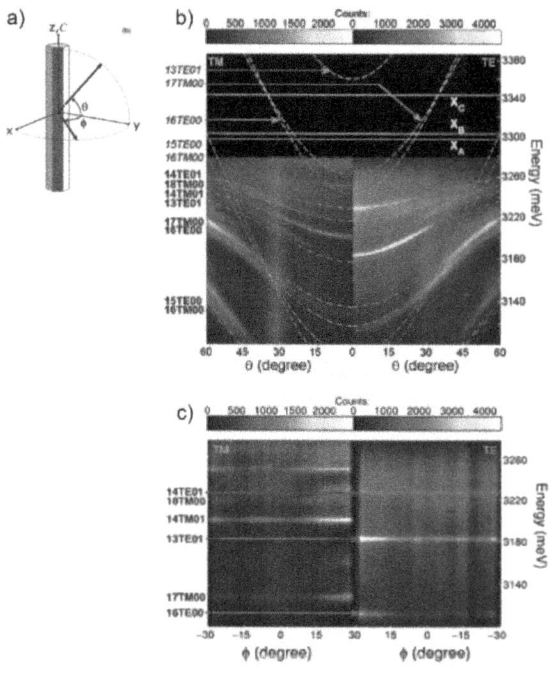

Figure 1.3. *Polariton dispersion in a single ZnO microwire. Angle-resolved far-field photoluminescence measured at 300 K from a single ZnO microwire. a) The two angles θ and ϕ characterizing the microwire far-field emission are one-to-one related to wave vectors kz along the c-axis and kx in the cross-section plane by $k_z = E \sin(\theta)/hc$ and $k_x = E \sin(\phi)/hc$, respectively. b) Far-field photoluminescence along the angle θ analyzed for the polarization TM (left) and TE (right). White solid lines at ~ 3.3 eV represent the three exciton resonances X_A, X_B, X_C; white dashed lines are bare photon modes, and white dashed-dotted lines polariton WGMs. c) Same as in b) but for the ϕ angle. (Reprinted with permission from [TRI 11] Copyright2011 American Physical Society). For a color version of this figure, see www.iste.co.uk/consonni/nanowires1.zip*

In summary, the absence of phonon broadening in GaN and ZnO NW polaritons is an invaluable benefit for the development of all-optical integrated circuits and spintronics operating at an elevated temperature [CER 13, BAL 13, NGU 13, STU 14, DEV 08, ESP 13]. It should also allow studies of the very rich 1D Bose physics [CAZ 11] over an exceptionally wide range of temperatures 4-500 K.

1.3. High efficiency single photon sources

The development of efficient solid-state SPSs is an important prerequisite for the implantation of large scale photonic quantum information processing [BUC 12]. An ideal SPS would emit exactly one photon in response to an external trigger. Currently, advanced SPSs are based on semiconductor QDs as emitters. A single QD would be embedded in a photonic structure designed to feed all photons emitted by the QD into a directed optical beam in the freespace, preferentially of Gaussian-type for subsequent coupling to optical fibers. The figure of merit for photon feeding into a given photonic mode can be defined by the factor $\beta = \Gamma/(\Gamma + \gamma)$, where Γ is the QD recombination rate into this selected mode and γ is the recombination rate into all other radiative modes.

There are basically two different schemes to achieve the ideal case of $\beta = 1$. The traditional scheme makes use of the Purcell effect in a photonic resonant cavity to accelerate the QD spontaneous emission rate into the cavity mode, making $\Gamma >> \gamma$ [MOR 01, HEI 10, GIE 13, GAZ 13]. It requires the QD to be in resonance with the cavity mode within $\Delta\lambda/\lambda \sim 1/Q < 10^{-3}$ (where Q is the cavity quality factor), a severe constraint for practical implementation. A more recent scheme relies on dielectric screening in photonic wires to inhibit any QD coupling to the unwanted optical modes, i.e. $\gamma \rightarrow 0$ [BLE 11, CLA 13]. Here, the constraint is much less demanding with $\Delta\lambda/\lambda \sim 0.1$ and, therefore, we will discuss the photonic wire concept in more detail.

Consider a QD inserted along the z-axis of a cylindrical wire of diameter d (see Figure 1.4).

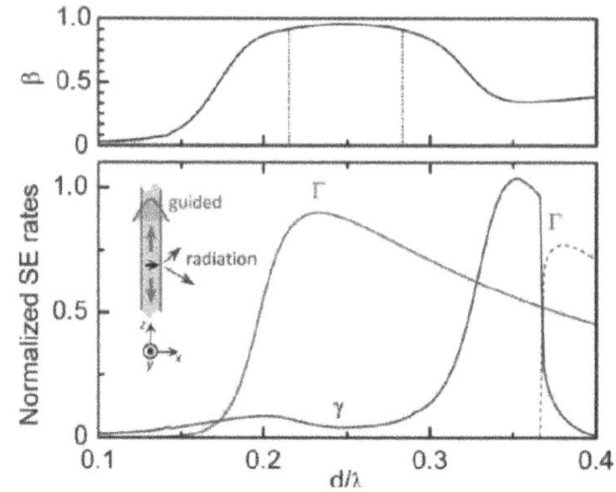

Figure 1.4. *Spontaneous emission of a transverse dipole embedded in a monomode photonic wire [BAR 08]. The dipole is on-axis and oriented along x. Bottom: Spontaneous emission rates of the dipole normalized to that in bulk environment as a function of the reduced wire diameter d/λ, using GaAs parameters: 3.45 for the wire refractive index, vacuum wavelength 900 nm. G is the spontaneous emission rate into the fundamental guided mode HE11, and the spontaneous emission rate into all other radiative modes. Top: Calculated factor $\beta = G/(G + \lambda)$ as a function of d/λ. The two dotted vertical lines represent the diameter range for which $\beta > 0.90$. For a color version of this figure, see www.iste.co.uk/consonni/nanowires1.zip*

Radiative recombination of electrons and holes confined in the QD can be modeled by a dipole moment parallel to the QD base plane (or (x,y) plane), as reported in GaN- [BAR 08, AML 12] and GaAs-based QDs [SIL 03]. This dipole can couple to the surrounding electromagnetic field formed by either guided modes in the wire or radiation modes in the free space. Usually, a guided mode is selected to collect the QD spontaneous emission because it offers a better control of out-coupling into free space. Then Γ and γ defined above are the QD recombination rates into this selected guided mode and into all other modes, respectively. For sufficiently small diameters, i.e. $d/\lambda << 1$ (λ being the mode wavelength in vacuum), photonic wires support a single guided mode, the fundamental hybrid HE11 mode. This mode features strong electric field distribution in the transverse (x,y) plane, fully compatible for coupling to the transverse QD dipole moment. On the other hand, dielectric screening effect in thin wires comes into play, inhibiting coupling of transverse dipole to free space radiation modes [BLE 11, CLA 13], leading to $\gamma/\Gamma << 1$, and consequently $\beta \sim 1$. According to modeling of GaAs-based photonic wire displayed in Figure 1.4, we would get $\beta > 0.90$ over a wide range of wavelengths $\lambda = 900 \pm 100$ nm and NW diameters $d/\lambda \sim 0.25 \pm 0.03$. This broad coupling bandwidth of the photonic wire concept is a crucial advantage when considering the inherent dispersion of QD emission wavelength. In the alternative approach based on the Purcell effect in a resonant cavity [MOR 01, HEI 10, GIE 13, GAZ 13], the coupling bandwidth $\sim 1/Q < 10^{-3}$ is too narrow for practical implementation.

GaAs-based photonic wires have been fabricated using either a top-down [CLA 10, MUN 13] or a bottom-up approach [REI 12]. They include a bottom mirror to redirect backward emitted photons in the forward direction as well as a designed top for out coupling into free space with minimum reflection loss and maximum emission directivity. Out coupling of the guided mode is initially realized with a cone-like top to smoothly increase the mode spatial expansion ("adiabatic" out coupling), which has permitted a record efficiency of 0.72 of single photons collected by standard optics with N.A. ~ 0.75 [CLA 10]. However it was recently shown that similar efficiency can be obtained with a trumpet-like top (Figure 1.5, [MUN 13]), which features additional benefits of Gaussian far-field distribution, of major interest for further manipulation with optical fibers, and weaker sensitivity to geometrical imperfections.

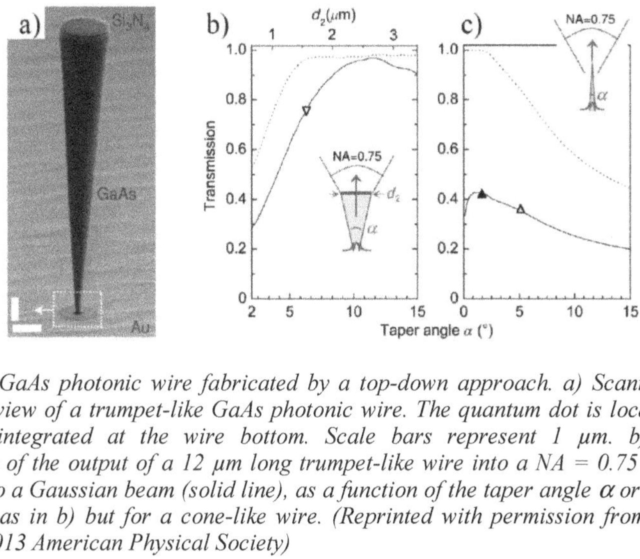

Figure 1.5. *GaAs photonic wire fabricated by a top-down approach. a) Scanning electron microscope view of a trumpet-like GaAs photonic wire. The quantum dot is located near the Au mirror integrated at the wire bottom. Scale bars represent 1 μm. b) Calculated transmission of the output of a 12 μm long trumpet-like wire into a NA = 0.75 lens (dashed line) and into a Gaussian beam (solid line), as a function of the taper angle α or top diameter d_2. c) Same as in b) but for a cone-like wire. (Reprinted with permission from [MUN 13]. Copyright 2013 American Physical Society)*

Last but not the least, the purity of the single photon emission is particularly high with photonic wires as assessed by their intensity autocorrelation function g2 (0) < 0.008 [CLA 10] and < 0.02 [MUN 13].

Interests of wide band gap semiconductors

Nitrides are more attractive than oxides because they can emit over a wider range of emission wavelengths, from deep (ultraviolet) UV wavelengths for solar blind transmission in free space to infrared wavelengths for optical fiber transmission. Furthermore, growth and processing technologies of nitrides are far more advanced.

In fact, basic building blocks for nitride SPSs are already available, such as QDs embedded in NWs [KIM 13, REN 08, REN 09, SON 11, DES 13, KAK 06, CHO 13a, CHO 13b]. Figures 1.6(a) and (b) show scanning and transmission electron microscopy images of GaN photonic wires fabricated by top-down [KIM 13] and bottom-up techniques [REN 09], respectively. The top-down approach consists of a two-step process: first, formation of nano-obelisks by chemical etching of a thick GaN layer in vapor phase HCl at high temperature; next, MOCVD growth of InGaN/GaN QDs on top of nano-obelisks [KIM 13]. In the bottom-up approach, nitride QDs are directly inserted in NWs using (molecular beam epitaxy) MBE [REN 08, REN 09, SON 11, DES 13] or MOCVD [KAK 06, CHO 13a, CHO 13b].

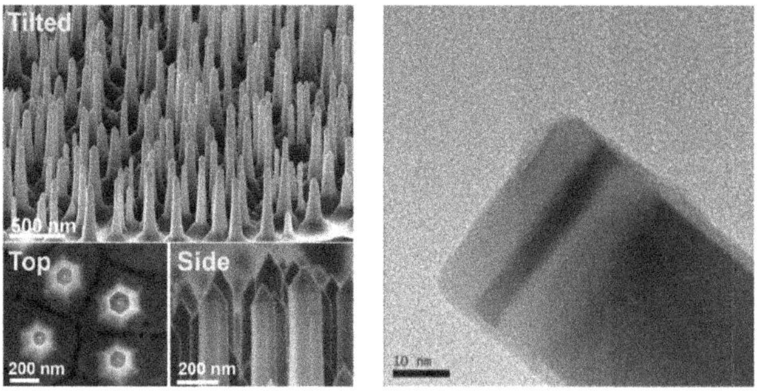

Figure 1.6. *Nitride quantum dots in nanowires. a) Scanning electron microscopy images of InGaN QDs in GaN nano-obelisks fabricated by top-down etching followed by MOCVD growth. (Reprinted with permission from [KIM 13], Copyright 2012 Nature Publishing Group). b) High-resolution TEM image of a GaN/AlN QD MBE grown on top of a GaN NW. GaN appears in dark and AlN in bright in the image. QD height ~ 4.5 nm. Scale bar = 10 nm. (Reprinted with permission from [REN 09]. Copyright 2008 American Chemical Society)*

Optical properties of nitride QDs in NWs are very promising. Figure 1.7 displays photoluminescence results obtained for an ensemble of 1 nm thick GaN/AlN QDs similar to that shown in Figure 1.6(b) [REN 09].

The QD photoluminescence intensity decreases by half between low temperature and room temperature, suggesting an internal quantum efficiency ~ 50% if the radiative recombination process was dominant at low temperature. However, the photoluminescence decay time of these QDs remains nearly constant ~ 300 ps, independent of the temperature. These optical data point to the existence of a thermally activated non-radiative process, which becomes comparable to the radiative process at room temperature. This loss channel could originate from the thin barrier separating the QD from the NW surface (Figure 1.6(b)). Figure 1.8 presents another type of QDs in NWs grown by site-controlled MOCVD [CHO 13b]. These small GaN/AlGaN QDs (height ~ 0.5-1 nm, diameter ~ 10 nm) are inserted in the cone-shaped tops of NWs, surrounded by a thick lateral barrier. They emit at ~4.4 eV, with typical exciton decay time ~ 300 ps at 4 K. Most interestingly, a record binding energy ~ 52 meV is found for the biexciton (Figure 1.8(c)). This is larger than the exciton broadening at room temperature (~ 45 meV), which could ensure reasonably pure single photons with exciton emission at room temperature. Finally, it should be noted that an electrically driven SPS has been demonstrated up to 150 K using InGaN QDs inserted in GaNp-*n* junction NWs grown by MBE on (111) Si [DES 13]. All these know-hows form a solid foundation for the development of bright SPSs based on nitride materials.

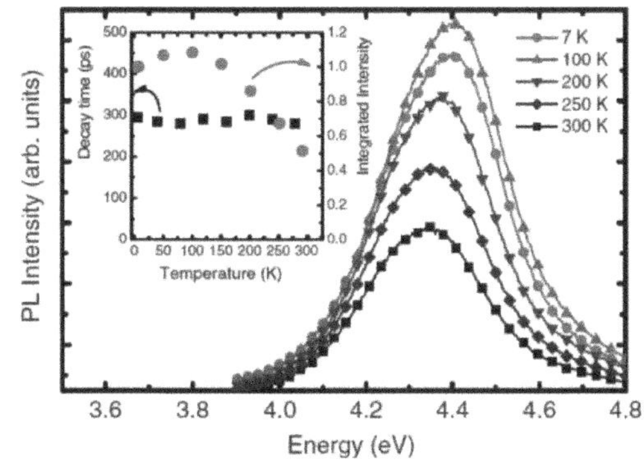

Figure 1.7. *GaN/AlN quantum dots in NWs grown by MBE. a) Temperature dependence of photoluminescence spectra of ensemble of quantum dots similar to the one shown in Figure 1.6(b). Inset displays their integrated intensity (circles) and decay time (black squares) as a function of the temperature. Reprinted with permission from [REN 09]. Copyright 2008 American Chemical Society. For a color version of this figure, see www.iste.co.uk/consonni/nanowires1.zip*

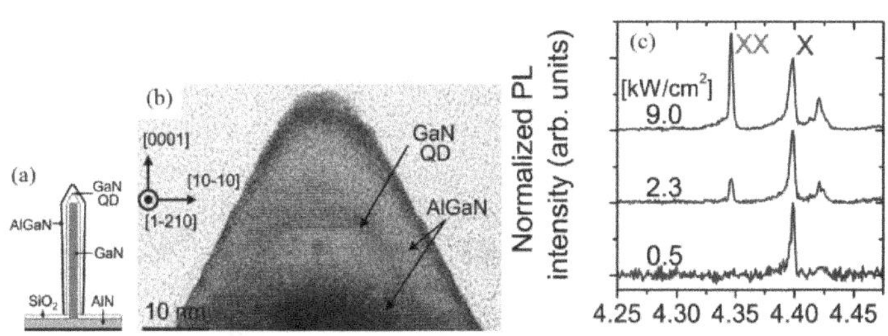

Figure 1.8. *GaN/AlGaN quantum dots in NWs grown by MOCVD. a) Sketch of the quantum dot in nanowire structure. b) High-resolution TEM image of the GaN/AlGaN quantum dot on top of the GaN NW. c) Microphotoluminescence spectra from a single GaN/AlGaN quantum dot as a function of the excitation power measured at ~ 4 K. X and XX denotes radiative recombination lines of exciton and biexciton, respectively. (Reprinted with permission from [LEI 13]. Copyright 2013 American Physical Society). For a color version of this figure, see www.iste.co.uk/consonni/nanowires1.zip*

1.4. High efficiency solar photovoltaics

Solar photovoltaics refers to the conversion of sun radiation into electrical power. The potential capacity of this renewable source of energy would largely exceed the global demand. The first generation of solar cells is based on single *p-n* junctions of bulk Si materials, with an average power conversion efficiency ~ 15%–20%. This is far from the theoretical Shockley–Queisser limit of 34%, mostly due to reflection loss at the front air–Si interface and non-optimum Si band gap. The second generation of solar cells aims at reducing fabrication and materials costs by using thin films of stronger absorbent materials, e.g. CdTe, copper indium gallium selenide (CIGS) or amorphous Si. Nowadays, a third generation of solar cells is emerging with new concepts encompassing nanotechnologies, novel materials (including QDs, NWs and organic materials), and novel concepts (multi-junction (MJ) cells and hot-carrier cells), that could boost efficiency beyond the Shockley–Queisser limit.

Efficiency in single *p-n* junction solar cells is mostly limited by the non-absorption of photons of energy below the band gap and thermal relaxation of hot carriers generated by the absorption of photons of energy above the band gap. These losses could be minimized in an MJ solar cell, each junction tuned to some specific spectral band of the solar spectrum. In theory, an infinite-junction cell would achieve an efficiency ~87%. The most common approach to the MJ cell structure is to stack subcells by order of decreasing band gaps, with the highest band gap on the top front surface, and connect them together in series by low resistance tunnel junctions. Modeling shows that efficiencies greater than 51% could be achieved with a three-junction solar cell built around InGaAs (0.94 eV), InGaAsP (1.39 eV) and InAlAs (1.93eV) [LEI 13].

Another concept for high efficiency solar cells is to replace conventional thin film devices by NW arrays [KAY 05, LAP 11]. Recent reports on NW-based solar cells strongly support their superior PV performance [KRO 13, WAL 13]. In the following, we examine the various arguments that could justify interest in developing NW solar cells made of wide band gap semiconductors. The reader may also refer to Chapter 10 in Volume 2 for further details on the subject.

1.4.1. *Potential photovoltaic benefits of the nanowire geometry*

Reflection loss, light absorption and carrier transport are key factors of PV performance. In planar solar cells, the large refractive index of the semiconductor induces important reflection loss, which has to be minimized by subwavelength surface texturing and antireflection coating. In NW arrays, insertion loss is usually lower due to their lower effective refractive index. However, the interplay between minimum reflection and maximum absorption of solar irradiation is complex because NW diameter D, length L and pitch p are of wavelength size. Thus, wave optics modeling is required to properly account for interference effects in NW arrays. Figure 1.9(a) displays electrodynamic modeling of reflection loss at the front air–NW interface of InP NWs as a function of the NW diameter and pitch assuming complete absorption of light into the NW array [ANT 13]. The inset shows the reflection loss calculated for an NW array (solid line) and a planar structure (dashed line). As intuitively expected, NW arrays are intrinsically less reflective: reflection loss steadily decreases with decreasing D and increasing p. Figure 1.9(b) shows absorption in an InP NW array as a function of the incident light wavelength, calculated for fixed NW pitch p = 680 nm, length L = 2 μm, and various diameters D = (i) 100 nm, (ii) 177 nm, (iii) 221 nm and (iv) 441 nm. The absorption behavior with increasing NW diameter is clearly not monotonic, particularly for long wavelengths $\lambda > 500$ nm. In fact, light absorption in NW arrays is fundamentally different from that in planar structures due to the periodic dielectric modulation at the wavelength scale.

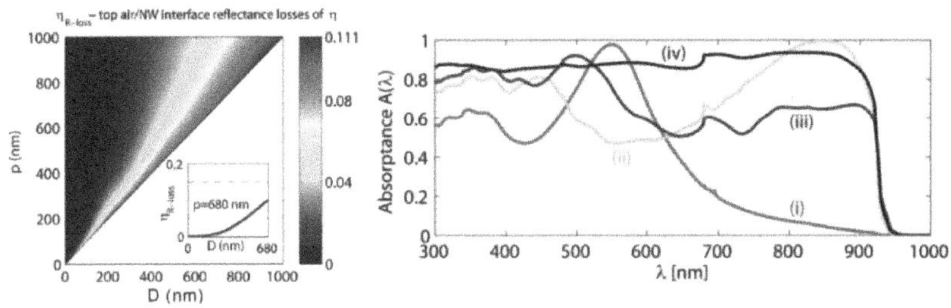

Figure 1.9. *Modeling of optical properties of an InP nanowire array. a) Reflection loss as a function of the NW diameter and the pitch p [DON 09]. The inset shows the reflection loss of the InP array as a function of the diameter for a fixed pitch p = 680 nm, compared to that of a planar InP structure (broken line). b) Absorption spectra of an InP NW array on top of an InP substrate, with fixed pitch p = 680nm, NW length L = 2 μm, diameter D = (i) 100 nm, (ii) 177 nm, (iii) 221 nm and (iv) 441 nm. The incident light is normal to the array. (Reprinted with permission from [ANT 13]. Copyright 2013 Optical Society of America). For a color version of this figure, see www.iste.co.uk/consonni/nanowires1.zip*

Figure 1.10 displays a theoretical study of absorption in InP NW arrays with D/p = 0.5 and length L = 2 μm, for an incident plane wave light [KUP 10]. Light is mostly coupled into two types of optical modes: modes confined in NWs, which can be seen as wave guiding modes, and modes confined in freespace between NWs. The top half of the figure shows the electric field intensity of these modes in NWs of 180 nm diameter and 360 nm pitch as a function of the wavelengths 400–900 nm. For short wavelengths (<700 nm), light is absorbed by the wave guiding modes (B in the figure), while for longer wavelengths absorption in NWs is mediated by the freespace modes (A in the figure). This complementary behavior is behind the high absorption efficiency of this NW array, as shown by the solid green curve in the bottom half of Figure 1.10. A comparison is given with absorption in a thin film coated with a perfect antireflection layer of equivalent volume (392 nm thickness, dashed black curve). Even in the absence of reflection loss, absorption performance is poorer in thin films than in optimized NW arrays.

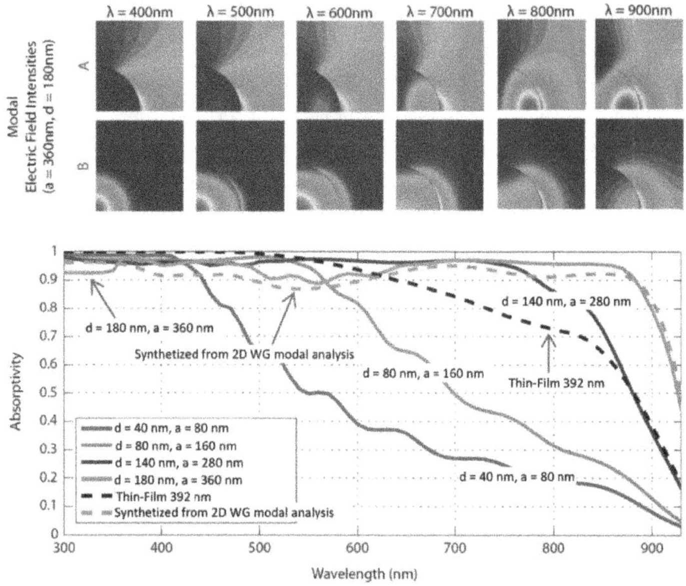

Figure 1.10. *Modeling of optical properties of an InP nanowire array. The (diameter d/p a) is kept fixed and equal to 0.5 and length L = 2 μm. a) Distribution of electric field intensity for various wavelengths for the two most relevant absorption modes: (A) modes confined in free space between NWs; (B) modes confined in NWs or wave guiding modes. NW diameter d = 180 nm, pitch a = 360 nm. b) Absorption as a function of wavelength for various NW diameters and pitches. The thin-film curve (dashed black curve) refers to absorption calculated for an InP thin-film of equal active volume, assuming no reflection loss at the air/semiconductor interface. (Reprinted with permission from [KUP 10]. Copyright 2010 Optical Society of America). For a color version of this figure, see www.iste.co.uk/consonni/nanowires1.zip*

Another benefit of NWs is the possibility of decoupling light absorption from carrier collection. In a planar solar cell structure, light absorption and carrier transport take place along the same material path. For most semiconductor PV materials, the light absorption length (~ $1/\alpha > 1$ μm, α being the material absorption coefficient) is usually longer than the minority carrier diffusion length, and a compromise has to be found for the material thickness. However, these two essential PV functionalities can be decoupled in NWs by growing a radial *p-n* junction: light would be absorbed along the NW length and photogenerated carriers rapidly separated and collected along the radial direction [KAY 05, LAP 11]. Fundamental processes in NW photovoltaics, e.g. carrier collection or surface passivation, can be investigated with single NW devices without being interfered by ensemble averaging or electrical shorting [KRO 13, DON 09, TIA 09, HOL 13], while only NW arrays can provide insights into light scattering and absorption in practical devices [WAL 13]. For a recent review of nanowire photovoltaics, see [LAP 13, GAR 11].

1.4.2. *Interests of wide band gap semiconductor photovoltaics*

Up until now, the role of ZnO NWs in photovoltaics has been rather limited, usually serving as an anode in dye-sensitized solar cells [LAW 05, PEN 11] (see Chapter 10 by J. Baxter in Volume 2). However, InGaN alloys were developed in the early 2000s as solar cell absorbers. These materials are well suited for MJ cells with a band gap tunability between 0.7 and 3.4 eV, *n* and *p* dopability, and large absorption coefficients 10^5 cm^{-1}. This would ensure complete light absorption in most of the solar radiation spectrum with active thickness ~ 0.5 µm, which is interesting in terms of material costs and ecological footprint.

Various InGaN-based solar cell structures have been studied, including InGaN homojunctions, InGaN/GaN heterojunctions, multiquantum wells and superlattices. Performances are still modest, e.g. V_{OC} = 1.81 V, I_{SC} = 4.2 mA/cm^2, fill factor = 75.3% under AM0 illumination of a heterojunction solar cell *p*-GaN/InGaN (2.95 eV band gap)/*n*-GaN (no antireflection coating) [NEU 08]. Of major concern in these planar structures are the difficulty in growing InGaN films with high In concentrations and the presence of high densities of threading dislocations generated by growth on sapphire substrates. For recent reviews on InGaN solar cells, see [BHU 12, MCL 13].

Material quality is expected to be much better in InGaN NWs thanks to their small aerial footprints and large strain accommodation at their free surfaces. Furthermore, Peidong Yang's group showed that high quality InGaN NWs can be grown for the entire alloy composition by low-temperature halide chemical vapor deposition [KUY 07]. PV devices have been fabricated using either single InGaN NWs [DON 09, TIA 09] or NW arrays [NGU 11, WIE 12], both yielding useful and complementary guidelines for optimum device designs. In [WIE 12], the authors first used a top-down approach to fabricate a periodic array of core *n*-type GaN NWs with minimal distribution in height and diameter. This size control is important for the device quality in terms of electrical contacting. Next, they overgrew a shell of InGaN/GaN multiquantum-wells by MOCVD and completed the structure with a *p*-type InGaN canopy by "coalescent growth" to ensure electrical *p*-contact on tops and sides of all NWs. The device absorbs at 2.1 eV but, as in planar devices, performance is far from promise with V_{oc} ~0.5 V, I_{sc} ~1 mA/cm^2, fill factor = 54%

under AM 1.5G solar illumination. At this early development stage, detailed characterization works, such as scanning photocurrent microscopy performed in [HOW 13], are needed to identify the efficiency limiting factors.

1.5. Conclusion

GaN and ZnO NWs feature remarkable optical properties deriving from the strong exciton binding energy and large oscillator strength of wide band gap semiconductors and the natural wave guiding effect of the NW geometry, largely exploited in optoelectronics devices (such as lasers, light-emitting diodes (LEDs) and photodetectors as discussed in the related chapters in Vol. 2 of this book). In this chapter, we have discussed three nanophotonics topics, which are less popular, but nevertheless very promising in terms of breaking new ground: (1) all-optical integrated circuits with Bose exciton polaritons, involving 1D photon modes interacting with bulk excitons in the so-called strong coupling regime of the light-matter interaction; (2) high-efficiency single photon sources (SPSs) for quantum information processing, based on single QDs axially embedded in photonic wires; (3) high-efficiency photovoltaics with core-shell InGaN NW arrays. As always, the most challenging task is in the control of NW growth, e.g. growth of photonic wire with trumpet-like top, or NW array with uniform size and radial doping profile.

1.6. Bibliography

[ALE 03] ALEKSIEJUNAS R., SUDZIUS M., MALINAUSKAS T., *et al.*, "Determination of free carrier bipolar diffusion coefficient and surface recombination velocity of undoped GaN epilayers", *Applied Physics Letters*, vol. 83, p. 1157, 2003.

[AML 12] AMLOY S., KARLSSON K.F., ANDERSSON T.G., *et al.*, "On the polarized emission from exciton complexes in GaN quantum dots", *Applied Physics Letters*, vol. 100, p. 021901, 2012.

[ANT 13] ANTTU N., XU H.Q., "Efficient light management in vertical nanowire arrays for photovoltaics", *Optics Express*, vol. 21, no. S3, pp. A558–A575, 2013.

[BAL 13] BALLARINI D., DE GIORGI M., CANCELLIERI E., *et al.*, "All-optical polariton transistor", *Nature Communications*, vol. 4, no. 1778, 2013.

[BAR 08] BARDOUX R., GUILLET T., GIL B., *et al.*, "Polarized emission from GaN/AlN quantum dots: single-dot spectroscopy and symmetry-based theory", *Physical Review B*, vol. 77, p. 235315, 2008.

[BHU 12] BHUIYAN G., SUGITA K., HASHIMOTO A., *et al.*, "InGaN solar cells: present state of the art and important challenges", *IEEE Journal of Photovoltaics*, vol. 2, pp. 276–293, 2012.

[BLE 11] BLEUSE J., CLAUDON J., CREASEY M., *et al.*, "Inhibition, enhancement and control of spontaneous emission in photonic nanowires", *Physical Review Letters*, vol. 106, p. 103601, 2011.

[BOR 00] BORRI P., JENSEN J.R., LANGBEIN W., *et al.*, "Direct evidence of reduced dynamic scattering in the lower polariton of a semiconductor microcavity", *Physical Review B*, vol. 61, p. R13377, 2000.

[BUC 12] BUCKLEY S., RIVOIRE K., VUCKOVIC J., "Engineered quantum dot single photon sources", *Reports on Progress in Physics*, vol. 75, p. 126503 (27pp), 2012.

[CAZ 11] CAZALILLA M.A., CITRO R., GIAMARCHI T., *et al.*, "One dimensional bosons: from condensed matter systems to ultracold gases", *Reviews of Modern Physics*, vol. 83, pp. 1405–1466, 2011.

[CER 13] CERNA R., LÉGER Y., PRAÏSO T.K., *et al.*, "Ultrafast tristable spin memory of a coherent polariton gas", *Nature Communications*, vol. 4, no. 2008, 2013.

[CHO 13a] CHOI K., ARITA M., KAKO S., *et al.*, "Site-controlled growth of single GaN quantum dots in nanowires by MOCVD", *Journal of Crystal Growth*, vol. 370, p. 328, 2013.

[CHO 13b] CHOI K., KAKO S., HOLMES M., *et al.*, "Strong exciton confinement in site-controlled GaN quantum dots embedded in nanowires", *Applied Physics Letters*, vol. 103, p. 171907, 2013.

[CHR 08] CHRISTMANN G., BUTTÉ R., FELTIN E., *et al.*, "Room temperature polariton lasing in a GaN/AlGaN multiple quantum well microcavity", *Applied Physics Letters*, vol. 93, p. 051102, 2008.

[CLA 10] CLAUDON J., BLEUSE J., SINGH MALIK N., *et al.*, "A highly efficient single-photon source based on a quantum dot in a photonic nanowire", *Nature Photonics*, vol. 4, p. 174, 2010.

[CLA 13] CLAUDON J., GREGERSEN N., LALANNE P., *et al.*, "Harnessing light with photonic nanowires: fundamentals and applications to quantum optics", *ChemPhysChem*, vol. 14, p. 2393, 2013.

[CON 14] CONSONNI V., FEUILLET G., (eds.), *Wide Band Gap Semiconductor Nanowires 2: Heterostructuers and Optoelectronic Devices*, ISTE, London, and John Wiley & Sons, New York, 2014.

[DES 13a] DESHPANDE S., HEO J., DAS A., *et al.*, "Electrically driven polarized single photon emission from an InGaN quantum dot in a GaN nanowire", *Nat. Commun.*, vol. 4, p. 1675, 2013.

[DES 13b] DESHPANDE S., BHATTACHARYA P., "An electrically driven quantum dot-in nanowire visible single photon source operating up to 150 K", *Applied Physics Letters*, vol. 103, p. 41117, 2013.

[DEV 08] DEVEAUD-PLÉDRAN B., "Polaritronics in view", *Nature*, vol. 453, p. 297, 2008.

[DON 09] DONG Y., TIAN B., KEMPA T.J., *et al.*, "Coaxial group III-nitride nanowire photovoltaics", *Nano Letters*, vol. 9, p. 2183, 2009.

[ESP 13] ESPINOSA-ORTEGA T., LIEW T., "Complete architecture of integrated photonic circuits based on AND and NOT logic gates of exciton polaritons in semiconductor microcavities", *Physical Review B*, vol. 87, p. 195305, 2013.

[FRE 00] FREIXANET T., SERMAGE B., TIBERJ A., *et al.*, "In-plane propagation of excitonic cavity polaritons", *Physical Review B*, vol. 61, p. 7233, 2000.

[GAR 11] GARNETT E.C., BRONGERSMA M.L., CUI Y., *et al.*, "Nanowire solar cell", *Annual Review of Materials Research*, vol. 41, p. 269, 2011.

[GAZ 13] GAZZANO O., MICHAELIS DE VASCONCELLOS S., ARNOLD C., *et al.*, "Bright solid-state sources of indistinguishable single photons", *Nature Communications*, vol. 4, no. 1425, 2013.

[GER 98] GERARD J.M., SERMAGE B., GAYRAL B., *et al.*, *Physical Review Letters*, vol. 81, p. 1110, 1998.

[GIE 13] GIESZ V., GAZZANO O., NOWAK A.K., *et al.*, "Influence of the Purcell effect on the purity of bright single photon sources", *Applied Physics Letters*, vol. 103, p. 033113, 2013.

[GIL 97] GIL B., CLUR S., BRIOT O., "The exciton-polariton effect on the photoluminescence of GaN on sapphire", *Solid State Communications*, vol. 104, p. 267, 1997.

[HEI 10] HEINDEL T., SCHNEIDER C., LERMER M., *et al.*, "Electrically driven quantum dot-micropillar single photon source with 34% overall efficiency", *Applied Physics Letters*, vol. 96, p. 011107, 2010.

[HOL 13] HOLM J.V., JORGENSEN H.I., KROGSTRUP P., *et al.*, "Surface passivated GaAsP single-nanowire solar cells exceeding 10% efficiency grown on silicon", *Nature Communications*, vol. 4, no. 1498, 2013.

[HOP 65] HOPFIELD J.J., THOMAS D.G., "Polariton absorption lines", *Physical Review Letters*, vol. 15, p. 22, 1965.

[HOW 13] HOWELL S.L., PADALKAR S., YOON K., *et al.*, "Spatial mapping of efficiency of GaN/InGaN nanowire array solar cells using scanning photocurrent microscopy", *Nano Letters*, vol. 13, p. 5123, 2013.

[KAK 06] KAKO S., SANTORI C., HOSHINO K., *et al.*, "A gallium nitride single-photon source operating at 200 K", *Nature Materials.*, vol. 5, p. 887, 2006.

[KAL 07] KALITEEVSKI M.A., BRAND S., ABRAM R.A., *et al.*, "Whispering gallery polaritons in cylindrical cavities", *Physical Review B*, vol. 75, p. 233309, 2007.

[KAS 06] KASPRZAK J., RICHARD M., KUNDERMANN S., *et al.*, "Bose–Einstein condensation of exciton polaritons", *Nature*, vol. 444, p. 409, 2006.

[KAV 03] KAVOKIN A., MALPUECH G., *Cavity Polaritons*, Elsevier, Amsterdam, 2003.

[KAY 05] KAYES B.M., ATWATER H.A., LEWIS N.S., "Comparison of the device physics principles of planar and radial p-n junction nanorod solar cells", *Journal of Applied Physics.*, vol. 97, p. 114302, 2005.

[KIM 13] KIM J.H., KO Y.H., GONG S.H., *et al.*, "Ultrafast single photon emitting quantum photonic structures based on a nano-obelisk", *Scientific Reports*, vol. 3, p. 2150, 2013.

[KLI 07] KLINGSHIRN C., HAUSCHILD R., FALLERT J., *et al.*, "Room-temperature stimulated emission of ZnO: alternatives to excitonic lasing", *Physical Review B*, vol. 75, p. 115203, 2007.

[KLI 10] KLINGSHIRN C., WAAG A., HOFFMANN A., *et al.*, *Zinc Oxide: From Fundamental Properties towards Novel Applications*, Springer, 2010.

[KRO 13] KROGSTRUP P., JORGENSEN H.I., HEISS M., *et al.*, "Single-nanowire solar cells beyond the Shockley–Queisser limit", *Nature Photonics*, vol. 7, p. 306, 2013.

[KUP 10] KUPEC J., STOOP R.L., WITZIGMANN B., "Light absorption and emission in nanowire array solar cells", *Optics Express*, vol. 18, p. 27589, 2010.

[KUY 07] KUYKENDALL T., ULRICH P., ALONI S., *et al.*, "Complete composition tunability of InGaN nanowires using a combinatorial approach", *Nature Materials*, vol. 6, p. 951, 2007.

[LAP 11] LAPIERRE R.R., "Numerical model of current-voltage characteristics and efficiency of GaAs nanowire solar cells", *Journal of Applied Physics.*, vol. 109, p. 034311, 2011.

[LAP 13] LAPIERRE R.R., CHIA A.C.E., GIBSON S.J., *et al.*, "III-V nanowire photovoltaics: review of design for high efficiency", *Physica Status Solidi (RRL)*, vol. 7, p. 815, 2013.

[LAW 05] LAW M., GREENE L.E., JOHNSON J.C., *et al.*, "Nanowire dye-sensitized solar cells", *Nature Materials*, vol. 4, p. 455, 2005.

[LEI 13] LEITE M.S., WOO R.L., MUNDAY J.N., *et al.*, "Towards an optimized all lattice-matched InAlAs/InGaAsP/InGaAs multijunction solar cell with efficiency > 50%", *Applied Physics Letters*, vol. 102, p. 033901, 2013.

[LI 13] LI F., OROSZ L., KAMOUN O., *et al.*, "From excitonic to photonic polariton condensate in a ZnO-based microcavity", *Physical Review Letters*, vol. 110, p. 196406, 2013.

[LIN 06] LIN Y.J., LIN W.X., LEE C.T., *et al.*, "Changes in optical and electrical properties and surface recombination velocity of n-type GaN due to $(NH_4)_2S_x$ treatment, *Solid State Communications*, vol. 137, p. 257, 2006.

[LU 12] LU T.C., LAI Y.Y., LAN Y.P., *et al.*, "Room temperature polariton lasing vs. photon lasing in a ZnO-based hybrid microcavity", *Optics Express*, vol. 20, p. 5530, 2012.

[MCL 13] MCLAUGHLIN D.V.P., PEARCE J.M., "Progress in InGaN materials for solar photovoltaic energy conversion", *Metallurgical and Materials Transactions. A*, vol. 44, p. 1947, 2013.

[MOR 01] MOREAU E., ROBERT I., GÉRARD J.M., *et al.*, "Single mode solid-state single photon source based on isolated quantum dots in pillar microcavities", *Applied Physics Letters*, vol. 79, p. 2865, 2001.

[MOR 08] MORKOÇ H., ÖZGÜR U., *Zinc Oxide: Fundamentals, Materials and Device Technology*, John Wiley & Sons, 2008.

[MUN 13] MUNSCH M., MALIK N., DUPUY E., *et al.*, "Dielectric GaAs antenna ensuring an efficient broadband coupling between an InAs quantum dot and a Gaussian optical beam", *Physical Review Letters*, vol. 110, p. 177402, 2013.

[NEU 08] NEUFELD C.J., TOLEDO N.G., CRUZ S.C., *et al.*, "High efficiency InGaN/GaN solar cells with 2.95 eV band gap", *Applied Physics Letters*, vol. 93, p. 143502, 2008.

[NGU 11] NGUYEN H.P.T., CHANG Y.L., SHIH I., *et al.*, "InN p-i-n nanowire solar cells on Si", *IEEE Journal of Selected Topics in Quantum Electronics*, vol. 17, p. 1062, 2011.

[NGU 13] NGUYEN H.S., VISHNEVSKY D., STURM C., *et al.*, "Realization of a double-barrier resonant tunneling diode for cavity polaritons", *Physical Review Letters*, vol. 110, p. 236601, 2013.

[NOB 04] NOBIS T., KAIDASHEV E.M., RAHM A., *et al.*, "Whispering gallery modes in nanosized dielectric resonators with hexagonal cross section", *Physical Review Letters*, vol. 93, p. 103903, 2004.

[PEN 11] PENG Q., QIN Y., "ZnO nanowires and their application for solar cells", in HASHIM A. (ed.), *Nanotechnology and Nanomatetials – Nanowires – Implementation and Applications*, 2011. Available at http://www.intechopen.com/books/nanowires-implementations-and-applications/zno-nanowires-and-their-application-for-solar-cells.

[POR 10] PORTOLAN S., DI STEFANO O., SAVASTA S., *et al.*, "Emergent entanglement of microcavity polariton pairs", *Journal of Physics: Conference Series*, vol. 210, p. 012033, 2010.

[REI 12] REIMER M., BULGARINI G., AKOPIAN N., *et al.*, "Bright single-photon sources in bottom-up tailored nanowires", *Nature Communications*, vol. 3, p. 737, 2012.

[REN 08] RENARD J., SONGMUANG R., BOUGEROL C., *et al.*, "Exciton and biexciton luminescence from single GaN/AlN quantum dots in nanowires", *Nano Letters*, vol. 8, p. 2092, 2008.

[REN 09] RENARD J., SONGMUANG R., TOURBOT G., *et al.*, "Evidence for quantum-confined Stark effect in GaN/AlN quantum dots in nanowires", *Physical Review B*, vol. 80, p. 121305R, 2009.

[SAV 97] SAVONA V., PIERMAROCCHI C., "Microcavity polaritons: homogeneous and inhomogeneous broadening in the strong coupling regime", *Physica Status Solidi (a)*, vol. 164, p. 45, 1997.

[SIL 03] SILVERMAN K.L., MIRIN R.P., CUNDIFF S.T., *et al.*, "Direct measurement of polarization resolved transition dipole moment in InGaAs/GaAs quantum dots", *Applied Physics Letters*, vol. 82, p. 4552, 2003.

[SON 11] SONGMUANG R., KALITA D., SINHA P., *et al.*, "Strong suppression of internal electric field in GaN/AlGaN multi-layer quantum dots in nanowires", *App. Phys. Lett.*, vol. 99, p. 141914, 2011.

[SUN 08] SUN L., CHEN Z., REN Q., *et al.*, "Direct observation of whispering gallery mode polaritons and their dispersion in a ZnO tapered microcavity", *Physical Review Letters*, vol. 100, p. 156403, 2008.

[STU 14] STURM C., TANESE D., NGUYEN H.S., *et al.*, "All-optical phase modulation in a cavity-polariton Mach – Zehnder interferometer", *Nature Communications*, vol. 5, no. 3278, 2014.

[TAK 07] TAKAHASHI K., YOSHIKAWA A., SANDHU A., *Wide Bandgap Semiconductors*, Springer, 2007.

[TIA 09] TIAN B., KEMPA T.J., LIEBER C.M., "Single nanowire photovoltaics", *Chemical Society Reviews.*, vol. 38, p. 16, 2009.

[TRI 11] TRICHET A., SUN L., PAVLOVIC G., *et al.*, "One-dimensional ZnO exciton polaritons with negligible thermal broadening at room temperature", *Physical Review B*, vol. 83, p. 041302(R), 2011.

[TRI 12] TRICHET A., MÉDARD F., ZUNIGA-PÉREZ J., *et al.*, "From strong to weak coupling regime in a single GaN microwire up to room temperature", *New Journal of Physics.*, vol. 14, p. 073004, 2012.

[WAL 13] WALLENTIN J., ANTTU N., ASOLI D., *et al.*, "InP nanowire array solar cells achieving 13.8% efficiency by exceeding the ray optics limit", *Science*, vol. 339, p. 1057, 2013.

[WEI 92] WEISBUCH C., NISHIOKA M., ISHIKAWA A., *et al.*, "Observation of the coupled exciton-photon mode splitting in a semiconductor quantum microcavity", *Physical Review Letters*, vol. 69, p. 3314, 1992.

[WIE 12] WIERER JR J.J., LI Q., KOLESKE D.D., *et al.*, "III-nitride core-shell nanowire arrayed solar cells", *Nanotechnology*, vol. 23, p. 194007, 2012.

[ZHA 08] ZHAO Q.X., YANG L.L., WILLANDER M., *et al.*, "Surface recombination in ZnO nanorods grown by chemical bath deposition", *Journal of Applied Physics.*, vol. 104, p. 073526, 2008.

[ZHA 12] ZHANG S., XIE W., DONG H., *et al.*, "Robust exciton-polariton effect in a ZnO whispering gallery microcavity at high temperature", *Applied Physics Letters*, vol. 100, p. 101912, 2012.

2

Stress Relaxation in Nanowires with Heterostructures

2.1. Introduction

2.1.1. *Scope*

This chapter examines nanowires (NWs) comprising heterostructures between materials that present a *lattice mismatch*. If we call stress-free (SF) the elastic state of a homogeneous crystal not submitted to any external force (i.e. the intrinsic state of the bulk crystal), the crystals assembled in the heterostructure are supposed to have the same structure but different SF lattice parameters. Two main types of heterostructures can be formed in NWs: *axial* and *radial* heterostructures. If, during growth, we switch from material A to material B while maintaining growth along the NW axis, we obtain an axial heterostructure (Figure 2.1(a)). Conversely, if we initiate radial growth while at the same time switching from A to B, we form a radial (or core-shell) heterostructure, with B surrounding A (Figures 2.1(b) and (c)). For more details about these types of heterostructures in particular systems, refer to Chapters 1–6 of Volume 2 [CON 14]. Axial and radial heterostructures may also be combined to form, e.g. embedded inclusions with a double axial heterostructure and a protective shell, which may also confine charge carriers (see Chapter 1).

We are primarily concerned with wide band gap nitride and oxide semiconductors, such as GaN and ZnO and the compounds and alloys obtained by totally or partially substituting the cation by another element of the same column of the periodic table. These materials predominantly adopt the hexagonal wurtzite (WZ) crystal structure, which involves two lattice parameters a and c (Figure 2.1(c)). The relative lattice mismatches $\Delta a/a$ and $\Delta c/c$ between crystals A and B are usually different for these two parameters; even their signs may be different (case of the

---

Chapter written by Frank GLAS.

ZnO/(Zn,Mg)O heterostructures [PER 12]). It will, however, prove useful to examine not only the scarce results relative to these specific materials but also other families of semiconductors as well as generic models for NW heterostructures. The crystal structure is a diamond-type face centered cubic (fcc) for elementary semiconductors and WZ or fcc zinc blende (ZB) for compound semiconductors. The vast majority of heterostructures so far realized involve different materials with the same structure or else the same material with different structures (case of ZB/WZ III–V "crystal phase heterostructures", for which strain plays at most a minor role); however, hybrid heterostructures involving different materials with different structures have started to appear [HOC 12, CON 13]. In the models, the NWs are most often assumed to have a circular or a hexagonal section. The latter is that of most actual NWs, which grow along a <111> axis (cubic structure) or along the [0001] **c** axis (WZ structure)[1]. Figure 2.1 summarizes these geometries and indicates various coordinates systems employed in the text. We use standard notations for elastic constants (C_{ij}) and components of the strain (ϵ_{ij}) and stress (σ_{ij}) tensors [SAD 05]. Unless stated otherwise, the strains are understood as relative to the SF state of each constituent material. Recall that the WZ structure exhibits transverse elastic isotropy around the **c**-axis [SAD 05].

2.1.2. *Stress relaxation*

Let us first recall the different meanings of *stress relaxation* in heterostructures. An A/B interface is *coherent* if the homologous lattice planes not parallel to the interface are equally spaced in A and B and continuous across the interface. The lattice mismatch then makes it impossible for A and B to retain their SF state, simply because the spacings of these lattice planes are different for SF A and B. One of the materials at least, and both of them in the case of non-planar structures, is thus strained with respect to its bulk state. This is a first type of stress relaxation: the crystals strain *elastically* from their SF state to minimize the total elastic energy, given the constraint of coherency. Conversely, accommodation may be realized *plastically*, via the formation of networks of *misfit dislocations* at the interface, which becomes *incoherent*. The dislocations permit discontinuities of the lattice spacings across the interface. The nature and density of the dislocations may be such that they allow A and B to recover their SF lattice parameters. The crystals are then mainly strained in the vicinity of the dislocation cores but rapidly tend to their respective SF state away from the latter. In a dynamical growth situation, where the thickness of deposit B increases while that of A remains fixed, this usually occurs after a phase of coherent elastic relaxation. The formation of such networks is also called stress relaxation since it allows the elastic energy to decrease below that in the coherent state. If the density of dislocations is insufficient, the entire structure remains elastically strained, albeit less than in the fully coherent case. One major question

1 Note that in these structures, the directions "up" and "down" these axes are not equivalent.

regarding strain accommodation in heterostructures is to find out which factors determine the mode of relaxation (elastic or plastic).

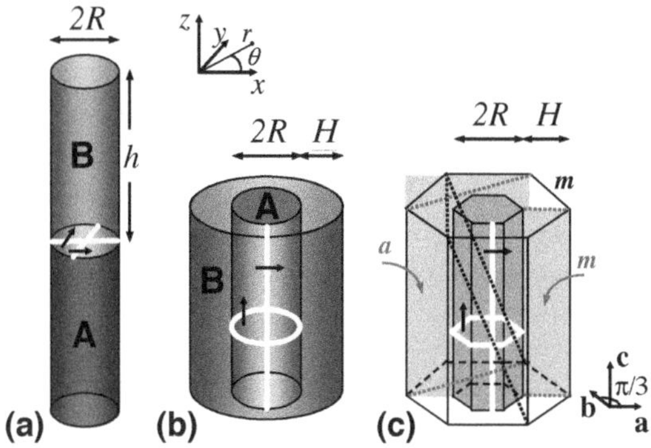

Figure 2.1. *Schematics of the two basic types of heterostructure in a NW (here with circular section): (a) axial and (b) radial. The thick white lines show the interfacial dislocations considered in simple models of plastic relaxation ((a) edge segments and (b) prismatic loop and axial dislocation), their Burgers vectors (BVs) being indicated by arrows. (c) A core-shell NW of WZ structure (axes indicated, with $a = b$) grown along* **c** *= [0001] with m-type sidewalls. R and H may be measured from the center toward the vertical edge (as shown) or toward the facet center. Two a and m prismatic planes and a pyramidal plane are marked by dotted outlines. Axial dislocation and basal loop with BVs as in (b). Various coordinates systems are indicated: orthonormal xyz axes with z along NW axis, cylindrical coordinates for NWs with circular section and unit cell axes for WZ structure. For a color version of this figure, see www.iste.co.uk/consonni/nanowires1.zip*

2.1.3. *Nanowire specificities*

The NW is bound by *free surfaces*, i.e. surfaces free of forces (Figure 2.1). Foremost among these are the NW sidewalls. The interfaces are usually close to these sidewalls (core-shell structure) and possibly intersect them (axial heterostructure without shell). During growth of an axial heterostructure, the top surface is also initially close to the axial interface. The cancellation of the forces $F_i = \sigma_{ij}n_j$, with $\mathbf{n}$ the unit normal to the free surface, imposes constraints on the stress tensor. However, these surfaces usually result in a decrease of the elastic energy with respect to the situation where material would have been present on both sides of the surface, because the material below the free surface has more latitude to relax elastically. Actually, one of the major interests of NWs is that their geometry allows one to form coherent heterostructures with higher mismatches or with thicker layers than in other geometries. Although this easier accommodation of stress in NWs compared with planar structures is often cited as an important advantage, the literature devoted only or principally to stress relaxation in NWs is rather scarce. Most of it deals with calculations and the first serious studies only recently appeared, aiming either at measuring elastic strains or at identifying the actual mechanisms of plastic relaxation.

After a brief introduction to the calculation and measurement of elastic strain in NWs (section 2.2), we will consider core-shell and axial heterostructures in sections 2.3 and 2.4, respectively. For each type of heterostructure, we will first review the calculations of coherent elastic relaxation and the main characters of the strain distribution, then the predictions about plastic relaxation, and confront these results to the few experiments that exist. Section 2.5 will consider other types of stress relaxation, involving a change of morphology of some NW bounding surfaces.

2.2. Calculation and measurement of elastic strain in nanowires

2.2.1. *Calculation of elastic strain*

The calculations of elastic relaxation in NWs are most often performed in the framework of linear elasticity. Because linear elasticity does not have any built-in length scale, some calculations performed long ago with macroscopic systems in mind may be applied to NWs, as long as relaxation remains coherent. The calculations may be analytical or numerical and may assume either isotropic or anisotropic elasticity. Analytical approaches usually involve more restrictive approximations (in particular with regard to NW geometry) whereas numerical methods are more flexible. A fair proportion of the WZ NWs so far treated by these methods is actually made up of those semiconductors that are cubic in their bulk form but tend to adopt the hexagonal structure when grown as NWs [GLA 07]. Nevertheless, the results so obtained may be useful for the GaN and ZnO families.

In the analytical calculations, one first derives differential equations verified by some quantities (e.g. the displacements or a stress function [SAD 05]), given the symmetries of the problem. The general solutions of these equations depend on a few parameters in each domain (core and shell, or core and axial deposit), which are found by applying the boundary conditions (interfacial coherency, free external sidewalls, free top surface for a NW of finite length, finite displacements, etc.).

Numerical methods give access to more realistic NWs. The typical hexagonal cross section and the difference of elastic constants between the different parts of the heterostructure may easily be taken into account. The calculations must, however, be performed for definite materials and the transferability of the results from one system to another is not guaranteed. The most common numerical method used in elasticity is probably the finite elements (FE) method, which finds an approximate solution for the equations of continuous elasticity after discretization of the domain of interest into elements of various sizes and shapes. Other methods rely on an atomistic description. The valence force field (VFF) method, which has been used for semiconductors for decades [KEA 66, MAR 70, GLA 90], minimizes the total energy of the system, taken as a sum of terms depending on the distortions of bond lengths and bond angles with respect to some ideal configuration (the SF state for simple and homogeneous materials). The VFF can handle millions of atoms, whereas the density functional theory (DFT) method is restricted to very small systems. It has, however, been applied to narrow Ge-Si core-shell NWs [MUS 05].

2.2.2. *Measurement of elastic strain*

Most experimental methods used to assess elastic relaxation in NWs measure strains rather than stresses. Since NWs are often fabricated as ensembles displaying variable geometries (and in particular a distribution of radii), the methods that can address individual NWs are particularly interesting. This is the case of several techniques of transmission electron microscopy (TEM). Electron diffraction provides easy access to the difference of lattice spacing in various directions [POP 11]. The spacings are, however, averaged over some volume of the NW (which may be as small as 1 nm in the transverse direction [BEC 11]) and, because the electron diffraction spots are broadened by various electron scattering processes, the accuracy is much worse than what can be achieved using X-ray diffraction (XRD). High-resolution TEM (HRTEM) methods provide atomic resolution, but it should be remembered that the information in each image pixel results from a dynamical scattering process that somehow averages the information in the thickness of the sample along the electron beam direction. Most often, the NWs are scraped off their substrate and dispersed on a thin membrane so that the NW axis is normal to the electron beam. Therefore, the interpretation may be complex for core-shell heterostructures (in parts of the NW, the electron beam traverses both the shell and the core) than for axial heterostructures. Simple Fourier filtering of extended areas of standard HREM images reveal the identity [NAZ 13] or on the contrary the difference [LAU 02] of lattice parameters between, for example, the core and the shell of an NW. The geometrical phase analysis (GPA) method has been widely used to measure displacement and strain fields in NWs with near-atomic resolution. This method is based on the filtering of an HREM image around a given lattice spatial frequency to obtain information about the displacement field of the corresponding planes, which locally modifies the phase of the electron wave function [HYT 98]. The method has been applied, in particular, to NWs with heterostructures of the nitride [BOU 10, BOU 09, LAN 10, HES 10, TOU 11, KEH 13] and II–VI [BEL 10] families.

XRD also gives valuable information on strained heterostructures. At variance with TEM, the NWs are usually kept on their substrate. Standard (high angle) XRD can then be used to measure the lattice spacings along the growth axis whereas grazing incidence XRD (GIXRD) gives access to the in-plane lattice spacings [EYM 07, KEP 09]. Coherent XRD is a rapidly progressing technique. High brightness synchrotron sources now make it possible to illuminate coherently single NWs with diameters below 100 nm. Scattering from this single object is recorded around one Bragg peak and the recorded three-dimensional (3D) data can be inverted to recover the shape of the scattering object [FAV 09] and, by comparison with calculations, the strain field (in case the NW is inhomogeneous) [FAV 10, HAA 13].

Micro-Raman spectroscopy is also able to probe strain in NWs. This has been done for core-shell heterostructures, in particular for nitride materials (ensembles of NWs [BOU 10, HES 10, CRO 13] or single NWs [LAN 11]) and for individual Si-(Si,Ge) [DIL 12] and GaAs-GaP [MON 10] NWs. Finally, a technique using incident particles with reduced penetration depth, medium energy ion spectroscopy, has been performed on ensembles of NWs to determine the strain in nanometric GaN insertions embedded in AlN sections deposited on top of GaN NWs [JAL 12].

2.3. Core-shell heterostructures

2.3.1. *Elastic relaxation in core-shell heterostructures*

In core-shell heterostructures, due to the geometric anisotropy of the NW, and even if elastic isotropy is assumed, we must consider that the misfit has two interfacial components, namely an *axial* component and an *in-plane tangential* component. In WZ NWs grown along the c-axis, these relate, respectively, to $\Delta c/c$ and $\Delta a/a$. In ZB or elastically isotropic NWs, the SF misfits are equal but their level of elastic accommodation and even their mode of plastic relaxation are *a priori* different. In calculations for core-shell heterostructures, the NW is most often considered as infinite along its axis. This is a good approximation as long as the length of the NW is large compared to its diameter, provided that one is not interested in the strains near the base of the NW or near its top facet. The elastic strains are then independent of z. Consequently, the axial dilatations ϵ_{zz} must be uniform in core (superscript c) and shell (superscript s), and such that their difference equals the relative lattice mismatch along this axis:

$$\epsilon^s_{zz} - \epsilon^c_{zz} = -\Delta c/c \qquad [2.1]$$

Analytical calculations of strain relaxation in axial heterostructures have been performed by several authors under various hypotheses. Starting with NWs of circular section, Warwick and Clyne considered materials with transverse elastic isotropy about the system axis [WAR 91], which includes WZ, as already mentioned. Trammel *et al.* [TRA 08], followed by Haapamaki *et al.* [HAA 12], treated cubic crystals [TRA 08] and Salehzadeh *et al.* compared cubic and hexagonal semiconductors [SAL 13]. Ovid'ko and coworkers, Aifantis *et al.* and Nazarenko *et al.* restrict themselves to isotropically elastic materials with identical Poisson ratios and identical [GUT 00, OVI 04, NAZ 13] or different [AIF 07] shear moduli[2]. In all the cases, all strain components are found to be uniform in the core, but not in the shell. At variance with these rigorous solutions, Raychaudhuri and Yu made the oversimplifying hypothesis that all strain components are uniform in core and *shell* (both being either WZ or cubic materials) [RAY 06b, RAY 06a]. Actually, if the shell is not too thick, the radial and in-plane tangential strains decrease continuously from the A/B interface to the NW sidewall [TRA 08], a characteristic instance of facilitated stress relaxation in NWs (see below).

With regard to numerical calculations, Søndergaard *et al.* [SØN 09] and Hocevar *et al.* [HOC 13] applied the FE method to NWs of cubic structure with hexagonal sections. Grönqvist *et al.* performed joint FE and VFF calculations for ZB crystals [GRO 09] in the same geometry and also for circular cylinders, and Boxberg *et al.* recently gave a comprehensive treatment of ZB and WZ NWs using the FE method [BOX 12]. Niquet and coworkers developed a VFF model for WZ and applied it to core-shell GaN/AlN NWs with hexagonal cross sections [CAM 10a, HES 10]. Kavanagh *et al.* applied molecular dynamics to InAs/GaAs core-shell structures, without however giving details of the calculations [KAV 12].

The broad features of elastic relaxation in core-shell heterostructures depend little on the method of calculation and on the precise geometry of the NW sidewalls. Let us summarize these features in the specific case of a GaN core of radius R and an AlN shell of thickness H (for more details about this particular system, refer to Chapters 1 and 6 of Volume 2). The shell then has smaller lattice parameters than the core, although the lattice misfits are rather different, namely $\Delta a/a = -2.5\%$ and $\Delta c/c = -4\%$. If $H \ll R$ (Figure 2.2), the situation is akin to the two-dimensional (2D) case, with most of the core close to its SF state and most of the strain localized in the shell, which is tangentially dilated (in both directions) and thus contracts radially: e.g. for the two xz-oriented sidewalls, $\epsilon^s_{xx} > 0$, $\epsilon^s_{zz} > 0$ whereas $\epsilon^s_{yy} < 0$. As H increases, the strain gets partitioned between the core and the shell (Figure 2.3). The core is now tangentially contracted by the shell and the shell is tangentially dilated by the core in the vicinity of the interface. The Poisson effect and free sidewalls impose the reverse signs on the strain normal to the sidewall. As already mentioned, ϵ_{zz} is uniform in core and shell and, in particular, independent of the radial position, whereas the in-plane components ϵ^s_{xx} and ϵ^s_{yy} decrease rapidly away from the interface over a distance of the order of R. Hence, the outer free surface of the NW and its particular geometry ensure efficient stress relaxation. As a consequence, the elastic energy stored in the NW per unit axial length, at fixed core radius, increases with H but rapidly saturates. This changes little if the shell becomes thick with respect to the shell. Thus, the strain is not confined in the core but still shared with the shell. This was pointed out as a significant difference from planar heterostructures [TRA 08] where, if one of the two layers is much thicker, it conserves its SF state everywhere and imposes its own lattice

2 Note that the second Ovid'ko reference corrects an error in the first one.

parameter to the thin layer in the interfacial plane. However, it is not specific to NWs. Any finite inclusion strains the matrix close to the interface. This is, for instance, manifest in the $H \to \infty$ limit case of a cylindrical inclusion in an infinite matrix [HAS 92].

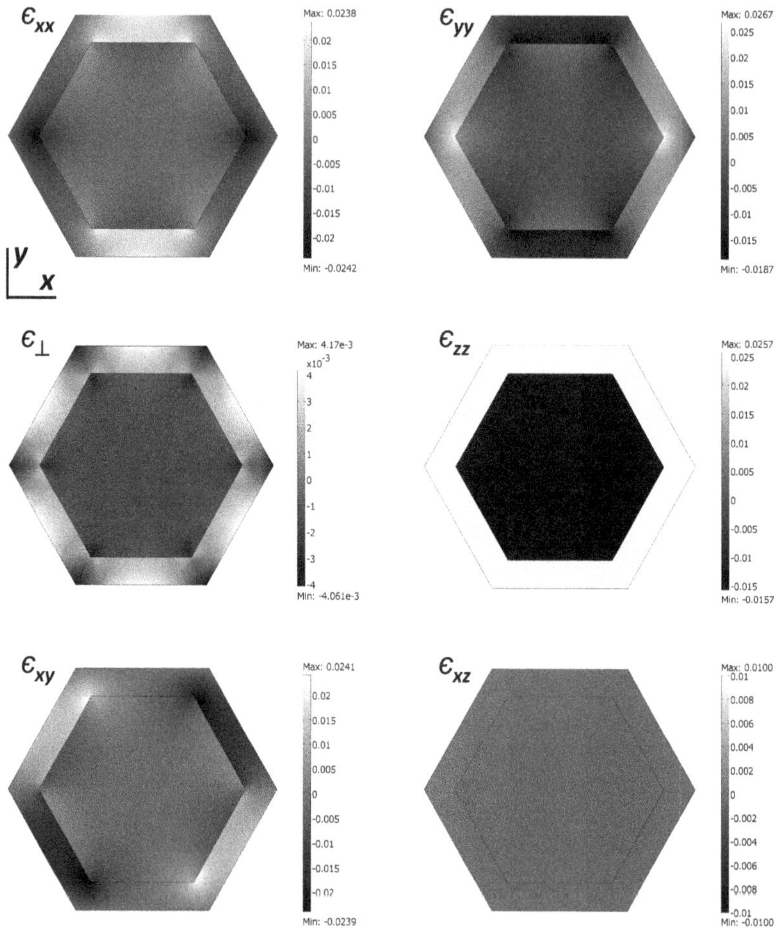

Figure 2.2. *Finite element calculations for a GaN core/AlN shell NW with $H/R = 0.3$. Maps of various components of the strain field in the xy plane (axes x and y as indicated) normal to the* $\mathbf{c}$ *(z) axis.* $\epsilon_\perp = \frac{1}{2}(\epsilon_{xx} + \epsilon_{yy})$. *For a color version of this figure, see www.iste.co.uk/consonni/nanowires1.zip*

The shear strain in the basal plane, ϵ_{xy}, is of the same order as the extensions ϵ_{ii} (Figure 2.2). This also manifests the increased latitude for relaxation of the NW due to its narrow lateral size. On the other hand, the shear strains ϵ_{iz} in the prismatic planes are zero. This was signaled as the major difference between the WZ and ZB structures by Boxberg *et al.*, who demonstrated it in the case of an NW with circular section [BOX 12]. Figure 2.2 shows that it is also true for a hexagonal section. Recall that for NWs of cubic structure, the shear strains ϵ_{iz} are of the same order as the other strain components [GRO 09].

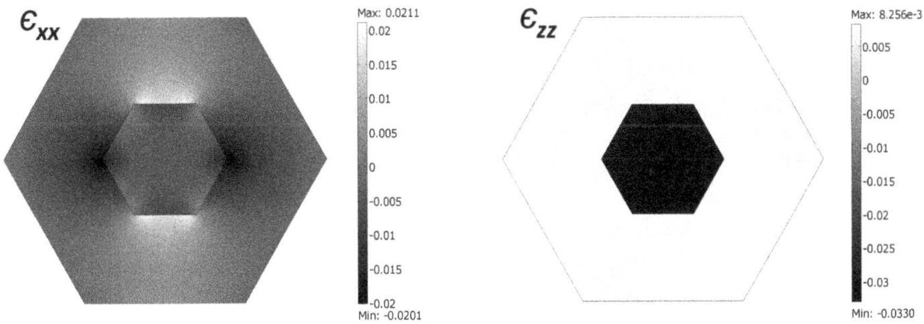

Figure 2.3. *Same as Figure 2.2 for* $H/R = 1.6$.
Only two strain components are shown. For a color version of this figure, see www.iste.co.uk/consonni/nanowires1.zip

Another remarkable property is that the in-plane strain $\epsilon_\perp = \frac{1}{2}(\epsilon_{xx} + \epsilon_{yy})$ is much smaller than its components ϵ_{xx}, ϵ_{yy} (Figure 2.2). This led Hestroffer *et al.* to propose a useful approximation [HES 10] that yields a good estimate of the uniform axial extensions in core and shell. Neglecting all strain components in the basal plane, the elastic energy is taken as $W_e = \frac{1}{2}\left(A_c C_{33}^c (\epsilon_{zz}^c)^2 + A_s C_{33}^s (\epsilon_{zz}^s)^2\right)$, where A_r and C_{33}^r are, respectively, the transverse area and the axial elastic constant of region r (c: core, s: shell). Minimizing W_e under constraint [2.1] readily yields the partition of the axial strain between ϵ_{zz}^c and ϵ_{zz}^s.

Because of the transverse elastic isotropy of crystals with a hexagonal structure, the orientation of the NW sidewalls is irrelevant as long as continuum elasticity is used, and the strain field is sixfold symmetric. In atomistic calculations where the termination of the sidewall facets matters, a subtle reduction of symmetry to order 3 may be detected [HES 10]. This may have a large effect for growth, to the point of producing nitride NWs with triangular sections [LIM 09].

2.3.2. *Plastic relaxation and critical parameters in core-shell heterostructures*

2.3.2.1. *Theoretical considerations*

The plastic relaxation of core-shell heterostructures has been the subject of several predictions since 2000, but the experiments remain scarce. The heterostructure is defined by its dimensions (R, H) (Figure 2.1) and by the core-shell misfits, which can vary continuously in the case of alloys. The bulk of theoretical work aims at determining possible critical values of these parameters, at which the system will undergo a transition from elastic to plastic relaxation (most often, one searches for critical dimensions at a given misfit). Following an approach common for planar heterostructures, this is usually done by comparing the energies of two states of the system, namely S_e, with purely elastic relaxation (section 2.3.1), and S_d, with one or several extended defects, most often interfacial dislocations. The onset of plastic relaxation is marked by the introduction of the first dislocation, which may *a priori* partially relieve either the axial misfit or the in-plane tangential misfit, or both.

Two main types of dislocations have been considered as candidates for stress relaxation in core-shell NWs (Figure 2.1): straight edge dislocations parallel to the NW axis, which relieve in-plane tangential misfit [GUT 00, OVI 06, RAY 06b, RAY 06a], and edge dislocation loops normal to the NW axis (prismatic loops), which relieve axial misfit [OVI 04, OVI 06, RAY 06b, RAY 06a, AIF 07, HAA 12]. Note that a single straight dislocation breaks the rotational symmetry of the system and a single prismatic loop breaks its translational symmetry.

One of the most striking features of stress relaxation in core-shell NWs is that, at fixed misfit, there exists a critical core radius R_c such that it should be possible to grow arbitrarily thick shells around an NW of radius $R \leq R_c$. The very existence of R_c can be understood by using simple dimensional arguments in line with those developed for axial heterostructures [GLA 06] (section 2.4.2.1). Recall that, for elastically strained shells thicker than the core, the in-plane strain components extend only over a distance of order R away from the interface (section 2.3.1 and Figure 2.3). Considering a length L of NW, the total energy in state S_e thus scales as LR^2, independently of H (assumed to be large). On the other hand, the energy in state S_d is surely larger than the energy of the dislocation core (in addition, there is an elastic energy). Since the core energy depends on the atomic structure of the dislocation [HIR 82], it scales as L and is independent of R. This immediately shows that, for narrow enough cores, S_e will have lower energy than S_d. A similar argument can be developed for a single interfacial loop. Indeed, if L is now a typical axial length over which the loop relieves stress, introducing the loop will procure a decrease of elastic energy scaling as LR^2 whereas the dislocation core energy scales as R.

For cores wider than $2R_c$, quite similarly to the planar case, there is a maximum shell thickness $H_c\,(R)$ above which dislocations are favored. This critical thickness will, of course, depend on R: it decreases from infinity at $R = R_c$ to the 2D critical thickness for very wide cores (Figure 2.4). Both critical dimensions are expected to decrease with increasing misfit (Figures 2.4 and 2.5), since the elastic energy scales with the square of the misfit whereas the dislocation core energy is independent of it.

Most models since the early work of Gutkin *et al.*, indeed, predict the existence of a radius-dependent critical thickness and of a critical radius as well as their decrease with increasing misfit [GUT 00, OVI 04, RAY 06b, RAY 06a, TRA 08,

HAA 12, SAL 13, NAZ 13][3]. However, quantitative estimates require evaluating the energy of state S_d. Introducing a dislocation modifies the strain and stress fields. In linear elasticity, the total fields ϵ^t_{ij}, σ^t_{ij}, are simply the sums of the initial fields ϵ^e_{ij}, σ^e_{ij} in state S_e (section 2.3.1) and of those induced by the dislocation, ϵ^d_{ij}, σ^d_{ij}. The latter are calculated independently of the "*e*" fields but should account for the geometry of the system, namely the free surfaces and the change of elastic parameters at the interface. The total energy is obtained by integrating the products $\epsilon^t_{ij}\sigma^t_{ij}$ over the whole system. This produces three terms. The first two terms, which involve only products $\epsilon^e_{ij}\sigma^e_{ij}$ and $\epsilon^d_{ij}\sigma^d_{ij}$, are simply the original elastic energy (state S_e) and the dislocation *self-energy*. The third term, which involves the cross products, is the *interaction energy*. An approximation consists in summing the self-energy of the dislocation and an elastic energy corresponding to a reduced misfit, allowing for partial accommodation by the dislocation. However, at variance with the planar case, this reduced misfit may not be easy to define in NWs. In any case, to these elastic contributions must be added the energy of the dislocation core. Before giving the main results about critical parameters, we briefly review the rare calculations of the energy of NWs with dislocations. With regard to straight dislocations, the full calculation of the elastic energy was performed by Gutkin *et al.* for isotropic materials with identical elastic constants [GUT 00] (the strain fields are not given and do not seem to have been calculated elsewhere). Raychaudhuri and Yu do treat WZ crystals but their calculations appear oversimplified: they used the reduced misfit approximation (once again with homogeneous strains in core and shell) and dislocation strain fields for infinite media, thereby neglecting relaxation at the free sidewalls [RAY 06b, RAY 06a]. With regard to prismatic loops, Raychaudhuri and Yu used the same crude approximations as for the straight dislocation and Haapamaki *et al.* also used the strain field of a loop in an infinite medium [HAA 12]. However, Ovid'ko and Sheinerman have calculated the elastic fields of the loop in an elastically isotropic cylinder [OVI 04] and Cai and Weinberger gave a semi-analytical calculation of the corresponding energy [CAI 09].

In addition, Chu *et al.* considered loops tilted with respect to the axis of an elastically isotropic NW with a circular section; these loops therefore adopt an elliptical shape in the interface [CHU 11]. This may look like a more realistic configuration than circular loops when one recalls that glide planes may not be normal to the NW axis (see below). The authors fitted extensive numerical calculations to a simple formula, taking into account the free surface by an image force method. However, at variance with other models (Figure 2.4), they predict an increase in the critical shell thickness with core radius and, consequently, no critical

3 Somewhat surprisingly, Ovid'ko and coworkers predicted that, at a given misfit, it is also possible to grow shells without dislocation *above* another critical value [GUT 00, OVI 06]. From the discussion by Raychaudhuri and Yu of their own results [RAY 06b], it appears that this might be an artifact induced by the consideration of a single dislocation only.

radius. The reasons for these discrepancies are not clear and this model will not be discussed further.

Comparing the energies of states S_e and S_d results in an equilibrium criterion which tells nothing about which dislocations may actually form. Since there are usually no dislocations in the core before shell growth, it is widely expected that dislocations form as half-loops at the surface of the shell and move toward the interface by glide or climb; dislocation reactions have also been invoked. These questions are important since the critical dimensions depend on the defects considered via the energy of S_d. We may go a step further by comparing states with different types of dislocations. There indeed exist speculations about which types of dislocations are possible and consequently about which part of the misfit might be relieved first. The predictions are, however, somewhat conflicting. Ovid'ko and coworkers [OVI 04] and Nazarenko *et al.* [NAZ 13] predicted, by simply comparing the total energies of the NW with either type of dislocations and leaving aside crystallography, that loops are generally easier to form than axial dislocations. Salehzadeh *et al.* also argued [SAL 13] that loops should appear first because they relieve axial strain which is present in both core and shell, whereas radial and in-plane tangential strains are much lower in the core (Figure 2.3). Raychaudhuri and Yu pointed out a difference between fcc and hexagonal crystals: in the former, dislocations along <111> (the NW axis) are not stable (according to Frank's criterion [HIR 82]) so that stress relief should occur first via loops, whereas in the latter, dislocations along **c** (with in-plane BVs along **a** or equivalent directions) are possible. Moreover, the calculations of these authors indicate that the critical dimensions relative to these axial dislocations are always lower than those relative to the loops so that they should appear first in WZ crystals. However, Raychaudhuri and Yu did not consider which slip systems might be activated to allow these dislocations to glide to the heterointerface. This question was treated in-depth by Perillat-Merceroz *et al.* [PER 12] in the case of WZ NWs with m-plane facets (Figure 2.1(c)). The authors noted that the glide planes activated depend on the location around the NW. Along the major part of each m facet, the stress state is nearly biaxial, close to what it would be in a 2D layer grown on this plane (Figure 2.2); glide of axial dislocations is then possible in prismatic planes not normal to the facet. Near the edges where two facets intersect, stress is not biaxial anymore and other slip systems may be activated, which possibly facilitates plastic relaxation. In any case, glide on the basal **c** plane is not allowed, which rules out the direct formation of the prismatic loops considered in the simple models. On the other hand, several pyramidal glide planes may be activated. Over and above these crystallographic limitations, the kinetic barriers to dislocation formation and movement, which are known to increase the effective critical thickness of 2D layers, must be kept in mind.

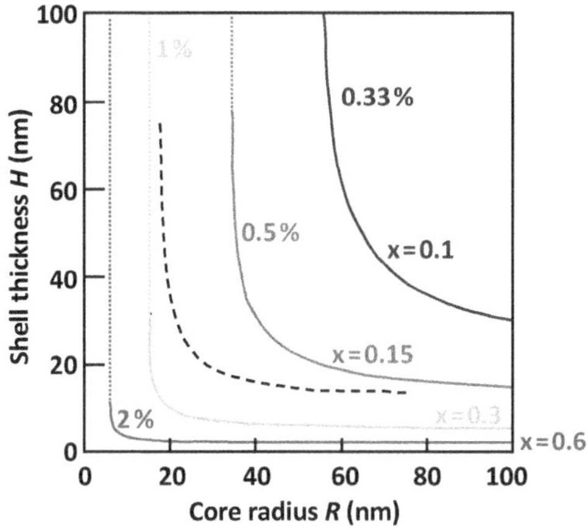

Figure 2.4. *Full lines and their dotted extensions: calculations of the critical dimensions of GaN/Al$_x$Ga$_{1-x}$N core-shell NWs by Raychaudhuri and Yu [RAY 06b]. Each curve corresponds to a different shell composition x and hence to the misfit indicated in % (average of $\Delta a/a$ and $\Delta c/c$), and gives the variation of the critical shell thickness H_c with core radius R. The asymptotes for $H \to \infty$ and $R \to \infty$ give, respectively, the critical radius R_c and the 2D critical thickness. The diagram may also be read as a stability map as a function of NW core radius R and shell thickness H: each curve separates the domain of elastic relaxation (state S_e) at left and bottom from the domain of plastic relaxation (state S_d) at right and top. Dashed line: the same for WZ InAs/InP NWs (3.2% misfit) as calculated by Salehzadeh et al. [SAL 13]. For a color version of this figure, see www.iste.co.uk/consonni/nanowires1.zip*

These considerations emphasize the limitations of the simple models that include prismatic loops and axial dislocations, and only these defects. It is nevertheless interesting to mention some quantitative results of these models. First, it is noteworthy that the predicted critical shell thickness usually transits from a value close to the 2D critical thickness to an infinite value over a small range of decreasing core radii. For most misfits, the critical shell thickness is thus either very small or quasi-infinite and the transition gets smoother only for small misfits (Figure 2.4). Second, various authors found different values of critical dimensions for a given misfit (Figures 2.4 and 2.5). The critical core radii calculated by Ovid'ko *et al.* [OVI 04], Aifantis *et al.* [AIF 07] (isotropic elasticity) and Raychaudhuri and Yu [RAY 06b] (nitrides) are of the same order whereas those calculated by Salehzadeh *et al.* (non-nitride WZ and ZB III–Vs) are nearly 10 times larger [SAL 13]. The critical radii found by Trammell *et al.* lie somewhere between these extremes whereas those of Haapamaki *et al.* [HAA 12] (InAs/Al$_x$In$_{1-x}$As) are extremely small and the value given by Nazarenko *et al.* (300 nm for a 2.7% misfit in In$_x$Ga$_{1-x}$As/GaAs) is extremely large. These discrepancies have not been explained or much commented so far, although they affect key portions of the parameter space. For instance, it is only for misfits less than approximately 1% that the first groups find critical radii approaching those of the narrowest existing NWs whereas, for Salehzadeh *et al.*, who claim that their calculations reproduce well the experimental results, the critical radii are still of about 20 nm for a misfit of 3% (this holds for ZB; for WZ crystals, the authors give slightly smaller values).

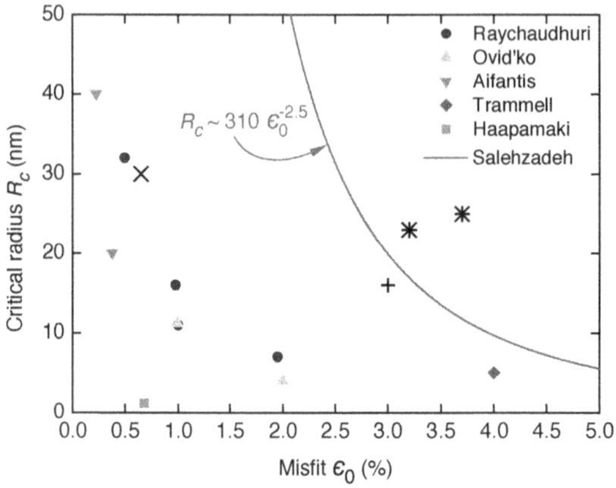

Figure 2.5. *Full symbols and line: critical radii calculated as a function of misfit* ϵ_0 *by Raychaudhuri and Yu [RAY 06b], Ovid'ko et al. [OVI 04], Aifantis et al. [AIF 07], Trammel et al. [TRA 08], Haapamaki et al. [HAA 12] and Salehzadeh et al. [SAL 13] for various material systems (see text). In the case of Raychaudhuri and Yu,* $\epsilon_0 = (1/2)(\Delta a/a + \Delta c/c)$ *and data for two nitride systems are included. The other symbols give upper bounds of* R_c *estimated by us from the observation of dislocations by Perillat-Merceroz et al. [PER 12] (X), Dayeh et al. [DAY 13] (+) and Salehzadeh et al. [SAL 13] (*). For a color version of this figure, see www.iste.co.uk/consonni/nanowires1.zip*

2.3.2.2. *Experiments*

There are convincing reports of misfitting core-shell heterostructures free of interfacial dislocations with shell thicknesses exceeding the 2D critical thickness in systems such as Ge/(Si,Ge) [LU 05, VAR 09, DAY 13], WZ InAs/InP [SAL 13], GaN/AlN [RIG 11] or ZnO/(Zn,Mg)O [PER 12].

The absence of dislocations for core-shell dimensions R,H means that $H_c(R) > H$, whereas the presence of dislocations implies both that $R_c < R$ and $H_c(R) < H$. The determination of the critical core radius appears difficult, since it should in theory involve the growth of very thick shells, which would in particular make difficult the observation of the interface. In practice, we might content ourselves with the growth of shells with thicknesses of the order of the core diameter, since the outer portion of thicker shells is effectively free of in-plane strain while most of the axial misfit is accommodated by the core (Figure 2.3). The first reports of plastic relaxation in these structures only appeared recently but it does not seem that any measurement of the critical core radius has been claimed so far.

Dislocations corresponding to the simple models discussed above, namely sets of prismatic edge loops and axial edge dislocations (Figures 2.1(b) and (c)), have apparently only been observed in WZ InAs/GaAs heterostructures [POP 11]. The NW cores being much thicker than any calculated critical radius for the very large SF misfit involved (about 8%), the prismatic loops, which are not expected as primary defects in WZ structures [PER 12], may have formed by dislocation interaction. For a much lower misfit ($\Delta a/a = +0.6\%$, $\Delta c/c = -0.7\%$), Perillat-Merceroz *et al.* have observed in WZ ZnO/(Zn,Mg)O NWs two families of half-loops lying, respectively, in prismatic and pyramidal planes and presenting segments threading through the shell. The loops of the first family relieve in-plane misfit and may later evolve in the axial dislocations of the model. The loops of the second family, which surround the core only partially, have BVs with both in-plane and axial components and therefore simultaneously relieve axial and in-plane tangential misfits [PER 12] (for more details about heterostructures in ZnO-based materials, see Chapter 3 of Volume 2). A quasiperiodic distribution of similar loops relieving both misfit components has been observed in cubic Ge/(Si,Ge) heterostructures by Dayeh *et al.* [DAY 13]. These loops lie in a $\{111\}$ plane not normal to the NW axis and have a standard $\mathbf{a}/2 < 110 >$ BV with both in-plane and axial components. Since they are closed and entirely located in the heterointerface, they represent a more mature stage of relaxation.

2.4. Axial heterostructures

2.4.1. *Elastic relaxation in axial heterostructures*

We now consider an NW of radius R, constituted of a foot of material A and an epilayer of material B with thickness h (Figure 2.1(a)). We assume that the foot is much longer than the NW diameter, thereby constituting an effective substrate of semi-infinite extension along the NW axis. In this case, we may forget about the actual bulk substrate which may support the NW foot, since it is located far away from the interface. If one is interested in the mechanisms of strain relaxation during growth, the most pertinent system involves a B layer with a free top surface. On the other hand, investigating the effect of strain on the electrical and optical properties of NWs may involve double A/B/A heterostructures, possibly wrapped by a shell (mixed axial/radial heterostructures).

The elastic state of purely axial heterostructures has been calculated by several authors. As in the case of core-shell structures, the differences between these calculations bear on the section of the NW, the elastic properties of the materials and the method of solution of the equations of elasticity. The problem is solved by imposing interfacial coherency and cancellation of the forces on all free surfaces. In the case of an infinitely long NW (the heterostructure then consists of two semi-infinite materials A and B), it suffices to cancel the radial and tangential forces on the sidewalls. Such infinite A/B NWs were treated by Ertekin *et al.* by way of a variational method, which consists of positing analytical forms for the elastic strains in the system, specified by a few parameters, calculating the total elastic energy and minimizing it as a function of the parameters [ERT 02, ERT 05]. Since the trial strain fields are not solutions of the equations of elasticity and do not produce free surfaces, this can hardly be used quantitatively. A more general problem was actually treated rigorously by Barton long before NWs were born [BAR 41, TIM 51]. However, in the framework of continuum linear elasticity (where the strain state of the system depends only on the ratios of its dimensions), Barton's solution for an infinitely long cylindrical bar subjected to a uniform pressure along part h of its length describes perfectly well an infinite NW A including a misfitting B section[4]. Barton's method consists of finding an analytical solution for an elementary sinusoidal axial modulation of SF parameter (assuming uniform elastic constants for an elastically isotropic material). Any distribution of SF parameter, and in particular any distribution of axial inclusions, can then be solved by Fourier synthesis. The same problem was more recently treated by other authors, who however overlooked Barton's work [KAS 04].

Barton's solution has been extended to WZ NWs with circular section and identical elastic constants. Zhong and Sun considered a single axial inclusion, assuming transverse elastic anisotropy [ZHO 02]. In Barton's spirit, Kaganer and Belov solved the problem of an arbitrary distribution of SF lattice parameter by deriving the elementary solution for a single Fourier component [KAG 12]. In the particular case of an axial superlattice (SL) consisting of alternating wells and barriers, Wölz *et al.* propose a "2D approximation" whereby, neglecting the presence of the free sidewalls, the in-plane lattice parameter is uniform in the whole structure [WOL 13]. The in-plane and axial strains (with respect to the SF states of A and B) are then uniform in well and barrier and proportional to the thicknesses of the barrier and well, respectively. This is indeed the solution for a laterally infinite SL without substrate. The authors show that the average axial strains so calculated are a good approximation if the total height of the SL is of the order of the NW diameter but deviate from the exact calculation for smaller heights. This suggests that strain engineering could be achieved in such structures by playing on the number of SL periods and ratio of well to barrier heights [WOL 13]. Boxberg *et al.* compared FE calculations of the strain fields induced by periodic distributions of axial inclusions and found that, as in the case of core-shell heterostructures, the main differences between ZB and WZ lie in the shear strains in the prismatic planes [BOX 12].

Turning now to atomistic approaches, Swadener and Picraux used an interatomic potential to find the equilibrium state of Si/Ge axial heterostructures [SWA 09]. The VFF method has also been used to compute strains for axial heterostructures in infinitely long NWs. Niquet and coworkers first considered ZB GaAs/InAs and

4 To recover a semi-infinite B layer, the $h \rightarrow \infty$ of Barton's "double heterostructure" solution may be taken.

InAs/InP SLs [NIQ 06, NIQ 08] and then the case of GaN nanodisks in an AlN NW [CAM 10b, LAN 10, JAL 12].

Mixed heterostructures comprising an axial insertion and a shell have been treated by several authors [NIQ 07, FUR 11, RAJ 12]. The shell is usually assumed to be of the same material as the semi-infinite NW foot and cap, and the final structure has a uniform diameter. As expected, the shell impedes elastic relaxation in the region of the inclusion. The elastic state of the system is very sensitive to the aspect ratio of the inclusion and the shell thickness [RAJ 12].

The model of an infinite NW is not relevant for growth, where one is interested in the evolution of thin epilayers with time. The case of a layer B of finite height with a free surface (Figure 2.1(a)) was treated by Glas for an isotropically elastic continuum [GLA 06]. The strains and the elastic energy were determined numerically in two ways, either by FE calculations (Figure 2.6) or by using Barton's solution. In the latter case, to find out the solution for the A/B heterostructure with a free top surface, an infinitely long NW, composed of a semi-infinite A foot, the B layer of interest and a semi-infinite A cap in which a small number of additional misfitting layers of finite thicknesses are inserted, is considered. Superimposing Barton's solution for each layer of this composite NW ensures that the lateral surfaces remain free. The number, positions, lengths and misfits of the additional inclusions are optimized to cancel as best as possible the forces on the top facet of layer B (which is now an internal interface of the system). The two methods give very similar results, except for the thinnest epilayers [GLA 06]. The second method has the advantage of providing analytical expressions for the elastic fields. FE calculations for heterostructures involving specific materials have also been performed [HAN 07], some of them including the effect of a rigid substrate relatively close to the heterostructure (finite foot height) [YE 09a, YE 09b].

Although different in details, these calculations confirm the main features expected for an uncapped axial NW heterostructure. When $h \ll R$, the system is close to a thin 2D B layer deposited on a semi-infinite A substrate and the elastic relaxation is the same as in this laterally infinite system, with no gain in energy, except at the very edge of the NW. When h increases (Figure 2.6 (a)), the B layer starts relaxing in all directions (because the top and lateral surfaces are free) and the substrate and epilayer share the strains even though the NW substrate is semi-infinite. As a rule of thumb, this happens when $h \sim 2R$, although Figure 2.6(a) shows that the top of the B layer is already quasi-strain-free for an aspect ratio $\rho = h/(2R)$ well below 1. When h increases further, any additional B material is deposited on an effectively strain-free surface and can thus adopt its own SF state (Figure 2.6(b)) so that the elastic energy does not increase anymore. This is the key to the existence of a critical radius.

This effect is conveniently quantified by function $f(\rho) = W_e^{NW}/W_e^{2D}$, where W_e^{NW} is the total elastic energy stored in the whole NW with a coherent interface (state S_e) and W_e^{2D} is the energy stored in the same volume cut in a planar heterostructure with a B epilayer of the same thickness h coherently deposited on a bulk A substrate. Quite generally, function f must satisfy two constraints [GLA 06]. First, as noted above, $f(\rho) \to 1$ when $h \to 0$. Second, when $h \gg R$, the system is only strained over a height of order R (Figure 2.6); hence, $W_e^{NW} \sim R^3$ whereas W_e^{2D} scales with the volume $\pi R^2 h$ of the uniformly strained epilayer, so that $f(\rho) \sim \rho^{-1}$ at large ρ. If A and B are elastically isotropic with the same Young's modulus E and Poisson ratio ν, $W_e^{2D} = \pi R^2 h E \epsilon_0^2/(1-\nu)$ and function f is independent of E (we denote it by f_ν). Figure 2.7 illustrates the variations of f_ν and shows that it depends only weakly on ν. We found that a good approximation is $f_\nu(\rho) \simeq \frac{1}{1+P_\nu \rho}$ ($P_\nu \simeq 27.3$ for $\nu = 1/3$) [GLA 06]. Since $P_\nu \gg 1$, the elastic energy is considerably reduced with respect to the 2D case even for modest aspect ratios, for instance, to about a quarter of its 2D value for $\rho = 0.1$. An even better approximation (that reproduces the non-trivial variations of W_e^{NW} around $\rho \simeq 0.3$; Figure 2.7) was proposed by Glas and Daudin in the form $f_\nu(\rho) = [1 + \tau_1 \rho \exp(-\tau_3 \rho)]/(1 + \tau_2 \rho)$, τ_i being ν-dependent parameters [GLA 12].

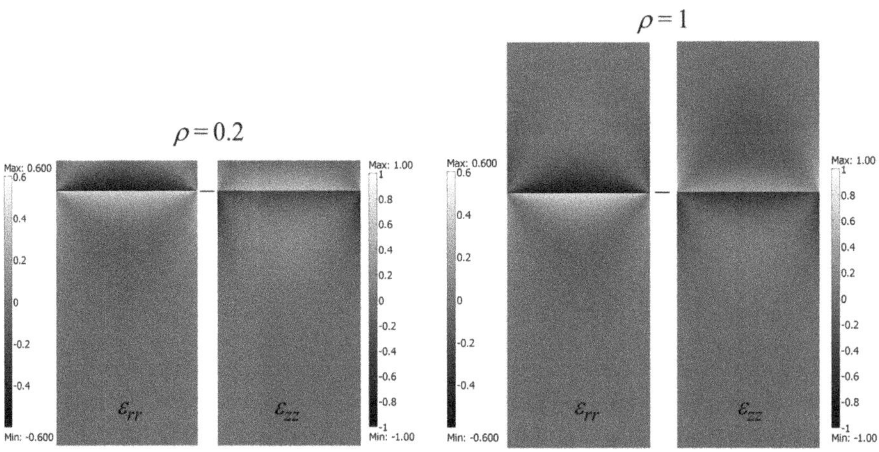

Figure 2.6. *Finite element calculation of the radial (ϵ_{rr}) and axial (ϵ_{zz}) dilatations in axial heterostructures for epilayer aspect ratios $\rho = 0.2$ and $\rho = 1$. The scale gives the strains normalized to the intrinsic misfit ϵ_0 between the semi-infinite foot of the NW and the epilayer, the interface between which is marked by a short segment. Note that even for $\rho = 0.2$, the strain is partitioned between the two materials. The materials are elastically isotropic with Poisson ratio $\nu = 0.33$ (Adapted from [GLA 13]). For a color version of this figure, see www.iste.co.uk/consonni/nanowires1.zip*

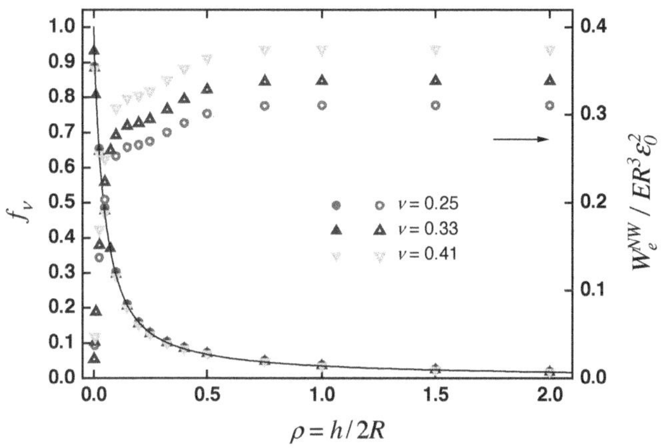

Figure 2.7. *Variation with B layer aspect ratio of the total energy W_e^{NW} stored in the coherent NW with A/B axial heterostructure (open symbols, right scale) and of the ratio f_ν of this energy to the energy stored in the same volume cut in a 2D system with identical epilayer thickness (full symbols, left scale), for three values of the Poisson ratio ν. The full line is a fit of the ratio with the second function f_ν given in the text [GLA 12] for $\nu = 0.33$. For a color version of this figure, see www.iste.co.uk/consonni/nanowires1.zip*

2.4.2. *Critical dimensions for axial heterostructures*

2.4.2.1. *Theoretical considerations*

With regard to plastic relaxation, axial heterostructures present the same two basic features as core-shell heterostructures, namely a radius-dependent critical thickness and a critical radius, both dependent on misfit.

For an A/B axial heterostructure of given radius R (Figure 2.1(a)), we may adopt the same criterion as in the 2D and core-shell cases, namely that the critical B layer thickness $h_c(R)$ is the thickness at which the energy of the fully coherent state S_e (section 2.4.1) becomes larger than that of state S_d, with interfacial misfit dislocations[5]. We are still interested in the introduction of the first dislocations, but the situation is different from the core-shell case, in which the stress could surely not be released fully by a single pair of dislocations of each type (whether they be prismatic loops, since the interface is infinite along the NW axis, or axial dislocations, due to the cylindrical geometry). Here, the interface is finite and has a single orientation. However, it remains necessary to accommodate the misfit along two independent directions in this plane. To do so, while preserving at least partly the

5 For the sake of completeness, we mention that Kästner and Gösele adopted another approach by considering the kinetics of dislocation formation and glide, albeit in an infinitely long wire, from which they derived values of critical thicknesses and radii for A/B/A axial heterostructures [KAS 04].

symmetry of the system for NWs of circular section with isotropic or cubic elasticity, the simplest is to posit a couple of orthogonal dislocations with in-plane edge components (Figure 2.1(a)). The few models so far developed have indeed considered a single pair of orthogonal dislocations of maximal length (i.e. passing through the center of the interfacial disk) [ERT 02, ERT 05, GLA 06, YE 09a, YE 09b]. As noted by Kästner and Gösele, this is somehow justified by the fact that, since the misfit-induced forces that act on the dislocation are radial, this is an equilibrium position for an interfacial dislocation [KAS 04]. This scheme is obviously not adapted to NWs with hexagonal cross sections.

A possible calculation method is to approximate the energy of state S_d by a sum of three terms (as outlined for the core-shell case in section 2.3.2.1), namely dislocation self-energy, core energy and elastic energy corresponding to the misfit reduced by introducing the dislocation. However, even with such a simple dislocation configuration, the calculation poses some problems. First, the strain field of a dislocation segment normal to the axis of a cylinder has not been calculated analytically so far. Second, in the case of a small finite interface, it is not clear which part of the misfit is accommodated by the dislocation pair. Notwithstanding these difficulties, Glas used this method to calculate the critical dimensions [GLA 06]. If ϵ_0 is the SF misfit, the reduced misfit is $\epsilon_0 - \epsilon_a$, ϵ_a being that part of the misfit which is accommodated by the dislocations. ϵ_a is taken as the misfit that would be accommodated in the 2D case by a dislocation array with the same length of dislocation per unit area. Building on an idea proposed previously for NWs lying on a substrate [OVI 02], the dislocation self-energy is evaluated by using the 2D formula [FIT 91] with a layer thickness equal to the distance $\bar{h}$ between the dislocation and the closest free surface. Since the distance to the NW sidewall obviously varies along the dislocation, a possibility is to take its average $\bar{d} = 2R/\pi$. For h small, $\bar{h} = h$, whereas $\bar{h} = \bar{d}$ for h large. Using this approach and the first expression of f_ν given above, Glas derived an implicit equation from which the radius-dependent critical thickness $h_c(R)$ may be calculated for given material parameters [GLA 06].

The variations of this critical thickness with R are shown by full lines in Figure 2.8 for several values of the misfit, assuming plastic relaxation by 60° dislocations of the fcc structure with $b = 0.4$ nm, a value pertaining to ZB GaAs and a good approximation for many cubic semiconductors. For each misfit, the $h_c(R)$ curve separates, in the (R, h) plane of layer dimensions, the coherency domain (below, left) from the domain of plastic relaxation (above, right). The critical thickness is seen to increase rapidly as the NW radius decreases. Moreover, for each misfit, $h_c(R) \to \infty$ when R tends to some finite value. This value is of course the critical radius $R_c(\epsilon_0)$, which will be discussed below.

Glas' approach was refined by Ye and coworkers, who performed FE calculations of the energy in state S_d. The effect of the edge component of the misfit dislocation was simulated by introducing an extra atomic half plane in the epilayer

[YE 09a, YE 09b]. In this way, the elastic energy is calculated in a single step; it then suffices to add the core energy of the dislocations. The counterpart is that this can only be done for particular values of the elastic parameters. The critical thicknesses determined in this way (dashed lines in Figure 2.8) are comparable to those determined by Glas but they vary less abruptly with NW radius.

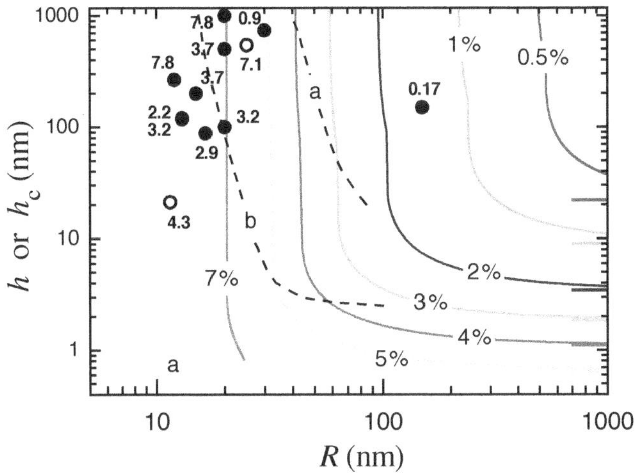

Figure 2.8. *Full lines: variations of the critical thickness h_c of a misfitting layer growing on top of an NW as a function of the NW radius, for various values of the misfit (given on each curve), as calculated by Glas for elastically isotropic materials with $\nu = 1/3$ [GLA 06]. In the plastically relaxed state, misfit is accommodated by 60° dislocations ($b = 0.4$ nm). Each segment at the right gives the asymptote of the curve immediately above for $R \to \infty$, i.e. the 2D critical thickness. Dashed lines: critical thickness calculated by Ye et al. assuming pure edge dislocations with $b = 0.3$ nm for misfits of 2.12% (a) and 4.78% (b) [YE 09b]. Symbols give the radii and thicknesses of layers grown without (disks) or with (circles) dislocations for various values of the misfit (indicated in percent near each point): 0.17% [WU 02], 0.9% [SVE 05], 2.2% and 2.9% [GLA 06], 3.2% [BJO 02b, LAR 07], 3.7% [VER 06, ZHA 10], 4.3% [TOU 11], 7.1% [HIR 96], 7.8% [GUO 06, JEP 08]. For a color version of this figure, see www.iste.co.uk/consonni/nanowires1.zip*

Since they model infinite A/B NWs, Ertekin *et al.* cannot calculate any critical thickness. These authors were, however, the first to determine a critical radius by comparing the energies of the infinite NW with and without a single pair of misfit dislocations [ERT 02, ERT 05]. In the former case, they evaluated the dislocation self-energy by replacing the layer thickness in the 2D formula [FIT 91] by the NW radius R and the residual elastic energy by setting $\epsilon_a = b/(2R)$. Ye *et al.* used their FE simulations to determine the critical radii for a small set of heterostructures [YE 09b]. In this case also, the approach of Glas has the advantage of providing an analytical solution. Glas indeed derived a second implicit equation which gives the critical radius for given misfit and material parameters [GLA 06]. These various estimates of critical radii are shown in Figure 2.9.

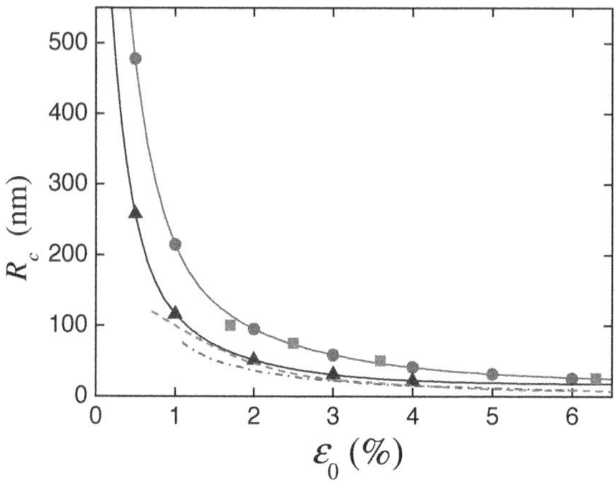

Figure 2.9. *Symbols and full lines: critical NW radius R_c under which arbitrarily thick misfitting layers should grow coherently, as a function of misfit ϵ_0, as calculated by Glas [GLA 06]. For $R > R_c$, plastic relaxation is assumed to occur via edge dislocations (triangles) or 60° dislocations (disks). b = 0.4 nm and $\nu = 0.33$. Dashed and dashed-dotted lines: calculations of Ye et al. and Ertekin et al. for pure edge dislocations with b = 0.3 nm [YE 09b, ERT 05]. Squares: critical radii measured by Chuang et al. for NWs on misfitting substrates [CHU 07]. For a color version of this figure, see www.iste.co.uk/consonni/nanowires1.zip*

As in the case of core-shell heterostructures, the very existence of a critical radius can be asserted independently of the particular expressions chosen for the energies involved. We saw in section 2.4.1 that the elastic energy W_e^{NW} in state S_e scales with R^3 if $\rho \geq 1$. It also limits when $\rho \to \infty$ (Figure 2.7). On the other hand, the energy of state S_d is surely larger than the energy of the dislocation cores, which is proportional to the dislocation length and hence to R (neglecting further relaxation at the intersections with the lateral surface, which involves a few atoms at most) and independent of layer height; this remains true irrespective of the precise location of the defects in the interface and even of their number. From these different power-law dependences upon R, it follows that, for infinitely thick layers, the dislocation energy always dominates, provided that R is small enough. In other words, there exists a critical radius R_c below which plastic relaxation is forbidden and coherent layers of arbitrary thickness should be obtainable. The general features of the $h_c(\epsilon_0, R)$ curves (Figure 2.8), in particular the horizontal (2D critical thickness) and vertical (critical radius) asymptotes, are thus independent of the expressions used in the calculations.

Let us now examine the quantitative estimates of the critical dimensions. First, similar to the core-shell case (Figure 2.4), at high misfit, the critical epilayer thickness $h_c(R)$ transits from values close to its 2D limit to a quasi-infinite value over a small range of decreasing NW radii (Figure 2.8), whereas for smaller misfits, the transition gets smoother. Second, the values of the critical dimensions found by various authors are comparable (Figures 2.8–2.9). The values of critical dimensions calculated by Ye *et al.* [YE 09b] tend to be lower than those of Glas [GLA 06], but the BVs considered are different. These similarities might arise partly from the fact that all authors consider the same dislocation configuration[6]. There is, however, an important difference between core-shell and axial heterostructures. We noted that in the former case, the calculated critical radii are small and they reach the order of typical NW radii only at low misfits (less than 1%, except for a few authors [SAL 13, NAZ 13]). For a given misfit, the critical radius of an axial heterostructure is about 10 times larger (compare Figures 2.5 and 2.9). Hence, even for misfits of several percent, the NW radius may be below the critical radius.

2.4.2.2. *Experiments*

To date, there is no systematic experimental determination of critical thicknesses in NWs with axial heterostructures and there is only a few mentions of the state (coherent or not) of the interface in such systems. In most strained axial heterostructures fabricated so far, the layers have been safely kept below their 2D critical thicknesses [POO 03, LAU 04]. However, there are reports of layers grown beyond this limit (Figure 2.8), most of which claim the absence of misfit dislocations [BJO 02a, WU 02, SVE 05, VER 06, GUO 06, GLA 06, JEP 08, LAR 07, CLA 08, ZHA 10]. There is an early report by Hiruma *et al.* of misfit dislocations in a cubic GaAs/InAs axial heterostructure [HIR 96]. More recently, Tourbot *et al.* observed by TEM a single edge dislocation with a BV $\mathbf{b} = 1/3 < \bar{1}2\bar{1}0 >$, lying in a GaN/ $Ga_{0.57}In_{0.43}N$ interface [TOU 11]. Interestingly, the dislocation is located midway between the sidewalls, as considered in the simple models (Figure 2.1(a)). In most cases (although not in this last one), the calculations of Glas [GLA 06] and Ye *et al.* [YE 09b] appear to predict correctly the state (coherent or not) experimentally observed (symbols in Figure 2.8). Probably for the reasons already mentioned in the case of core-shell heterostructures, there seems to be no experimental determination of the critical radii of axial heterostructures in NWs. However, here again, due to the rapid damping of the strains as the epilayer thickness increases (Figure 2.6), we might content ourselves with the absence of dislocations in layers with $h \simeq R$ to determine R_c.

6 The calculated critical diameters are larger than the spacing of an array of misfit dislocations that would relieve the same misfit in the 2D case. This suggests that the energy of the plastically relaxed state might be further reduced by introducing more than one pair of interfacial dislocations. The calculated critical radii would then be overestimated for large misfits.

2.5. Other possible modes of stress relaxation in nanowires with heterostructures

It has been known for decades that a mismatched material deposited on a planar substrate may accommodate the epitaxial stress by developing a non-planar surface. If, at least to start with, this occurs via a smooth surface corrugation, the phenomenon is usually considered as an instance of the Asaro–Tiller–Grinfeld (ATG) instability [ASA 72, GRI 86, SRO 89] of the planar free surface of a solid submitted to uniaxial or biaxial stress. For high misfit, growth may proceed via the formation of faceted islands without any initial planar growth stage (Volmer–Weber growth) or after a thin planar wetting layer has grown (Stranski–Krastanow growth). In any case, the initial relaxation is entirely elastic and the driving force for the morphological instability is the reduction of elastic energy afforded by relaxation at the lateral free surfaces, be they the flanks of the corrugation or the island facets. Since, as recalled in the previous sections, such lateral relaxation, which is made possible by the non-planarity of the structure, is the very reason why NWs relax stress efficiently, one may however wonder whether any significant extra elastic relaxation can be attained by forming a shell or an axial deposit of non-uniform thickness. The answer seems to be positive.

The spontaneous corrugation of the outer surface of a core-shell NW with a circular section was treated in detail by Schmidt *et al.* [SCH 08]. The authors performed a linear stability analysis in the spirit of that carried out for the standard ATG instability, where transport by surface diffusion allows a uniformly thick shell to develop a corrugation, if it is energetically favorable [SRO 89, SPE 92]. They tested the stability of the cylindrical surface against an elementary corrugation with a sinusoidal profile along the NW axis and around it, thereby producing localized mounds rather than ridges (except for zero wave numbers). The corrugation wavelength around the NW was supposed to be a harmonic $2\pi R/n$ of the base period, namely the shell perimeter. The authors argued that surface stress [MUL 04], which relates to the change of energy due to straining the surface (as opposed to that due to change of area, described by surface tension), must be included in the analysis due to the nanometric size of the structure. This, in turn, induces a behavior that depends on the sign of the misfit. The authors found that core-shell NWs may indeed be morphologically unstable. As usual in such analyses, they concentrated on the perturbation that develops most rapidly. The main effects of NW dimensions on this perturbation are that its wavenumber n gets smaller as R and H become smaller and that its growth rate decreases as H increases. The same effect was studied by other authors [DUA 08, WAN 08, CAO 13]. Some experiments do indeed reveal surface islanding in strained core-shell Si-Ge NWs, which might represent a later stage of the process; the islands' positions along the axis are in antiphase for opposite facets [PAN 05], which might correspond to the $n = 2$ harmonic of the theory [SCH 08]. InAs islands have also been grown on the side facets and at the edges of the side facets of AlAs-capped GaAs NWs [UCC 10] and GaN islands at the edges of AlN-capped GaN NWs [ARB 12]. It is however not clear whether strain plays an important role in these cases.

In turn, the possibility of the spontaneous formation of a non-uniformly thick axial deposit was investigated by Glas and Daudin [GLA 12]. The authors considered the formation of a cylindrical island that is narrower than the NW stem. To this end, they numerically calculated the total energy of the system (which comprises elastic and surface contributions) for a given deposit volume. They could thereby determine the optimal shape of the deposit as a function of misfit, NW radius and nominal deposit thickness. If the values of any two of these parameters are set, then there is a critical value of the third one above which the formation of a genuine island that is narrower than the stem is favored compared to that of a disk of equal volume covering entirely the top facet. These critical values are easily accessible in current NW systems and, at high misfit, islanding may lead to a large reduction of the total energy provided that R is not too small. Most often, the optimal island has a radius notably smaller than the NW; it is only for values of NW radius and deposit thickness just above their respective critical values that the deposit tends to the shape of a disk of radius R [GLA 12]. For semiconductor NWs, islanding is likely to occur primarily in the case of catalyst-free growth, since in the case of vapor–liquid–solid growth, each new ML tends to nucleate at the triple phase line [GLA 07] or in its vicinity before rapidly spreading over the whole top facet of the NW. It is possible that the spontaneous formation of an In-rich core surrounded by a Ga-rich shell which has been observed when an (In,Ga)N alloy is deposited on top of a GaN NW [CHA 10, TOU 11] is triggered by the formation of such an island. In this case, the shell would form after the core. The formation of faceted (In,Ga)N islands with bases narrower than the NW, observed in the same system, seems even a clearer evidence of the mechanism [TOU 12] (for more details, see Chapter 2 of Volume 2). However, core-shell structures might also form if all successive MLs adopted the NW radius but with a radially modulated composition, as studied theoretically by Niu *et al.* [NIU 12], a mechanism in which strain also plays an important role.

2.6. Summary and conclusions

By selecting the appropriate material system, growth method and growth conditions, it is possible to fabricate axial or core-shell heterostructures (and combinations thereof) in semiconductor NWs. The NW geometry offers great flexibility for associating in such heterostructures materials that differ significantly in terms of lattice constants, without introducing extended defects. This arises mainly from the ease with which the internal misfit stress may relax elastically at the NW sidewalls, which constitute a free boundary. This allows the grower to associate materials that cannot be assembled coherently not only in 2D planar geometry, but even as quantum dots on a substrate.

More precisely, for either core-shell or axial heterostructures, there exists a critical deposit thickness below which no interfacial dislocation could form. This is, of course, also the case for planar deposits; however, for a given misfit, the values for a NW are systematically larger. Even more noteworthy is the existence of a critical NW radius for each type of heterostructure, below which arbitrary thick deposits may be grown without interfacial defects being generated. This results from the fact that a thick deposit is strained only over a distance from the interface of the order of the NW radius, the rest of it being effectively free of stress. Although they tend to the 2D values only when the radius becomes infinite, the critical deposit thicknesses decrease rapidly when the NW radius increases above its critical value, unless the misfit is quite small. Critical thicknesses and critical radii also decrease when misfit increases. Experiments have confirmed that thicker axial or shell layers may be grown without defects on a NW than in the case of a planar heterostructure.

Beyond these similarities, axial and core-shell heterostructures present important differences, at least in the present state of our knowledge. For axial heterostructures, the few calculations of critical thicknesses and critical radii that exist yield values of the same order, which also appear compatible with the experiments. There are more calculations of critical dimensions for core-shell heterostructures, but the values found by various authors may differ widely. As yet, the reasons for these discrepancies are not clear and they cannot be explained solely by differences of material systems. Because misfit must be relieved along the axial and in-plane tangential directions of the core-shell interface, it is often assumed that two different types of dislocations (straight lines and loops, of edge character) must form. The joint consideration of the NW geometry and the possible slip systems for a given crystal structure shows that the actual situation might be more complex. The first experiments on plastic relaxation indeed show the "textbook" situation in some cases but sometimes also the presence of tilted dislocation loops that relieve both components of the misfit. Especially for core-shell structures, plastic relaxation thus calls for more theoretical and experimental investigations.

Since strain affects the electronic and optical properties of semiconductor NWs (see Chapter 1), the studies reviewed in this chapter provide useful guidelines for the choice of the geometry and material parameters likely to produce heterostructures with specific strain-related physical properties, without compromising the physical properties of the structure by introducing extended defects.

2.7. Bibliography

[AIF 07] AIFANTIS K.E., KOLESNIKOVA A.L., ROMANOV A.E., "Nucleation of misfit dislocations and plastic deformation in core/shell nanowires", *Philosophical Magazine*, vol. 87, pp. 4731–4757, 2007.

[ARB 12] ARBIOL J., MAGEN C., BECKER P., *et al.*, "Self-assembled GaN quantum wires on GaN/AlN nanowire templates", *Nanoscale*, vol. 4, pp. 7517–7524, 2012.

[ASA 72] ASARO R.J., TILLER W.A., "Interface morphology development during stress corrosion cracking: part I. Via surface diffusion", *Metallurgical Transactions*, vol. 3, pp. 1789–1796, 1972.

[BAR 41] BARTON M.V., "The circular cylinder with a band of uniform pressure on a finite length of the surface", *Journal of Applied Mechanics*, vol. 8, pp. A97–A104, 1941.

[BEC 11] BÉCHÉ A., BOUGEROL C., COOPER D., *et al.*, "Measuring two dimensional strain state of AlN quantum dots in GaN nanowires by nanobeam electron diffraction", *Journal of Physics nference Series*, vol. 326, p. 012047, 2011.

[BEL 10] BELLET-AMALRIC E., ELOUNEG-JAMROZ M., BOUGEROL C., *et al.*, "Epitaxial growth of ZnSe and ZnSe/CdSe nanowires on ZnSe", *Physica Status Solidi C*, vol. 7, pp. 1526–1529, 2010.

[BJO 02a] BJÖRK M.T., OHLSSON B.J., SASS T., *et al.*, "One-dimensional heterostructures in semiconductor nanowhiskers", *Applied Physics Letters*, vol. 80, pp. 1058–1060, 2002.

[BJO 02b] BJÖRK M.T., OHLSSON B.J., SASS T., *et al.*, "One-dimensional steeplechase for electrons realized", *Nano Letters*, vol. 2, pp. 87–89, 2002.

[BOU 09] BOUGEROL C., SONGMUANG R., CAMACHO D., *et al.*, "The structural properties of GaN insertions in GaN/AlN nanocolumn heterostructures", *Nanotechnology*, vol. 20, p. 295706, 2009.

[BOU 10] BOUGEROL C., SONGMUANG R., CAMACHO D., *et al.*, "Structural properties of GaN nanowires and GaN/AlN insertions grown by molecular beam epitaxy", *J. Phys. Conf. Ser.*, vol. 209, p. 012010, 2010.

[BOX 12] BOXBERG F., SØNDERGAARD N., XU H.Q., "Elastic and piezoelectric properties of zincblende and wurtzite crystalline nanowire heterostructures", *Advanced Materials*, vol. 24, pp. 4692–4706, 2012.

[CAI 09] CAI W., WEINBERGER C.R., "Energy of a prismatic dislocation loop in an elastic cylinder", *Mathematics and Mechanics of Solids*, vol. 14, p. 192, 2009.

[CAM 10a] CAMACHO D., NIQUET Y.M., "Application of Keating's valence force field model to non-ideal wurtzite materials", *Physica E*, vol. 42, pp. 1361–1364, 2010.

[CAM 10b] CAMACHO D., NIQUET Y.-M., "Stark effect in GaN/AlN nanowire heterostructures: Influence of strain relaxation and surface states", *Physical Review B*, vol. 81, p. 195313, 2010.

[CAO 13] CAO Y.Y., OUYANG G., WANG C.X., *et al.*, "Physical mechanism of surface roughening of the radial Ge-core/Si-shell nanowire heterostructure and thermodynamic prediction of surface stability of the InAs-core/GaAs-shell nanowire structure", *Nano Letters*, vol. 13, pp. 436–443, 2013.

[CHA 10] CHANG Y.-L., WANG J.L., LI F., *et al.*, "High efficiency green, yellow, and amber emission from InGaN/GaN dot-in-a-wire heterostructures on Si(111)", *Applied Physics Letters*, vol. 96, p. 013106, 2010.

[CHU 07] CHUANG L.C., MOEWE M., CHASE C., *et al.*, "Critical diameter for III-V nanowires grown on lattice-mismatched substrates", *Applied Physics Letters*, vol. 90, p. 043115, 2007.

[CHU 11] CHU H.J., WANG J., ZHOU C.Z., *et al.*, "Self-energy of elliptical dislocation loops in anisotropic crystals and its application for defect-free core/shell nanowires", *Acta Materialia*, vol. 59, pp. 7114–7124, 2011.

[CLA 08] CLARK T.E., NIMMATOOR P., LEW K.-K., *et al.*, "Diameter dependent growth rate and interfacial abruptness in vapor-liquid-solid Si/$Si_{1-x}Ge_x$ heterostructure nanowires", *Nano Letters*, vol. 8, pp. 1246–1252, 2008.

[CON 13] CONESA-BOJ S., DUNAND S., RUSSO-AVERCHI E., *et al.*, "Hybrid axial and radial Si-GaAs heterostructures in nanowires", *Nanoscale*, vol. 5, pp. 9633–9639, 2013.

[CON 14] CONSONNI V., FEUILLET G., *Wide Band Gap Semiconductor Nanowires 2: Heterostructures and Optoelectronic Devices*, ISTE, London and John Wiley & Sons, New York, 2014.

[CRO 13] CROS A., MATA R., HESTROFFER K., *et al.*, "Ultraviolet Raman spectroscopy of GaN/AlN core-shell nanowires: core, shell, and interface modes", *Applied Physics Letters*, vol. 102, p. 143109, 2013.

[DAY 13] DAYEH S.A., TANG W., BOIOLI F., *et al.*, "Direct measurement of coherency limits for strain relaxation in heteroepitaxial core/shell nanowires", *Nano Letters*, vol. 13, pp. 1869–1876, 2013.

[DIL 12] DILLEN D.C., VARAHRAMYAN K.M., CORBET C.M., *et al.*, "Raman spectroscopy and strain mapping in individual Ge-Si_xGe_{1-x} core-shell nanowires", *Physical Review B*, vol. 86, p. 045311, 2012.

[DUA 08] DUAN H.L., WEISSMÜLLER J., WANG Y., "Instabilities of core-shell heterostructured cylinders due to diffusions and epitaxy: spheroidization and blossom of nanowires", *Journal of the Mechanics and Physics of Solids*, vol. 56, pp. 1831–1851, 2008.

[ERT 02] ERTEKIN E., GREANEY P.A., SANDS T.D., *et al.*, "Equilibrium analysis of lattice-mismatched nanowire heterostructures", in KLIMOV V., BURIAK J.M., WAYNER D. D.M., *et al.* (eds.), *Quantum Confined Semiconductor Nanostructures*, Mat. Res. Soc. Proc., vol. 737, Pittsburgh, p. F10.4, 2002.

[ERT 05] ERTEKIN E., GREANEY P.A., CHRZAN D.C., "Equilibrium limits of coherency in strained nanowire heterostructures", *Journal of Applied Physics*, vol. 97, p. 114325, 2005.

[EYM 07] EYMERY J., RIEUTORD F., FAVRE-NICOLIN V., *et al.*, "Strain and shape of epitaxial InAs/InP nanowire superlattice measured by grazing incidence X-ray techniques", *Nano Letters*, vol. 7, pp. 2596–2601, 2007.

[FAV 09] FAVRE-NICOLIN V., EYMERY J., KOESTER R., *et al.*, "Coherent-diffraction imaging of single nanowires of diameter 95 nanometers", *Physical Review B*, vol. 79, p. 195401, 2009.

[FAV 10] FAVRE-NICOLIN V., MASTROPIETRO F., EYMERY J., *et al.*, "Analysis of strain and stacking faults in single nanowires using Bragg coherent diffraction imaging", *New Journal of Physics*, vol. 12, p. 035013, 2010.

[FIT 91] FITZGERALD E.A., "Dislocations in strained-layer epitaxy: theory, experiment, and applications", *Materials Science Reports*, vol. 7, pp. 87–142, 1991.

[FUR 11] FURTMAYR F., TEUBERT J., BECKER P., *et al.*, "Carrier confinement in GaN/Al$_x$Ga$_{1-x}$N nanowire heterostructures ($0 < x \leq 1$)", *Physical Review B*, vol. 84, p. 205303, 2011.

[GLA 90] GLAS F., GORS C., HÉNOC P., "Diffuse scattering, size effect and alloy disorder in ternary and quaternary III/V compounds", *Philosophical Magazine B*, vol. 62, pp. 373–394, 1990.

[GLA 06] GLAS F., "Critical dimensions for the plastic relaxation of strained axial heterostructures in free-standing nanowires", *Physical Review B*, vol. 74, p. 121302, 2006.

[GLA 07] GLAS F., HARMAND J.C., PATRIARCHE G., "Why does wurtzite form in nanowires of III-V zinc blende semiconductors?", *Physical Review Letters*, vol. 99, p. 146101, 2007.

[GLA 12] GLAS F., DAUDIN B., "Stress-driven island growth on top of nanowires", *Physical Review B*, vol. 86, p. 174112, 2012.

[GLA 13] GLAS F., "Heterostructures and strain relaxation in semiconductor nanowires", WANG S., (ed.), *Lattice engineering: Technologies and applications*, Chapter 5, Pan Stanford, Singapore, pp. 189–228, 2013.

[GRI 86] GRINFEL'D M.A., "Instability of the separation boundary between a non-hydrostatically stressed elastic body and a melt", *Soviet Physics – Doklady, or: Soviet Physics Doklady*, vol. 31, pp. 831–834, 1986.

[GRO 09] GRÖNQVIST J., SØNDERGAARD N., BOXBERG F., *et al.*, "Strain in semiconductor core-shell nanowires", *Journal of Applied Physics*, vol. 106, p. 053508, 2009.

[GUO 06] GUO Y.N., ZOU J., PALADUGU M., *et al.*, "Structural characteristics of GaSb/GaAs nanowire heterostructures grown by metal-organic chemical vapor deposition", *Applied Physics Letters*, vol. 89, p. 231917, 2006.

[GUT 00] GUTKIN M.Y., OVID'KO I.A., SHEINERMAN A.G., "Misfit dislocations in wire composite solids", *J. Phys.: Condens. Matter*, vol. 12, pp. 5391–5401, 2000.

[HAA 12] HAAPAMAKI C.M., BAUGH J., LAPIERRE R.R., "Critical shell thickness for InAs-Al$_x$In$_{1-x}$P core-shell nanowires", *Journal of Applied Physics*, vol. 112, p. 124305, 2012.

[HAA 13] HAAG S.T., RICHARD M.-I., FAVRE-NICOLIN V., *et al.*, "In situ coherent X-ray diffraction of isolated core-shell nanowires", *Thin Solid Films*, vol. 530, pp. 113–119, 2013.

[HAN 07] HANKE M., EISENSCHMIDT C., WERNER P., *et al.*, "Elastic strain relaxation in axial Si/Ge whisker heterostructures", *Physical Review B*, vol. 75, p. 161303, 2007.

[HAS 92] HASEGAWA H., LEE V.-G., MURA T., "The stress fields caused by a circular cylindrical inclusion", *Journal of Applied Mechanics*, vol. 59, pp. S107–S114, 1992.

[HES 10] HESTROFFER K., MATA R., CAMACHO D., *et al.*, "The structural properties of GaN/AlN core-shell nanocolumn heterostructures", *Nanotechnology*, vol. 21, p. 415702, 2010.

[HIR 82] HIRTH J.P., LOTHE J., *Theory of Dislocations*, Wiley, New York, 1982.

[HIR 96] HIRUMA K., MURAKOSHI H., YASAWA M., *et al.*, "Self-organized growth of GaAs/InAs heterostructure nanocylinders by organometallic vapor phase epitaxy", *Journal of Crystal Growth*, vol. 163, pp. 226–231, 1996.

[HOC 12] HOCEVAR M., IMMINK G., VERHEIJEN M., *et al.*, "Growth and optical properties of axial hybrid III-V/silicon nanowires", *Nature Communications*, vol. 3, p. 1286, 2012.

[HOC 13] HOCEVAR M., GIANG L.T.T., SONGMUANG R., *et al.*, "Residual strain and piezoelectric effects in passivated GaAs/AlGaAs core-shell nanowires", *Applied Physics Letters*, vol. 102, p. 191103, 2013.

[HYT 98] HŸTCH M.J., SNOECK E., KILAAS R., "Quantitative measurement of displacement and strain fields from HREM micrographs", *Ultramicroscopy*, vol. 74, pp. 131–146, 1998.

[JAL 12] JALABERT D., CURÉ Y., HESTROFFER K., *et al.*, "Strain state of GaN nanodisks in AlN nanowires studied by medium energy ion spectroscopy", *Nanotechnology*, vol. 23, p. 425703, 2012.

[JEP 08] JEPPSSON M., DICK K.A., WAGNER J.B., *et al.*, "GaAs/GaSb nanowire heterostructures grown by MOVPE", *Journal of Crystal Growth*, vol. 310, pp. 4115–4121, 2008.

[KAG 12] KAGANER V.M., BELOV A.Y., "Strain and x-ray diffraction from axial nanowire heterostructures", *Physical Review B*, vol. 85, p. 125402, 2012.

[KAS 04] KÄSTNER G., GÖSELE U., "Stress and dislocations at cross-sectional heterojunctions in a cylindrical nanowire", *Philosophical Magazine*, vol. 84, pp. 3803–3824, 2004.

[KAV 12] KAVANAGH K.L., SAVELIEV I., BLUMIN M., *et al.*, "Faster radial strain relaxation in InAs-GaAs core-shell heterowires", *Journal of Applied Physics*, vol. 111, p. 044301, 2012.

[KEA 66] KEATING P.N., "Effect of invariance requirements on the elastic strain energy of crystals with application to the diamond structure", *Physical Review*, vol. 145, pp. 637–645, 1966.

[KEH 13] KEHAGIAS T., DIMITRAKOPULOS G.P., BECKER P., *et al.*, "Nanostructure and strain in InGaN/GaN superlattices grown in GaN nanowires", *Nanotechnology*, vol. 24, p. 435702, 2013.

[KEP 09] KEPLINGER M., MÅRTENSSON T., STANGL J., *et al.*, "Structural investigations of core-shell nanowires using grazing incidence X-ray diffraction", *Nano Letters*, vol. 9, pp. 1877–1882, 2009.

[LAN 10] LANDRÉ O., CAMACHO D., BOUGEROL C., *et al.*, "Elastic strain relaxation in GaN/AlN nanowire superlattice", *Physical Review B*, vol. 81, p. 153306, 2010.

[LAN 11] LANEUVILLE V., DEMANGEOT F., PÉCHOU R., *et al.*, "Double strain state in a single GaN/AlN nanowire: probing the core-shell effect by ultraviolet resonant Raman scattering", *Physical Review B*, vol. 83, p. 115417, 2011.

[LAR 07] LARSSON M.W., WAGNER J.B., WALLIN M., *et al.*, "Strain mapping in free-standing heterostructured wurtzite InAs/InP nanowires", *Nanotechnology*, vol. 18, p. 015504, 2007.

[LAU 02] LAUHON L.J., GUDIKSEN M.S., WANG D., *et al.*, "Epitaxial core-shell and core-multishell nanowire heterostructures", *Nature*, vol. 402, pp. 57–61, 2002.

[LAU 04] LAUHON L.J., GUDIKSEN M.S., LIEBER C.M., "Semiconductor nanowire heterostructures", *Philosophical Transactions of the Royal Society A*, vol. 362, pp. 1247–1260, 2004.

[LIM 09] LIM S.K., BREWSTER M., QIAN F., *et al.*, "Direct correlation between structural and optical properties of III-V nitride nanowire heterostructures with nanoscale resolution", *Nano Letters*, vol. 9, pp. 3940–3944, 2009.

[LU 05] LU W., XIANG J., TIMKO B.P., *et al.*, "One-dimensional hole gas in germanium/silicon nanowire heterostructures", *Proceedings of the National Academy of Sciences*, vol. 102, pp. 10046–10051, 2005.

[MAR 70] MARTIN R.M., "Elastic properties of ZnS structure semiconductors", *Physical Review B*, vol. 1, pp. 4005–4011, 1970.

[MON 10] MONTAZERI M., FICKENSCHER M., SMITH L.M., *et al.*, "Direct measure of strain and electronic structure in GaAs/GaP core-shell nanowires", *Nano Letters*, vol. 10, pp. 880–886, 2010.

[MUL 04] MÜLLER P., SAÚL A., "Elastic effects on surface physics", *Surface Science Reports*, vol. 54, pp. 157–258, 2004.

[MUS 05] MUSIN R.N., WANG X.-Q., "Structural and electronic properties of epitaxial core-shell nanowire heterostructures", *Physical Review B*, vol. 71, p. 155318, 2005.

[NAZ 13] NAZARENKO M.V., SIBIREV N.V., NG K.W., *et al.*, "Elastic energy relaxation and critical thickness for plastic deformation in the core-shell InGaAs/GaAs nanopillars", *Journal of Applied Physics*, vol. 113, p. 104311, 2013.

[NIQ 06] NIQUET Y.M., "Electronic and optical properties of InAs/GaAs nanowire superlattices", *Physical Review B*, vol. 74, p. 155304, 2006.

[NIQ 07] NIQUET Y.M., "Effects of a shell on the electronic properties of nanowire superlattices", *Nano Letters*, vol. 7, pp. 1105–1109, 2007.

[NIQ 08] NIQUET Y.-M., CAMACHO D., "Quantum dots and tunnel barriers in InAs/InP nanowire heterostructures: electronic and optical properties", *Physical Review B*, vol. 77, p. 115316, 2008.

[NIU 12] NIU X., STRINGFELLOW G.B., LEE Y.-J., *et al.*, "Simulation of self-assembled compositional core-shell structures in $In_xGa_{1-x}N$ nanowires", *Physical Review B*, vol. 85, p. 165316, 2012.

[OVI 02] OVID'KO I.A., "Relaxation mechanisms in strained nanoislands", *Physical Review Letters*, vol. 88, p. 046103, 2002.

[OVI 04] OVID'KO I.A., SHEINERMAN A.G., "Misfit dislocation loops in composite nanowires", *Philosophical Magazine*, vol. 84, pp. 2103–2118, 2004.

[OVI 06] OVID'KO I.A., SHEINERMAN A.G., "Misfit dislocation in nanocomposites with quantum dots, nanowires and their ensembles", *Advances in Physics*, vol. 55, pp. 627–689, 2006.

[PAN 05] PAN L., LEW K. K., REDWING J.M., *et al.*, "Stranski-Krastanow growth of Germanium on silicon nanowires", *Nano Letters*, vol. 5, pp. 1081–1085, 2005.

[PER 12] PERILLAT-MERCEROZ G., THIERRY R., JOUNEAU P.-H., *et al.*, "Strain relaxation by dislocation glide in ZnO/ZnMgO core-shell nanowires", *Applied Physics Letters*, vol. 100, p. 173102, 2012.

[POO 03] POOLE P.J., LEFEBVRE J., FRASER J., "Spatially controlled nanoparticle-free growth of InP nanowires", *Applied Physics Letters*, vol. 83, pp. 2055–2057, 2003.

[POP 11] POPOVITZ-BIRO R., KRETININ A., VON HUTH P., *et al.*, "InAs/GaAs core-shell nanowires", *Cryst. Growth Design*, vol. 11, pp. 3858–3865, 2011.

[RAJ 12] RAJADELL F., ROYO M., PLANELLES J., "Strain in free standing CdSe/CdS core-shell nanorods", *Journal of Applied Physics*, vol. 111, p. 014303, 2012.

[RAY 06a] RAYCHAUDHURI S., YU E.T., "Calculation of critical dimensions for wurtzite and cubic zinc blende coaxial nanowire heterostructures", *Journal of Vacuum Science and Technology B*, vol. 24, pp. 2053–2059, 2006.

[RAY 06b] RAYCHAUDHURI S., YU E.T., "Critical dimensions in coherently strained coaxial nanowire heterostructures", *Journal of Applied Physics*, vol. 99, p. 114308, 2006.

[RIG 11] RIGUTTI L., JACOPIN G., LARGEAU L., *et al.*, "Correlation of optical and structural properties of GaN/AlN core-shell nanowires", *Physical Review B*, vol. 83, p. 155320, 2011.

[SAD 05] SADD M.H., *Elasticity*, Elsevier, Amsterdam, 2005.

[SAL 13] SALEHZADEH O., KAVANAGH K.L., WATKINS S.P., "Geometric limits of coherent III-V core/shell nanowires", *Journal of Applied Physics*, vol. 114, p. 054301, 2013.

[SCH 08] SCHMIDT V., MCINTYRE P.C., GÖSELE U., "Morphological instability of misfit-strained core-shell nanowires", *Physical Review B*, vol. 77, p. 235302, 2008.

[SØN 09] SØNDERGAARD N., HE Y., FAN C., *et al.*, "Strain distributions in lattice-mismatched semiconductor core-shell nanowires", *Journal of Vacuum Science and Technology B*, vol. 27, pp. 827–830, 2009.

[SPE 92] SPENCER B.J., VOORHEES P.W., DAVIS S.H., "Morphological instability in epitaxially strained dislocation-free solid films: linear stability theory", *Journal of Applied Physics*, vol. 73, pp. 4955–4970, 1992.

[SRO 89] SROLOVITZ D.J., "On the stability of surfaces of stressed solids", *Acta Metallurgica*, vol. 37, pp. 621–625, 1989.

[SVE 05] SVENSSON C.P.T., SEIFERT W., LARSSON M.W., *et al.*, "Epitaxially grown GaP/$GaAs_{1-x}P_x$/GaP double heterostructure nanowires for optical applications", *Nanotechnology*, vol. 16, pp. 936–939, 2005.

[SWA 09] SWADENER J.G., PICRAUX S.T., "Strain distributions and electronic property modifications in Si/Ge axial nanowire heterostructures", *Journal of Applied Physics*, vol. 105, p. 044310, 2009.

[TIM 51] TIMOSHENKO S.P., GOODIER J.N., *Theory of Elasticity*, McGraw-Hill, New York, 1951.

[TOU 11] TOURBOT G., BOUGEROL C., GRENIER A., *et al.*, "Structural and optical properties of InGaN/GaN nanowire heterostructures grown by PA-MBE", *Nanotechnology*, vol. 22, p. 075601, 2011.

[TOU 12] TOURBOT G., BOUGEROL C., GLAS F., *et al.*, "Growth mechanism and properties of InGaN insertions in GaN nanowires", *Nanotechnology*, vol. 23, p. 135703, 2012.

[TRA 08] TRAMMELL T.E., ZHANG X., LI Y., *et al.*, "Equilibrium strain-energy analysis of coherently strained core-shell nanowires", *Journal of Crystal Growth*, vol. 310, pp. 3084–3092, 2008.

[UCC 10] UCCELLI E., ARBIOL J., MORANTE J.R., *et al.*, "InAs quantum dot arrays decorating the facets of GaAs nanowires", *ACS Nano*, vol. 4, pp. 5985–5993, 2010.

[VAR 09] VARAHRAMYAN K.M., FERRER D., TUTUC E., *et al.*, "Band engineered epitaxial Ge-Si_xGe_{1-x}core-shell nanowire heterostructures", *Applied Physics Letters*, vol. 95, p. 033101, 2009.

[VER 06] VERHEIJEN M.A., IMMINK G., DE SMET T., *et al.*, "Growth kinetics of heterostructured GaP-GaAs nanowires", *Journal of the American Chemical Society*, vol. 128, pp. 1353–1359, 2006.

[WAN 08] WANG H., UPMANYU M., CIOBANU C.V., "Morphology of epitaxial core-shell nanowires", *Nano Letters*, vol. 8, pp. 4305–4311, 2008.

[WAR 91] WARWICK C.M., CLYNE T.W., "Development of composite coaxial cylinder stress analysis model and its application to SiC monofilament systems", *Journal of Materials Science*, vol. 26, pp. 3817–3827, 1991.

[WOL 13] WÖLZ M., RAMSTEINER M., KAGANER V.M., *et al.*, "Strain engineering of nanowire multi-quantum well demonstrated by Raman spectroscopy", *Nano Letters*, vol. 13, pp. 4053–4059, 2013.

[WU 02] WU Y., FAN R., YANG P., "Block-by-block growth of single-crystalline Si/SiGe superlattice nanowires", *Nano Letters*, vol. 2, pp. 83–86, 2002.

[YE 09a] YE H., LU P., YU Z., *et al.*, "Critical lateral dimension for a nanoscale-patterned heterostructure using the finite element method", *Semiconductor Science and Technology*, vol. 24, p. 025029, 2009.

[YE 09b] YE H., LU P., YU Z., *et al.*, "Critical thickness and radius for axial heterostructure nanowires using finite-element method", *Nano Letters*, vol. 9, pp. 1921–1925, 2009.

[ZHA 10] ZHANG G., TATENO K., GOTOH H., *et al.*, "Structural, compositional, and optical characterizations of vertically aligned AlAs/GaAs/GaP heterostructure nanowires epitaxially grown on Si substrate", *Jpn. Journal of Applied Physics*, vol. 49, p. 015001, 2010.

[ZHO 02] ZHONG Z., SUN Q.P., "Analysis of a transversely isotropic rod containing a single cylindrical inclusion with axisymmetric eigenstrains", *International Journal of Solid Structures*, vol. 39, pp. 5763–5765, 2002.

3

Surface-related Optical Properties of GaN-Based Nanowires

3.1. Introduction

Among semiconductor nanowires (NWs), nanocolumns or nanorods, those based on wide band gap compounds are being considered as interesting building blocks for visible light emitting devices (for more details, refer to Chapter 5 in Volume 2 [CON 14]) or photovoltaic converters. One advantage that is often put forward in favor of *self-organized* nitride NWs (diameters between 10 and 100 nm, typically) is the extremely low density of defects such as threading dislocations, at significant variance from the usual case of two-dimensional (2D) layers that contain a lot of those non-radiative defects (for more details, refer to Chapter 8). Therefore, much effort has been made in the past decade in order to promote the utilization of dense arrays of nitride-based NWs for efficient visible-light emitting devices. In addition to the absence of dislocations, the NW configuration is, indeed, expected to favor higher incorporation of indium (for more details, refer to Chapter 2 in Volume 2), lower quantum confined Stark effect and better light extraction from the device. After some spectacular demonstrations in Kishino and coworkers [SEK 10, NAK 12], the current trend is to leave self-organized growth aside, in favor of selective-area growth (SAG, for more details, refer to Chapter 5), which is foreseen to permit better control of the position, diameter, composition, contacting and periodicity of arrays of nano-LEDs. The essential differences between the latter SAG and self-organized growth modes are: (1) significantly larger NW diameters (above ~100 nm, typically) in case of SAG and (2) the loss, with SAG, of the crystalline perfection of self-organized growth.

Chapter written by Pierre LEFEBVRE.

Now, as far as optical properties are concerned, the feature that is common to all NW growth modes is the large developed surface of the samples, mainly consisting of the lateral m-planes, since the growth axis is the (0001) axis in the vast majority of cases. For example, with diameters of 50 nm, heights of 1 μm and NW density of $10^{10}cm^{-2}$, we obtain a developed surface of ~30 cm^2, per cm^2 of sample. It is therefore logical to expect that the general properties of GaN-based NWs are more sensitive to surface-related phenomena than in the case of 2D layers. A renowned demonstration of this has been made by Calarco *et al.* [CAL 05] who proved the sensitivity of electronic conduction in GaN NWs to their diameter thus emphasizing the importance of the pinning of the Fermi level at (lateral) surface states, which they modeled via a continuous-medium approach of near-surface band curvature in an n-doped semiconductor with donor density N_D.

The purpose of this section is to review and discuss the recent observations of the impact of surface effects on the optical properties of GaN-based NWs. It is worth mentioning that most of the discussions can be directly transposed to the case of ZnO-based NWs (for more details, refer to Chapter 4 in Vol. 1). Restricting ourselves to radiative recombination of electron–hole pairs, we will consider three aspects: (1) the existence of specific exciton and donor quantum states related to surfaces; (2) the intrinsic non-radiative activity of surfaces; (3) the impact on electron–hole pair recombination dynamics of the physical and chemical environment of the NWs. We will not discuss optical properties related to lattice vibrations, such as surface optical phonons.

3.2. Specific exciton and donor states related to surfaces

Refined emission spectroscopy of pure GaN NWs revealed excitonic features that were clearly related to defects (e.g. stacking faults) stemming from the coalescence of adjacent NWs during their growth [CON 09, LEF 11]. Such defects occupy the entire section of the NW, including its core, forming inclusions of cubic GaN in the wurtzite matrix, acting as quantum wells at which the excitons are trapped. They induce a series of low-temperature (low-T) photoluminescence (PL) lines at lower energies than the dominant neutral-donor-bound exciton line (I_2-line), such as the well-known line near 3.42 eV (for more details, refer to Chapter 6).

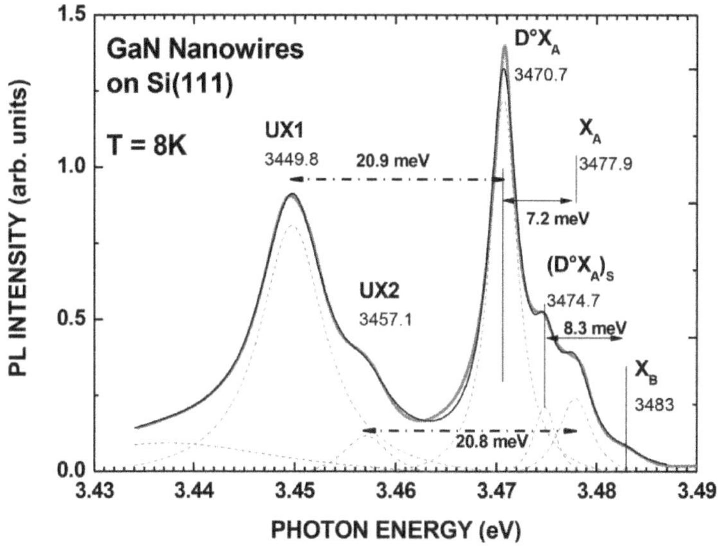

Figure 3.1. *Typical PL spectrum of a dense ensemble of GaN NWs grown by molecular beam epitaxy on Si(111) substrate*

However, some PL lines, uniquely observed in GaN NWs, were proven to be intrinsic to *isolated* NWs, [LEF 11] and recognized as being related to surface effects. This is the case of the doublet of lines near 3.45 eV [CAL 00]. Indeed, contrary to the lines arising from cubic inclusions, the intensity of this doublet increases relatively to the I_2 line, when the NW density (and also their diameter) is reduced and therefore when the NW coalescence is suppressed. Figure 3.1 shows one example of deconvolution by Lorentzian line-shapes of a typical continuous-wave (CW) PL spectrum of GaN NWs grown by plasma-assisted molecular beam epitaxy. Two free exciton lines X_A and X_B are associated with the two upper valence bands (A and B) of wurtzite GaN and lie exactly at the energies that they have in unstrained GaN. The I_2 line (labeled $D°X_A$) dominates the spectrum, as it results from the efficient trapping of A-excitons, at low T, by residual donors (likely silicon atoms). UX1 and UX2 denote the two "unknown exciton" lines that compose the doublet.

The identification of the UX lines is still controversial, though there is a current consensus on its excitonic nature and on its connection with surfaces. During the past few years, most investigations of the UX lines were based on relative intensity behaviors of the different lines upon changing doping, excitation conditions, T, NW environment, etc. They were tentatively assigned to inversion domain boundaries [ROB 07], to point to defects such as Ga vacancies [FUR 08], or to "some abundant surface defect" [BRA 10]. However, the latter group implicitly questioned

[PFU 10b] the relation of UX lines to surfaces, since their intensity seemed to be less affected by changes in the near-surface electric field than the I_2 line (see discussion on those fields below). A thorough review of the literature on the 3.45 eV luminescence is available in the recent paper by Sam-Giao *et al.* [SAM 13], along with a critical discussion of the different hypotheses. Below, we give more details about one of them, which will allow us to emphasize the role of the – indisputable – alteration of excitonic and donor quantum states by the proximity of the surface.

Upon its first observation by Calleja *et al.* [CAL 00], the energetic similarity of the UX doublet with the so-called "two-electron satellite" (TES) of the I_2 line was noticed. But this interpretation was discarded because of the abnormally large relative intensity of the doublet. In high-quality GaN, the TES lines are, indeed, ~50–100 times less intense than I_2 [KOR 99, PAS 07]. They correspond to recombination of a neutral-donor-bound exciton (D°X), which leaves, as a final state, the donor in one of its excited states (2s, 2p, 3s, 3p, 3d, etc.), instead of the ground state (1s), like for the I_2 line. Therefore, the TES band is in fact composed of a large number of lines, depending on the donor (e.g. O or Si) and on the final state involved. The lines involving the 2s and 2p states are generally the most intense. This is why the energy separation between I_2 and the main TES is of ~21–22 meV, close to 3/4 of the silicon donor binding energy in GaN, i.e. the splitting between n = 1 and n = 2 donor states. Paskov *et al.* [PAS 07] also pointed out the existence of a second family of TES, lying ~7 meV above the main TES. The second line corresponds to an initial state that is none of the D°X bound states but, instead, a state in the continuum of unbound (scattering) states: an unbound exciton interacts with the donor without binding to it and promotes the donor into one of its excited states. This yields a PL line red-shifted from the free exciton line by ~3/4 of the donor binding energy.

A point that has never been observed up to now is illustrated in Figure 3.1: the splittings marked therein strongly suggest that UX2 could be related to the latter "unbound exciton TES" whereas UX1 could correspond to the bound-exciton TES. This suggestion is reinforced by the T dependent relative intensities of UX1 and UX2 (Figure 3.2), which are nearly identical to the intensities exhibited by the I_2 and FX_A lines: the collapsing of the lower lying transition results from the thermally activated de-trapping of the exciton from the donor into the continuum. However, as we already noted, for GaN NWs with diameters in the 20–100 nm range, transitions UX1 and UX2 are much more intense than the TES usually are in bulk GaN.

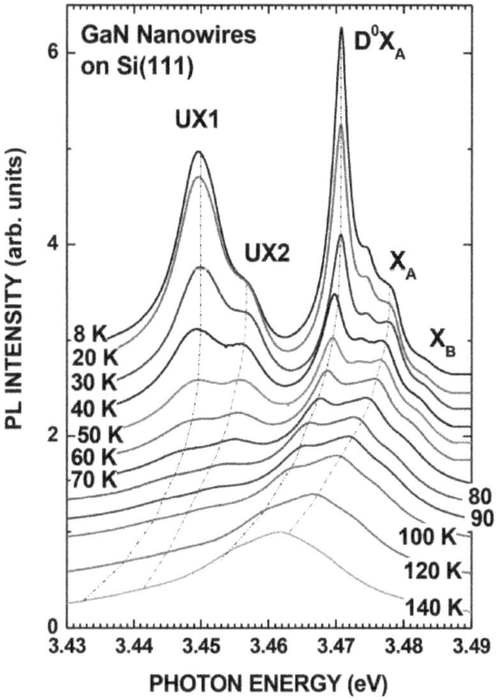

Figure 3.2. *Temperature dependence of the PL spectrum of a dense ensemble of GaN NWs grown by molecular beam epitaxy on Si(111) substrate*

Microscopic PL (μPL) of NWs detached from their substrates provides interesting information regarding the various excitonic lines that can be observed and their relative intensities. Examples are shown in Figure 3.3: not all examined excitation spots reveal sharp PL lines, as illustrated by Figure 3.3(a) [PFU 11], and in some cases only very weak PL – and even *no line* at all – is seen around 3.47 eV, whereas a group of intense, narrow lines appear near 3.45 eV (*see,* e.g. the lower right spectrum in Figure 3.3(a). The same observation was reported by Sam-Giao [SAM 12], as shown in Figure 3.3(b), in an attempt to correlate the nature of the substrate on which the NWs are dispersed and the aspect of the μPL spectrum: the 3.45 eV PL is obviously composed by the superposition of several distinct, narrow lines which are, in some cases, the only lines observable. We wish to make an important remark, here: the absence of I_2 lines in some of these spectra *does not* rule out the assignment of UX lines to TES of the D°X recombination. Indeed, as discussed below, symmetry consideration can very well explain why, for certain positions of the donors in the NW, the I_2 line would lose its intensity to the benefit of the TES.

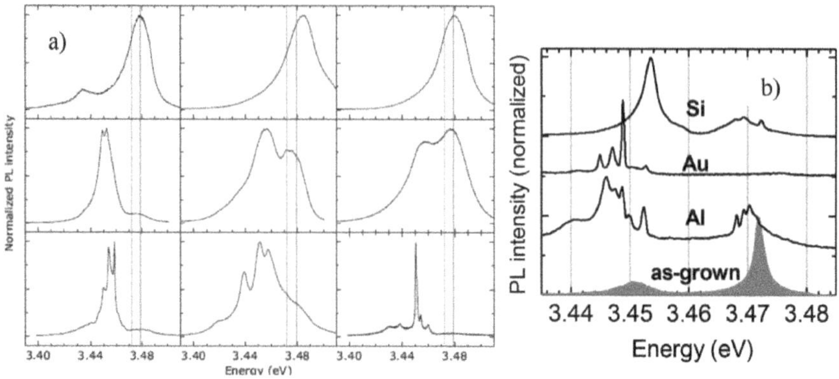

Figure 3.3. *A few examples of microPL spectra of isolated GaN NWs dispersed on various substrates: a)[PFU 11], b) [SAM 12]*

It can be useful, at this stage, to figure out what controls the relative intensities of all the donor-related excitonic lines. For finite T between 2 and 15 K, Paskov *et al.* [PAS 07] made the remarkable demonstration that the possible initial state for both I_2 and TES lines *is not* the only ground state of the D°X complex, but rather one among the various "excited rotational" states that can be seen as excited orbital states for the hole, attracted by the two electrons and repelled by the D^+ nucleus [NEU 99]. These so-called (a), (b), (c), etc. excited states of the D°X can be thermally populated at 10–15 K, because their energy splitting scales with the exciton binding energy and remains in the range of ~1 meV in GaN (rather ~2 meV for ZnO) [GIL 07]. Even more importantly, Paskov *et al.* demonstrated that the PL intensity of a given TES line is really a matter of compatible symmetries between the initial and final states involved, in addition to the population of the initial state.

Based on such symmetry arguments, Corfdir, Lefebvre and coworkers [COR 09, LEF 11] suggested that the UX lines do correspond to TES lines, but that their abnormal intensities result from the tremendous distortion undergone by the wave functions of all involved states when the donor is close to the NW surface. Due to the vicinity of the surface, initial donor-bound (ground and excited) exciton states, on the one hand, and final neutral donor (ground and excited) states, on the other hand, are affected by the impossibility for the carriers to exist outside the semiconductor. Hence, depending on the exact position of the donor from the surface, not only all transition energies (I_2, TES, etc.) are altered but, above all, the spatial distributions of electron and hole presence probabilities are profoundly different from the highly symmetrical bulk situation. In other words, for donor nuclei lying within a surface layer of, say, 15 nm, both the energies and relative probabilities of the various donor-related excitonic transitions are extremely position-dependent.

A detailed calculation supporting this argumentation is still lacking, for obvious reasons of complexity: we need to compute the position-dependent exact energies and wave functions of the ground *and* excited states of *both* the D°X complex and of the D°, inside a hexagon-shaped GaN NW of varying size. Nevertheless, restricting ourselves to the only final states of the transitions, we can convince ourselves of the sensitivity of the donor states to the donor distance from a flat surface.

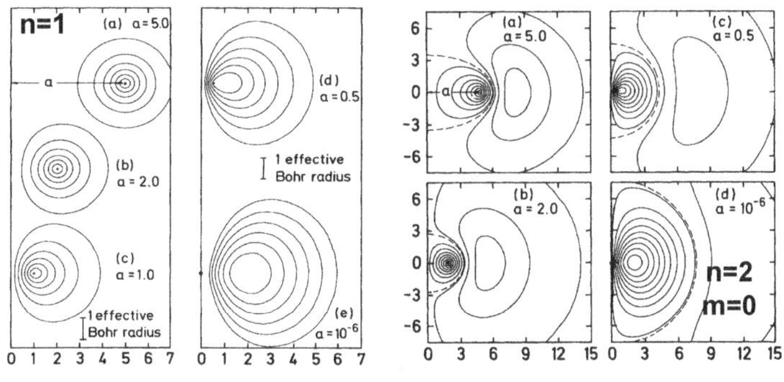

Figure 3.4. *Calculated probability densities for the ground states (n=1) and one of the first two excited states (n=2,m=0) of a donor, for different distances, a, of the donor nucleus from the surface of the semiconductor. (Adapted figures with permission from [SAT 83] – copyright 1983, the American Physical Society)*

In fact, the calculation of the n = 1 and n = 2 states was performed 30 years ago by Satpathy [SAT 83], who treated the hypothetical case of a position-dependent exciton close to a surface, but who set the hole mass to infinity: therefore the case treated was really the one of an effective-mass donor. In anticipation of the discussions in the next paragraphs, we wish to emphasize the simplicity of the approach: no dielectric mismatch between the semiconductor and its surrounding medium is considered and no intrinsic surface states are introduced, yet. The vacuum is seen here as a sharp, infinitely high-potential barrier. Satpathy's calculations were presented in units of the exciton Bohr radius (Figure 3.4) and binding energy. Therefore, we can transpose his results to the case of silicon donor in GaN by setting the Bohr radius to 3 nm and the binding energy to 30 meV. From Figure 3.4, we can readily infer the following: (1) the wave-function of the n = 1 state is distorted for the donor lying within a first, ~6 nm thick, surface region (SR1); (2) the wave functions of the two non-degenerate n=2 (m = 0, m = 1) states are distorted for the donor lying not only in SR1 but also in an extended surface region (SR2), up to ~15 nm from the surface; (3) for electrons weakly bound to near-surface donors, the wave functions – including the one of the ground state – present a much larger spatial extension than in the bulk case.

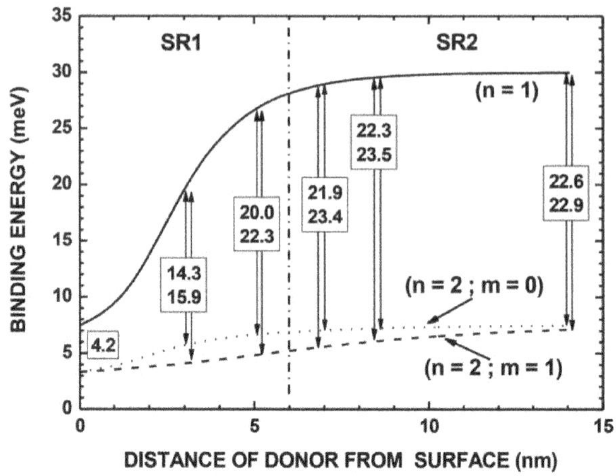

Figure 3.5. *Existence of two surface regions (SR1 and SR2): binding energies of the ground and first excited states for a shallow donor in GaN versus the distance of the nucleus from the surface (energy differences indicated in meV)*

Figure 3.5 displays the binding energies of the (n = 1), (n = 2,m = 0) and (n = 2,m = 1) states of the donor, extrapolated from the dimensionless calculation by Satpathy [SAT 83] (given, therein, in the form of numerical tables) as a function of the distance, *a*, from an infinite-potential wall. Note that the two symmetries m=0 and m=1 exhibit different sensitivities to the presence of the surface. We can distinguish two cases:

1) For 4 $<a<$ 15 nm, the energy difference between n =1 and n = 2 states remains in the range of 20–24 meV. However, displacing the donor within the same range of positions results in an extremely large variety of situations in terms of respective wave functions of the same n = 1 and n = 2 states (Figure 3.4).

2) For 0 $<a<$ 4 nm, all the energies vary drastically, in particular the n=1 binding energy which covers the 5–20 nm range.

Generally, excited states being more extended in space, they will be influenced by the surface from larger distances. This is expected to be the case, too, for D°X states, which involve two electrons attracted by the nucleus and repelling each other and a hole attracted by the two electrons but repelled by the nucleus. In terms of symmetry, reducing the distance *a* will make the D°X wave function of the ground state resemble more and more the wave functions of the less symmetrical (1), (2), etc., rotational excited states, just like the ground state of the donor looks more and more similar to a 2p state. It is even possible, given the small binding energy of the exciton to the neural donor (7 meV), that the D°X be unstable for a small enough *a*. In such a case, neither the I_2 line nor any TES would be observed.

For a certain range of a< 4 nm, both binding energies of the D°X and of the D° are certainly positive but drastically reduced, compared to the bulk situation. This should lead to I_2 lines *above* the usual line. This is precisely what has been observed by Brandt *et al.* [BRA 10] by μPL on selected, strain-free, individual GaN NWs. Sharp transitions were indeed observed in the region of 3.475 meV and were assigned to surface donor-bound excitons, denoted by $(D°X_A)_S$. Interestingly, a peak at this energy is also observed by cw-PL in high-quality GaN NW ensembles, as shown in Figure 3.1. Brandt *et al.* [BRA 10] also discovered that the donors may also be able to migrate toward the NW surface, as proved by the blue-shift of the emission energy of one $(D°X_A)_S$ after one month of natural ageing.

For 4 <a< 15 nm, both the I_2 and main satellites are found more or less at their "normal" energies. For the larger a values, in particular, the situation is quite close to the one in bulk GaN, except for a slight a-dependence of the transition energy for the I_2 transition. This is now recognized as the intrinsic origin of the broadening of the I_2 line in ensembles of GaN NWs: it result from the statistical distribution of distances between donor sites and surfaces [COR 09, BRA 10]. Statistical treatment of micro-PL results also reveals [PFU 10a] that unintentionally doped GaN NWs, which generally contain a very small number of donors per NW and even no donor at all in a significant percentage of cases. This point is extremely important not only regarding the mechanisms of generation, capture and recombination of excitons, but also because it questions the modeling of surface band curvatures in GaN NWs by the classical approach based on continuous space charges arising from homogeneous densities of donors. This approach may not be transposable to the nanoscale. For nuclei lying in a certain range of a, between 4 and 6 nm, the distortion of the D°X and D° wave functions is such that the satellite recombination would become the more probable, for reasons of *compatible symmetries*, but still keeping energies close to the "normal" ones. The latter point is really the one that is still under debate.

Recently, Sam-Giao *et al.* [SAM 13] used polarization-resolved μPL and magneto-PL on ensembles of GaN NWs to study the UX lines in self-organized GaN NWs. They discussed in particular the importance of birefringence and intrinsic polarization effects in a highly anisotropic effective medium, together with the sensitivity of emitted light polarization to the emission angle, which had never been explicitly accounted for in previous studies. Based on this, they argue against the direct, quantitative use of relative PL intensities, without a proper modeling of the electromagnetic modes as a function of NWs densities and diameters. They eventually refute the TES hypothesis, based on two main arguments:

1) The I_2 and UX lines appear to be counter-polarized with a significant amount of linear polarization of the UX line along the (0001) growth axis of the NWs. It is

argued that since the D°X and TES lines have a common initial state, built upon the A-valence band, they should have identical polarization features.

2) The shifts and splitting (in fact, *no* splitting, here) of the UX line under applied magnetic fields both in Voigt and Faraday configurations are not those observed for TES lines in bulk GaN, whereas the I_2 line of GaN NWs does show similar behavior as in the bulk. Based on the hypothesis that the only hole band involved would be the A-valence-band, it is argued that, since both I_2 and TES transitions have identical initial states, this observation rules out the possibility that the UX line would be a two-electron satellite of the $D°X_A$ transition.

We believe that this conclusion is insufficiently supported: the initial state involved in the intense I_2 line of the cw-PL spectrum *of an ensemble* of NWs is *not the same* as the initial state involved in the intense UX doublet. They most certainly pertain to different donors in a given NW and most probably even to different NWs. Some NWs exhibit solely the UX emission, others the only I_2 line, some are even dominated by the free exciton emission [PFU 10a]. It is the complex, statistical distribution of situations, including donor positions, charge transfer and recombination dynamics, optical out-coupling scheme, etc., which results in a given final cw-PL spectrum. In terms of modeling the symmetry, we should consider the D°X problem in a GaN NW as one of the hole bound to a D^- complex, close enough to the curved surface of a cylinder of radius comprised between 20 and 100 nm. A natural axial symmetry of the problem appears along the c-axis, due to this curvature. In addition, preventing the hole to penetrate into nearly half of the space usually available in the bulk case cannot be considered as a small perturbation and therefore treating the problem without mixing the three valence-bands of wurtzite GaN is most certainly too strong an approximation.

We see that surface-donor based explanation of the UX lines is at the same time the most satisfying regarding the *parsimony principle* (Occam's razor) and the most difficult to handle from the theoretical point of view. As a matter of fact, in addition to all the ingredients already mentioned (multiple particles, valence-band mixings, curved surfaces, ground and excited states, etc.), the dielectric mismatch between the NW and its surrounding medium, was proved [COR 12, DIA 07, SLA 07] to have a significant impact on impurity and exciton eigenstates in NWs. Last but not least, the exact near-surface band bending scheme, discussed in the sections, certainly should be accounted for. Now, the resulting radial electric field is another argument in favor of the distortion of donor and exciton wave functions.

In summary, (1) surface-related excitonic transitions have been identified around 3.475 eV as resulting from the presence of donors close to the lateral m-planes of GaN NWs; (2) the distance of donors from these planes is recognized as the cause of slight energy variations of the main D°X (I_2) PL line; (3) most authors associate the 3.45 eV UX doublet to surface-related excitons, but its relation to regular shallow donors is still debated.

3.3. Non-radiative surface recombination

From the perspective of improving visible light emitting devices, great expectations have been placed on arrays of NWs to bypass the recurrent problem of high densities of threading dislocations that limit the radiative efficiency of classical, 2D heterostructures. However, in spite of the superior – even quasi-perfect – crystal quality, e.g. of self-organized NWs, it has become quite clear, along the past decade, that the room-temperature (RT) radiative efficiency is limited by significant losses. The latter are very often assigned to surface recombination processes that are also detrimental to photovoltaic applications.

In the case of pure GaN NWs, the PL spectra in Figure 3.2 show that excitons tend to delocalize from donors and to become free to move throughout the entire NW above ~100 K. When further increasing T, the overall PL intensity usually decreases quite rapidly, with eventual intensity at 300 K, 2 to 4 orders of magnitude smaller that the low-T one. It is logical to assign this behavior to the capture of carriers by non-radiative defects. In the absence of structural extended or point defects, and given the small lateral sizes, "surface states" are the usual suspects. The situation is rather similar when confining layers, namely quantum discs (QDisks) are grown along the c-axis: both GaN/(Al,Ga)N and (In,Ga)N/GaN QDisks exhibit significant reduction of both effective lifetime and PL intensity when T is increased up to 300 K [NAK 12]. Alike in quantum wells, though, (In,Ga)N based systems can exhibit quite small intensity drops due to strong carrier localization at potential fluctuations induced by disorder in (In,Ga)N.

To elucidate the mechanisms at work, the electrical engineers' methodology is to measure LED external quantum efficiencies versus current density and T, and then to analyze them by the so-called ABC (or ABf) models in order to extract the coefficients, respectively, representative of Auger effects and of Shockley-Reed-Hall (SRH) carrier capture by non-radiative defects. With this approach, surface recombination was proposed to be the main (logical) cause of efficiency losses in nitride-based NW devices [NGU 12a, NGU 12b], with relatively much less Auger effect, but with no further microscopic description. The physicists' method consists of measuring the T-dependence of both the PL intensity and the effective PL lifetime, in order to extract the internal quantum efficiency. Rough estimations of the latter are often calculated as the ratio between PL intensities at 300 K and at low T. This is based on the strong assumption that this efficiency would be equal to 1 at low T, which might not be always the case for NW systems, due to small diameters. This method was applied, in a pragmatic way, to LEDs containing InGaN QDisks, in order to evaluate the possible passivation of surface recombination [JAH 13] by using either Si_3N_4 or parylene layers, eventually claiming for a slight improvement of efficiency. But no real model of the physical mechanism of passivation of non-radiative recombination processes was discussed.

To get a better insight into the physics of surface recombination mechanisms, it is worthwhile to go through what has been proposed in the past two decades in studies of NWs based on other semiconductors [YON 12, REI 07], even though we should bear in mind essential differences such as much larger exciton Bohr radii and smaller binding energies in GaAs or InP, for example, than in GaN. GaAs NWs are described as intrinsic semiconductors, with mid-band gap Fermi level in the core of the NW and substantial upward band bending toward the surfaces, owing to the presence of surface states that are essentially electron-deficient in character [YON 12] and on which the Fermi level is pinned. The resulting radial electric field is therefore responsible for the separation of the electron–hole pair and thus for the loss of excitonic radiative recombination. This situation can be changed by passivating the surface states, e.g. via polymers having the proper ionization potential [YON 12], or by growing a core-shell structure, the shell being made in this case by epitaxial (Al,Ga)As, with wider band gap. It is important to notice that, depending on the intensity of optical excitation, the field-induced charge separation can counteract the initial band bending (screening effect) so that a nearly flat-bad situation can be produced [REI 07], with consequences on the respective dynamics of recombination and trapping to surface states. A phenomenological model has been proposed for narrow GaAs NWs [OGA 92]: the non-radiative lifetime related to surface recombination is simply given by $\tau_{NR}=d/2S$, where d is the NW diameter and S the surface recombination velocity. The latter depends exponentially the surface band bending energy, Φ, resulting from charged surface states. If S_0 is the ideal, flat-band velocity (the one that one may reach under strong photoexcitation), then the actual velocity is given by: $S= S_0 \exp(-\Phi / k_BT)$.

In other words, the surface band bending results in radial fields that would be able to dissociate excitons or prevent them from forming, on the one hand, but it is also, on the other hand, what prevents the carriers from reaching non-radiative recombination centers at the surface.

We can see that in the case of GaN NWs, it is crucial to know about energy and density-of-states of the intrinsic surface states that may be present on lateral m-planes (the majority of encountered cases). Then, accounting for small doping densities and for the *essential discontinuity* of the fixed charge landscape, one needs to devise a reasonable model for potential energies, band curvatures, and eventually exciton dynamics. Such a model is still lacking, in particular because the doping has, so far, been treated [CAL 05, RES 09] by a continuous density of charge, with donor densities estimated, by fitting photocurrent data, to be in the range of several

$10^{17}cm^{-3}$ whereas it has been recently estimated as a few $10^{16}cm^{-3}$ [PFU 10a, PFU 10b], i.e. a few donors par NW, and even no donor in 25% of the NWs.

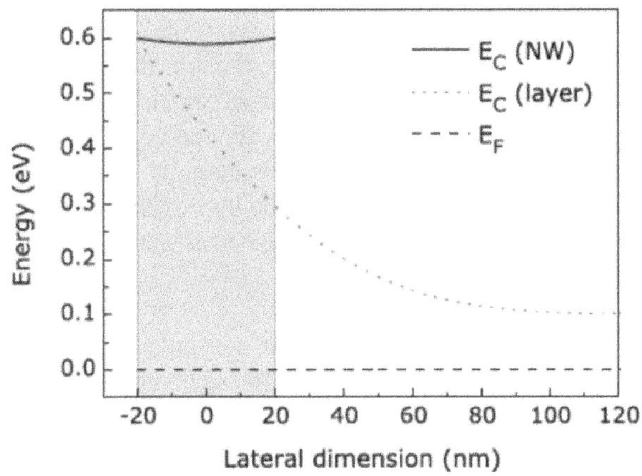

Figure 3.6. *Classically calculated conduction band profile (solid line) in a 40nm-wide GaN slab (gray area) compared to the case of a semi-infinite medium (dotted line). (Reprinted figure with permission from [PFU 10b] – copyright 2010, the American Physical Society)*

Figure 3.6 displays the classical calculation of the conduction band bending [PFU 10b] in a GaN slab, accounting for a residual donor concentration of $5 \times 10^{16}cm^{-3}$ and for the pinning of the Fermi level 0.6 eV below the minimum conduction band. Clearly, for a width of 40 nm, the conduction band profile hardly presents any curvature, at least if we assume a continuous, positive fixed charge density and the asymptotic pinning of the Fermi level to the shallow-donor level, far from the surface. Accounting for ~60 donors in a 1µm long NW, we should reexamine the question of the Fermi level position, both at the surface and in the core of the NW.

The existence and the energy of surface states on m-planes still has to be ascertained and therefore the magnitude of Φ and the possibility for surface fields to ionize the strongly bound excitons of GaN, still have to be verified. Scanning Kelvin-probe microscopy of GaN *non-polar* a-plane surfaces concluded to an upward band-bending of 1.1 V [CHE 06], comparable to what had be obtained for the *polar* c-planes [RES 04] and assigned to charged surface states, namely an effective density of surface acceptors, lying between 1.2 and 1.6 eV below the conduction band. It was concluded [CHE 06] that possible spontaneous polarization effects are widely surpassed by those surface states, with typical densities of $10^{12}cm^{-2}$, in the explanation of the observed band bending. Photoconductivity-mode deep-level optical spectroscopy was used on GaN NWs with triangular sections, grown under Ni catalyst along the (11–20) direction, allowing for identification of similar surface states, 2.6 eV above the valence band. [ARM 10] These states correspond to one {0001} facet of unknown polarity and to two {1–101} facets, and not to the more common {1–100} (m-plane) non-polar facets of hexagonal NW grown along the (0001) direction. Importantly, though, the pragmatic approach in [ARM 10] regarding surface states consisted of passivating them by growing core-shell structures, i.e. GaN cores surrounded by with epitaxial (Al,Ga)N. In this case, the signature of surface states disappeared. However, up to now, it has been very difficult to find, in the literature, results about the improvement of room-temperature PL efficiency of GaN NWs by such core-shell heterostructures, even though refined low-T PL spectroscopy is available [JAC 12].

To summarize this section, we can say that non-radiative recombination at the surface of nitride NWs is generally believed to occur and that the band bending resulting from the pinning of Fermi level on surface states is generally believed to control access of carriers to those surface states, whereas it would also efficiently separate photo-generated electrons and holes. In particular, holes are believed to be driven toward the surface. In section 3.4, we will see that the photochemical activity of this surface, most certainly related to photocarrier transport from and toward surface states, has quite strong effects on the optical properties of the NWs.

3.4. Influence of surface photochemical activity on nitride nanowire optical properties

Wide band gap NWs are already used as elements in gas sensors based on the alteration of conductivity upon exposure to various gases (see, e.g. the review by Wright *et al.* [WRI 10]). Optical properties of those NWs are affected, too, by the chemical environment. For example, the response of photodetectors based on GaN NWs was proven to be sensitive to their environment (vacuum or air) [GON 13].

As an obvious manifestation of such effects, Figure 3.7 illustrates the dependence of the room-T PL intensity of a dense ensemble of GaN NWs [LEF 12] to sudden exposure to air, under constant above-band gap laser excitation. Within a few minutes, the PL intensity drops by one order of magnitude, with no change in the PL spectrum, which is dominated by recombination of unbound excitons, at 300 K. This effect is reversible: upon evacuation of the atmosphere, the initial PL signal is recovered. Experiments conducted with pure nitrogen, helium and dry air allowed concluding in favor of an effect essentially related to oxygen.

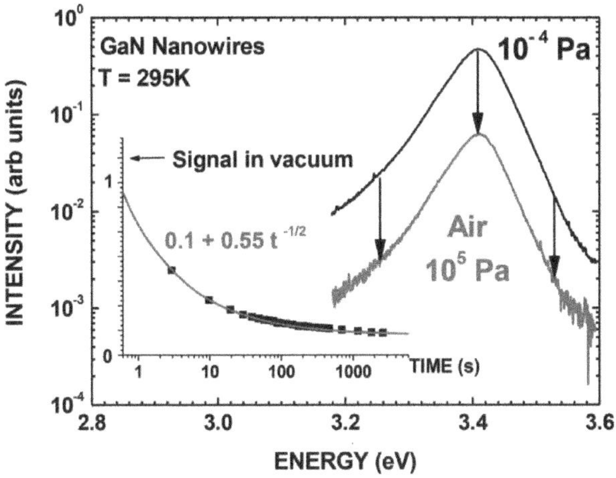

Figure 3.7. *Room-T PL spectrum of a dense ensemble of GaN NWs: when moving from vacuum to air, the intensity is (reversibly) divided by ~10, following the dynamics in inset*

To interpret this result (totally reproducible on numerous GaN NW samples) at the microscopic scale, we can refer to the detailed PL study performed *at T = 10 K* by Pfüller *et al.* [PFU 10b] on GaN NWs placed in vacuum: the slow increase of PL intensities of D°X and UX lines was assigned to the photoinduced desorption of oxygen from the NW sidewalls. Readsorption of oxygen to nearby NWs was proved, via a modeling of the dynamics of PL change. Time-resolved PL results pointed to an increase of quantum efficiency by enhancement of the radiative decay rate of excitons. Since the energies of the donor-bound excitons remain constant during the desorption process, Pfüller *et al.* concluded that a radial electric field within the NW is responsible for the relatively slow exciton recombination. Oxygen desorption would increase the recombination probability by reducing this near-surface field. The latter being determined by the surface Fermi level pinning and the doping of the NW, the observed effect was assigned to the unpinning of the Fermi level.

Let us remark that the latter results, for T=10 K, involve excitons that are bound to donors – thus localized – whereas experiments at room-T instead concern delocalized, unbound excitons, for which the mechanism of PL quenching, upon oxygen adsorption, may be slightly different. In fact, exactly the same effects have been studied for free excitons in ZnO NWs at 300 K, using variable atmospheres and laser excitation intensities [RES 09]. Qualitatively, the model proposed is identical to the one discussed in section 3.3: an upward band bending at lateral surfaces of the NWs is induced by chemisorption of oxygen, resulting in natural negatively charged surface. A depletion region would exist near the surface, where photocreated electron–hole pairs are dissociated. The subsequent accumulation of holes at the surface then reduces the band bending so that, in steady-state conditions, a balance is reached between the holes flowing to the surface and the electrons transferring from the NW core to the surface, over the barrier. This scheme is consistent with the observation of weaker quenching effects between vacuum and air ambient, when laser excitation is increased, the latter reducing both the surface electric field and the width of the depletion region. We see that the essential mechanism for exciton PL quenching, here, is the impossibility for excitons to exist in the region of strong electric field, near the surface. Still, they necessarily *can* exist at the center of the NW, since an extremum of potential essentially exists, there.

The conclusion of [RES 09], which we may transpose to GaN NWs [LEF 12], is that adsoption of oxygen is the principal cause of upward band bending. The increase of PL signal upon evacuation is then just a matter of relative volume in which excitons can exist and the latter depends on the photodesorption of oxygen by UV excitation, via the reduction of the near-surface barrier and the depletion region width. Desorption itself results from the transfer of photogenerated holes from the core to the surface, thus neutralizing negatively charged ionosorbed oxygen.

The latter is a well-known property of semiconductor surfaces, since the late 1950s, as noted by Pfüller *et al.* [PFU 10b]. It has been studied in GaN only for c-plane samples [FOU 09] with different n-type doping, by Kelvin-probe microscopy and PL. In ambient air the upward band bending was found to increase under UV excitation. The increase of negative surface charge results from the transfer of electrons from surface states or bulk to oxygen physisorbed at the GaN surface. In vacuum, the upward band bending decreases under UV excitation, due to the photoinduced desorption of oxygen. The photoadsorption and photodesorption of negatively charged species caused the surface depletion region to increase and decrease, respectively. This explained the significant quenching of PL intensity in air ambient, compared to vacuum, for a sample with wide depletion region (residual doping of 3×10^{16} cm^{-3}). The change of PL intensity between air and vacuum was only of 20% for a sample with wide depletion region (n-doped to 7×10^{18} cm^{-3}). Once again, in those planar c-plane samples, the quenching was assigned to change in depletion region width.

The latter results qualitatively support the above explanations of the strong effects observed in GaN NWs. Quantitatively, though, the reported doping levels of GaN NWs (a few 10^{16} cm^{-3}) lead to typical depletion lengths largely exceeding the NW diameters (Figure 3.6). In addition, the above-mentioned essential discontinuity of the fixed charge landscape is incompatible with a classical treatment in terms of continuous band profiles. The proposed explanations therefore need to be better supported by quantitative models, all the more that realistic numerical parameters *hardly lead to any potential variation* across the NW diameter.

We will finish this discussion with a more pragmatic approach, which successfully permitted us to fabricate opto-chemical gas sensors based on GaN NWs, using PL intensity changes induced by exposure to various gases [TEU 11, PAU 12]. Oxygen exposure induces a signal drop while hydrogen exposure induces a signal increase on condition that Pt coating is used, in the latter case. The physical principle is believed, here, to arise from the control of surface-related non-radiative recombination by change of dipole situation and electric fields at surfaces. The same group [WAL 12] applied similar principles for pH sensing, with the same explanation (band bending control) and subsequent detection based on non-radiative surface recombination. (for more details, refer to Chapter 1 in Volume 2)

PL intensity in GaN NWs is clearly a very sensitive probe of the chemical environment. Whatever the detailed mechanism, this certainly results from the interplay of creation/relaxation/recombination processes with the adsorption of charged species at lateral m-planes. Carrier localization also can play a role, in particular if (In,Ga)N NWs are considered [LEF 12]. In fact, these surface processes can also be used in a reversed way: it has been shown recently [KIB 13] that water splitting under visible-light irradiation can be achieved using InGaN/GaN NW heterostructures as photocatalyst, with significant improvement when decorated with Rh/ Cr_2O_3 core-shell nanoparticles. Therefore, the photochemical activity of nitride NW surfaces make them interesting candidates for solar-powered artificial photosynthesis for the production of hydrogen and other solar fuels.

3.5. Summary

The size reduction in nitride based nanowires with usual diameters hardly leads to any quantum confinement and certainly has a complex influence on optical mode confinement. However, several optical properties are still very much influenced by the nano-size effects that are induced by the proximity between the recombining excitons and lateral surfaces. A great deal of modeling and more systematical experiments still remain to be performed to fully understand the influence of those surfaces on excitonic properties. Clearly, intrinsic or extrinsic surface states, photoadsorption and desorption of charged species, along with a correct description of the discontinuous fixed charge landscape will be the necessary ingredients of this understanding. Nevertheless, strong effects have been demonstrated and are already being used as principles for active opto-chemical devices.

3.6. Bibliography

[ARM 10] ARMSTRONG A., LI Q., LIN Y., *et al.*, "GaN nanowire surface state observed using deep level optical spectroscopy", *Applied Physics Letters*, vol. 96, p. 163106, 2010.

[BRA 10] BRANDT O., PFÜLLER C., CHEZE C., *et al.*, "Sub-meV linewidth of excitonic luminescence in single GaN nanowires: direct evidence for surface excitons", *Physical Review B*, vol. 81, p. 045302, 2010.

[CAL 00] CALLEJA E., SÁNCHEZ-GARCÍA M.A., SÁNCHEZ F.J., *et al.*, "Luminescence properties and defects in GaN nanocolumns grown by molecular beam epitaxy", *Physical Review B*, vol. 62, no. 24, pp. 16826–16834, 2000.

[CAL 05] CALARCO R., MARSO M., RICHTER T., *et al.*, "Size-dependent photoconductivity in MBE-Grown GaN–Nanowires", *Nano Letters*, vol. 5, p. 981, 2005.

[CHE 06] CHEVTCHENKO S., NI X., FAN Q., *et al.*, "Surface band bending of a-plane GaN studied by scanning Kelvin probe microscopy", *Applied Physics Letters*, vol. 88, p. 122104, 2006.

[CON 09] CONSONNI V., KNELANGEN M., JAHN U., *et al.*, "Effects of nanowire coalescence on their structural and optical properties on a local scale", *Applied Physics Letters*, vol. 95, p. 241910, 2009.

[COR 09] CORFDIR P., LEFEBVRE P., RISTIĆ J., *et al.*, "Time-resolved spectroscopy on GaN nanocolumns grown by plasma assisted molecular beam epitaxy on Si substrates", *Journal of Applied Physics*, vol. 105, p. 013113, 2009.

[CON 14] CONSONNI V., FEUILLET G., (eds), *Wide Band Gap Semiconductor Nanowires 2: Heterostructures and Optoelectronic Devices*, ISTE, London, and John Wiley & Sons, New York, 2014.

[COR 12] CORFDIR P., LEFEBVRE P., "Role of the dielectric mismatch on the properties of donors in semiconductor nanostructures bounded by air", *Journal of Applied Physics*, vol. 112, p. 106104, 2012.

[DIA 07] DIARRA M., NIQUET Y.-M., DELERUE C., *et al.*, "Ionization energy of donor and acceptor impurities in semiconductor nanowires: importance of dielectric confinement", *Physical Review B*, vol. 75, p. 045301, 2007.

[FOU 09] FOUSSEKIS M., BASKI A.A., RESHCHIKOV M.A., "Photoadsorption and photodesorption for GaN", *Applied Physics Letters*, vol. 94, p. 162116, 2009.

[FUR 08] FURTMAYR F., VIELEMEYER M., STUTZMANN M., *et al.*, "Optical properties of Si- and Mg-doped gallium nitride nanowires grown by plasma-assisted molecular beam epitaxy", *Journal of Applied Physics*, vol. 104, p. 113506, 2008.

[GIL 07] GIL B., BIGENWALD P., LEROUX M., *et al.*, "Internal structure of the neutral donor-bound exciton complex in cubic zinc-blende and wurtzite semiconductors", *Physical Review B*, vol. 75, p. 085204, 2007.

[GON 13] GONZÁLEZ-POSADA F., SONGMUANG R., DEN HERTOG M., *et al.*, "Environmental sensitivity of n-i-n and undoped single GaN nanowire photodetectors", *Applied Physics Letters*, vol. 102, p. 213113, 2013.

[JAC 12] JACOPIN G., RIGUTTI L., BELLEI S., *et al.*, "Photoluminescence polarization in strained GaN/AlGaN core/shell nanowires", *Nanotechnology*, vol. 23, p. 325701, 2012.

[JAH 13] JAHANGIR S., BANERJEE A., BATTACHARYA P., "Carrier lifetimes in green emitting InGaN/GaN disks-in-nanowire and characteristics of green light emitting diodes", *Physica Status Solidi (c)*, vol. 10, no. 5, pp. 812–815, 2013.

[KIB 13] KIBRIA Md G., NGUYEN H.P.T., CUI K., *et al.*, "One-step overall water splitting under visible light using multi-band inGaN/GaN nanowire heterostructures", *ACS Nano*, 2013.

[KOR 99] KORNITZER K., EBNER T., GREHL M., *et al.*, "High-resolution photoluminescence and reflectance spectra of homoepitaxial GaN layers", *Physica Status Solidi (b)*, vol. 216, no. 1, pp. 5–9, 1999.

[LEF 11] LEFEBVRE P., FERNÁNDEZ-GARRIDO S., RISTIĆ J., *et al.*, "Radiative defects in GaN nanocolumns: correlation with growth conditions and sample morphology", *Applied Physics Letters,* vol. 98, no. 8, p. 083104, 2011.

[LEF 12] LEFEBVRE P., ALBERT S., RISTIĆ J., *et al.*, "Oxygen photo-adsorption related quenching of photoluminescence in group-III nitride nanocolumns", *Superlattices and Microstructures,* vol. 52, p. 165, 2012.

[NAK 12] NAKA J., INOSE Y., KUNUGITA H., *et al.*, "Optical properties of InGaN/GaN nanocolumns in yellow-to-red region", *Physica Status Solidi (c)*, vol. 9, no. 12, pp. 2477–2480, 2012.

[NEU 99] NEU G., TEISSEIRE M., BEAUMONT B., *et al.*, "Near-band gap selective photoluminescence in Wurtzite GaN", *Physica Status Solidi (b)*, vol. 216, no. 1, pp. 79–83, 1999.

[NGU 12a] NGUYEN H.P.T., CUI K., ZHANG S., *et al.*, "Controlling electron overflow in phosphor-free InGaN/GaN nanowire white light-emitting diodes", *Nano Letters*, vol. 12, pp. 1317–1323, 2012.

[NGU 12b] NGUYEN H.P.T., DJAVID M., CUI K., *et al.*, "Temperature-dependent nonradiative recombination processes in GaN-based nanowire white-light-emitting diodes on silicon", *Nanotechnology*, vol. 23, p. 194012, 2012.

[OGA 92] OGAWA K., HARAGUCHI K.-I., HIRUMA K., *et al.*, "Spectral and temporal features of photoluminescence of gallium arsenide quantum-wire crystals", *Journal of Luminescence*, vol. 53, p. 387, 1992.

[PAS 07] PASKOV P.P., MONEMAR B., TOROPOV A., *et al.*, "Two-electron transition spectroscopy of shallow donors in bulk GaN", *Physica Status Solidi (c)*, vol. 4, p. 2601, 2007.

[PAU 12] PAUL S., HELWIG A., MÜLLER G., *et al.*, "Opto-chemical sensor system for the detection of H_2 and hydrocarbons based on InGaN/GaN nanowires", *Sensors and Actuators B*, vol. 173, pp. 120–126, 2012.

[PFU 10a] PFÜLLER C., BRANDT O., FLISSIKOWSKI T., *et al.*, "Statistical analysis of excitonic transitions in single, free-standing GaN nanowires: probing impurity incorporation in the poissonian limit", *Nano Research*, vol. 3, no. 12, pp. 881–888, 2010.

[PFU 10b] PFÜLLER C., BRANDT O., GROSSE F., *et al.*, "Unpinning the Fermi level of GaN nanowires by ultraviolet radiation", *Physical Review B*, vol. 82, p. 045320, 2010.

[PFU 11] PFÜLLER C., Optical properties of single GaN nanowires and nanowire ensembles, PhD Thesis, Humbold University of Berlin, Germany, June 2011. Available at http://edoc.hu-berlin.de/dissertationen/pfueller-carsten-2011-06-27/PDF/pfueller.pdf.

[REI 07] REITZENSTEIN S., MÜNCH S., HOFMANN C., *et al.*, "Time resolved microphotoluminescence studies of single InP nanowires grown by low pressure metal organic chemical vapor deposition", *Applied Physics Letters*, vol. 91, p. 091103, 2007.

[RES 04] RESHCHIKOV M.A., SABUKTAGIN S., JOHNSTONE D.K., *et al.*, "Transient photovoltage in GaN as measured by atomic force microscope tip", *Journal of Applied Physics,* vol. 96, no. 5, pp. 2556–2560, 2004.

[RES 09] RESHCHIKOV M.A., BEHRENDS A., BAKIN A., *et al.*, "Photoluminescence from ZnO nanowires", *Journal of Vacuum Science and Technology B*, vol. 27, no. 3, pp. 1688–1692, 2009.

[ROB 07] ROBINS L.A., BERTNESS K.A., BARKER J.M., *et al.*, "Optical and structural study of GaN nanowires grown by catalyst-free molecular beam epitaxy. II. Sub-band-gap luminescence and electron irradiation effects", *Journal of Applied Physics,* vol. 101, p. 113506, 2007.

[SAM 12] SAM-GIAO D., Optical study of GaN nanowires and GaN/AlN microcavities, PhD Thesis, Joseph Fourier University, Grenoble, France, November 2012. Available at http://www.researchgate.net/publication/235323521_PhD_thesis

[SAM 13] SAM-GIAO D., MATA R., TOURBOT G., *et al.*, "Fine optical spectroscopy of the 3.45 eV emission line in GaN nanowires", *Journal of Applied Physics,* vol. 113, p. 043102, 2013.

[SAT 83] SATPATHY S., "Eigenstates of Wannier excitons near a semiconductor surface", *Physical Review B*, vol. 28, pp. 4585–4592, 1983.

[SEK 10] SEKIGUCHI H., KISHINO K., KIKUCHI A., "Emission color control from blue to red with nanocolumn diameter of InGaN/GaN nanocolumn arrays grown on same substrate", *Applied Physics Letters*, vol. 96, p. 231104, 2010.

[SLA 07] SLACHMUYLDERS A.F., PARTOENS B., MAGNUS W., *et al.*, "Excitons and trions in cylindrical nanowires with dielectric mismatch", *Physica Status Solidi (c)*, vol. 5, no. 7, pp. 2416–2419, 2008.

[TEU 11] TEUBERT J., BECKER P., FURTMAYR F., *et al.*, "GaN nanodiscs embedded in nanowires as optochemical transducers", *Nanotechnology*, vol. 22, no. 27, p. 275505, 2011.

[WAL 12] WALLYS J., TEUBERT J., FURTMAYR F., *et al.*, "Bias-enhanced optical pH response of group III–Nitride nanowires", *Nano Letters*, vol. 12, no. 12, pp. 6180–6186, 2012.

[WRI 10] WRIGHT J.S., LIM W., NORTON D.P., *et al.*, "Nitride and oxide semiconductor nanostructured hydrogen gas sensors", *Semiconductor Science and Technology*, vol. 25, p. 024002, 2010.

[YON 12] YONG C.K., NOORI K., GAO Q., *et al.*, "Strong carrier lifetime enhancement in GaAs nanowires coated with semiconducting polymer", *Nano Letters*, vol. 12, no. 12, pp. 6293–6301, 2012.

4

Surface Related Optical Properties of ZnO Nanowires

4.1. Introduction

One of the most fascinating aspects of nanotechnology is the fact that many electronic, thermal and optical properties become a function of the nanostructures' size because of quantum-confinement effects. However, for quantum confinement in semiconductors to show a significant impact on the material properties, the lateral dimensions of the nanostructures have to be comparable with, or smaller than, the de-Broglie wavelength of the electrons, the Bohr radius of the excitons, or some other characteristic length scale associated with the spatial extension of particles or quasi-particles. Depending on the material under consideration, this requires nanostructures smaller than some tens of nanometers, in the case of wide-gap materials such as ZnO and GaN even smaller than about 10 nm [STI 07]. The diameters of semiconductor nanowires that are most often studied in the context of functional electronic or optoelectronic devices and that have been grown in a vapor–liquid–solid (VLS) or wet-chemical process, however, are typically one order of magnitude larger. This shows that in the vast majority of cases, quantum-confinement effects can be completely neglected when studying the optical properties of VLS ZnO nanowires.

The ratio of the nanowire surface area to its volume increases with ~$1/d$ as the nanowire diameter d is reduced. To be specific, let us assume a cylindrical wire with length l and diameter d. With the surface area:

$$s = \pi \cdot d \cdot l + 2 \cdot \tfrac{\pi d^2}{4} = \pi \cdot d(l + \frac{d}{2}) \qquad [4.1]$$

Chapter written by Tobias VOSS and Jürgen GUTOWSKI.

and the volume

$$V = \frac{\pi \cdot d^2}{4} \cdot l \qquad [4.2]$$

we directly arrive at

$$\frac{S}{V} = \frac{4}{d} + \frac{2}{l}. \qquad [4.3]$$

With typical aspect ratios of $l/d > 50$ of ZnO nanowires, the second summand on the right side of equation [4.3] can be neglected in a good approximation. A similar relation holds if we not only consider the two-dimensional surface area but also calculate the ratio of the volume V_{SL} of a surface layer with a finite thickness t to the remaining inner volume V_i of the wire associated with the diameter d-$2t$. If we again assume a wire with a large aspect ratio, we can neglect the contributions from the top and bottom facets and obtain

$$\frac{V_{SL}}{V_i} = \frac{dt - t^2}{d^2/4 - dt + t^2}. \qquad [4.4]$$

Also, in this case, we see that the ratio roughly scales as $\sim 1/d$ for $d >> t$, thus demonstrating that the relative influence of the electronic and optical properties of near-surface layers becomes increasingly important for semiconductor crystals with nanowire morphology. As an example, for a nanowire with diameter $d = 200$ nm and a surface layer thickness of $t = 20$ nm, we calculate a ratio $V_{SL}/V_i = 0.56$; if we reduce the diameter to $d = 50$ nm, we see that the ratio substantially increases to $V_{SL}/V_i = 24$.

The above analysis demonstrates that electronic and optical properties of semiconductor nanowires will usually exhibit a significant dependence on their diameter d, although d is often more than one order of magnitude larger than the corresponding (quasi-)particle dimensions in the material system under consideration. Surface effects will bring forth strong deviations from the bulk properties of the material because the periodicity of the crystal lattice is broken at the surface. Adsorption of atoms or molecules at the nanowire surface, reconstructions of the crystal structure at the surface, surface defects and other processes will alter the band structure in the near-surface layer. In the following sections, we will discuss specific processes and properties of ZnO nanowires that are directly related to their increased surface-to-volume ratio compared to larger bulk crystals.

4.2. Surface excitons in ZnO nanowires

Being a wide- and direct-gap semiconductor, ZnO has a substantially large exciton binding energy of E_X = 60 meV [KLI 07] (compared to E_x = 4.8 meV for GaAs). When optically excited with photons that have an energy larger than the band gap (approximately 3.4 eV at room temperature), electron-hole pairs are generated in the vicinity of the Γ point (if the photon energy is not too large). In particular at low temperatures, these electron-hole pairs will form excitons due to their Coulomb attraction. The excitons dominate the optical emission spectra of ZnO at low and moderate temperatures and excitation densities. Even at room temperature, excitonic emission processes can still be identified, and excitonic ultraviolet (UV) electroluminescence has been demonstrated for a single-nanowire ZnO light emitting diode (LED) [ZIM 09]. Because of the significant overlap of the electron and hole wave functions, excitons typically show large oscillator strengths and lead to efficient and fast radiative recombination processes.

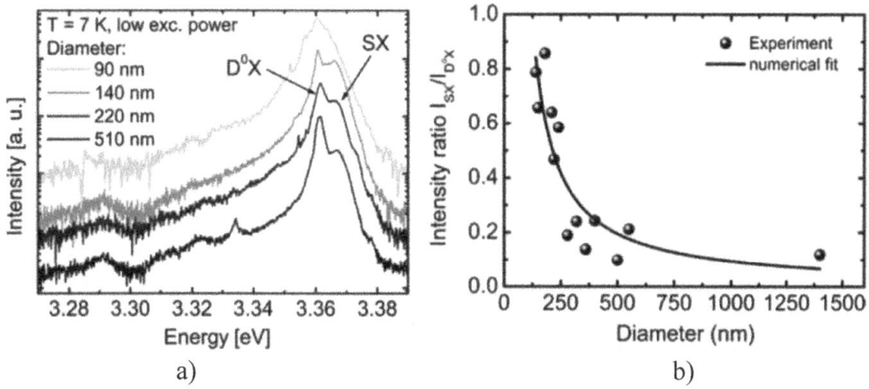

Figure 4.1. *a): photoluminescence spectra of individual wires with different diameters. The spectra are normalized to the intensity of the D^0X peak and shifted in intensity for a better overview [WIS 06a]; b): diameter dependence of the intensity ratio SX/D^0X for VLS nanowires fitted with a model with a surface layer of finite thickness as described in the text [VOS 10]*

Photoluminescence (PL) spectroscopy of single ZnO nanowires already demonstrated in 2006 [WIS 06a] that, apart from the well-known exciton related emission lines, surface related features can quite easily be identified by studying the PL spectra of nanowires with different diameters *d* (Figure 4.1(a)) The PL spectra are dominated by the emission line at 3.365 eV, which originates from the radiative recombination of excitons bound to specific neutral donors (D^0X). The broad peak SX at 3.367 eV, however, exhibits a line shape being uncommon for D^0X or D^+X emission lines that are typically observed in this spectral region [MEY 04]

(D^+X: exciton bound to an ionized donor). A clear change in the intensity ratio of the SX and D^0X peaks is observed for nanowires with different diameters. Figure 4.1(b) shows this intensity ratio as a function of the wire diameter [VOS 08, VOS 10]. The solid line is a fit according to equation [4.4], which includes a surface layer with finite thickness t. The fitting procedure yields $t < 25$ nm with the rather large uncertainty being due to possible errors in the determination of the wire diameters and the possibly varying donor concentration in wires with different diameters. These results strongly suggest that the SX emission line is related to recombination processes involving excitons in a surface layer where the local potential landscape is substantially modified by the presence of surface states.

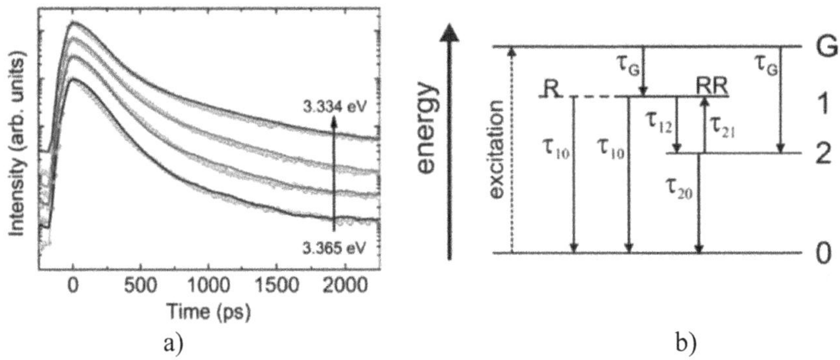

Figure 4.2. *a) comparison of the measured transients (gray dots) and the results of the rate-equation model (solid lines) as function of detection energy; b) schematic energy diagram describing the excitation of excitons (state G) and their relaxation or recombination pathways via the surface states 1 and 2. State 1 is separated into weakly localized excitons that are able to relax and recombine (RR: solid line) and strongly localized excitons that only recombine (R: dashed line) [VOS 10]. For a color version of this figure, see www.iste.co.uk/consonni/nanowires1.zip*

A critical step toward an understanding of the microscopic nature of the recombination processes associated with the SX line was the analysis of its dynamics [WIS 06b, VOS 10]. Figure 4.2(a) shows the results of time-resolved PL spectroscopy of a nanowire sample grown by vapor-phase epitaxy (ensemble with wire lengths from 1 to 6 µm and diameters d from 40 to 130 nm). For the experiments, a system for time-correlated single-photon counting (TCPC) was used that provided a spectral resolution of 2.4 meV and a time resolution of about 30 ps. A rate-equation model was used to simulate the experimentally obtained decay dynamics of the SX emission band [WIS 06b]. The basic idea is visualized in Figure 4.2(b). Excitons are generated by optical pumping from the ground state 0 (no exciton) into state G (exciton in its ground state). The free excitons will be captured with a characteristic time constant τ_G into the SX state (1). Due to the inhomogeneity of the surface layer, some excitons will be captured in deep and strongly localizing local potential minima, others may be captured in potential minima at the same energy level but with the possibility of relaxing into lower SX energy states (2). From these weakly localizing centers, the SX can either show relaxation or radiative recombination (level RR). For the strongly localized near-surface exciton level (R), only radiative recombination is possible. The superposition of the emission from both weakly and strongly localized excitons at the same energy position leads to the experimentally observed non-monoexponential decay dynamics. The model can explain the shape of the SX PL transients for different detection energies (changing fraction of weakly and strongly localized excitons), different temperatures (thermal activation of weakly localized excitons), and different excitation fluences (saturation of the limited number of SX states) [WIS 06b]. The solid lines in Figure 4.2(a) are numerical fits to experimental data using this rate equation model. The excellent agreement demonstrates that the model captures all essential aspects of the recombination dynamics of near-surface excitons.

Surface related excitonic effects in ZnO, CdS and CdSe bulk crystals were already described in the literature as early as 1990 [TRA 90b, TRA 90a, TEN 92]. The probably most conclusive evidence of a surface-related property was given by Travnikov and coworkers who studied PL spectra of ZnO bulk crystals immersed in liquid helium ($T = 2$ K) [TRA 90b].

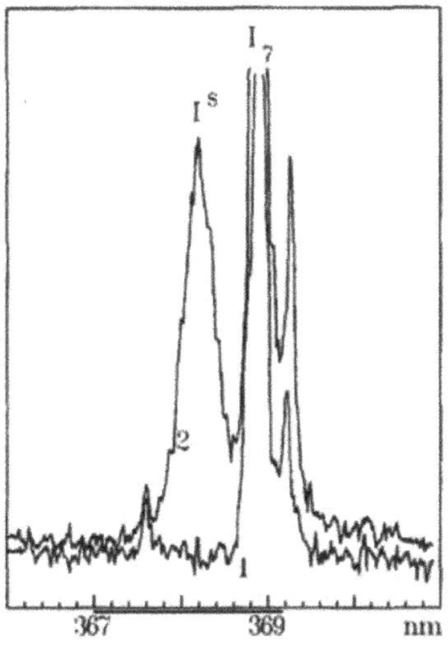

Figure 4.3. *Photoluminescence spectra of a ZnO crystal cleaved in liquid helium; 1: spectrum obtained directly after the cleaving; 2: spectrum obtained after exposure of the cleaved sample to air for several hours at room temperature. (Reprinted from [TRA 90a], with permission from Elsevier)*

Figure 4.3 shows two luminescence spectra of a ZnO single crystal: spectrum 1 was taken directly after the crystal was cleaved in liquid helium ($T = 2$ K); spectrum 2 was again taken in liquid helium but after the cleaved crystal had been exposed to air at room temperature for several hours. It is striking that the broad and asymmetric peak I_S is absent directly after cleaving but returns after exposure to air. By its spectral position and shape it can be identified as the SX emission band seen in the nanowire samples (note that the spectra in Figure 4.1 are shown on a logarithmic intensity scale while the spectra in Figure 4.3 chooses a linear scale). The results let us directly conclude that the surface-related excitonic emission band is related to the adsorption of molecules, which can be found on the ZnO surface in normal air. Therefore, the SX band is not directly connected with the clean, reconstructed ZnO surface, but contaminations of the surface result in additional (charged or uncharged) localization centers. The energy distribution of these localization centers is directly reflected in the line shape of the SX band. A further proof of this interpretation was given by Travnikov *et al.*, [TRA 90b] who also demonstrated that the SX emission at $T = 2$ K (in their case labeled as I_S) quenches under illumination of the sample with a 250 W high-pressure Hg lamp for more than 1 h. This result can be interpreted by assuming photo-induced desorption of atoms or molecules from the ZnO surface that lead to the SX (or I_S) emission line.

After SX related PL emission bands had been found for ZnO nanowires and other ZnO nanostructures independently by two groups [GRA 05, WIS 06a], several experiments were designed and carried out to clarify the underlying microscopic origin. On the one hand, plasma treatment (plasma immersion ion implantation) of ZnO nanocomb samples resulted in a quenching of the SX emission band [YAN 07] when a 5 nm thick surface layer was modified by the ion implantation process. Coating of the nanowires with different kinds of polymers, on the other hand, strongly enhanced the relative intensity of the SX emission line compared to the bulk related D^0X features [RIC 08b]. A similar enhancement of the SX emission was found after conformally covering ZnO nanowires with a thin (10 nm) amorphous Al_2O_3 shell by atomic-layer deposition [RIC 08a]. Therefore, it was concluded that surface passivation of ZnO nanowires with a large variety of different dielectric materials (organic and inorganic) changes the electronic band structure in the near-surface region in such a way that the SX emission intensity becomes stronger. The following model was proposed to explain this effect: (1) fermi-level pinning at the surface of *n*-doped ZnO nanowires (induced by surface states) results in an upward band bending in the near-surface region and the formation of an electron depletion layer. This is a direct result of negatively charged surface states that repel electrons from the surface but attract holes by electrostatic interactions. The negative surface charge is compensated by the positive charges of ionized donor centers in the near-surface layer. (2) Surface-coating with dielectric materials will passivate the surface states (through the formation of chemical bonds) and, in addition, dielectrically screen the charge states of the remaining surface states. Both of these processes will reduce the band bending in the near-surface layer and the width of the electron depletion layer. (3) The reduced band bending will increase the overlap of electron and hole wave functions in the near-surface layer as the repulsive force for electrons and the attractive force for holes are reduced. This favors the formation and subsequent radiative recombination of near-surface excitons in the coated nanowire samples, thus enhancing the intensity of the SX emission band relative to the bulk-related D^0X bands.

Experiments by Biswas *et al.*, [BIS 11] have conclusively shown that the SX emission processes are most probably related to OH species adsorbed on certain surfaces of ZnO and its nanostructures. The authors point out that the nanocrystal morphology, grain boundaries and other planar defects will have a significant influence on the density of preferential adsorption sites for OH molecules and strongly affect their binding and localization energies. Therefore, both the crystal structure and the adsorption of molecules to the surface of ZnO nanowires will determine the relative intensity of the SX emission band compared to the bulk-related D^0X lines. This also explains why the relative strength of the SX emission line strongly varies from sample to sample and is typically difficult to observe in conventional bulk crystals.

We now want to conclude the section about surface excitons in ZnO nanowires with an example which demonstrates the importance of the SX emission processes for device applications. In [KIM 10], the authors fabricated field-effect transistors (FETs) from single ZnO nanowires where the nanowires had been grown under oxygen-rich (OR) and oxygen-poor (OP) environments. They determined the electronic transfer characteristics of the different FETs and found a correlation with the relative intensity of the SX emission in the respective nanowires. Specifically, the transfer characteristics of wires grown under OP conditions showed a modulation ratio of the drain current (on/off ratio) of 10^7 while only 10^4 was found for wires grown under OR conditions. At the same time, the OP wires showed a stronger SX emission than the OR wires (normalized to the D^0X emission band and measured at about 10 K). These results can be interpreted by considering the donor-like nature of many point defects that are created during the nanowire growth. Under OP conditions, a larger density of oxygen vacancies (V_O) and related defect complexes is generated that leads to a larger unintentional *n*-type doping concentration of the nanowires. This consequently results in the improved transfer characteristics of the wires and, at the same time, reduces the width of the surface depletion layer by reducing the band bending. As discussed above, this will result in an increase of the relative SX emission intensity of the OP nanowires.

Taking all these experimental results into consideration, we have to conclude that two processes determine the SX recombination process: (1) the adsorption of molecules on the ZnO nanowire surface (most probably OH related species) and

(2) the band bending in the near surface region. Together, these two processes yield a particular potential landscape that electrons, holes and excitons experience in the near-surface region. The characteristic distribution of localization energies leads to the special line shape and recombination dynamics of the surface-exciton SX emission. In this respect, this luminescence band can also be used as a sensitive probe to study surface modifications of ZnO nanowires.

4.3. Surface-related defect luminescence in ZnO nanowires

The PL spectra of ZnO samples are commonly divided into at least two distinct regions: the near-band-edge emission (NBE) in the near-UV spectral region, which is related to excitonic transitions, and the defect luminescence in the visible spectral region (DLE), which is attributed to radiative electronic transitions involving defect levels deep in the band gap (energy offset relative to the conduction-band minimum or valence-band maximum larger than about 500 meV). The chemical and structural nature of these defects is still under strong debate because a multitude of different defect complexes can usually be formed which typically show spectrally overlapping emission bands. This also holds for ZnO nanowires where, however, the situation is further complicated by the increasing influence of surface effects. Whereas the previous section dealt with surface effects in the NBE region of ZnO nanowires, we will now take a look at properties of the DLE that are related to the increased surface-to-volume ratio of ZnO nanowires.

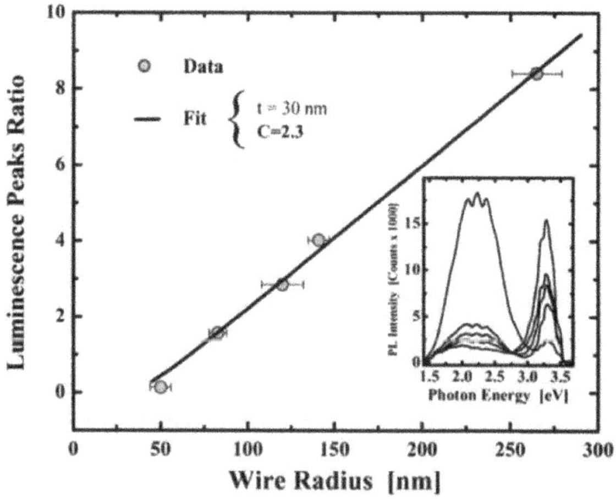

Figure 4.4. *Photoluminescence peak intensity ratio I_{NBE}/I_{DLE} as a function of the average wire radius. Solid line: numerical fit according to equation [4.4]. From [SHA 04](copyright 2004, The American Physical Society)*

There are at least two different ways in which surface effects contribute to DLE processes: (1) radiative recombination processes via surface defects and (2) activation of bulk defects via electrons or holes in the near-surface region. The first case was analyzed in [SHA 04] where the ratio of the visible to near-UV luminescence in ZnO nanowires was studied for different wire diameters. The results (see Figure 4.4) demonstrate that the ratio of the luminescence intensities $I_{NBE}/I_{DLE} = V_i/V_{SL}$ increases with the wire radius according to the inverse of equation [4.4]. A numerical fitting procedure yields the thickness t of the surface layer from which the photogenerated electrons can recombine radiatively via surface traps. In [SHA 04], the authors determined a value of $t = 30$ nm. However, they point out that this value is not a general characteristic of ZnO nanowires as, among other factors, the density of surface states will strongly change from sample to sample. Still, the results demonstrate that there is typically a strong contribution to the DLE from the radiative recombination via surface states in ZnO nanowires. It is also instructive to note that the thickness of the relevant surface layer in this experiment is comparable to that obtained in the studies of the near-surface exciton emission band SX. This points to a closely related origin of these two processes. However, we have to bear in mind that in the investigations many other relevant material properties such as doping level, density of bulk defects and faceting of the nanowires were not taken into account. The fact that they could be neglected for the interpretation of the discussed experiments definitely does not imply that they could not be influencing the surface-related optical properties of ZnO nanowires in general.

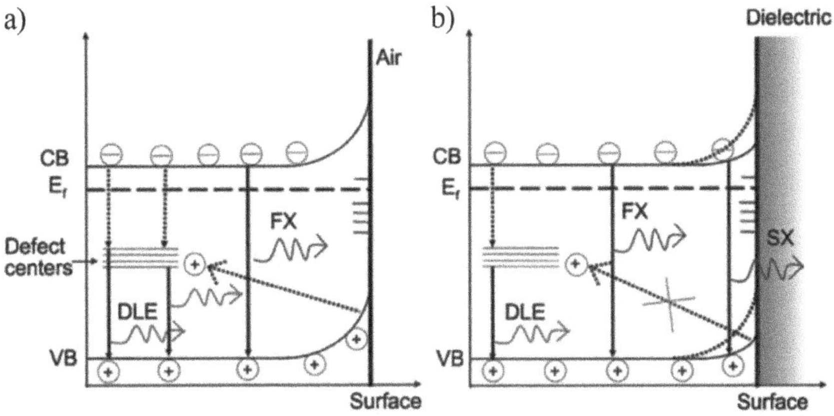

Figure 4.5. *Model of the surface-related activation of the DLE in ZnO nanowires [VOS 10]*

The second case, namely the activation of bulk defects by electrons or holes in the surface layer, was discussed in [VOS 10] based on a model developed for nanoparticles in [DIJ 00]. While studying the influence of dielectric coatings on the relative intensity of the SX emission band, the authors found that, in addition to the increased relative intensity of the SX emission, the green deep-level emission was significantly reduced for nanowires covered with polymer films [RIC 08b] or Al_2O_3 layers [RIC 08a]. For the interpretation of the results, they assumed that V_O [LEI 03] are among the most likely candidates for the microscopic origin of the DLE band in their nanowire samples. These vacancies need to be activated by a hole since they are located far below the Fermi level (see Figure 4.5(a)). The holes required for this activation process can be provided by the surface layer of the nanowires. Due to the pinning of the Fermi level at the surface of the *n*-type ZnO nanowires, an upward band bending of both the valence and conduction bands is induced [LIN 06] yielding a hole accumulation near the surface. These holes may preferably tunnel into the DLE states. In a further step, Al_2O_3 or polymer shells around the ZnO nanowire core screen the charge carriers located in the surface states thereby reducing the density of accumulated holes in the surface layer and the band bending. These effects result in an increased energy level offset between the holes trapped in the surface layer and the (V_O) (see Figure 4.5(b)), thus reducing the probability for tunneling into an (V_O). This process will result in a reduction of the relative intensity of the green defect luminescence in ZnO nanowires coated with dielectric materials. Exactly this behavior was observed in the above mentioned experiments.

Related experiments were also discussed in [WAN 12] where the authors studied the PL of ZnO nanowires before and after annealing treatment under OR and OP conditions, respectively. They emphasize that the experiments clearly demonstrate that the annealing treatment does not change the V_O concentration in the nanowires but leads to either chemisorption or desorption of oxygen at the nanowire surface and subsequent changes in the band bending. After annealing in oxygen atmosphere, the larger density of oxygen chemisorbed at the surface increases the band bending due to the trapping of additional electrons in surface states. In this case, the relative intensity of the green DLE band was found to be stronger, in perfect agreement with the SX related experimental results discussed above. Subsequent annealing under OP conditions removes the oxygen from the nanowire surface, decreases the density of electrons trapped in surface states and thereby reduces the band bending. The result is a lower relative intensity of the green DLE.

The results discussed in this section clearly show the importance of surface effects for the defect luminescence bands of ZnO nanowires in the visible spectral region. Surface defects directly contribute to the relevant recombination processes, and the band structure in the near-surface layer affects the activation of bulk defects. Still, it has to be emphasized that a precise attribution of certain defect luminescence bands to specific microscopic processes is extremely difficult for ZnO in general due to the simultaneous presence of a large variety of different defects and defect complexes.

4.4. Surface functionalization of ZnO nanowires with colloidal quantum dots

Optical sensing and solar energy conversion are two of the most relevant fields of application for ZnO nanowires (for more details, refer to Chapter 10, Volume 2 [CON 14]). For both of these applications, it is desirable to benefit from the enhanced surface-to-volume ratio of nanowires compared to larger bulk crystals. In addition, the possibility for large scale bottom–up synthesis of high-quality nanowire arrays can be exploited for the design and fabrication of sensors with high sensitivity and/or the parallelized detection of a large variety of different chemical species. For solar energy conversion, the active area per wafer footprint can be significantly increased if core–shell nanowire geometries are synthesized. Combined with the possibility of low-temperature nanowire synthesis from aqueous solutions in a chemically mild environment, cheap and flexible substrates can allow for the fabrication of low-cost, medium-efficiency nanowire solar cells.

Obviously, applications as solar cells and optical sensors (for more details, refer to Chapter 8, Volume 2) require a functionalization of the ZnO nanowire surface because ZnO is transparent throughout the whole visible spectral region. For sensing applications, surface functionalization is required to achieve a selective response to certain chemicals. Organic chemistry provides a variety of methods to realize a selective functionalization of the ZnO surface. In particular, carboxylic acid groups can be employed as chemical bridges to achieve a binding of various substances to the oxide surface. This strategy has been used to achieve photosensitization by covering ZnO nanowires with CdSe quantum dots for potential application in photovoltaic devices [LES 07].

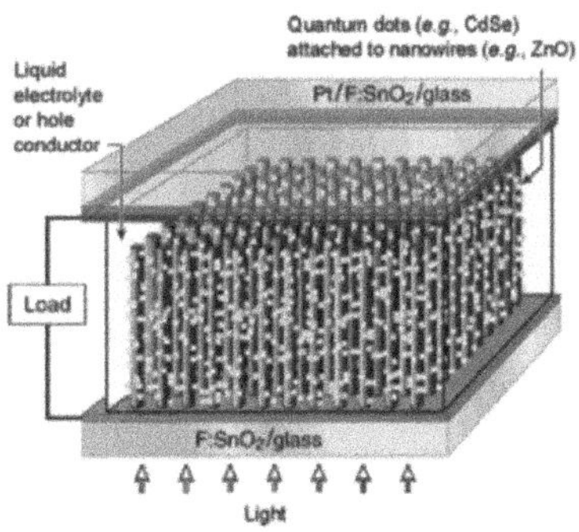

Figure 4.6. *Schematic of a quantum-dot-sensitized solar cell. An array of ZnO nanowires, grown vertically on an F-doped SnO_2/glass substrate and decorated with CdSe quantum dots, serves as the photoanode. A second F-doped SnO_2/glass substrate, coated with a 10 nm layer of Pt, serves as photocathode. The space between the two electrodes is filled with a liquid electrolyte. The cell is illuminated from the bottom. (Reprinted with permission from [LES 07]. Copyright 2007, American Chemical Society). For a color version of this figure, see www.iste.co.uk/consonni/nanowires1.zip*

Figure 4.6 shows a schematic of the thus created nanowire-based solar-cell device. Two glass slides are coated with conductive layers (the bottom one is transparent) and form the top and bottom electrodes of the solar cell. *n*-conductive ZnO nanowires were grown on the transparent bottom electrode, and the space between the two electrodes was filled with a liquid electrolyte. To achieve photoactivity in the visible spectral region, surface functionalization of the nanowires with colloidal CdSe quantum dots was performed. The quantum dots are surrounded by a shell of mercaptopropionic acid which stabilizes them during their synthesis and prevents agglomeration. The sulfur ends of the molecules bind selectively to the quantum-dot surface, the carboxylic acid end will eventually attach to the ZnO nanowire surface. This way, the organic linker molecules introduce a small separation between the surface of the nanowires and the quantum dots which can be tuned by changing the length of the carbon chain of the linker molecule.

The mode of operation of the device is as follows: upon excitation with visible light, electron-hole pairs are generated in the excited states of the CdSe quantum dots provided the photon energy is larger than the lowest energy gap of the dots. The electrons will tunnel into the conduction band of the ZnO nanowires which is located energetically below the excited electronic state in the quantum dots (see Figure 4.8) while the holes remain in the dot. This way, a spatial separation of the photoexcited electrons and holes is introduced so that a direct recombination is no longer possible. Through the electrolyte (typically containing a redox couple), the remaining hole is transported to the platinum-coated top electrode, and the electron will flow through the *n*-conductive nanowire to the bottom electrode. This way, the photon energy is converted into electrical energy which can be provided to the external load. In [LES 07], the authors report a short circuit current of 1–2 mA/cm^2, an open circuit voltage of 0.5–0.6 V (under illumination with 100 mW/cm^2 simulated AM1.5 solar spectrum), and internal quantum efficiencies as high as 50–60%.

Taking into account the importance of surface band bending and oxygen adsorption at the nanowire surface as discussed in the previous sections, it is quite natural to raise the question: to what extent do these surface-related properties and processes influence the electron-transfer dynamics and the overall performance of quantum-dot-sensitized ZnO-nanowire solar cells? Details of the electron-transfer dynamics in such a device were published in [HOU 12]. In this work, the authors studied a device similar to the device depicted in Figure 4.6 with the main difference being the absence of the electrolyte. In addition, the device was operated in different atmospheres to study the dynamics of gas adsorption and desorption processes on and from the surface of the nanowire/quantum-dot assembly. The quantum dots were linked to the ZnO surface also via mercaptopropionic acid molecules. In Figure 4.7, the photoconductivity of nanowire/quantum-dot hybrid devices is shown for operation in different gas environments. The authors measured the current flowing through the nanowire array under an applied constant external bias of $U_B = 1.0$ V.

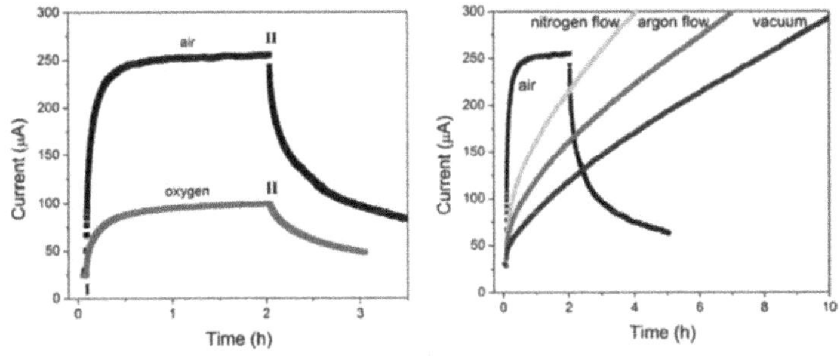

a) b)

Figure 4.7. *a): photoconductivity of the NW/QD assembly in air and pure oxygen atmosphere under Ar laser irradiation (1.0 mW cm^{-2}) and with bias voltage $U_B = 1.0$ V. I: laser on, II: laser off; b): photoconductivity dynamics of the NW/QD assembly measured in four different gas environments under the same excitation conditions. (Reprinted with permission from [HOU 12]. Copyright 2012, American Chemical Society)*

The results demonstrate that upon illumination with an argon ion laser (cw, photon energy 2.71 eV, i.e. below the band gap of the ZnO nanowires) the photoconductivity of the device operated in ambient air significantly rises on a time scale of some tens of minutes. After blocking the laser beam, the device slowly returns to its initial state in a few hours, which shows that no irreversible changes were induced by the laser irradiation. Compared to the operation in ambient air, the maximum current reached under the same excitation conditions but in pure oxygen atmosphere is substantially smaller whereas the dynamics is not significantly changed in this case. However, the same experiment carried out in oxygen-free environments (nitrogen flow, argon flow and vacuum conditions 10^{-6} mbar) always systematically shows a much slower increase of the photoconductivity, and a significantly higher saturation current is reached.

From the results of these experiments it was concluded that the adsorption and desorption of oxygen related molecular species from both the nanowire and the quantum-dot surfaces plays a key role for understanding the electron-transfer dynamics in these hybrid systems. The essential elements of the model are depicted in Figure 4.8. In the as-prepared devices, oxygen molecules and related species (such as O_2, O, OH, OH^-, etc.) passivate surface states on the ZnO nanowires and the CdSe quantum dots. For the quantum dots, this passivation is crucial for suppressing a rather fast and efficient non-radiative recombination channel for photoexcited electron-hole pairs involving surface states.

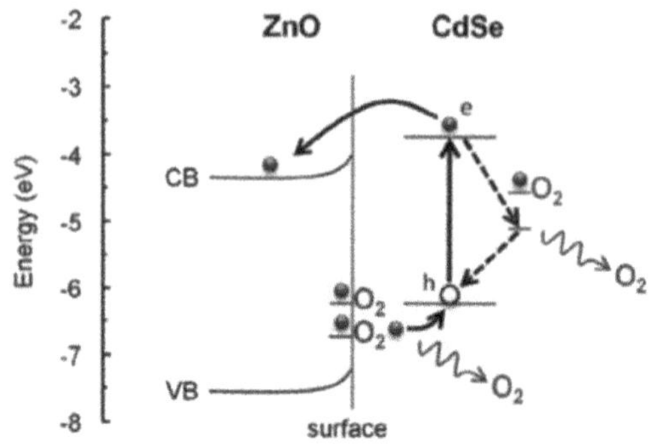

Figure 4.8. *Diagram of the energy level alignment, the electron transfer, as well as the surface oxygen desorption from the ZnO nanowire surface in the photoconductivity measurement of the NW/QD assembly. (Reprinted with permission from [HOU 12]. Copyright 2012, American Chemical Society)*

After photoexcitation of one electron-hole pair in a quantum dot, the electron can tunnel into the conduction band of the ZnO nanowire. As no hole-conductive electrolyte is present, the hole will stay in the quantum dot. The dot can be refilled with an electron through the interaction with negatively charged surface states of the nanowire or the quantum dot, respectively. This way, surface oxygen will be released from the nanowire or the quantum dot. If this happens on the nanowire, the whole excitation process has effectively removed a localized electron from an oxygen related surface state and transferred it into an extended conduction-band state while releasing the neutral oxygen molecule from the surface. This whole process therefore increases the density of electrons in the conduction band of the nanowire and increases the width of the conductive nanowire core by reducing the band bending (and thus the width of the electron depletion layer). The overall effect is an increased photoconductivity that was indeed measured for the functionalized nanowires.

At the same time, however, the continuous photoexcitation of electron-hole pairs in the quantum dots can release oxygen from the quantum-dot surface thereby activating non-radiative recombination channels. These channels trap photogenerated electrons rather efficiently and substantially reduce the probability of electron tunneling into the conduction band of the nanowire. Therefore, this competing process will slow down the dynamics of the photoconductivity enhancement of the nanowire/quantum-dot assembly. Under typical experimental conditions as reported in [HOU 12], steady-state conditions for the assembly placed in vacuum (10^{-6} mbar) under photoexcitation with an argon ion laser (photon energy 2.71 eV) could not be reached even after several days.

The results discussed in this section demonstrate the importance of (negatively charged) surface states, band bending and adsorbed atomic or molecular species the nanowire surface for the electronic and photonic properties of simple devices based on functionalized nanowires. Suitable passivation strategies must be developed in order to achieve full control over the nanowire surface states. In particular, the adsorption and desorption processes of oxygen and OH related molecular species will be of crucial importance for optimizing the performance of nanowire-based devices and ensuring their long-term stability.

4.5. Other surface-related effects in ZnO nanowires

The previous sections have dealt with surface-related luminescence properties of ZnO nanowires, emphasized the importance of band bending and illustrated the importance of adsorption and desorption processes at the nanowire surface. Of course, there exist far more surface-related properties of ZnO nanowires which cannot all be covered in the scope of this chapter. Still, a glimpse onto some further surface-related properties will be given in this last section.

Already the pure ZnO surface, without any further functionalization, can be used for the fabrication of gas sensing devices [WAN 04] albeit with very limited selectivity. The operating principle is very similar to the surface-related phenomena discussed in the previous sections. In oxide ceramics or thin polycrystalline films, surface defects at grain boundaries are typically passivated by negatively charged oxygen molecules. The result is again an upward band bending in the near-surface region which consequently leads to relatively broad depletion layers and thus high resistivity of the device under ambient conditions. Exposure to a reducing gas at moderate temperatures removes the oxygen from the grain boundaries, thus increasing the conductivity of the device. As reported in [WAN 04], the use of ZnO nanowires with diameters of about 25 nm for gas sensing applications can, therefore, substantially increase the sensitivity of such sensors because the conductivity of the whole nanowire will be strongly modulated when its surface is oxidized or reduced,. In particular, the authors find a sensitivity of $S = R_a/R_g$=47 to 200 ppm ethanol gas at an operating temperature of $T = 300°C$ which is more than one order of magnitude higher than the value of $S< 2$ reported for ZnO ceramic gas sensors [RAO 00]. The response time of their sensor is on the order of 10 s for the "on"-signal, the recovery needs roughly 120 s.

Fast response times and a high sensitivity are required for sensing applications. However, for stable operation of nanowire FETs, processes favoring those properties have to be completely suppressed. In [CHA 06], the authors applied a (*complementary metal-oxide semiconductor*) CMOS-compatible surface passivation treatment (a SiO_2/Si_3N_4 bilayer coating patterned by a polyimide shadow mask) to

ZnO nanowire FETs. Prior to passivation, the nanowire devices were placed in a vacuum chamber at 10^{-6} torr to remove surface adsorbates. The authors report substantially improved FET characteristics (subthreshold swing, on/off ratio and mobility) obtained with the passivated ZnO nanowires. In particular, they find that the field-effect mobility of ZnO nanowire FETs is increased by about two orders of magnitude after the passivation process.

Apart from passivation treatments, certain applications of ZnO nanowires may also require their surface to exhibit either a hydrophobic or hydrophilic character. In [KWA 09], the authors demonstrated that the treatment of chemically synthesized ZnO nanowires (hydrothermal method, assynthesized nanowire array shows a hydrophilic surface) with fatty acids of different chain lengths allows them to tune the nanowire surface energy over a wide range. Their experiments demonstrated that the wettability of the ZnO nanowire array was monotonically converted from hydrophilicity to hydrophobicity by increasing the carbon chain length of the chemisorbed fatty acid. The maximum water contact angle of 167° was achieved with stearic acid (C18). Upon illumination with UV light, the contact angle monotonically decreased due to light induced decomposition of the organic monolayer. The authors used this UV-stimulated conversion to fabricate a photo induced pattern composed of super-hydrophobic and super-hydrophilic areas on the nanowire array.

Theoretical [HE 08] and experimental studies [CHE 06] have indicated that the elastic properties of ZnO nanowires are strongly influenced by surface effects, too. In particular, the Young's modulus in [0001] oriented ZnO nanowires was experimentally found to depend on the nanowire diameter [CHE 06]. Below about 120 nm, it becomes substantially larger with decreasing nanowire diameter, whereas thicker nanowires show Young's moduli comparable to those of bulk ZnO. The authors explain their findings on the basis of a core–shell model where surface stiffening effects arise from the bond length becoming gradually shorter when going from the bulk to the surface of the nanowire.

4.6. Conclusion

We conclude the whole chapter by stating that ZnO nanowires show an extremely rich variety of properties that are directly related to their surface. At the same time, they can be used as single crystalline channels for electronic or photonic transport with the surface effects providing means for tuning the transport properties or selectively interacting with the electrons or photons in the transport channel. This combination makes semiconductor nanowires unique among other nanostructures and allows for fundamental and applied research that cannot be carried out with other nanomaterials. The past decade has shown remarkable progress in the controlled growth of nanowires and provided us with a thorough understanding of their basic physical properties. Yet, the richness of surface effects in ZnO nanowires has still to be fully understood, controlled, and carefully made use of in future nanowire-based electronic and optoelectronic devices.

4.7. Bibliography

[BIS 11] BISWAS M., JUNG Y.S, KIM H.K., *et al.*, "Microscopic origins of the surface exciton photoluminescence peak in ZnO nanostructures", *Phys. Rev. B*, vol. 83, p. 235320, 2011.

[CHA 06] CHANG P.-C, FAN Z., CHIEN C.-J., *et al.*, "High-performance ZnO nanowire field effect transistors", *Appl. Phys. Lett.*, vol. 89, no.13, p. 133113, 2006.

[CHE 06] CHEN C.Q., SHI Y., ZHANG Y.S., *et al.*, "Size dependence of young's modulus in ZnO nanowires", *Phys. Rev. Lett.*, vol. 96, p. 075505, 2006.

[CON 14] CONSONNI V., FEUILLET G., (eds), *Wide Band Gap Semiconductor Nanowires 2: Heterostructures and Optoelectronic Devices*, ISTE, London, and John Wiley & Sons, New York, 2014.

[DIJ 00] VAN DIJKEN A., MEULENKAMP E.A., VANMAEKELBERGH D., *et al.*, "The kinetics of the radiative and nonradiative processes in nanocrystalline ZnO particles upon photoexcitation", *J. Phys. Chem.*, vol. 104, no. 8, pp. 1715–1726, 2000.

[GRA 05] GRABOWSKA J., MEANEY A., NANDA K.K., *et al.*, "Surface excitonic emission and quenching effects in ZnO nanowire/nanowall systems: limiting effects on device potential", *Phys. Rev. B*, vol. 71, p. 115439, 2005.

[HE 08] HE J., LILLEY C.M., "Surface effect on the elastic behavior of static bending nanowires", *Nano Letters*, vol. 8, no. 7, pp. 1798–1802, 2008.

[HOU 12] HOU D., DEV A., FRANK K., *et al.*, "Oxygen-controlled photoconductivity in ZnO nanowires functionalized with colloidal CdSe quantum dots", *J. Phys. Chem C*, vol. 116, pp. 19604–19610, 2012.

[KIM 10] KIM D.S., RICHTERS J.-P., SCHOLZ R., *et al.*, "Modulation of carrier density in ZnO nanowires without impurity doping", *Appl. Phys. Lett.*, vol. 96, p. 123110, 2010.

[KLI 07] KLINGSHIRN C., "ZnO: material, physics and applications", *ChemPhysChem*, vol. 8, no. 6, pp. 782–803, 2007.

[KWA 09] KWAK G., SEOL M., TAK Y., *et al.*, "Superhydrophobic ZnO nanowire surface: chemical modification and effects of UV irradiation", *J. Phys. Chem. C*, vol. 113, pp. 12085–12089, 2009.

[LEI 03] LEITER F., ALVES H., PFISTERER D., *et al.*, "Oxygen vacancies in ZnO", *Physica B*, vol. 340, pp. 201–204, 2003.

[LES 07] LESCHKIES K.S., DIVAKAR R., BASU J., *et al.*, "Photosensitization of ZnO nanowires with CdSe quantum dots for photovoltaic devices", *Nano Letters*, vol. 7, no. 6, pp. 1793–1798, 2007.

[LIN 06] LIN Y.-J., TSAI C.-L., "Changes in surface band bending, surface work function, and sheet resistance of undoped ZnO films due to $(NH_4)_2S_x$ treatment", *J. Appl. Phys.*, vol. 100, p. 113721, 2006.

[MEY 04] MEYER B.K. *et al.*, "Bound exciton and donor-acceptor pair recombinations in ZnO", *Phys. Stat. Sol. (b)*, vol. 241, no. 2, pp. 231–260, 2004.

[RAO 00] RAO B.B., "Zinc oxide ceramic semi-conductor gas sensor for ethanol vapour", *Mater. Chem. Phys.*, vol. 64, p. 62, 2000.

[RIC 08a] RICHTERS J.-P., VOSS T., KIM D.S., *et al.*, "Enhanced surface-excitonic emission in ZnO/Al_2O_3 core-shell nanowires", *Nanotechnology*, vol. 19, p. 305202, 2008.

[RIC 08b] RICHTERS J.-P., VOSS T., WISCHMEIER L., *et al.*, "Influence of polymer coating on the low-temperature photoluminescence properties of ZnO nanowires", *Appl. Phys. Lett.*, vol. 92, p. 011103, 2008.

[SHA 04] SHALISH I., TEMKIN H., NARAYANAMURTI V., "Size-dependent surface luminescence in ZnO nanowires", *Phys. Rev. B*, vol. 69, p. 245401, 2004.

[STI 07] STICHTENOTH D., RONNING C., NIERMANN T., *et al.*, "Optical size effects in ultrathin ZnO nanowires", *Nanotechnology*, vol. 18, p. 435701, 2007.

[TEN 92] TENNE R., NABUTOVSKY V.M., LIFSHITZ E., *et al.*, "Unusual photoluminescence of porous CdS (CdSe) crystals", *Solid State Communications*, vol. 82, no.9, pp. 651–654, 1992.

[TRA 90a] TRAVNIKOV V.V., "Surface radiative recombination in CdS crystals", *Journal of Crystal Growth*, vol. 101, pp. 579–583, 1990.

[TRA 90b] TRAVNIKOV V.V., FREIBERG A., SAVIKHIN S.F., "Surface excitons in ZnO crystals", *Journal of Luminescence*, vol. 47, pp. 107–112, 1990.

[VOS 08] VOSS T., WISCHMEIER L., "Recombination dynamics of surface-related excitonic states in single ZnO nanowires", *J. Nanosci. Nanotechnol.*, vol. 8, no.1, pp. 228–232, 2008.

[VOS 10] VOSS T., RICHTERS J.-P., DEV A., "Surface effects and nonlinear optical properties of ZnO nanowires", *Phys. Stat. Sol. (b)*, vol. 247, no. 10, pp. 2476–2487, 2010.

[WAN 04] WAN Q., LI Q.H., CHEN Y.J., *et al.*, "Fabrication and ethanol sensing characteristics of ZnO nanowire gas sensors", *Appl. Phys. Lett.*, vol. 84, p. 3654, 2004.

[WAN 12] WANG D., REYNOLDS N., "Photoluminescence of zinc oxide nanowires: the effect of surface band bending", *ISRN Condensed Matter Physics*, vol. 2012, p. 950354, 2012.

[WIS 06a] WISCHMEIER L., VOSS T., BÖRNER S., *et al.*, "Comparison of the optical properties of as-grown ensembles and single ZnO nanowires", *Appl. Phys. A*, vol. 84, pp. 111–116, 2006.

[WIS 06b] WISCHMEIER L., VOSS T,. RÜCKMANN I., *et al.*, "Dynamics of surface-excitonic emission in ZnO nanowires", *Phys. Rev. B*, vol. 74, p. 195333, 2006.

[YAN 07] YANG Y., *et al.*, "Quenching of surface-exciton emission from ZnO nanocombs by plasma immersion ion implantation", *Appl. Phys. Lett.*, vol. 91, p. 071921, 2007.

[ZIM 09] ZIMMLER M.A., VOSS T., RONNING C., *et al.*, "Exciton-related electroluminescence from ZnO nanowire light-emitting diodes", *Appl. Phys. Lett.*, vol. 94, p. 241120, 2009.

5

Doping and Transport

5.1. Introduction

The understanding of semiconductor doping has been one of the main issues of semiconductor research since the appearance of electronic devices in the 1940s. Since then, semiconductor physics and electronic device performances have been closely linked to doping. When a new semiconductor emerges, the control of $n-$type and $p-$type doping are generally the main targets for researchers growing or studying such materials. As an example, doping of wide band gap semiconductors, like GaN and ZnO, has been intensively studied during these last three decades and, up to now, major issues are still unsolved. Concerning GaN, the $n-$type was easily mastered whereas the $p-$type was more difficult to achieve because of the depth of Mg acceptor (245 meV from the topmost of the valence band). However, $p-$type doping is now controlled and it is therefore possible to fabricate efficient bipolar electronic devices dedicated to a large market like light-emitting diodes (LED). ZnO is a more problematic issue since $n-$type residual doping is generally high in this semiconductor. This currently hampers the reproducible control of $p-$type doping process, making impossible the fabrication of bipolar devices like LED.

Since the beginning of the 2000s, the emergence of nanowires (NW) and microwires (MW) introduced a new situation for GaN and ZnO doping. Nanowires and microwires are attractive systems due to their particular geometry. The large surface to volume ratio of NW and MW makes them good candidates for efficient and sensitive devices (light emission, light absorption, gas detection, etc.) as well as for low-dimensional physics (0D or 1D systems, topological insulators, etc.). GaN and ZnO are interesting semiconductors because of their physical properties (direct wide band gap, large exciton binding energy) as well as the ability to grow wires

Chapter written by Julien PERNOT, Fabrice DONATINI and Pierre TCHOULFIAN.

(NW and MW). The understanding of the doping mechanism and transport properties in GaN and ZnO wires is a major task currently under investigation.

Today, some important questions concerning GaN and ZnO wires doping remain open. Of particular interest is the spatial distribution of dopants both axially along the growth axis (from the substrate to the top), and radially, from the core to the surface. The possible inhomogeneity could modify the physical properties of the wires (for example, the surface band bending) and should be taken into account when analyzing their electrical properties. The distribution of dopants must be controlled in order to develop electronic devices based on GaN and ZnO NW and MW. However, it implies that two practical challenges must first be solved: (1) how to connect electrically these objects (at single or ensemble levels) and (2) how to perform reliable measurements of electrical properties of doped NW and MW.

This chapter will present an exhaustive method to study the transport properties of doped GaN and ZnO NW and MW in light of recent results published in the literature. The chapter is organized into three main sections. In section 5.2, an original lithography process perfectly well suited to connect ZnO NW and GaN MW will be described. Section 5.3 will be dedicated to the electrical transport techniques, which can be used in the case of NW and MW. An analysis of the carrier mobility extracted from such measurements will be discussed and compared to bulk case. In section 5.4, a local probe technique based on cathodoluminescence will be used to investigate exciton diffusion length and space charge region extension in NW. Finally, section 5.5 will conclude the chapter with a summary and perspectives.

5.2. Advanced lithography processes for direct wide band gap nanowire and microwire devices

Making electrical contacts on single nanowires is of crucial importance for the understanding of the transport in nanowires at single level but also at ensemble level. This could be done via the direct connection of probes driven by nanomanipulators [LOR 13], by direct ion or electron beam-induced deposition (IBID or EBID) of metal electrodes [HER 06] or by deposition of metal electrodes through a mask made of resist patterned using electron beam lithography (EBL) [LAN 06]. The direct connection is possible on as grown NW or MW (only one probe with a nano-positioning system is needed at the top end of the NW or MW if the substrate could be used as the second contact) or on dispersed NW or MW on insulating substrate. However, obtaining ohmic contacts with the direct connection technique remains challenging. With regard to the IBID, EBID and EBL techniques, the dispersion of the nanowires on an insulating substrate is required. The drawbacks of IBID and EBID compared to EBL are the limited number of precursors, the slow deposition rate, the presence of impurities in the deposit and the difficulty of obtaining steep deposition profiles. Recently, an original *in situ* EBL process with voltage contrast imaging has been successfully used for the fabrication of single ZnO nanowire devices [LON 09]. However, its main drawback is that high-voltage contrast is only obtained at low ebeam energy (below 4 keV) whereas lithography writing needs higher ebeam energy (above 10 keV).

Here, we will focus on a new *in situ* EBL lithography technique based on cathodoluminescence (CL) imaging [DON 10], which is perfectly well suited for the practical case of GaN and ZnO. It has been demonstrated that beyond optical characterization at the nanometer scale, CL could be involved in single nanowire device fabrication (this section) and transport measurements on specific devices (section 5.4). In a cathodoluminescence system (see Figure 5.1), the energy of the ebeam (up to 30 keV) of the scanning electron microscope (SEM) is large enough to excite, at nanometer scale, the luminescence in semiconductors including wide band gap semiconductors.

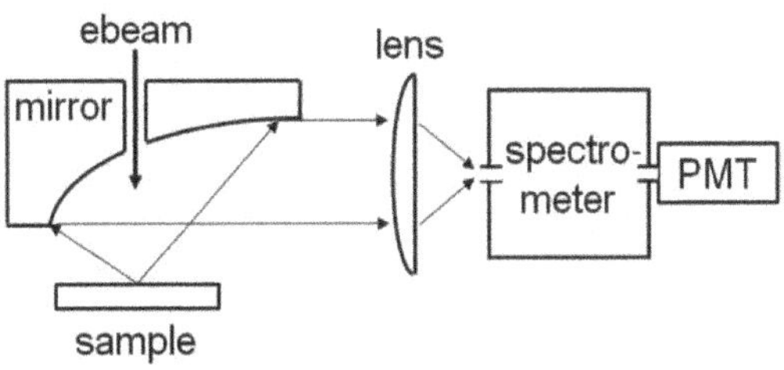

Figure 5.1. *CL imaging setup. (Reproduced with permission from [DON 10]. © IOP Publishing. All rights reserved)*

Even if silicon is an indirect band gap semiconductor, many other semiconducting materials have a direct band gap, such as GaN and ZnO. Moreover, single GaN and ZnO NW and MW could exhibit large cathodoluminescence emission even under low electron beam excitation. Therefore, for these wide band gap semiconductors, the CL signal enables us to acquire a CL mapping of nanowires buried under resist with an ebeam dose much lower than the writing (exposure) dose of the resist. Panchromatic or monochromatic mappings are obtained using a photomultiplier tube (PMT) associated to a spectrometer. *In situ* EBL could then be considered in the wavelength range of the resist optical transparency. The transmission of the poly methyl methacrylate (PMMA) is 90% for the near-band-edge emission of GaN and ZnO as shown in Figure 5.2. Figure 5.3 illustrates the ability to map out the location of a single ZnO nanowire buried under 600 nm PMMA thick resist using cathodoluminescence imaging technique without resist writing.

The first step of this *in situ* EBL technique is to locate, by scanning the substrate surface, an isolated nanowire beneath the resist with a low ebeam dose. In this read step, the quality of the nanowire could be checked by monitoring the brightness and uniformity of the CL (with or without spectral filtering) along the nanowire. Once an appropriate nanowire has been located, the second step is dedicated to the writing of the pattern associated with the nanowire (one connection at one end of the nanowire, for example, as shown in Figure 5.4). In the write step, the ebeam dose is set to the writing dose of the resist (for example 250 μC/cm^2 for PMMA resist). Read and write cycles could be performed from place to place until the substrate area has been scanned. This *in situ* technique is less time consuming than the classical technique since the latter needs substrates with alignments marks and an additional SEM imaging step to locate nanowires with respect to these marks.

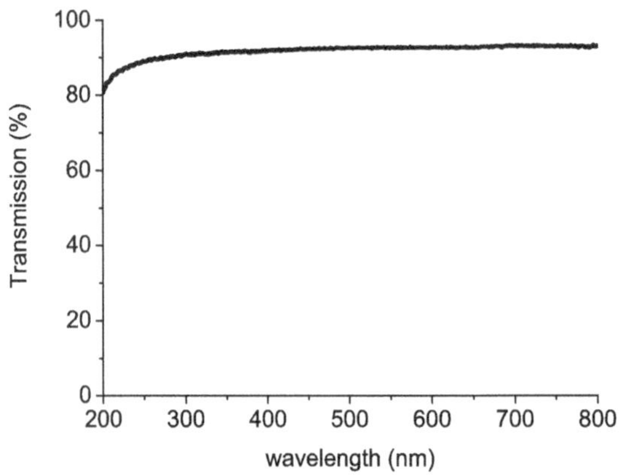

Figure 5.2. *Optical transmission of a 600 nm thick PMMA layer in the near-UV to near-IR range. (Reproduced with permission from [DON 10]. © IOP Publishing. All rights reserved)*

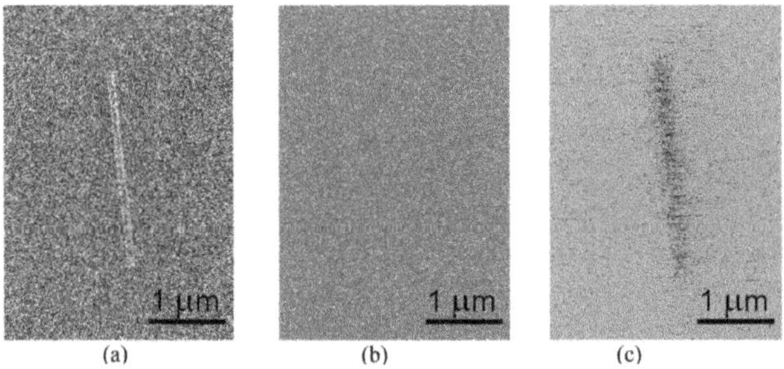

Figure 5.3. *Observation of the same nanowire using the same 5 μC.cm^{-2} e-beam dose: a) SEM image of the nanowire on the Si/SiO$_2$ substrate; b) SEM image of the nanowire under 600 nm thick PMMA resist and c) CL image of the nanowire under 600 nm thick PMMA resist at 5 K. (Reproduced with permission from [DON 10]. © IOP Publishing. All rights reserved)*

Since temperature plays an important role on the radiative efficiency of semiconductors (higher radiative efficiency at lower temperature as illustrated in

Figure 5.5 in the case of ZnO single nanowire without PMMA resist), cryogenic EBL (down to liquid helium temperature) has been studied. A shift of the writing threshold of the PMMA resist has been clearly observed toward larger ebeam dose (see Figure 5.6 where the relative thickness is the ratio between the written thickness to the total PMMA thickness). Thus, in cryogenic *in situ* EBL with CL imaging, a much better signal-to-noise ratio is obtained on CL mappings at the read step. Moreover, the optical selection of nanowires could be achieved more accurately by looking at one specific wavelength in the nanowire CL emission range. No degradation of the spatial resolution has been seen when EBL patterns are realized at low temperature.

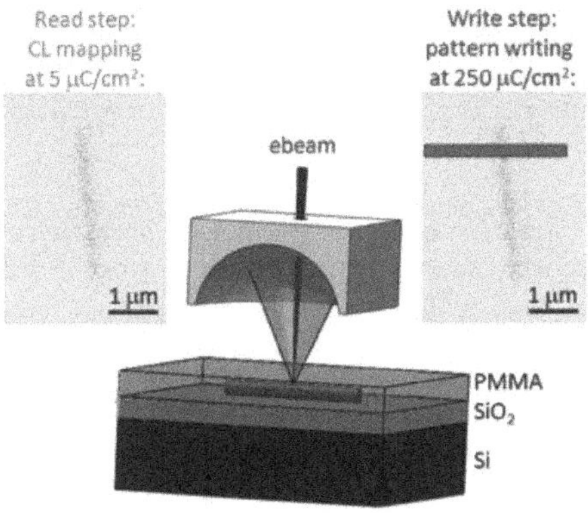

Figure 5.4. *Schematic of the in situ EBL with CL imaging. For a color version of this figure, see www.iste.co.uk/consonni/nanowires1.zip*

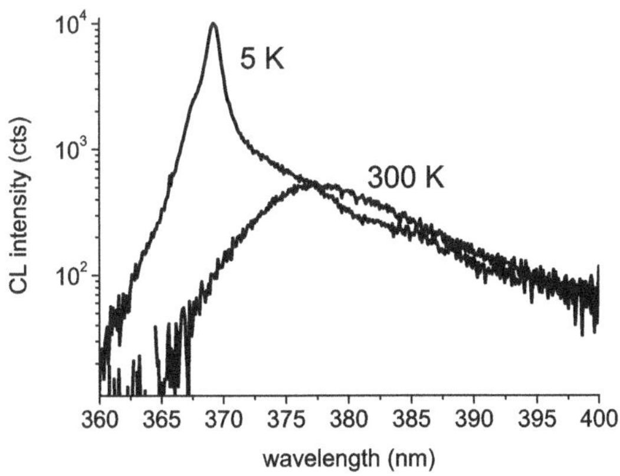

Figure 5.5. *CL spectra of the same nanowire under the same e-beam dose at room temperature and 5 K. (Reproduced with permission from [DON 10]. © IOP Publishing. All rights reserved)*

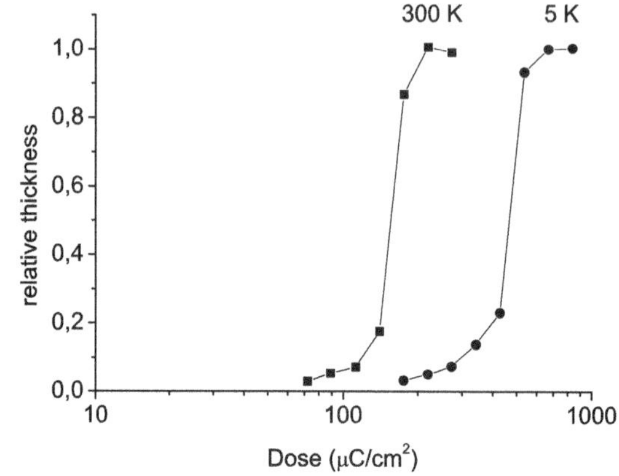

Figure 5.6. *Sensitivity of the PMMA at room temperature and 5 K. (Reproduced with permission from [DON 10]. © IOP Publishing. All rights reserved)*

EBL (CL-based or not) is routinely used as a suitable technique to electrically contact nanowires with typical diameters up to 600 nm. Recently, GaN-based microwires (diameter around 1–2 μm) grown by MOVPE have, however, attracted considerable interest because of their desirable intrinsic properties such as their core-shell structures.

With regard to these structures, the standard EBL technique is still considered appropriate to write the contact pattern on a single MW, even though two additional issues arise in this case: the loss of spatial resolution and the poor step coverage during the classical metal deposition. First, a resist layer a lot thicker than the wire itself should be spin-coated. Taking as an example a 1 μm large microwire, the necessary 2 μm thick resist deteriorates the achievable spatial resolution because beam electrons diffuse further away from the exposure landing point as the resist gets thicker. The achievable pattern size is then close to the μm range, which could still be fine on particularly long wires. The second issue arises during the metal deposition step. As shown in Figure 5.7(a), the directional nature of the evaporated metal atoms gives poor step coverage. This shadowing effect makes the contacting of hexagonal MW difficult. We can get rid of the poor coverage by using either conformal sputtering but subsequent lift-off is difficult or by performing incident deposition and constantly rotating the sample holder during the metal deposition as demonstrated in Figure 5.7(b) [TCH 13b].

To achieve a high resolution, an ideal solution is to use a lifting layer in order to planarize the device [ZIM 08, JAC 12, STO 12, TCH 13a]. The lifting layers are based on either spin-on glass (H-silsequioxane) [ZIM 08, JAC 12]] or photoresist (S1805)

[STO 12, TCH 13b]. After spin-coating the planarizing layer, a hard-baking step is necessary to make the lifting layer insoluble to acetone in order for it to sustain the lift-off treatment. Access to the lying wires surface is obtained by wet or dry etching. Finally, a standard PMMA-based EBL can be performed on this flat device with a high resolution. This technique has been proved to be useful due to its auto-alignement feature and high resolution as shown in Figure 5.7(c).

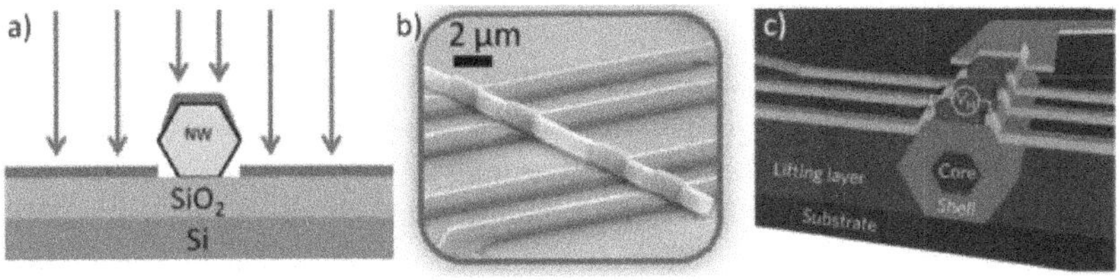

Figure 5.7. *a) Shadowing effect during metal deposition. b) Microwire connected using the incident-deposition technique. (Reprinted with permission from [TCH 13b]. Copyright 2013, AIP Publishing LLC). c) Schematic of a device using a self-aligned lifting layer in order to achieve high resolution. (Reprinted with permission from Macmillan Publishers Ltd, [STO 12], copyright 2012). For a color version of this figure, see www.iste.co.uk/consonni/nanowires1.zip*

The advanced lithography process for direct wide band gap semiconductor presented here allows us to connect wires with diameters ranging from tens of nanometers up to 2 μm. In the next section, some examples of single wire device connection via this technique will be used for electrical or thermoelectric measurements.

5.3. Electrical transport properties of single wire: ZnO nanowire and GaN microwire

5.3.1. *Electrical transport measurements*

5.3.1.1. *General remarks about near-surface band bending*

Due to their large surface-to-volume ratio, NW and MW need special attention concerning the electrical transport measurement and analysis. Some effects, generally neglected in the case of bulk or thin films, need to be taken into account in the case of NW and MW. For example, the band bending from the surface to the core of the wire could create an insulating (depletion) or a conductive (accumulation) subsurface layer, making difficult the analysis of the electrical measurements performed on the wire. These situations are schematically described in Figure 5.8 where different Fermi-level pinnings at the surface are considered (accumulation, flat band or depletion) for both $n-$type and $p-$type materials. To simplify the picture, the doping level (acceptors or donors) is considered to be homogeneous in the whole wire, except on the surface where surface doping could be one of the reasons for band bending.

The wire in the middle column of Figure 5.8 corresponds to the generally accepted assumption in bulk material: no influence of the near-surface region on the electrical conduction of the wire and therefore no band bending (flat band at the surface Figure 5.8(c) for the n−type and (d) for the p−type). In that case, the carrier density is constant from the core to the surface of the wire (Figure 5.8(b)) and usual bulk-like laws can be used to determine the basic properties of the wire. In this discussion, we neglect confinement effect in the wires.

The situation is more complicated when the band bending results in an accumulation of majority carrier population below the surface, as shown on the two left columns of Figure 5.8. In that case, the high carrier density under the surface should induce an additional conductive path (which could be even metallic) in parallel to the bulk-like conduction at the core of the wire. In that case, the conductance of the wire is generally larger than in the flat-band case (considering the same bulk-like conductivity values at the core of the wire). It must be noted that this situation could be inverted in the case of low carrier mobility in the accumulation layer; then the conductivity value of the accumulation layer could be below that of the core.

In the last situation, the band bending induces a depletion of majority carrier under the wire surface as shown in the two right columns, thereby reducing the actual section of the electrical conduction. A so called effective radius r_0 is defined to quantify this effect. In that case, the conductance of the wires described by the two right columns of Figure 5.8 will be lower than the middle one. The wire could even behave like an insulator in case of full depletion (last column of Figure 5.8).

The Fermi-level pinning could originate either from an inhomogeneous doping level in the wires (see, for example, in Ge NW [PER 09] or InN NW [ZHA 12]) or because of the interaction between the material and its surrounding. ZnO and GaN surfaces strongly interact with species in atmosphere, changing significantly the electrical transport in wires. Numerous papers reported the interaction mechanisms of ZnO wire surfaces with oxygen and hydrogen species [KIN 02, LIN 06, HU 09, PAR 09, GUR 14]. We may refer to section 1.4, Chapter 1, for further reading on the subject of surface effects in ZnO nanowires. When adsorbed to the surface of the wire, the species create an electron depletion layer (for oxygen) or an electron accumulation layer (for hydrogen or surface doping). These effects are therefore directly dependent on the species present in the surrounding atmosphere of the wires. The extent and time response of the band bending are determined by the adsorption mechanism and the amount of species (and thus to the vacuum level). It is therefore difficult to control the surface state in ZnO wire. UV illumination or electron beam bombardment could break the bonds between the ZnO NW and the species [HU 10, SOU 12] and could even create persistent conduction in ZnO wires [BAO 11, JOH 11, GUR 14]. It must be noted that persistent photoconduction (PPC) is a phenomenon already reported in bulk ZnO. These surface-related phenomena make the analysis of conduction properties of ZnO wires difficult [SCH 08]. The case of GaN is simpler since the pinning of the Fermi level at the surface creates a depletion layer (no accumulation layer observed) [RIC 08, GUR 11, SAN 13]. This surface depletion layer together with the induced spatial separation of photogenerated carriers is at the origin of the potential use of GaN (and ZnO) nanowires for high-gain photodetectors (Chapter 8 of Volume 2 [CON 14], for further information). To conclude, measurement in GaN and ZnO NW and MW strongly depends on the measurement conditions (illumination, vacuum, heating, NW surrounding (oxygen gas, hydrogen gas), etc.). Precaution is needed in order to determine reliable resistivity or doping/mobility values.

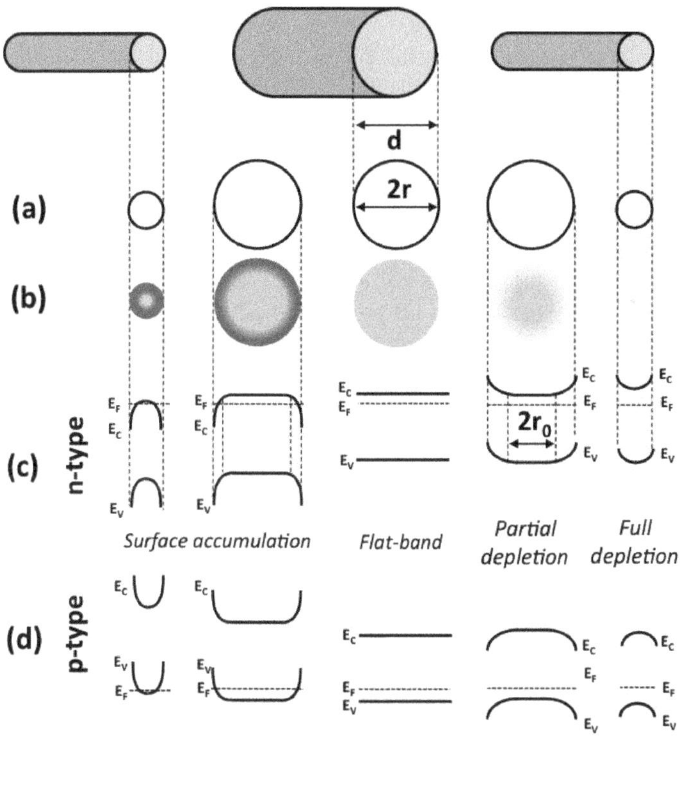

Figure 5.8. *Band bending along the radius of the wire and its effect on the spatial distribution of majority carriers in the wire: a) sections of the wire considered here; b) radial distribution of majority carriers in the wire; c) and d) band diagrams of the wire in case of n–type and p–type semiconductors, respectively. The five columns correspond to different Fermi-level pinning at the surface and wire diameter: accumulation in the case of small-diameter wire, accumulation in the case of large-diameter wire, flat band, partial depletion and full depletion.*

5.3.1.2. *Resistivity measurement: two-probe and four-probe configurations*

Four-probe measurement is a useful technique for the determination of the sheet resistance in thin film or bulk semiconductor. For wires, the principle is to use two outer ohmic contacts to source the current and two other inner contacts to probe the voltage. This technique permits us to infer the resistance of a sample without the influence of serial contact resistances. Two-probe measurements are often reported in the literature in order to determine the resistivity of a wire considering negligible contact resistances. The underlying assumption, which consists of neglecting the two contact resistances, can lead to an overestimation of the wire resistivity. In the case of omega-shaped contacts fabricated around a wire as shown in Figure 5.7(b), the contact resistance is proportional to the inverse of the radius $1/r$ of the wire and the wire resistance itself is inversely proportional to the section of the wire, i.e. proportional to $1/r^2$. In the case of a two-probe measurement, these resistances are in series. Considering a homogeneously doped wire, the ratio between the cylindric wire resistance $R_W = \frac{\rho l}{\pi r^2}$ (where ρ is the resistivity, l is the gap between the voltage probe contacts and $S = \pi r^2$ is the conducting section of the wire) and the contact resistance $R_C = \frac{\rho_C}{2\pi r d}$ (where d is the width of the electrical contact), $R_W/R_C \sim \rho/\rho_C \times 2d/r$, tends to ∞ when r tends to 0. In other words, the influence of the contact resistance will be larger for large-diameter wires; therefore, simple two-probe measurement is more relevant in the case of NW than MW. Obviously, apart from the geometrical aspect of contacts, the ratio ρ/ρ_C plays an important role. In any case, the four-probe method is the most reliable technique, assuming that the two inner voltage probes are sufficiently narrow to limit their influence on the equipotential lines in the wires. This requirement is more stringent in the wire case as compared to bulk one.

The four-probe technique allows us to infer the resistivity of the homogeneous doped semiconductor wire using the standard formula $R_W = \rho l/S$, where R is the measured resistance, ρ is the resistivity, l is the gap between the voltage probe contacts and S is the measured section of the wire. The analysis is much more complicated when the situations described in Figure 5.8 occur (except for the flat-band regime, which is the homogeneous case). Indeed, in these cases, the conductance of the near-surface region or surface layer must be taken into account. In case of depleted area, the region below the surface insulates and the conductive section becomes $S = \pi r_0^2$, where r_0 is the effective radius. The thickness of the depleted region below the surface is directly related to the pinning of the Fermi level at the surface (density and energy of the surface levels) and to the doping level of the sample, which determines the band bending curvature. For low doping and/or small radius, the wire could be in full depletion regime as evidenced, for example, by Richter *et al.* in GaN [RIC 08]. The critical doping concentration and/or depleted thickness can be evaluated using the formula given in [RIC 08, GUR 11]. For negligible depleted thickness (highly doped NW or a large radius), the standard formula is valid, the ratio $R/l = \rho/S = \rho/(\pi r^2)$ follows a $1/r^2$ law in case of bulk-like resistivity ρ. The experimental data in

Figure 5.9(a) illustrate this effect in highly Si-doped GaN MW (diamond) [TCH 13b], GaN NW (open circles) [GUR 11] and ZnO NW (full circles) [LOR 13]. In these cases, the doping levels of the wires are relatively high (larger than 10^{18} cm^{-3}). The dotted lines represent simulations using the parameters summarized in Table 5.1.

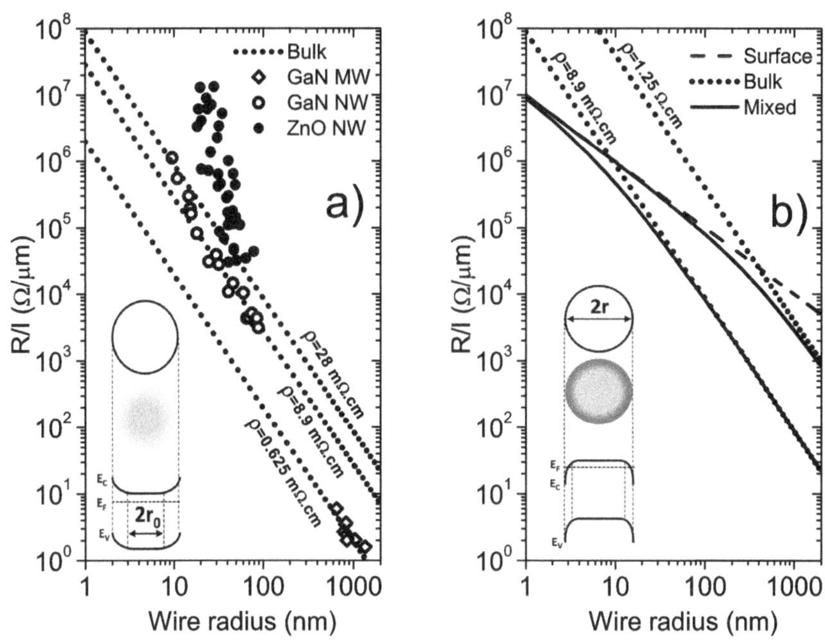

Figure 5.9. *Ratio between the resistance of the wire R and the gap length l between the two inner contacts of a four-probe geometry versus the radius of the wire r. Dotted lines correspond to simulations of bulk-like conduction within the whole section of the wire (πr^2) assuming a constant resistivity value (see Table 5.1). a) The symbols are experimental data: Si-doped GaN MW (diamond) [TCH 13b], GaN NW (open circles) [GUR 11] and ZnO NW (full circles) [LOR 13]. b) The dashed line corresponds to simulation for a surface conduction all around the wire perimeter ($2\pi r$). The full lines correspond to a mixed conduction with two parallel paths: the surface conduction and bulk-like conduction (core). It must be noted that no quantum effects are taken into account here leading to incorrect values for a small radius (<10 nm)*

For a small radius NW ($r < 20$ nm for GaN and $r < 50$ nm for ZnO), i.e. for depleted thickness of the order of magnitude of a wire radius, the wire is partially depleted and the resulting resistivity could be overestimated if such depletion is not taken into account. This is illustrated in Figure 5.9(a) where the R/l values deviate from the dotted line corresponding to a constant resistivity (ρ = 8.9 mΩ.cm for GaN and ρ = 28 mΩ.cm for ZnO).

Resistivity ρ (mΩ.cm)	Density n or p (cm^{-3})	Mobility μ (cm^2/Vs)
1.25×10^3	10^{16}	500
28	10^{18}	225
8.9	2.24×10^{18}	350
0.625	2×10^{20}	50

Table 5.1. *Parameters used in the simulation shown in Figure 5.9. It must be noted that the resistivity is the only parameter used for this plot. The carrier density and mobility are given as examples corresponding to the resistivity values*

However, if the wire surface is in an accumulation regime, the resistance of the wire core (flat-band region) is in parallel with the resistance of the wire surface (electron gas due to accumulation regime) leading to a mixed conduction. If the surface conductance is higher than the core conductance, the resistivity cannot be directly determined without knowing the properties of the surface electron gas. A full line of Figure 5.9(b) takes into account a mixed conduction including the core conduction with $\rho = 8.9$ mΩ.cm and the surface conduction $R/l = 1/(2\pi r n_S e\mu)$ (for example $n_S = 10^{12}$ cm^{-3}; μ = 100 cm^2/Vs in the range of typical values reported by Allen *et al.* [ALL 10] for bulk ZnO polar and non-polar surfaces). This simulation, where no quantum effect is taken into account, shows that the radius of ZnO wire must be of the order of the accumulation thickness (few nm [ALL 10]) to influence the resistance of the wire measured by the four-probe technique when the core doping level is in the range of 10^{18} cm^{-3}. However, for low doped wire core (around 10^{16} cm^{-3}), this critical radius could reach a value as high as hundreds of nanometers as shown by the other full line. This phenomenon has never been evidenced up to now but should be observed when the residual $n-$type doping level of ZnO wire will be lowered.

5.3.1.3. *Field-effect transistor characteristics*

Similar to the planar epitaxial case, the control of doping is essential to obtain high-performance wire-based devices. The knowledge of the electrical properties (carrier concentration and mobility) in wire-based semiconductor devices is therefore crucial and is obtained primarily by measuring wires at the single-wire level. The field-effect transistor (FET) is the most common device in order to easily estimate the carrier concentration and the mobility in doped nanowires. Figure 5.10(a) shows a typical device configuration where a GaN nanowire (NW) is reported on an oxidized substrate and subsequently electrically contacted using EBL technique [HUA 02].

As shown from the linear $I_{sd} - V_{sd}$ curves in Figure 5.10(b), the source and drain metal contacts are ohmic and the silicon back gate controls the amount of charges in the nanowire. The effect of applied gate voltage on I_{sd} current is actually a common method to infer the dopant type (n or p) of the studied nanowire. $n-$type is inferred here because the conductance increases as the positive gate voltage increases. Figure 5.10(c) shows the $I_{sd} - V_g$ curve from which the threshold voltage required to deplete the wire is found to be V_{th}=-8 V. The total charge in the wire is $Q_{tot} = C \times V_{th}$, with the capacitance commonly modeled as $C = \frac{2\pi\epsilon L}{cosh^{-1}(\frac{r+h}{r})} \approx \frac{2\pi\epsilon L}{ln(\frac{2h}{r})}$ if $h \gg r$ where ϵ is the dielectric constant, h is the thickness of the SiO_2 dielectric, L is the length and r is the NW radius. The electron concentration $n = \frac{Q_{tot}}{2\pi r^2 L}$ was found to be in the range 10^{18} to 10^{19} cm^{-3}. It must be noted here that the value measured by this technique is the donor density N_D and not the carrier concentration (here the approximation is correct because the residual donor level in GaN is shallow leading to the equality $n = N_D$). The mobility μ of the carriers is then estimated from the transconductance plot where $\frac{dI_{sd}}{dV_g} = \mu\frac{C}{L^2}V_{sd}$. Same methods have been used in the case of ZnO NW [PAR 04, GOL 05, CHA 06, YAN 08]. However, the accuracy of this approach relies on the difficult estimate of both the gate capacitance and interface charges. By comparing Hall-effect and FET results on InAs nanowires [BLO 12], it has been evidenced recently that the carrier concentration (respectively, the mobility) could be overestimated (underestimated) by a factor of 4 in the FET configuration. Using ZnO nanostructures, Likovich *et al.* [LIK 09] also concluded that the FET approach gives a poor estimation of the mobility and carrier concentration (overestimation of the mobility and underestimation of the carrier concentration in their case). Moreover, the FET approach is not suitable for studying microwires or highly doped material because of the poor electrostatic control of the back-gate in these cases. For a thorough estimation of the electrical properties of the wires, other characterization techniques such as thermoelectric and Hall-effect measurements can be performed.

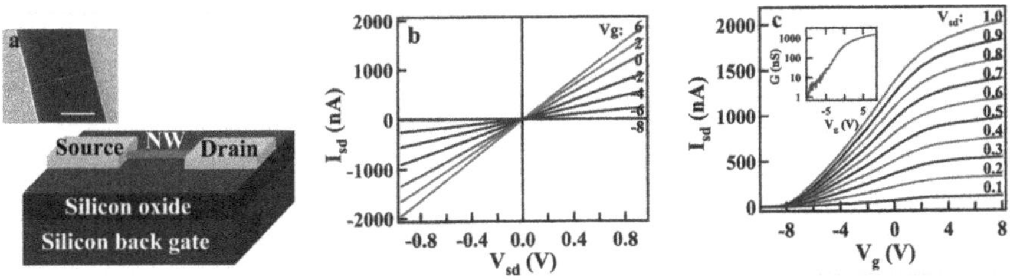

Figure 5.10. *a) Schematic of an NW FET, and (inset) FESEM image of a GaN NW FET. Scale bar is 2 μm. b) Gate-dependent $I_{sd} - V_{sd}$ data recorded on a 17.6 nm diameter GaN NW. The gate voltages for each $I_{sd} - V_{sd}$ curve are indicated. c) $I_{sd} - V_g$ data recorded for values of V_{sd}= 0.1-1 V. (inset) Conductance, G, versus gate voltage. (Reprinted with permission from [HUA 02]. Copyright 2002, American Chemical Society). For a color version of this figure, see www.iste.co.uk/consonni/nanowires1.zip*

5.3.1.4. *Thermoelectric measurement: Seebeck effect*

In the well-known thermoelectric probe experiment, the contact of a hot and a cold probe on a semiconductor gives rise to a thermovoltage whose sign allows us to determine the majority carrier type. It points out the interest for thermoelectric measurement since this experiment could be readily applied to ZnO nanowires where p–doping is still challenging. More generally, thermoelectric measurements are well adapted to doped semiconductors as they are based on majority carrier distribution asymmetry at the Fermi level and allow us to probe small changes in the scattering rate and the density of states at Fermi level.

Measurement of the Seebeck coefficient is actually a way to probe the Fermi-level energy and then estimate the carrier concentration n of the nanowire material. An experimental work has reported such measurements in both GaN and ZnO nanowires [LEE 09]. In addition, it is known that the resistivity ρ combines the effect of carrier concentration n and mobility μ through the expression ρ=1/(enμ). Therefore, by combining thermoelectric and resistivity measurements at the same location on the same part of a wire, the mobility μ can be deduced. This combined approach has been recently undertaken to study heavily Si-doped GaN wires. [TCH 13a]

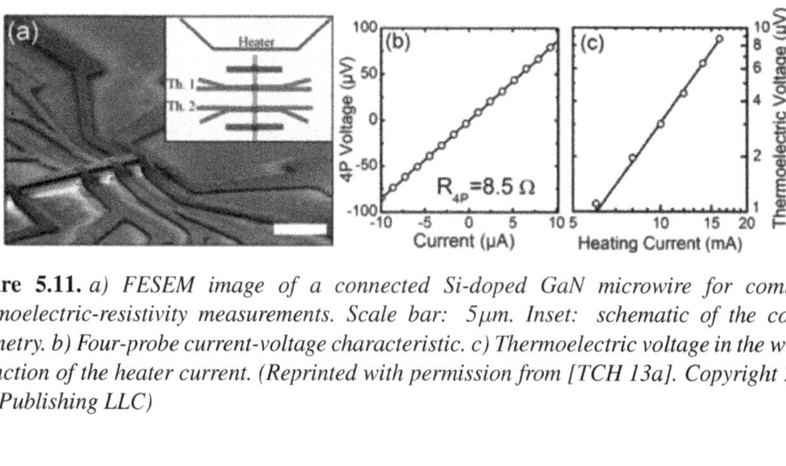

Figure 5.11. *a) FESEM image of a connected Si-doped GaN microwire for combined thermoelectric-resistivity measurements. Scale bar: 5μm. Inset: schematic of the contact geometry. b) Four-probe current-voltage characteristic. c) Thermoelectric voltage in the wire as a function of the heater current. (Reprinted with permission from [TCH 13a]. Copyright 2013, AIP Publishing LLC)*

Figure 5.11(a) depicts the most common configuration to perform combined thermoelectric resistivity measurements in the same region of a wire. EBL is needed to define the metallic contacts, among which are found a micro-fabricated heater and two sets of thermometers (thermometers 1 and 2), which simultaneously serve as electrodes for electrical measurements. The sourcing of a current I_h in the heater creates a thermal gradient ΔT along the structure with isothermal lines parallel to the heater line. ΔT is quantified by probing the change in four-probe resistivity of the *thermometer* metallic lines (Figure 5.11(b)). The generated thermoelectric voltage V_{th} across the two thermometers is measured as reported in Figure 5.11(c). Joule heating from the heater is responsible for the generated V_{th} as shown by the almost quadratic variation of V_{th} versus I_h (power coefficient is 2.1–2.6) as shown in Figure 5.11(c). Neglecting the Seebeck coefficient of the metal contacts, the Seebeck coefficient S for the material under study is simply written as $S = -\frac{V_{th}}{\Delta T}$.

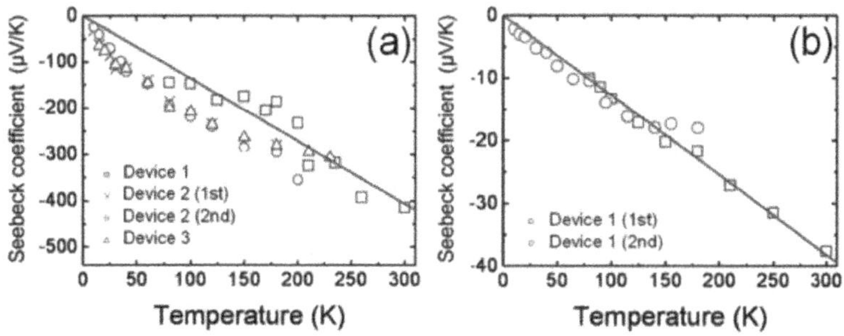

Figure 5.12. *Seebeck coefficient values as a function of temperature for individual: a) ZnO nanowires and b) GaN nanowires. (Reprinted with permission from [Lee 09]. Copyright 2009, AIP Publishing LLC). For a color version of this figure, see www.iste.co.uk/consonni/nanowires1.zip*

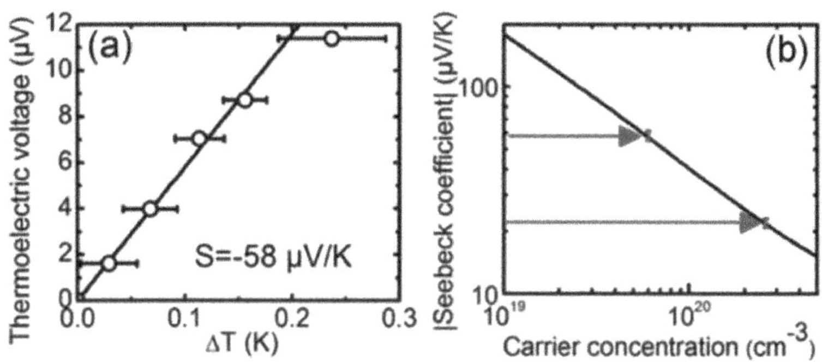

Figure 5.13. *a) Thermoelectric voltage versus generated temperature difference ΔT in a Si-doped GaN wire showing a linear behavior. b) Seebeck coefficient versus carrier concentration in $n-$GaN for two Si-doped wires (Reprinted with permission from [TCH 13a]. Copyright 2013, AIP Publishing LLC)*

In the framework of Boltzmann's transport theory, where the relaxation time dependence on energy is defined as to follow a power-law, a general description of the Seebeck coefficient can be written as [NOL 01]:

$$S = -\frac{1}{cT} \times \left[E_F - \frac{\int_0^\infty g(E)\tau(E)E^2 \frac{\partial f_0(E)}{\partial(E)}}{\int_0^\infty g(E)\tau(E)E \frac{\partial f_0(E)}{\partial(E)}} \right] \qquad [5.1]$$

where E is the kinetic energy of the electron, E_F is the Fermi energy, g is the density of states, τ is the relaxation time, f_0 is the Fermi distribution function, T is the temperature and e is the electron charge. This description is valid irrespective of the doping properties of the material within the effective mass approximation. Once appropriate density of states and dominant scattering mechanism are chosen, this description could be simplified and could even yield analytical formulations in the case of non-degenerated ($\frac{E_C - E_F}{k_b T} \gg 1$) or highly degenerated material ($\frac{E_C - E_F}{k_b T} \ll 1$), where E_C is the bottom of the conduction band and k_b is the Boltzmann's constant. In the intermediate doping range, numerical calculations are, however, necessary to find out the relevant E_F. Under high degeneracy condition, the model is known as the Mott equation which, if non-parabolicity effects are taken into account, is written as :

$$S = -\frac{8\pi(\frac{\pi}{3})^{\frac{5}{3}} k_b^2}{|e|h^2} \times \frac{T}{n^{\frac{2}{3}}} \times [m^*(s + 1 - \lambda)]_{E=E_F} \qquad [5.2]$$

where h is Planck's constant, s is the scattering parameter associated with the dominant scattering mechanism and $\lambda = 3n/m^* \frac{dm^*}{dn}$ is the effect of the non-parabolicity of the conduction band. The non-parabolicity is included by using the momentum effective mass $m^*(E) = m_0^*[1 + 2M_1E + 3M_2E^2 + 4M_3E^3]$ whose values in GaN are $m_0^* = 0.183m_0$ (where m_0 is the electron mass), $M_1 = 0.177$, $M_2 = -0.0169$ and $M_3 = 0.460$ [SHO 08]. As far as highly doped GaN material is concerned, ionized impurity scattering is known to be the main scattering mechanism. In order to calculate the value for the scattering parameter s, the relevant theory (Brooks–Herring for ionized impurity scattering) should be employed.

Figure 5.12 shows that the linear temperature dependence of the Seebeck coefficient described in the Mott equation has been verified experimentally in both doped ZnO and GaN wires [LEE 09]. Experimentally, the existence of the Seebeck coefficient is evidenced in Figure 5.13(a) by the linear relation between the thermoelectrical voltage and the differential temperature.

Finally, the carrier concentration is estimated from the measured Seebeck coefficient as shown in Figure 5.13(b) and the mobility is deduced from the measured resistivity combined to Seebeck coefficient at the same location in the wire. The combined thermoelectric-resistivity measurements are therefore effective to infer the electrical properties of semiconducting wires over a wide range of diameters and doping. The same procedure has been more recently used on InAs NW [SCH 14].

5.3.2. *Mobility versus doping*

An optoelectronic nanowire device, such as an LED, could be efficient only when $p-$ and $n-$type doping are both achieved and controlled. Among the range of possible characterization techniques, two approaches (FET and combined thermoelectric-resistivity) have been described in this chapter in order to infer the electrical properties of nanowires. In the past few years, several groups have managed to estimate the mobility and carrier density of $n-$doped GaN wires using either the two described approaches or other measurement techniques such as Raman spectroscopy, photoconductivity or resistivity.

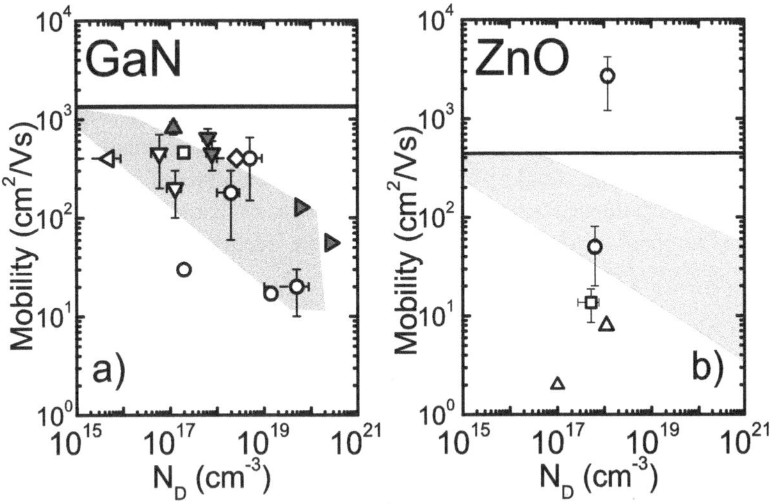

Figure 5.14. *a) Mobility as a function of doping concentration for non-intentionally doped (NID) (empty symbols) and Si-doped GaN (filled symbols) wires with different radii. Experimental data are obtained by different techniques: combined resistivity-photovoltage (diamond) [GUR 11], combined resistivity-thermoelectric (right-pointing triangles) [TCH 13a], FET measurements (circles)[MOT 07, STE 05, HUA 02], Raman spectroscopy (square) [JEG 09], Combined Raman-Thz photoconductivity (up-pointing triangle) [PAR 12], photoconductivity (down-pointing triangle) [BER 11], space-charge limited current (left-pointing triangle) [TAL 10]. b) Mobility as a function of doping concentration for ZnO measured by FET approach: square [GOL 05], circles [CHA 06] and triangles [YAN 08]. The light gray areas of each figure represent the commonly obtained mobility for $n-$doped planar layers. The black line represents the maximum mobility achievable in each material when considering intrinsic mobility exclusively limited by phonon scattering (for GaN [LOO 01] and for ZnO [MAK 05]). For a color version of this figure, see www.iste.co.uk/consonni/nanowires1.zip*

Figure 5.14 (a) shows the reported electron mobility as a function of doping concentration in $n-$doped GaN wires measured by different techniques. First, electron mobility in GaN wires with different radii follows the trend existing in planar layers illustrated by the gray area. Experimental data points are found to span over the full range of $n-$doping. The appeal for using nanowires has been evidenced, for example, by Parkinson *et al.* [PAR 12]. The authors claimed an improved mobility of 820 cm^2/Vs in nanowires, as compared to 590 cm^2/Vs for similarly doped layers ($n \approx 10^{17} cm^{-3}$). Surface scattering mechanism has also been evidenced to play a significant role in decreasing the mobility as the NW gets thinner [MOT 07]. Today, the main issue toward efficient optoelectronics GaN NW devices remains the mastering of $p-$type doping. As a proof, resistivity values have never been reported on $p-$type nanowire material either on a bare $p-$type wire or on the $p-$part of an axial or radial $p - n$ junction. Thus, there is currently little information about the electrical properties of the $p-$GaN used in NWs. This difficulty is not surprising since the $p-$type doping of GaN has been a well-known technological barrier in planar layers grown by metal organic chemical vapor deposition until the 1990s. Additionally, part of the difficulty to obtain an ohmic contact on a possibly poor $p-$doped nanowire could also come from the less matured nanowire contacting technology. This bottleneck must be overcome or new characterization techniques must be developed in order to provide valuable feedback about the electrical properties of $p-$GaN to NW growers.

Figure 5.14(b) shows the reported electron mobility as a function of doping concentration in $n-$doped ZnO wires measured by the FET approach [GOL 05, CHA 06, YAN 08]. These typical data illustrated the present situation in ZnO. Residual doping of ZnO is larger than 10^{17} cm^{-3}. Since the FET approach is the only technique that has been used up to now for mobility measurement, no data exist for highly doped ZnO wires (more than $10^{19} cm^{-3}$). Electron mobility values deduced by this technique are highly dispersed (more than three decades for the same doping-level range). Some of them are even much larger than the maximum mobility achievable in ZnO when considering intrinsic mobility exclusively limited by phonon scattering [MAK 05]. This comparison shows that the FET approach certainly leads to erroneous values of mobility.

5.4. Local probe and mapping of the electric field: cathodoluminescence

The accurate determination of some physical properties like minority carrier or exciton diffusion lengths requires high spatial resolution techniques based on electron beam microscopy: electron beam-induced current (EBIC) and CL, and specific devices: $p - n$ junction [GUT 12], Schottky barrier [HWA 11] or heterojunction [BOL 11]. Moreover, a precise mapping of the internal electric field could also be useful for the position of a heterostructure [GUS 07] or for the direct visualization of the carrier depletion and its extension with the external bias.

Concerning ZnO nanowires, the *in situ* EBL technique with CL imaging has been successfully involved for the realization of a single nanowire Schottky device [DON 10]. The main advantages of this kind of device are the creation of a depletion area on the Schottky side and the ability to tune the position of the frontier between the depletion area and the neutral part of the nanowire with an external reverse bias. Optical far-field [GUS 07] or near-field [SOU 10] techniques facilitate the direct visualization of the depletion area and the measurement of exciton or minority carrier diffusion lengths since the depleted area acts as a sink for both exciton and minority carrier. As shown in Figure 5.15, the internal electric field in the depletion region drifts apart electron and hole and may also dissociate the electron and hole of an exciton diffusing from the neutral part of the nanowire toward the depletion area. Using the CL mapping [HWA 11] rather than pure optical techniques leads to a better spatial resolution of the mappings and thus a more accurate measurement of the depletion width and diffusion length. Figure 5.15 gives the measurement principle and Figure 5.16 shows that the acquisition of CL mappings while increasing the reverse bias allows first to measure the extension of the depleted area and then to probe the diffusion length at different positions along the NW axis.

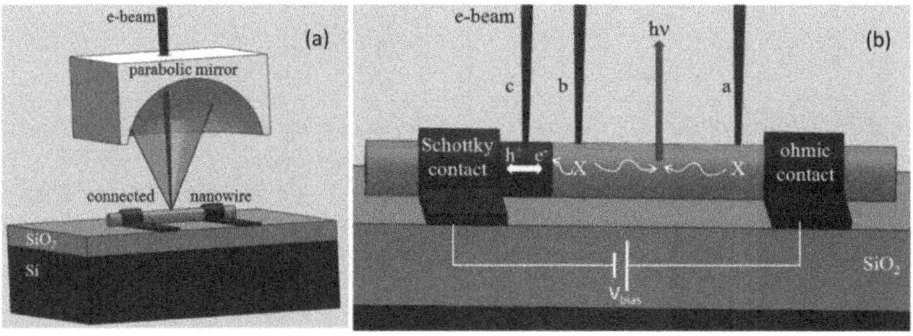

Figure 5.15. *a) Schematic of the CL setup for the study of a single ZnO nanowire Schottky diode. b) Schematic of the measurement of exciton (X) diffusion for different e-beam locations (a, b, c) along the nanowire axis. (Reproduced with permission from [HWA 11]. © IOP Publishing. All rights reserved). For a color version of this figure, see www.iste.co.uk/consonni/nanowires1.zip*

It has been confirmed, as shown in Figure 5.17(b), that in the case of ZnO nanowires the extension of the depleted area is proportional to the reverse bias and deviates from the bulk behavior (proportional to the square root of the reverse bias). The constant value of the diffusion length along the NW axis confirms the uniform quality of the NW already visible on the CL mapping during the read step of the *in situ* EBL (homogeneous CL emission in the neutral part). Unfortunately, the extension of the depleted area could not be readily used to estimate the carrier concentration since the dependance deviates from the bulk case. In the near future, the simultaneous determination of both minority carrier diffusion length and life time using local probe techniques will provide minority carrier mobility values in GaN and ZnO nanowires.

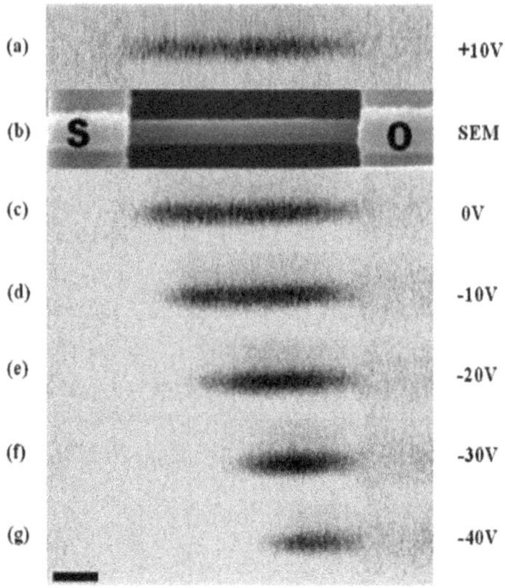

Figure 5.16. *CL mappings at 5 K (intensity scale increasing from white to black). a) CL mapping under the forward bias condition +10 V. b) SEM image of the ZnO NW Schottky device with one Schottky (S) and one Ohmic (O) contact. c) CL mapping without bias (0 V). d)–g) CL mappings under reverse bias conditions from -10 to -40 V. The scale bar is 300 nm. (Reproduced with permission from [HWA 11]. © IOP Publishing. All rights reserved)*

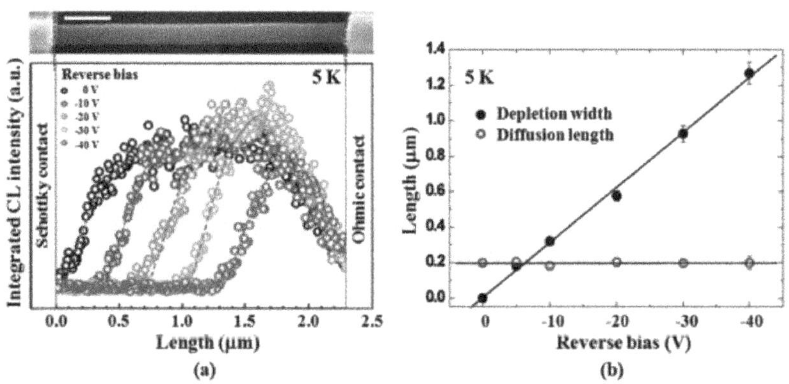

Figure 5.17. *a) CL intensity profiles along the NW as a function of the reverse bias condition (0 to -40 V) on a linear scale with the dashed lines representing the fitting using equation [5.1]. The scale bar is 400 nm. b) Diffusion length of the excitons and carrier depletion width in a ZnO NW as a function of the reverse bias condition. The values are extracted from the CL intensity profiles in (a). (Reproduced with permission from [HWA 11]. © IOP Publishing. All rights reserved). For a color version of this figure, see www.iste.co.uk/consonni/nanowires1.zip*

5.5. Conclusion and perspectives

GaN and ZnO wires are promising elementary building blocks for the fabrication of wire-based devices, like LEDs for example. The need to understand and control the doping mechanism of such objects is crucial for fabricating efficient devices. A large research effort has been carried out during the last decade for the improvement of electrical characterization of wide band gap semiconductor wires. In this chapter, we first described an original lithography process well suited to connecting ZnO and GaN wires. Then, the electrical properties of single wire were investigated and discussed in detail. The different near-surface regimes (depletion, flat band and accumulation) induced by the radial band bending were summarized and their impacts on the electrical properties were highlighted. ZnO wire surfaces exhibited depletion or accumulation depending on the experimental environment (mainly species in surrounding environment) while, only surface depletion has been observed in GaN wires. The electrical transport properties induced by multiple (core and shell) or partial (reduced core) conduction paths were discussed and quantified in the case of four-probe resistance measurement. The possibilities offered by the FET approach to measure the carrier concentration and mobility as well as its limitations were detailed. An original approach combining resistivity (four probe) and carrier concentration (thermoelectric Seebeck effect) measurements has been used to deduce the carrier mobility in individual wire. This method applied to Si-doped GaN MW enabled us to demonstrate that the larger conductivity of GaN MW compared to thin film is due to a combination of a mobility value (around 50 cm^2/Vs) and a large carrier concentration (around $n = 2 \times 10^{20}$ cm^{-3}). In the coming years, spatially the resolved Hall effect measurement [STO 12, BLO 12] should be applied to GaN and ZnO wires. Even if it is time-consuming, this technique will be powerful and informative for the understanding of doping mechanism in GaN and ZnO wires. However, special care will be needed to deduce carrier concentration from the Hall voltage measurement. Indeed, the combination of this experimental technique with some finite element models will be mandatory to determine the current path through the wire. More generally, the knowledge of the current density distribution within the section of a wire where band-bending exists (surface band-bending, core-shell $p-n$ junction or, more generally, axial or radial doping gradient) is one of the challenges for the coming years. This relies on an accurate experimental determination of axial and radial dopant distribution within a whole wire. In this context, local probes/spatially resolved techniques such as CL-based measurement, energy-dispersive X-ray spectroscopy, electron holography, EBIC, secondary electron potential contrast [LU 13, TCH 13c] and atom probe tomography [RIL 13, HOW 13, RIG 14] are attractive methods to be further developed in the case of GaN and ZnO wires. The doping process understanding and control will eventually allow us to fabricate efficient wire-based devices.

5.6. Bibliography

[ALL 10] ALLEN M.W., SWARTZ C.H., MYERS T.H., *et al.*, "Bulk transport measurements in ZnO: the effect of surface electron layers", *Physical Review B*, vol. 81, no. 7, p. 075211, February 2010.

[BAO 11] BAO J., SHALISH I., SU Z., *et al.*, "Photoinduced oxygen release and persistent photoconductivity in ZnO nanowires", *Nanoscale Research Letters*, vol. 6, no. 1, p. 404, May 2011.

[BER 11] BERTNESS K., SANFORD N., DAVYDOV A.V., "GaN nanowires grown by molecular beam epitaxy", *IEEE Journal of Selected Topics in Quantum Electronics*, vol. 17, no. 4, pp. 847–858, July 2011.

[BLO 12] BLOMERS C., GRAP T., LEPSA M.I., *et al.*, "Hall effect measurements on InAs nanowires", *Applied Physics Letters*, vol. 101, no. 15, p. 152106, October 2012.

[BOL 11] BOLINSSON J., MERGENTHALER K., SAMUELSON L., *et al.*, "Diffusion length measurements in axial and radial heterostructured nanowires using cathodoluminescence", *Journal of Crystal Growth*, vol. 315, no. 1, pp. 138–142, January 2011.

[CHA 06] CHANG P.-C., FAN Z., CHIEN C.-J., *et al.*, "High-performance ZnO nanowire field effect transistors", *Applied Physics Letters*, vol. 89, no. 13, p. 133113, September 2006.

[CON 14] CONSONNI V., FEUILLET G., (eds.), *Wide Band Gap Semiconductor Nanowires 2: Heterostructures and Optoelectronic Devices*, ISTE, London, and John Wiley & Sons, New York, 2014.

[DON 10] DONATINI F., DANG L.S., "A single-step electron beam lithography of buried nanostructures using cathodoluminescence imaging and low temperature", *Nanotechnology*, vol. 21, no. 37, p. 375303, September 2010.

[GOL 05] GOLDBERGER J., SIRBULY D.J., LAW M., *et al.*, "ZnO nanowire transistors", *The Journal of Physical Chemistry B*, vol. 109, no. 1, pp. 9–14, January 2005.

[GUR 11] GURWITZ R., SHALISH I., "Method for electrical characterization of nanowires", *Nanotechnology*, vol. 22, no. 43, p. 435705, October 2011.

[GUR 14] GURWITZ R., COHEN R., SHALISH I., "Interaction of light with the ZnO surface: photon induced oxygen breathing, oxygen vacancies, persistent photoconductivity, and persistent photovoltage", *Journal of Applied Physics*, vol. 115, no. 3, p. 033701, January 2014.

[GUS 07] GUSTAFSSON A., BJÖRK M.T., SAMUELSON L., "Locating nanowire heterostructures by electron beam induced current", *Nanotechnology*, vol. 18, no. 20, p. 205306, May 2007.

[GUT 12] GUTSCHE C., NIEPELT R., GNAUCK M., *et al.*, "Direct determination of minority carrier diffusion lengths at axial GaAs nanowire p-n junctions", *Nano Letters*, vol. 12, no. 3, pp. 1453–1458, March 2012.

[HER 06] HERNANDEZ-RAMIREZ F., RODRIGUEZ J., CASALS O., *et al.*, "Characterization of metal-oxide nanosensors fabricated with focused ion beam (FIB)", *Sensors and Actuators B: Chemical*, vol. 118, no. 1–2, pp. 198–203, October 2006.

[HOW 13] HOWELL S.L., PADALKAR S., YOON K., *et al.*, "Spatial mapping of efficiency of GaN/InGaN nanowire array solar cells using scanning photocurrent microscopy", *Nano Letters*, vol. 13, no. 11, pp. 5123–5128, November 2013.

[HU 09] HU Y., LIU Y., LI W., *et al.*, "Observation of a 2D electron gas and the tuning of the electrical conductance of ZnO nanowires by controllable surface band-bending", *Advanced Functional Materials*, vol. 19, no. 15, p. 2380–2387, 2009.

[HU 10] HU Y., ZHOU J., YEH P.-H., *et al.*, "Supersensitive, fast-response nanowire sensors by using Schottky contacts", *Advanced Materials*, vol. 22, no. 30, p. 3327–3332, 2010.

[HUA 02] HUANG Y., DUAN X., CUI Y., *et al.*, "Gallium nitride nanowire nanodevices", *Nano Letters*, vol. 2, no. 2, pp. 101–104, February 2002.

[HWA 11] HWANG J.-S., DONATINI F., PERNOT J., *et al.*, "Carrier depletion and exciton diffusion in a single ZnO nanowire", *Nanotechnology*, vol. 22, no. 47, p. 475704, November 2011.

[JAC 12] JACOPIN G., BUGALLO A.D.L., LAVENUS P., *et al.*, "Single-wire light-emitting diodes based on GaN wires containing both polar and nonpolar InGaN/GaN quantum wells", *Applied Physics Express*, vol. 5, no. 1, p. 014101, 2012.

[JEG 09] JEGANATHAN K., DEBNATH R.K., MEIJERS R., *et al.*, "Raman scattering of phonon-plasmon coupled modes in self-assembled GaN nanowires", *Journal of Applied Physics*, vol. 105, no. 12, p. 123707, June 2009.

[JOH 11] JOHANNES A., NIEPELT R., GNAUCK M., *et al.*, "Persistent ion beam induced conductivity in zinc oxide nanowires", *Applied Physics Letters*, vol. 99, no. 25, p. 252105, December 2011.

[KIN 02] KIND H., YAN H., MESSER B., *et al.*, "Nanowire ultraviolet photodetectors and optical switches", *Advanced Materials*, vol. 14, no. 2, p. 158–160, 2002.

[LAN 06] LANGFORD R.M., WANG T.-X., THORNTON M., *et al.*, "Comparison of different methods to contact to nanowires", *Journal of Vacuum Science & Technology B*, vol. 24, no. 5, pp. 2306–2311, September 2006.

[LEE 09] LEE C.-H., YI G.-C., ZUEV Y.M., *et al.*, "Thermoelectric power measurements of wide band gap semiconducting nanowires", *Applied Physics Letters*, vol. 94, no. 2, p. 022106, January 2009.

[LIK 09] LIKOVICH E.M., RUSSELL K.J., PETERSEN E.W., *et al.*, "Weak localization and mobility in ZnO nanostructures", *Physical Review B*, vol. 80, no. 24, p. 245318, December 2009.

[LIN 06] LIN Y., WANG D., ZHAO Q., *et al.*, "Influence of adsorbed oxygen on the surface photovoltage and photoluminescence of ZnO nanorods", *Nanotechnology*, vol. 17, no. 9, p. 2110, May 2006.

[LON 09] LONG R., CHEN J., LIM J.-H., *et al.*, "Precise voltage contrast image assisted positioning for in situ electron beam nanolithography for nanodevice fabrication with suspended nanowire structures", *Nanotechnology*, vol. 20, no. 28, p. 285306, July 2009.

[LOO 01] LOOK D.C., SIZELOVE J.R., "Predicted maximum mobility in bulk GaN", *Applied Physics Letters*, vol. 79, no. 8, pp. 1133–1135, August 2001.

[LOR 13] LORD A.M., MAFFEIS T.G., WALTON A.S., *et al.*, "Factors that determine and limit the resistivity of high-quality individual ZnO nanowires", *Nanotechnology*, vol. 24, no. 43, p. 435706, November 2013.

[LU 13] LU Y.-J., LU M.-Y., YANG Y.-C., *et al.*, "Dynamic visualization of axial p-n junctions in single gallium nitride nanorods under electrical bias", *ACS Nano*, vol. 7, no. 9, pp. 7640–7647, September 2013.

[MAK 05] MAKINO T., SEGAWA Y., TSUKAZAKI A., *et al.*, "Electron transport in ZnO thin films", *Applied Physics Letters*, vol. 87, no. 2, p. 022101, July 2005.

[MOT 07] MOTAYED A., VAUDIN M., DAVYDOV A.V., *et al.*, "Diameter dependent transport properties of gallium nitride nanowire field effect transistors", *Applied Physics Letters*, vol. 90, no. 4, p. 043104, January 2007.

[NOL 01] NOLAS G.S., SHARP J., GOLDSMID H.J., *Thermoelectrics: Basic Principles and New Materials Developments,* Springer series in materials science, vol. 45, Springer, Berlin, London, 2001.

[PAR 04] PARK W.I., KIM J.S., YI G.-C., *et al.*, "Fabrication and electrical characteristics of high-performance ZnO nanorod field-effect transistors", *Applied Physics Letters*, vol. 85, no. 21, pp. 5052–5054, November 2004.

[PAR 09] PARK W., HONG W.-K., JO G., *et al.*, "Tuning of operation mode of ZnO nanowire field effect transistors by solvent-driven surface treatment", *Nanotechnology*, vol. 20, no. 47, p. 475702, November 2009.

[PAR 12] PARKINSON P., DODSON C., JOYCE H.J., *et al.*, "Noncontact measurement of charge carrier lifetime and mobility in GaN nanowires", *Nano Letters*, vol. 12, no. 9, pp. 4600–4604, September 2012.

[PER 09] PEREA D.E., HEMESATH E.R., SCHWALBACH E.J., *et al.*, "Direct measurement of dopant distribution in an individual vapour-liquid-solid nanowire", *Nature Nanotechnology*, vol. 4, no. 5, pp. 315–319, May 2009.

[RIC 08] RICHTER T., MEIJERS H. L.R., CALARCO R., *et al.*, "Doping concentration of GaN nanowires determined by opto-electrical measurements", *Nano Letters*, vol. 8, no. 9, pp. 3056–3059, September 2008.

[RIG 14] RIGUTTI L., BLUM I., SHINDE D., *et al.*, "Correlation of microphotoluminescence spectroscopy, scanning transmission electron microscopy, and atom probe tomography on a single nano-object containing an InGaN/GaN multiquantum well system", *Nano Letters*, vol. 14, no. 1, pp. 107–114, January 2014.

[RIL 13] RILEY J.R., PADALKAR S., LI Q., *et al.*, "Three-dimensional mapping of quantum wells in a GaN/InGaN core-shell nanowire light-emitting diode array", *Nano Letters*, vol. 13, no. 9, pp. 4317–4325, September 2013.

[SAN 13] SANFORD N.A., ROBINS L.H., BLANCHARD P.T., *et al.*, "Studies of photoconductivity and field effect transistor behavior in examining drift mobility, surface depletion, and transient effects in Si-doped GaN nanowires in vacuum and air", *Journal of Applied Physics*, vol. 113, no. 17, p. 174306, May 2013.

[SCH 08] SCHLENKER E., BAKIN A., WEIMANN T., *et al.*, "On the difficulties in characterizing ZnO nanowires", *Nanotechnology*, vol. 19, no. 36, p. 365707, September 2008.

[SCH 14] SCHMIDT V., MENSCH P. F.J., KARG S.F., *et al.*, "Using the Seebeck coefficient to determine charge carrier concentration, mobility, and relaxation time in InAs nanowires", *Applied Physics Letters*, vol. 104, no. 1, p. 012113, January 2014.

[SHO 08] SHOKHOVETS S., AMBACHER O., MEYER B.K., *et al.*, "Anisotropy of the momentum matrix element, dichroism, and conduction-band dispersion relation of wurtzite semiconductors", *Physical Review B*, vol. 78, no. 3, p. 035207, July 2008.

[SOU 10] SOUDI A., DHAKAL P., GU Y., "Diameter dependence of the minority carrier diffusion length in individual ZnO nanowires", *Applied Physics Letters*, vol. 96, no. 25, p. 253115, June 2010.

[SOU 12] SOUDI A., HSU C.-H., GU Y., "Diameter-dependent surface photovoltage and surface state density in single semiconductor nanowires", *Nano Letters*, vol. 12, no. 10, pp. 5111–5116, October 2012.

[STE 05] STERN E., CHENG G., CIMPOIASU E., *et al.*, "Electrical characterization of single GaN nanowires", *Nanotechnology*, vol. 16, no. 12, p. 2941, December 2005.

[STO 12] STORM K., HALVARDSSON F., HEURLIN M., *et al.*, "Spatially resolved Hall effect measurement in a single semiconductor nanowire", *Nature Nanotechnology*, vol. 7, no. 11, pp. 718–722, November 2012.

[TAL 10] TALIN A.A., LÉONARD F., KATZENMEYER A.M., *et al.*, "Transport characterization in nanowires using an electrical nanoprobe", *Semiconductor Science and Technology*, vol. 25, no. 2, p. 024015, February 2010.

[TCH 13a] TCHOULFIAN P., DONATINI F., LEVY F., *et al.*, "Thermoelectric and micro-Raman measurements of carrier density and mobility in heavily Si-doped GaN wires", *Applied Physics Letters*, vol. 103, no. 20, p. 202101, November 2013.

[TCH 13b] TCHOULFIAN P., DONATINI F., LEVY F., *et al.*, "High conductivity in Si-doped GaN wires", *Applied Physics Letters*, vol. 102, no. 12, p. 122116, March 2013.

[TCH 13c] TCHOULFIAN P., DONATINI F., LEVY F., *et al.*, "Direct imaging of p–n Junction in Core–Shell GaN Wires", *Nano Lett.*, May 2014.

[YAN 08] YANG W.Q., DAI L., MA R.M., *et al.*, "Back-gate ZnO nanowire field-effect transistors each with a top shaped Au contact", *Applied Physics Letters*, vol. 93, no. 3, p. 033102, July 2008.

[ZHA 12] ZHAO S., FATHOLOLOUMI S., BEVAN K.H., *et al.*, "Tuning the surface charge properties of epitaxial InN nanowires", *Nano Letters*, vol. 12, no. 6, pp. 2877–2882, June 2012.

[ZIM 08] ZIMMLER M.A., STICHTENOTH D., RONNING C., *et al.*, "Scalable fabrication of nanowire photonic and electronic circuits using spin-on glass", *Nano Letters*, vol. 8, no. 6, pp. 1695–1699, June 2008.

6

Microstructure of Group III-N Nanowires

6.1. Introduction

The epitaxial growth of group III-N films on dissimilar substrates is accompanied by the formation of a high density of threading dislocations and other extended defects affecting their physical properties and limiting their successful application in optoelectronic devices such as lasers, detectors or solar cells. An alternative approach to the planar epitaxy technology that has been discussed for several years is the realization of perfectly aligned and laterally well-ordered nanowires. Because of the small footprint to the substrate, those free-standing nitride nanowires are expected to grow with minimized defect density and residual strain.

Similar to the case of planar semiconductor heterostructures, the knowledge of the nanowire's microstructure, i.e. defect type, density and spatial distribution, is a prerequisite for the basic understanding of the mechanical and optoelectronic properties. The small scale as well as the three-dimensional (3D) shape of nanowires requires electron microscopy techniques with high spatial resolution to obtain detailed structural information in the desired quality.

In this chapter, we review the structural properties of III-N nanowires. First, we will basically introduce the wurtzite and zinc-blende crystal structures, the building block concept and the close relationship along the close-packed planes of both structures. Following this, the shape and morphology of III-N nanowires are described as well as their influences on macroscopic (nanowire ensemble) and microscopic strain states measured by x-ray diffraction. The polarity of hexagonal nanowires is an important physical parameter and its determination with high spatial

Chapter written by Achim TRAMPERT, Xiang KONG, Esperanza LUNA, Javier GRANDAL and Bernd JENICHEN.

resolution is proven by channeling-enhanced electron energy-loss spectroscopy in combination to transmission electron microscopy (TEM).

The small interface area between substrate and lattice mismatched nanowire and the free surfaces of the wires produce a misfit dislocation network located at the interface plane and a strain relaxation process that is confined at the nanowire base. As a consequence, threading dislocations are not a main feature of the nitride nanowire's microstructure. On the other hand, the building principle for both the wurtzite and zinc-blende structure is based on the accurate stacking of the close-packed planes. Disorder in the stacking sequence will form intrinsic or extrinsic stacking faults of low energy. As the predominant extended defect type, their particular nature will be described and their detection in nanowires is demonstrated by high-resolution TEM imaging. Besides the formation of stacking faults, nitride nanowires grown along the polar c-direction can be affected by the presence of inversion domains, which are defined by a polarity switch across the domain boundary. The determination of inversion domains in GaN nanowires by dark-field TEM is demonstrated and the reason for the formation of this defect is indicated.

In the last section the interface structure between III-N nanowires and Si or AlN buffer layers grown on Si substrates is systematically described depending on the amount of lattice mismatch. A coincidence site lattice model for large mismatched systems is introduced and its capability is illustrated by selected examples. Besides the structural classification in coherent and semi-coherent interfaces, the relevance of the chemical element distribution or composition profile across interfaces in axial nanowire heterostructures is discussed and the complexity of an accurate measurement is demonstrated for the (In,Ga)N/GaN case.

6.2. Structural properties

6.2.1. *Crystal structure*

Group III-nitride compound semiconductors crystallize in either the hexagonal wurtzite or the cubic zinc-blende structure. The first-mentioned is the thermodynamic stable phase.

The wurtzite crystal structure has a hexagonal unit cell with two lattice parameters *a* and *c* in the ratio $c/a = \sqrt{8/3} = 1.633$. As shown in Figure 1(a), this structure is composed of two hexagonal close-packed sublattices, which are shifted with respect to each other along the three-fold c-axis by the amount of $u=3/8$ in fractional coordinates. The internal parameter *u* is defined as the bond length along the *c*-axis (anion–cation bond length) divided by the *c* lattice constant. Both sublattices are occupied by one atomic species resulting in only four atoms per unit cell. Locally, every anion (cation) is surrounded by four cations (anions), which are arranged at the edges of a tetrahedron (coordination number 4). The symmetry of the wurtzite lattice is given by the space group *P6₃mc*, and the two inequivalent atom positions are $\left(\frac{1}{3},\frac{2}{3},0\right)$ and $\left(\frac{1}{3},\frac{2}{3},u\right)$. For group III-nitrides GaN, InN and AlN, the axial ratios deviate considerably from the ideal value for hexagonal close packing of spheres, and hence, each lattice is distorted from the ideal geometry to some extent. As a result, there are two slightly different bond lengths defined by [YEH 96]

$$R(1)=u\cdot c \qquad \text{and} \qquad R(2)=a\sqrt{\frac{1}{3}+\left(\frac{1}{2}-u\right)^2\left(\frac{c}{a}\right)^2}$$

In the case of an ideal wurtzite crystal structure, *R(1)* is equal to *R(2)*. The degree of deviation for the different nitrides is related to their phase stability in the sense that a critical deviation exists when the wurtzite structure becomes unstable [LAW 72]. In Table 6.1, structural data of the binary nitrides are summarized and compared to the ideal case including the lattice constants, the internal parameter and the *c/a* ratio. Because the *c/a* ratio correlates with the difference in the electronegativities of the constituents, the compounds with the largest differences show the largest discrepancy from the ideal *c/a* ratio.

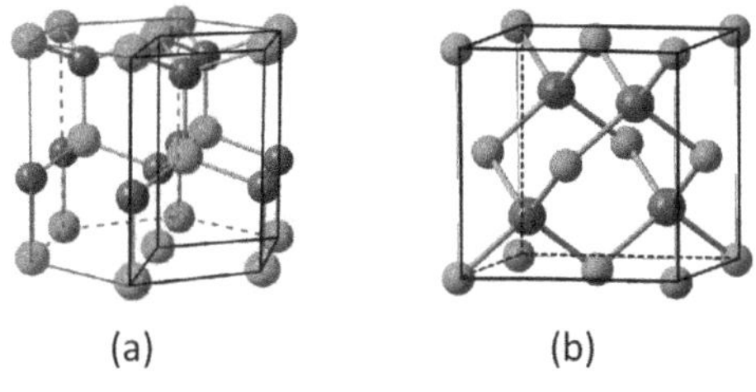

Figure 6.1. *a) The wurtzite crystal structure: group-III atoms are light gray (blue), N atoms are dark gray (green). The unit cell is marked by black lines; b) the zinc-blende structure. For a color version of this figure, see www.iste.co.uk/consonni/nanowires1.zip*

The zinc-blende structure belongs to space group $F\bar{4}3m$ and consists of two interpenetrating and identically oriented face-centered cubic lattices offset from one another by one quarter of a body diagonal as shown in Figure 6.1(b). There are four atoms/unit cells, the coordinates of one atom type being (0,0,0), (0,½,½), (½,0, ½), (½,½,0) and those of the other type being (¼,¼,¼), (¼,¾,¾), (¾,¼,¾), (¾,¾,¼). As a result, each anion (cation) is surrounded by four cations (anions) forming a regular tetrahedron (coordination number 4). The overall equivalent bond length amounts to $R=\sqrt{3}a/4$.

Because of the tetrahedral coordination of both wurtzite and zinc-blende structures, the short range order of these structures is virtually identical. Neglecting slight deviations from the ideal spherical closed packing in the wurtzite structure, each atom is tetrahedrally surrounded by four nearest neighbors with the same bond distance as in the zinc-blende modification. Additionally, the twelve next-nearest neighbors are at the same distance in both structures. Significant differences are found only for the third next-nearest neighbors. This difference can be simply described by regarding the handedness of the fourth interatomic bond along the <111> and <0001>[1], respectively, closed packed chain corresponding to the third nearest neighbor relation in both structures (see Figure 6.2(a)). In the zinc-blende structure, triangularly arranged atoms in the {111} plane are in a "staggered" configuration with the triangularly arranged unlike atoms in the third nearest neighbor distance, that is, the latter are rotated 60° with respect to the former (Figure 6.2(b)). In contrast, triangularly arranged atoms in the {0001} closed packed basal plane are in an "eclipsed" (i.e. identical) configuration in the wurtzite structure. Taking the different arrangements in closed-packed planes as the building principle for both crystal structures, one finally ends up with the following stacking sequence:

.... AaBbAaBbAaBbAaBbAaBb ...

for wurtzite and

... AaBbCcAaBbCcAaBbCcAaBbCc ...

for zinc-blende type that completely characterizes both structure types as shown in Figure 6.2. Lower and uppercase letters stand for the two different kinds of constituents (i.e. the group III element and N). A fault in the stacking sequence can transform one structure into the other, as will be discussed in section 6.4.1, if neglecting the material-depending deviation from the ideal close-packing of hard spheres.

	GaN		**InN**		**AlN**	
	ideal	exp.	ideal	exp.	ideal	exp.
a [Å]		3.189		3.545		3.110
c/a	1.633	1.626	1.633	1.609	1.633	1.601
u	0.375	0.377	0.375	0.379	0.375	0.382

Table 6.1. *Calculated (for ideal crystal) and experimentally observed structural parameters for GaN, InN and AlN, e.g. [VUR 03, AMB 02]*

1 Note here the use of the four-index notation [*hkil*] with $i=-(h+k)$.

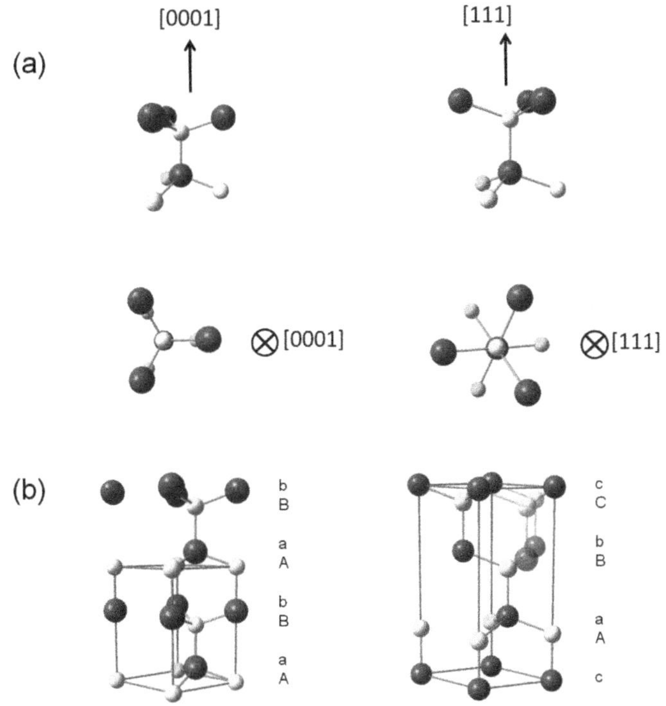

Figure 6.2. *a) Handedness of the fourth interatomic bond along the closed-packed direction for wurtzite (left) and zinc-blende (right). b) Schematic of the layer sequence and the corresponding three-dimensional representation. For a color version of this figure, see www.iste.co.uk/consonni/nanowires1.zip*

6.2.2. *Nanowire morphology*

There are several names used in the literature for one-dimensional (1D) nitride nanostructures: nanowires, nanocolumns, nanopillars or nanorods, simply depending on their effective aspect ratio between length and diameter (see Figure 6.3). Basically, 1D group III-nitride nanostructures are grown by molecular beam epitaxy (MBE) or by metal organic chemical vapor deposition (MOCVD). The most common ways of growing these nanostructures are either in a self-assembled or in a selected area growth (SAG) fashion. The former does not require any substrate preparation prior to the growth and reduces the complexity of the fabrication while the latter – even it requires a pre-patterning with several technological processes involved – has the advantage of a better control of the nanowire's dimension and homogeneity and, generally, allows for a more direct approach to its characterization. Both fabrication techniques can be performed in a self-induced mode (vapor-solid) or by taking advantage of a catalyst (vapor-liquid-solid). The main difference in all these growth modes appears in the size and morphology of the nanowires. In general, self-induced nanowires grown by MBE present small diameters because the critical nuclei to induce the subsequent axial growth have a small size [CON 10]. In the following, only these 1D nanostructures with a high aspect ratio are considered [BEN 11, ALB 13].

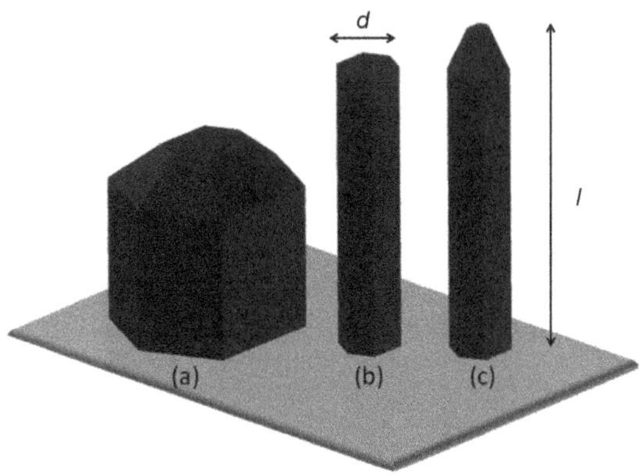

Figure 6.3. *Diagram of typical morphologies of one-dimensional III-N nanostructures: a) pillar with $l \geq d$ and pencil-like top, b) wire with $l \gg d$, c) wire with pencil-like top*

All these 1D nano-systems have a unique growth direction in common defining their linear structure and determining the specific physical properties. Due to the anisotropy in the surface energies of GaN and related binary compounds and alloys, the nanowires reveal a morphology and shape with hexagonal cross-section and well-defined crystallographic side facets. The growth direction (i.e. long wire axis) usually corresponds to the hexagonal c-axis, i.e. the <0001> crystal direction ([0001] or $[000\bar{1}]$ depending on the polarity), whereas the side facets ordinarily are of $\{1\bar{1}00\}$ m-plane type.

Figure 6.4(a) represents a typical (Al, Ga)N nanowire observed in bright-field TEM. The homogenous contrast indicates the absence of extended defects. The nanowire with a diameter of about 25 nm is embedded in epoxy glue for mechanical stabilization during the thin foil preparation and the following TEM observation [TRA 03]. The corresponding high-resolution (HR) TEM image taken along the $[11\bar{2}0]$ zone axis verifies the hexagonal – AaBb – stacking of the basal planes along the growth direction. The atomically sharp side facet is obviously visible and corresponds to the $\{1\bar{1}00\}$ plane according to the Fourier transform. (Figure 6.4(a) and (c)).

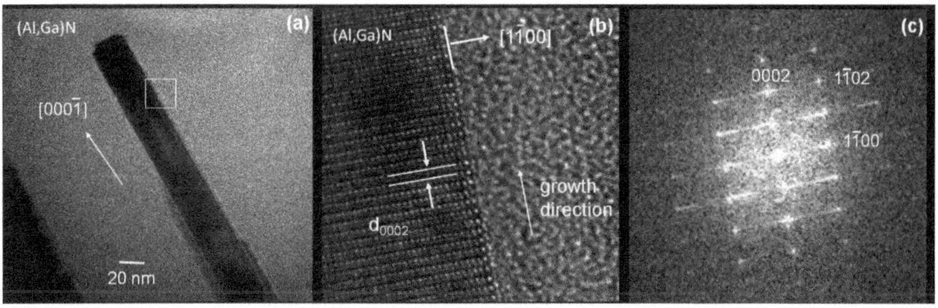

Figure 6.4. *a) Cross-sectional bright-field TEM micrograph of (Al,Ga)N nanowire, b) magnified part of the wire in HRTEM and c) corresponding Fourier transform of the HRTEM image*

The most direct way to determine the crystallography of side surfaces is to prepare a TEM specimen in plan-view geometry and to image the nanowire directly along the growth direction, i.e. along the wire *c*-axis. In this way, atomically resolved high-resolution TEM, or the combined use of diffraction and bright-field imaging will provide direct information about the facet crystallography (see Figure 6.5).

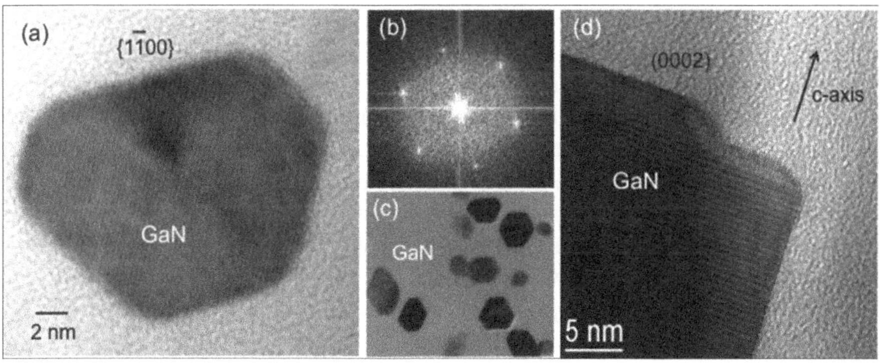

Figure 6.5. *a) Plan-view HRTEM of a GaN nanowire imaged along the <0001> direction and b) corresponding Fourier transform indicating the coincidence of* {1 $\bar{1}$00} *planes and surface facets; c) low magnification plan-view micrograph of nanowire ensemble; d) cross-sectional HRTEM micrograph of the GaN nanowire top indicating the flat surface terminated by (0002) lattice plane*

The growth front plane of GaN and InN nanowires is usually terminated by the {0002} basal plane ("flat top"), or alternatively, the top surface is composed of inclined {1 $\bar{1}$02} or {11$\bar{2}$3} planes forming a pyramidal form ("pencil-like top") or a truncated pyramid (see Figure 6.3). Both, flat and pyramidal top structures were observed in MBE grown nanowires depending on the growth conditions [BEN 11]. In general, the crystal habit in equilibrium is determined by their surface free energies (so-called "Wulff" construction [WUL 01]) favoring basal *c*-, prism *m*- and *a*-plane, respectively [SHA 05]. However, kinetic factors such as surface migration, segregation or atom adsorption as well as local stresses acting on the growth front can affect the actual shape of the nanowire and, additionally, be responsible for the generation of stacking faults.

Quite a common modification of the nanowire morphology is cross-sectional tapering or widening. This process is usually the result of a local variation in the conditions on the growth front. For instance, a small excess of element III in the growth front causes the nanowires to increase their diameter to keep local stoichiometry, generating a widening and/or roughening or undulation of the sidewalls. As a result, stacking faults and nanotwins are often observed in the immediate vicinity (see Figure 6.12). It is assumed that this kind of growth front roughening may generate local surface stress being responsible for the formation of such defects.

6.2.3. *Macroscopic and microscopic strain*

MBE grown GaN nanowires are often realized in two ways: by heteroepitaxy on Si or AlN-buffered Si(111) substrates or by homoepitaxy via selected area growth (on thick GaN epitaxial layers). During the early stage of heteroepitaxial growth, 3D GaN nuclei relieve strain elastically by undergoing several shape transitions before introducing misfit dislocations at the nanowire/substrate interface [CON10]. These dislocations do not propagate along the nanowires, but remain at the interface or bend toward the free sidewalls. Thus, residual strain should be negligibly small or can be released elastically thanks to the extreme aspect ratio of the nanowires. The macroscopic strain state of those GaN nanowires averaged over their length and over nanowire ensembles is accurately determined by x-ray diffraction lattice parameter measurements [JEN 11]. Using an analyzer crystal in front of the detector, the angular positions of the incident and diffracted beams are determined precisely in order to obtain the Bragg angle. Based on necessary calculations in dynamic approximation for the Si substrate, a lattice constant value of 0.54299 nm is obtained. Thanks to the small diameter of the nanowires, the kinematic approximation is sufficient for calculation of the angular position of their x-ray reflections. The position for the symmetrical in-plane and out-of-plane reflections in $\omega/2\Theta$ scans yield a = 0.31885 nm and c = 0.51854 nm for the GaN lattice parameters within the error margin of $\pm 5 \times 10^{-5}$ nm. These values perfectly agree with those reported in references [MOR 09, ROB 07] for strain-free and pure GaN. This means that the GaN nanowires are free of macroscopic strain on average. However, the widths of $\omega/2\Theta$ scans across the GaN *000.n* reflections with n = 2,4,6 increase linearly with the reflection order n. This result is expected for broadening governed by micro-strain, i.e. local variations in strain state. We assume that the breadth β_f of the measured profiles consists of size and strain components, β_s and β_D. The size component β_s does not depend on the diffraction angle and the strain component β_D grows linearly with the order of reflection. We apply the reciprocal lattice notation $|d^*| = 1/d = 2\sin(\Theta)/\lambda$ with $|d^*|$ as the absolute value of the reciprocal lattice vector and $\beta_f^* = \beta_f \cos(\Theta)/\lambda$.

For a Lorentzian peak shape the two contributions have to be added [LAN 00, WIL 53]:

$$\beta_f^* = \beta_s^* + \beta_D^* = \beta_s^* + 2ed^* \qquad [6.1]$$

where $e = \beta_D^*/2d^* = \Delta d/d$ is the micro-strain that can be interpreted as the variation of d spacing within the nanowires. The plot of β_f^* over d^* is expected to give a straight line and such a plot, as shown in Figure 6.6, is a Williamson-Hall plot in reciprocal lattice representation. The intersection with the ordinate axis allows us to determine β_D^* corresponding to the size $l = 1/\beta_D^*$. This is the size of an average domain in the direction of the diffraction vector probing the length along the wire. The characteristic size here is $l = 0.3$ µm, which is smaller than the nanowire length of about 1.6 µm. The micro-strain $e = 1.4 \times 10^{-4}$ is obtained from the slope of the Williamson-Hall plot [WIL 53]. This value describes the fluctuation of the lattice plane distances in the nanowires, not their average value. We estimate an error of $\pm 1 \times 10^{-4}$ in the determination of the micro-strain. The inhomogeneous micro-strain is caused by residual defects in the GaN wires, such as small angle grain boundaries including boundary dislocations arising from the nanowire coalescence [CON 09]. By implication, this means that for nanowire ensembles without defects creating long-range strain fields, macro- and micro-strain would be identical.

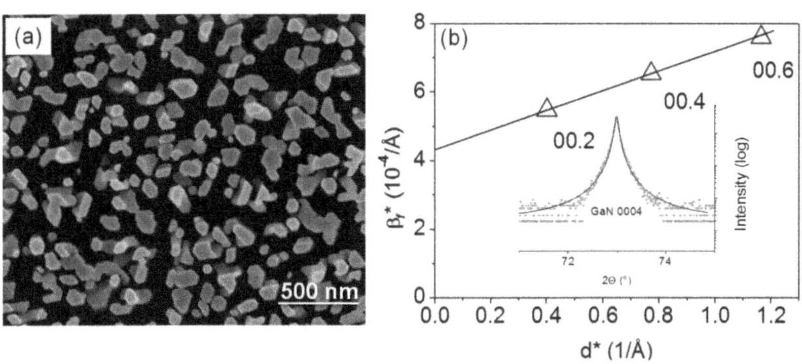

Figure 6.6. *Scanning electron microscope image of GaN nanowires (left) with low degree of coalescence and the corresponding results of the XRD (right)*

6.3. Polarity

Wurtzite and zinc blende-crystal structures do not possess a center of symmetry. This means that, when defining an atom position with coordinates (xyz), it is not invariant to the position $(\overline{xyz})$ since inversion results in the replacement of group III atoms by nitrogen atoms and vice versa. As a result of the lack of inversion center (symmetry element "$\bar{1}$"), the crystal structure exhibits crystal polarity[2]. Closed-packed (0001) basal planes differ from $(000\bar{1})$ planes, i.e. the <0001> direction is called polar direction. In the literature, the [0001] and $[000\bar{1}]$ polar directions are also referred to as Ga- and N-polar orientations. Beyond this primary polar plane, many other secondary planes exist (semi-polar planes) as well as pure non-polar directions, which have developed some practical importance in optoelectronic devices.

The polar orientation specifies the direction of the spontaneous polarization vector and thus determines the type of charge induced at the surface or interface. Along with the piezo-electric polarization, the polarization-induced charges influence the electrical and optical properties for the materials [HEL 98, SUM 04]. Furthermore, the crystal polarity strongly affects the spontaneous nucleation and growth of GaN nanowires grown along the c-axis on SiC (0001) and $(000\bar{1})$ substrates, respectively, by plasma-assisted molecular beam epitaxy [FER 12].

Compared to bulk crystals or planar films, the determination of polarity in nanowires is considerably demanding due to their small dimension and three-dimensional character. Transmission electron microscopy techniques are able to determine the polarity with high spatial resolution including convergent-beam electron diffraction (CBED) [PON 96, ROU 98, SPI 02], x-ray energy dispersive spectroscopy (EDS) [TAF 83, JIA 02] and electron energy-loss spectroscopy (EELS) [TAF 87, KON 02, KON 11], as well as the recently demonstrated high-resolution Z-contrast scanning TEM that is able to image [MKH 06, LIU 07, ROU 08] the group III and N atom columns directly.

Figure 6.7(a) is a cross-sectional TEM image of self-assembled GaN nanowires grown on Si(111) by catalyst-free MBE. Although the bright-field contrast is strongly affected by bend contours along the individual wires, the image obviously reflects that most of the wires are grown perpendicular to the Si substrate with a length up to 650 nm. The diameter varies between 30 and 40 nm only. For CBED measurements the sample is tilted out of the zone axis toward the systematic row diffraction condition with symmetric (0002) and $(000\bar{2})$ reflections (the so-called

2 Polarity is a bulk property, whereas "termination" describes a surface property.

Laue case). The experimental CBED pattern recorded at the center of the wire is shown in Figure 6.7(b) together with the CBED simulation, which is realized based on the Bloch wave method (JEMS program [STA 87]). A reasonable fit is obtained for a thickness of 28 nm, which is in good agreement with the nanowire geometry (Figure 6.7(b)). This CBED result appoints the $[000\bar{1}]$ direction as the growth direction of the wires. However, due to the missing dynamic contributions inside the diffraction discs, the so-called Kossel-Möllenstedt fringes, the exact imaging alignment is questionable and limits our polarity analysis. Alternative polarity determinations are performed by spatially resolved spectroscopic techniques as described in the following example of channeling enhanced EELS [KON 11].

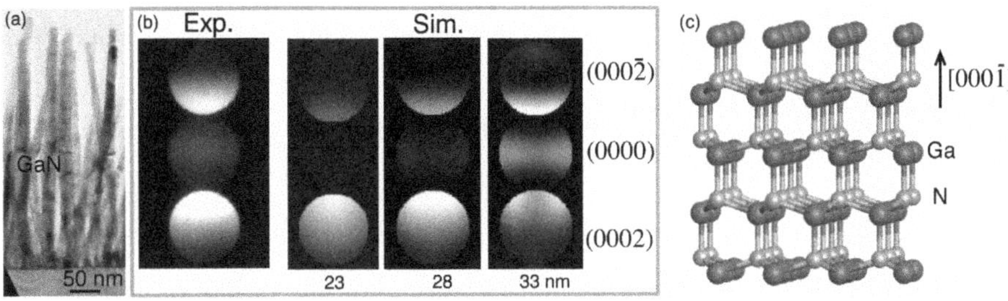

Figure 6.7. *a) Cross-sectional bright-field TEM image of GaN nanowires grown by MBE on Si(111); b) experimental and simulated CBED pattern taken from the center of a nanowire, simulations are presented for three thickness values, 23, 28 and 33 nm; c) schematic atomic arrangement for N-polarity*

Channeling-enhanced EELS

Fast electrons inside a thin crystal propagate as a superposition of Bloch waves [HIR 77]. In the case of two-beam conditions $(0, \vec{g}_{hkl})$, i.e. strong scattering from a single set of planes *(hkl)*, the electron distribution is highly localized at the special atom planes as a consequence of the interference between the Bloch waves. The thickness-averaged electron current intensity (TAECI) close to the Bragg condition $\vec{g}_{hkl}$ is given as [SPE 92, EGE 96]:

$$I(r) = \frac{1}{t}\int_0^t |\psi(z)|^2 dz = 1 - \frac{1}{\sqrt{w^2+1}}\frac{1-\cos(2\pi t')\sqrt{w^2+1}}{2\pi t'\sqrt{w^2+1}}\sin(2\pi\vec{g}\cdot\vec{r}+\varphi_g) \quad [6.2]$$

$$-\frac{w}{w^2+1}(1-\frac{\sin(2\pi t')\sqrt{w^2+1}}{2\pi t'\sqrt{w^2+1}})\sin(2\pi\vec{g}\cdot\vec{r}+\varphi_g)$$

Here, $\psi(z)$ is the electron wave function inside the crystal, z is the electron beam propagation depth and t the thickness of the crystal. The parameter w represents the orientation of the electron beam with respect to the crystal: $w = s_g \xi_g$ with s_g as the deviation parameter and ξ_g as the two-beam extinction distance, respectively. The atomic position inside the unit cell is given by $\vec{r}$ and φ_g describes the phase shift due to the absence of the center of symmetry [KON02]. When the sample is tilted at the exact two-beam condition ($w = 0$), the thickness-averaged intensity will be simplified to

$$\begin{aligned} I(\vec{r}) &= \frac{1}{t}\int_0^t |\psi(z)|^2 \, dz \\ &= 1 - \frac{1-\cos(2\pi t/\xi_g)}{2\pi t/\xi_g}\sin(2\pi \vec{g}\cdot\vec{r} + \varphi_g) = 1 - A(t)B(\vec{r}) \end{aligned} \qquad [6.3]$$

In equation [6.3], the component $B(\vec{r}) = \sin(2\pi \vec{g}\cdot\vec{r} + \varphi_g)$ indicates that the wave function oscillates within the unit cell reflecting the difference of electron intensity at different atomic planes (see Figure 6.8(a)). In case of the Bragg condition $\vec{g} = 0002$ (dashed line in (Figure 6.8(a)) a maximum in intensity is found at the N atomic plane, while the intensities at the Ga atomic plane are roughly equal. For this reason, the larger thickness-averaged intensity $I(\vec{r})$ is expected to correspond to a higher intensity of the N K edge in the EELS spectrum. Therefore, the intensity of the N K edge taken at the (0002) Bragg condition should be higher than that taken at the $(000\bar{2})$ condition if the spectra are normalized by using Ga L edges. The component $A(t)$ in equation [6.3] is only associated with sample thickness t and thus only influences the magnitude of the measured EELS signal difference without affecting the relative intensity at the atomic planes and thus the polarity determination. Additionally, it must be noticed that a small deviation w from the exact two-beam condition cannot be avoided due to the sample bending and beam convergent effects. The variation of TAECI with respect to small orientation deviations w for $\vec{g} = 0002$ and $\vec{g} = 000\bar{2}$ conditions is verified in Figure 6.8(c) revealing no remarkable change in the intensity difference.

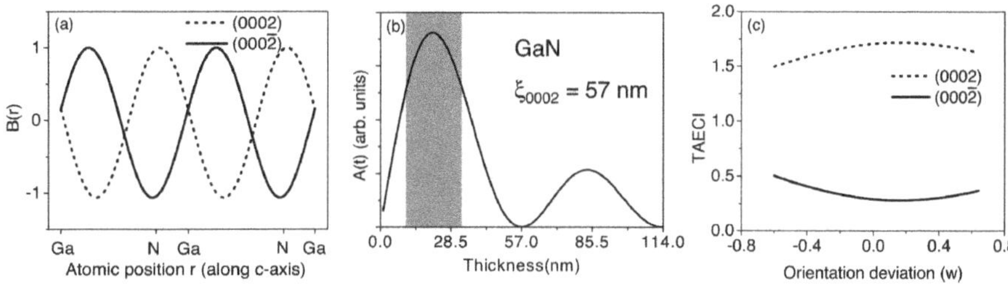

Figure 6.8. *a) Plot of $B(\vec{r})$ calculated for a specimen thickness of $0.4 \cdot \xi_{0002}$ and b) of A(t) components of the thickness averaged electron intensity. c) Variation of TAECI with crystal orientation w*

Two critical features regarding the sample thickness should be pointed out: (1) at the N-atomic plane, the difference of *I(r)* for $\vec{g} = 0002$ and $\vec{g} = 000\bar{2}$ does not change sign and is only damped with increasing sample thickness and turns out to be zero for thicknesses of $n \cdot \xi_{0002}$ (*n* as integer). (2) The main maximum of $A(t)$ is at the thickness of $0.4 \cdot \xi_{0002}$, which is the most suitable sample thickness for polarity determination (shadowed area in Figure 6.8(b)). The value of $0.4 \cdot \xi_{0002}$ for GaN is about 23 nm taking into account that the two-beam extinction distance ξ_{0002} for GaN is about 57 nm. For that reason, this method is complementary to CBED and particularly applicable for GaN, AlN (ξ_{0002}=79 nm) and InN (ξ_{0002}= 49 nm) nanostructures with thicknesses in the range between 5 and 35 nm.

For the polarity determination by channeling enhanced EELS analysis, the GaN nanowire is tilted into the two-beam condition, first with the (0002) and then with the (000$\bar{2}$) diffracted beam – far away from the zone axis. The exact alignment of the two-beam condition is less critical because of the large difference in EELS intensities. In order to enhance the signal-to-noise ratio, the EEL spectra, including N *K* edge and Ga *L* edges are collected by accumulating several spectra in an energy-loss window from 300 to 1640 eV. The spectra are acquired in the diffraction mode with a collective semi-angle of about 3.5 mrad in a 300 kV microscope. Before comparing the spectra taken at (0002) and (000$\bar{2}$) Bragg conditions, respectively, the background of the N *K* edge has been subtracted by a standard power law method [EGE 96] and the spectra are normalized at the Ga *L* edges. The results are shown in Figure 6.9. Based on the theoretical consideration, the N *K* edge peak at the (0002) orientation has a higher intensity compared to the (000$\bar{2}$) case, and thus, an unambiguous assignment of the {0002} reflections can be accomplished. In combination with the TEM image of the GaN nanowire fixing the growth direction, the EELS measurements determine a N-polarity of the nanowire, i.e. the $[000\bar{1}]$ growth direction.

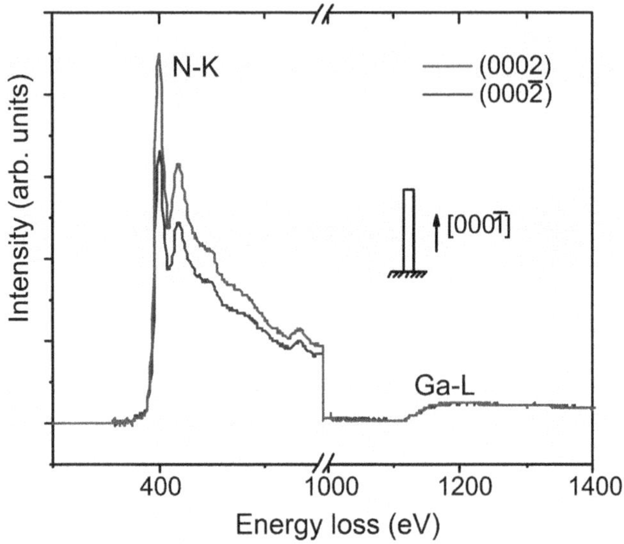

Figure 6.9. *A pair of EEL spectra of GaN nanowires acquired at (0002) and (000$\bar{2}$) Bragg conditions with the N K edge background subtracted. The inset indicates a N-polarity of the GaN nanowires on Si(111). (From [KON 11] © 2011 IOP Publishing. Reproduced by permission of IOP Publishing. All rights reserved). For a color version of this figure, see www.iste.co.uk/consonni/nanowires1.zip*

Another demonstration of the polarity determination by channeling-enhanced EELS is related to InN nanowires. The self-assembled nanowires were grown on bare Si(111) by plasma-assisted MBE [GRA 07]. In this sample containing nanowires of around 50 nm diameter the EELS method still displays its advantage. Figure 6.10 reveals a cross-sectional bright-field image of the nanowires and the corresponding EEL spectra taken at $\vec{g} = 0002$ and $\vec{g} = 000\bar{2}$, respectively. The EELS measurements show that the N K-edge intensity difference between (0002) and (000$\bar{2}$) conditions is much smaller compared to the aforementioned GaN nanowire, since the InN nanowire diameter is only slightly larger than ξ_{0002} with a small TAECI difference (compare Figure 6.8(b)). Nevertheless, using the In M edges as reference, the InN nanowire is accordingly determined to be N-polar.

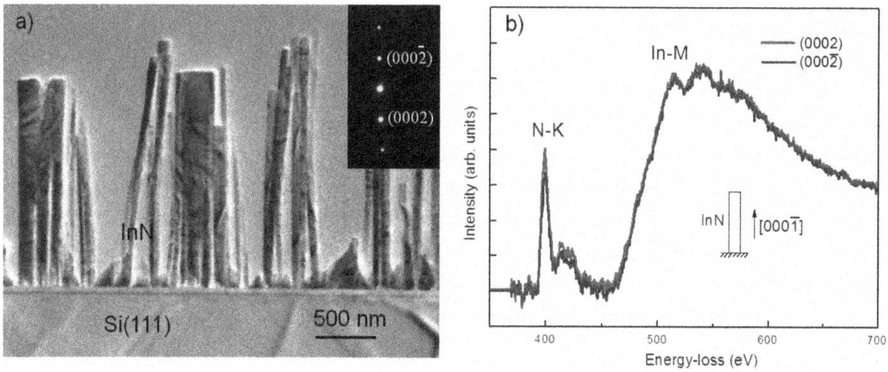

Figure 6.10. *a) TEM bright-field image of InN nanowires grown on Si(111), inset: selected area diffraction pattern indicating the growth direction; b) EEL spectra of nanowires taken at (0002) and (000$\bar{2}$) Bragg condition with the background subtracted, the schematic (inset) indicates the N-polarity. For a color version of this figure, see www.iste.co.uk/consonni/nanowires1.zip*

6.4. Extended defects in nanowires

Nanowires are often considered as ideal crystals with high structural perfection (similar to whiskers) without crystalline defects. The reasons thereto are discussed on one side based on the small footprint of nanowires to the substrate where they nucleate and, on the other side, on their free-standing character preventing the appearance of *external* stress sources. The first argument describes the fact that the epitaxial mismatch is only confined to a small area where strain release takes place including the formation of misfit dislocations. And the second reason points to stress-free growth as the requirement for defect-free crystal growth. Nevertheless, real single-phase nanowires can accidentally contain stacking faults or inversion domains, which will be discussed in the context of growth defects.

Another scenario for the generation of extended defects is the realization of lattice mismatched heterostructures in axial or core/shell geometry. Internal stress is applied because of the epitaxial alignment of the layers with different composition and thus a different lattice parameter – similar to the 2D planar case. Here, low dimensions and crystallographic edges of nanowires can modify the relaxation process (see section 6.5). At present, there is very little data about the formation of misfit dislocations and their line direction and Burgers vector, and additionally, their interplay with wire geometry, 3D stress state or interface roughness.

In this section the different types of stacking faults as well as inversion domain boundaries are described and experimentally verified. Misfit dislocations are discussed in section 6.5 in connection with interfaces.

6.4.1. *Stacking faults*

Stacking faults are the most frequently observed planar defects in nanowires and rods. Basal plane stacking faults can be considered as faults in the regular stacking sequence, preserving the original tetrahedral bonding (nearest-neighbor arrangement). There are three possible kinds of stacking faults, two intrinsic types I_1 and I_2 and one extrinsic type E [HUL 11]. Stacking fault I_1 is formed by the removal of a basal plane layer (for example *Aa*), which produces a high energy fault, followed by a slip of $1/3<1\bar{1}00>$ of the crystal above this fault (*Bb* and *Aa* bi-layers to *Cc* and *Bb*) to reduce the energy. The stacking sequence in these stages is:

… AaBbAaBbAaBb … → … AaBbAaBbBbAaBb … → … AaBb**AaBbCc**BbCc… (I_1)

The resultant displacement vector $\vec{R}$ is then given by

$$R = -\frac{1}{2} < 0001 > + \frac{1}{3} < 1\bar{1}00 > = \frac{1}{6} < 2\bar{2}0\bar{3} >,$$

and the fault is surrounded by a Frank–Shockley partial dislocation. As the removal of a double layer is required, this type of basal plane stacking fault cannot be created by the dissociation of perfect dislocations into partials in response to plastic deformation. On the contrary, it can be formed as an accident during crystal growth.

The stacking fault I_2 results from slip of $1/3<1\bar{1}00>$ only in a perfect crystal

… AaBbAaBbAaBbAaBb … → ….AaBb**AaBbCcAa**CcAa … (I_2)

and the displacement corresponds to the slip vector. Such a stacking fault is terminated by Shockley partial dislocations characterized with the Burgers vector $\vec{b}_{sp} = \frac{1}{3} < 1\bar{1}00 >$.

The extrinsic fault (E) is formed by inserting an extra plane, e.g. *Cc*,

… AaBbAaBbAaBb … → … AaBb**AaBbCcAaBb**AaBbAa … (E)

with the displacement vector $\vec{R} = +\frac{1}{2}[0001]$ and the fault E is bounded by Frank partial dislocations.

These stacking faults introduce into the wurtzite crystal a thin layer of zinc-blende stacking (***AaBbCc***) and, therefore, the main contribution to the fault energy

γ_{SF} arises from changes in the second neighbor sequence of the basal planes (defining a cubic or hexagonal phase). Consequently, one can argue on the basis of pure geometry that I_2 has twice as many fault layers compared to I_1 and E even three times more: $\gamma_E \approx 3/2\gamma_{I_2} \approx 3\gamma_{I_1}$ [HUL 11]. Of course, in real crystals the deviation from the ideal geometry – for instance the *c/a* ratio or the specific bonding character – must be taken into consideration for the precise stacking fault energy.

Using first-principles calculations, the atomic structure and corresponding formation energy of basal plane stacking faults in AlN, GaN and InN are investigated by C. Stampf and C. Van de Walle [STA 98]. The intrinsic type I_1 has been calculated with the lowest energy in each material followed by I_2 and E as expected.

On the other hand, the experimental determination of stacking fault energies turns out to be extremely difficult. A reliable estimation of the energy of intrinsic stacking faults is to evaluate by TEM the equilibrium width of dislocations dissociated into partial dislocations bounding the intrinsic fault I_2. The fault energy is calculated using the balance between the surface tension of the fault and the repulsive force between the partials. S. Takeuchi and K. Suzuki have introduced the dissociation of dislocations during the plastic deformation of crystalline powder for several nitride materials and further compound semiconductors [TAK 99]. The partial dislocations and their distances to each other are detected by weak-beam dark-field TEM or by lattice resolved high-resolution TEM. The results for GaN, InN and AlN are presented together with theoretical values in Table 6.2.

	γ (mJ/m^2) Experimental	I_2 (mJ/m^2) Theory	$\Delta E_{w\text{-}zb}$ (meV/atom)	c/a
AlN	220±70[a]	204[b]	-18.41[c]	1.601[d]
GaN	20±3[a]	43[b]	-9.88[c]	1.626[d]
InN	41±8[a]	65[b]	-11.44[c]	1.609[d]

a Takeuchi and Suzuki [TAK 99].
b Stampf and Van de Walle [STA 98].
c Yeh *et al.* [YEH 92].
d Vurgaftman *et al.* [VUR 03].

Table 6.2. *Values of the experimentally measured stacking fault energy γ_{I2}, the calculated stacking fault energies, the relaxed energy difference between wurtzite and zinc-blende phase, and the c/a ratio for AlN, GaN and InN*

It is remarkable that the trend in stacking fault energy for the binary nitrides follows the same trend as for the energy difference between the wurtzite and the zinc-blende phases as well as for the deviation in the *c/a* ratio from the ideal value of 1.633. The last statement also implies that the introduction of a stacking fault in wurtzite crystal results in a local stacking sequence of the cubic phase but this segment cannot equate with the corresponding zinc-blende crystal structure because of the difference in deviations of c/a-ratio as well as the lack of inherent symmetry elements.

The following examples of pure GaN and InN nanowires will demonstrate the interplay between wire morphology and the existence of planar defects. Assuming the absence of external stress and strain acting on nanowires or internal epitaxial stress based on heterostructures, the morphology results in a very regular shape with smooth sidewalls and, thus, only few stacking faults are detected as growth-induced defects. In this case, the incorporation into the crystal is driven by a minimization of the total energy that points to the formation of an intrinsic stacking fault I_1. An example of such a planar fault is presented in Figure 6.11. The high-resolution TEM image taken along the $[11\bar{2}0]$ zone axis reveals a direct observation of the stacking sequence and allows the determination of the stacking fault type. The fault crosses the entire nanowire and, consequently, it does not possess a partial dislocation. The stacking fault is characterized by a pure displacement of $\vec{R} = 1/6 < 2\bar{2}0\bar{3} >$ across the faulted plane of the upper nanowire part related to the lower part without producing any long-range strain field.

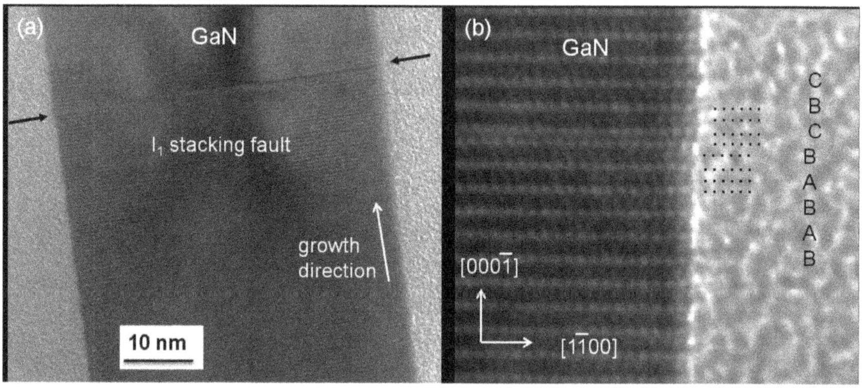

Figure 6.11. *a) Cross-sectional HRTEM image of a GaN nanowire including a I_1 type stacking fault; b) magnified part of the stacking fault identifying the sequence of the (0002) lattice planes across the fault*

Figure 6.12 presents a HRTEM image of an InN nanowire grown under non-optimized conditions that causes a sidewall roughening and wire tapering along the growth direction. All types of stacking faults – intrinsic I_1, I_2 and extrinsic E – are detectable in this nanowire region. Although the stacking fault energies for InN are considerably larger than the corresponding values for GaN, their occurrence is directly correlated with the tapering of the wire generating local variations in surface tension. This surface stress may act as the driving force for the formation of faults along the growth direction running through the entire nanowire.

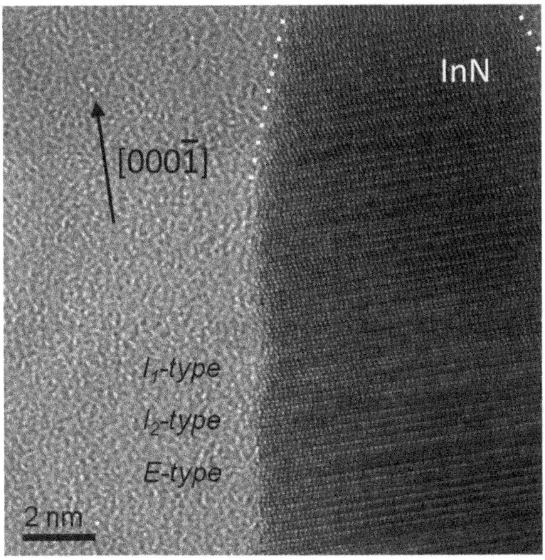

Figure 6.12. *Cross-sectional HRTEM image of InN nanowire close to the top displaying a high number of stacking faults (different types are indicated as I_1, I_2 and E type) associated with surface roughening (tapering is indicated by dotted white lines)*

6.4.2. *Inversion domain boundaries*

Another planar defect detected in nanowires is the inversion domain boundary, i.e. the boundary between domains of opposite polarity. This domain boundary is observed in GaN nanowires grown homoepitaxially by selected area growth on patterned GaN buffered sapphire or Si substrates [BEN 11].

Figure 6.13(a) shows a cross-sectional bright-field TEM image of an ordered (InGa)N/GaN axial nanowire heterostructure realized on a Ti-masked GaN template. The nanowire contains no threading dislocations and is free of stacking faults. However, a vertical narrow columnar domain is observed in the center of the wire passing through the whole nanowire structure (marked by a white arrow). The typical diameter of the domain is found in the range of 5 to 15 nm that is much thinner than the TEM sample thickness. Therefore, the polarity determination by CBED or EELS is affected by the signal overlap of matrix and domain and is thus inaccurate. On the other hand, these boundaries show contrast reversals in dark-field TEM if applying $\vec{g} = 0002$ and $\vec{g} = 000\bar{2}$ imaging conditions, that is a typical behavior of inversion domains. Based on the violation of Friedel's law in case of multiple diffraction conditions along the non-centrosymmetric $[1\bar{1}00]$ zone axis, the intensity distribution within the $(000\bar{2})$ electron diffraction disk cannot be obtained by simple symmetry operations of the intensity distribution within the (0002) disk [SER 73]. Therefore, the asymmetry between the (0002) and the $(000\bar{2})$ disk leads to contrast differences between adjacent regions of opposite polarities in the dark-field images as shown in Figure 6.13(b). Because the GaN buffer and the GaN nanowire matrix exhibit Ga-polarity (see CBED in Figure 6.13(a)), the inversion domain must reveal the N-polar direction.

The origin of the inversion domain is localized at the homo-junction between GaN nanowire and buffer layer. The HRTEM image of the boundary region between the inversion domain and GaN substrate in Figure 6.14 indicates a stacking disorder along the basal planes similar to an intrinsic stacking fault. However, the 0002 lattice plane spacing is remarkably reduced at the faulted plane compared to the regular plane distance, indicating the presence of Ti impurities. Figure 6.14(b) shows EELS spectra taken across the homo-junction (position 1,2,3) and next to the inversion domain (position 4) that verify the presence of Ti at the faulted plane only. The Ti *L*-edges located at about 455 eV allows a sensitive detection of Ti atoms even at low concentration levels.

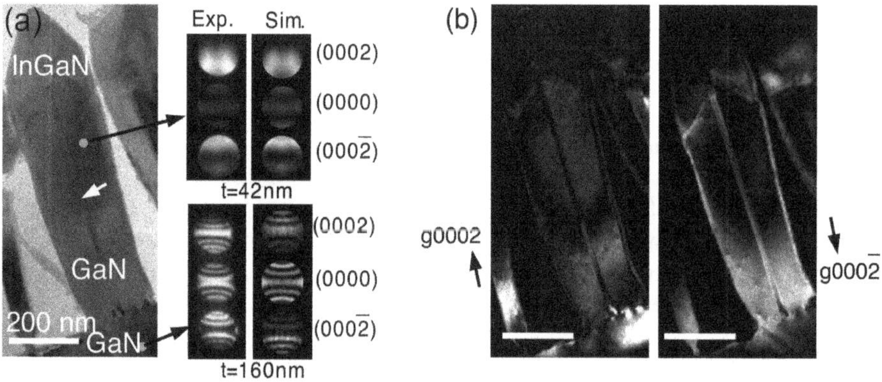

Figure 6.13. *a) Cross-sectional bright-field TEM image of an (In,Ga)N/GaN axial nanowire heterostructure with an inversion domain in the center (white arrow); the experimental and simulated CBED pattern indicate the Ga-polarity for GaN buffer and wire matrix. b) Dark-field TEM images recorded with (0002) and* $(000\bar{2})$ $\vec{g}$ *-vectors. Notice the contrast inversion between regions of inverted polarity (scale bar: 200 nm)*

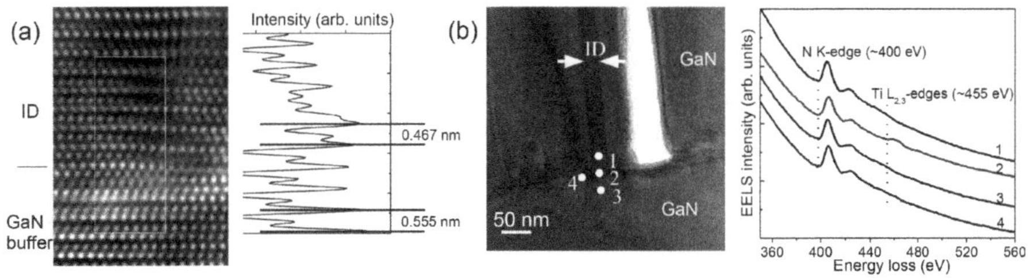

Figure 6.14. *a) Cross-sectional HRTEM image of the boundary between the inversion domain and the GaN buffer exhibiting an extra (0002) plane. b) TEM micrograph around the domain (ID) in the GaN nanowire. The numbers 1-4 indicate the different positions where EEL spectra have been acquired. Local EEL spectra confirm the presence of Ti impurities at the ID nucleating site*

6.5. Interfaces and heterostructures

A major contribution to the interest in III-N nanowires and their achievement potentials is given by the feasibility to fabricate them epitaxially aligned on Si substrates and to grow heterostructures. The structural description of epitaxial interfaces follows the same principles as in the case of planar layers. Depending on the epilayer thickness and the magnitude of the lattice misfit f_0, the epitaxial interface appears either coherent or semi-coherent. The lattice misfit is defined by $f_0 = (a_o - a_s)/a_s$, where a_o and a_s is the unstrained natural spacing of the corresponding atoms in overlayer and substrate, respectively. In the case of nanowires, the 3D shape including the nanowire radius has additionally to be taken into account for the determination of the coherent/semi-coherent transition, [GLA 06] see for instance. Generally, the minimization of the interface energy controls the exact atomic arrangement at interfaces. Because in pure semiconductor systems the bonding part of the interface energy can be neglected, the interface structure is specified only by the lattice misfit accommodation including the formation of misfit relieving dislocations.

6.5.1. *Interface between III-N nanowire and substrate*

In the case of low lattice misfits, the theories of van der Merwe and Matthews describing the interface formation during epitaxial growth are well established and experimentally verified for a large number of systems [e.g. VAN 73, MAT 75, FIT 91]. The epitaxy starts with the two-dimensional growth, and the lattice misfit is accommodated by a coherently strained interface associated with the continuity of lattice planes across the interface. At a critical thickness, the accumulated strain energy is decreased by a relaxation process via morphological instabilities and dislocation nucleation and glide. The ultimate result is a semi-coherent interface, which is defined by large areas of preserved coherency separated by narrow areas with localized misfit dislocations. Such a scenario is observed during the initial growth of GaN nanowires on AlN-buffered Si(111) substrate [CON 10]. The lattice misfit of about 2.5% is initially relieved elastically by the formation of coherent islands undergoing several strain-induced shape transitions before finally – at a critical radius – a misfit dislocation is created at the interface. If the substrate is assumed to be rigid and the strain is built up only within the nanowire, the total lattice misfit is described by

$$f_0 = \varepsilon_{res} + \varepsilon_{el} + \varepsilon_{pl}, \qquad [6.4]$$

where ε_{res} is the residual strain in the nanowire, ε_{el} represents the elastically degraded strain part thanks to the free surface of the nanowire, and ε_{pl} describes the strain accommodated by misfit dislocations. If the misfit dislocation is defined by the Burgers vector $\vec{b}$, which is identical to the dislocation type of the bulk material, then ε_{pl} is in first order approximation determined by the mean spacing D_f between the dislocation cores (neglecting the limited interface area):

$$\varepsilon_{pl} = b_f / D_f . \qquad [6.5]$$

Here, b_f is the Burgers vector component parallel to the interface and perpendicular to the dislocation line direction, which is responsible for the misfit strain relief. If the nanowire is strain-free far from the interface region, the residual strain is zero and the lattice constant has reached its bulk value, and simultaneously, a minimum distance $D_f^{\min} = b_f /(f_0 - \varepsilon_{el})$ has been ensured. The exact interface structure depends on the symmetry of the misfit dislocation network, the corresponding Burgers vectors and the surface of the 3D nanowire contributing to ε_{el}.

This description is only useful for low-misfit systems because the dislocations have to be well separated: if f_0 is larger than 10%, the dislocation distance will achieve values where the dislocation cores overlap and the linear elastic theory (equation [6.4]) is not appropriate anymore. Nevertheless, epitaxial interfaces in large-misfit systems must be alike connected with special configurations exhibiting minimal total interface energy. Assuming that the strain energy represents the most prominent part of the total interfacial energy for semiconductor heterostructures, the interface character can be described by a near coincidence site lattice model predicting points of best match and highest mismatch [see TRA 00]. Perfect coincidence between the epilayer lattice a_e and the substrate lattice a_s would occur when $a_e / a_s = m / n$, where m and n are positive integers. If $m = n+1$, there is one extra lattice plane in each unit cell of the coincidence site lattice, i.e. a geometrical edge type dislocation is generated. The name "geometrical" misfit dislocation is used to emphasize that its nature is defined by the coincidence site lattice and therefore, must not generally correspond to a bulk-like dislocation, which is able to glide to the interface. Accordingly, this means that the dislocation character is derived by the symmetry of the coincidence lattice and must not necessarily be an invariant vector of the bulk lattice.

In general, the lattice misfit at the interface between dissimilar materials is not expected to be at exact coincidence, and a coincidence lattice misfit F_0 expresses the deviation from perfect coincidence as

$$F_0 = (ma_s - na_e) / ma_s . \qquad [6.6]$$

This deviation induces strain associated with each coincidence unit that can be accommodated by (i) regularly introducing a coincidence unit with a larger or smaller ratio of lattice planes depending on the sign of F_0 and by (ii) elastic relaxation at the free surface of nanowire. The following two examples illustrate the applicability of this coincidence model on epitaxially aligned III-N nanowires grown on substrates with large lattice mismatch.

(a) (b)

Figure 6.15. *a) Cross-sectional HRTEM image of GaN nanowire epitaxially aligned to Si(111) substrate; b) magnified part of the interface indicating the coincidence site lattice with 5/4 and 6/5 ratios*

GaN nanowires on Si(111)

The epitaxial alignment of GaN nanowires on Si(111) is given by the following epitaxial relationship: GaN(0001) || Si(111) and GaN$[11\bar{2}0]$ || Si[110]. This relationship corresponds to an agreement of the in-plane symmetry of both close-packed planes, the GaN(0001) and Si(111) plane. Nevertheless, the orientation results in the large lattice misfit of about 17%. Figure 6.15 shows a cross-sectional HRTEM image of the interface region between an epitaxial aligned GaN nanowire and the Si(111) substrate taken along the Si [110] and GaN$[11\bar{2}0]$ zone axis, respectively. Although a significant amount of lattice disorder is detected at the interface, the continuity of GaN$\{1\bar{1}00\}$ and Si $\{111\}$ lattice planes across the interface is revealed as well as periodical interruptions at every fifth or sixth GaN plane (indicated by arrows). Furthermore, the bending of lattice planes around the core is obviously visible indicating the strong interfacial bonding and large coherency stresses. By taking the ratio of 6/5 planes into account, a deviation of about 0.3% remains if applying the corresponding bulk values for GaN (a = 0.3189 nm) and Si ($a_{in\text{-}plane}$ = 0.3840 nm). This small deviation leads to the formation of coincidence units with 5/4 ratio. Nevertheless, the small diameter of the nanowire of about 10 nm certainly involves additional elastic strain relaxation at the free surface.

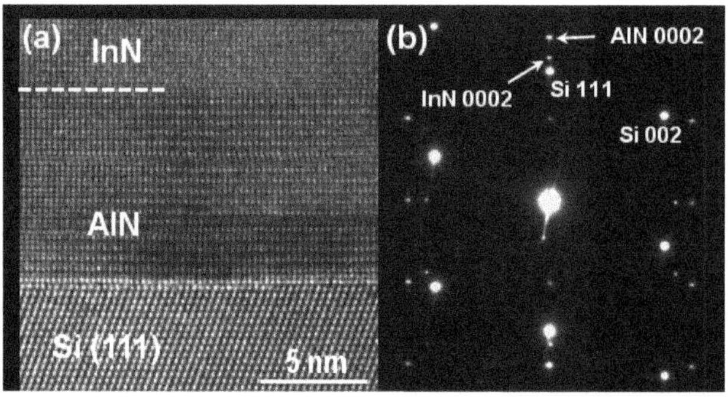

Figure 6.16. *a) Cross-sectional HRTEM image of the InN/AlN/Si(111) interface region and b) corresponding selected area electron diffraction pattern along the Si $[\bar{1}\bar{1}0]$ zone axis*

InN nanowires on AlN buffer

Figure 6.16(a) presents a further example of a large lattice-mismatched interface ($f_0 \approx 14.6\%$). It shows a cross-sectional HRTEM image of interface regions between an InN nanowire grown on AlN-buffered Si(111) proving the same orientation relationship as described in the previous case of GaN/Si(111) [GRA 07]. The AlN buffer layer of about 10 nm thickness reveals atomically smooth interfaces to the Si substrate as well as to the InN nanowire. Based on the selected area diffraction pattern, the AlN and the InN reflect bulk lattice constants within the measuring accuracy and using Si as reference. A higher magnified image of the InN/AlN interface in Figure 6.17 indicates the presence of extra planes in accordance with the coincidence site lattice model. Applying a Fourier-filtered lattice fringe image by selecting the $\{\bar{1}100\}$ and $\{1\bar{1}00\}$ spots only, the array of geometric misfit dislocations is detectable with lattice plane ratios between the expected 7/8 and 8/9 ratio. It is remarkable that the extra $\{1\bar{1}00\}$ plane in the filtered image does not characterize the Burgers vector but the center of the mismatched area.

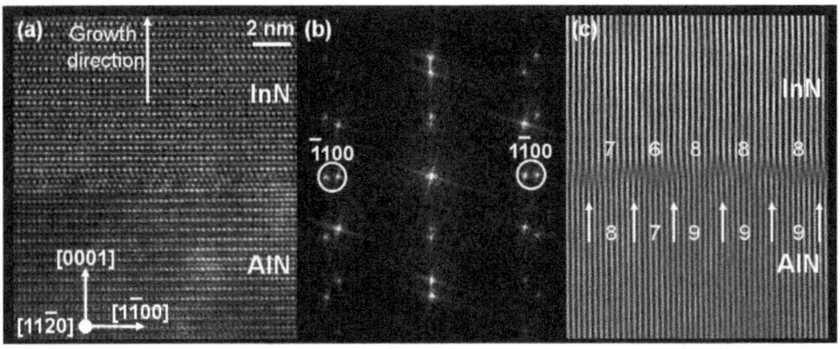

Figure 6.17. *(11$\bar{2}$0) cross-sectional HRTEM image of the InN/AlN interface, the corresponding Fourier transform with selected spots used for the filtered fringe image*

In order to observe the two-dimensional geometric misfit dislocation network or the corresponding coincidence site lattice directly, plan-view TEM is performed [LUN 14]. Figure 6.18(a) depicts a plan-view micrograph along the <0001> zone axis displaying the lattice image of AlN and a Moiré-like pattern due to the superposition with the InN lattice. The Moiré-like pattern with its three- or six-fold symmetry reflects the coincidence lattice and the local atomic arrangement. The periodic sequence of bright and dark contrast areas indicates the regions of perfect matching (coincidence sites, bright contrast) and regions of maximal lattice mismatch where the geometric dislocations are located (dark contrast). In Figure 6.18(b), a model of the coincidence site lattice is presented defining the areas of maximal lattice mismatch and thus the location of the misfit dislocations. Two networks are possible which only differ in a shift of lines in order to avoid the presence of a high-energy triple dislocation junction [SAN 07]. In both networks, the dislocation lines are running along $<11\bar{2}0>$ directions in agreement with the experimental observation based on the Moiré pattern. Furthermore, the arrangement of the dark spots in the HRTEM image indicates a configuration of the misfit dislocation network without triple junctions (see blue (i) and red (ii) networks in Figure 6.18(b)). Moreover, it is found that, towards the nanowire edges, the dislocation network is slightly distorted due to elastic surface relaxation based on image forces [HUL 11].

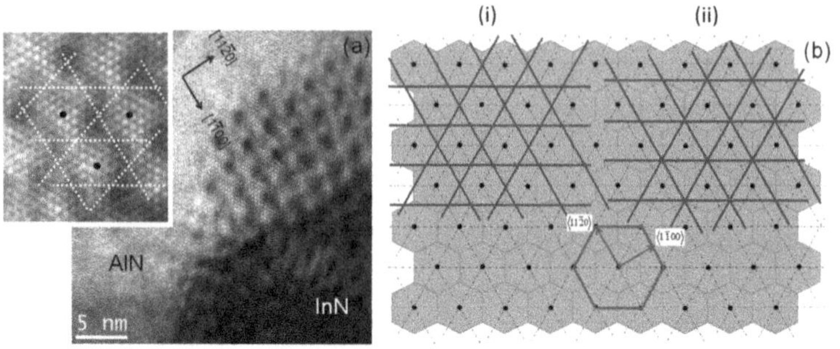

Figure 6.18. *a) Plan-view HRTEM image partly displaying the interface between InN and AlN as Moiré pattern (inset: higher magnification); b) schematic of the coincidence site lattice with two possible arrangements of misfit dislocations. For a color version of this figure, see www.iste.co.uk/consonni/nanowires1.zip*

6.5.2. *Axial nanowire heterostructure: (In,GaN)/GaN case study*

Besides the hetero-phase boundaries between nanowire and substrate, the interfaces in axial and radial nitride nanowire heterostructures are most relevant for applications. The basic physics of plastic relaxation and misfit dislocation generation is based on similar mechanisms as those described in the previous section. On the other hand, there are only a few investigations on dislocations formed in core-shell systems, in particular, if the systems are characterized by a large aspect ratio combined with a diameter of less than 100 to 200 nm. A review about misfit dislocations in nanowire heterostructures can be found in [KAV 10].

A further important aspect for achieving a complete interface analysis of III-N heterostructures is related to the quantification of the interface roughness and chemical width [LUN 12]. The measurement of composition profiles in axial heterostructures is complex compared to the corresponding planar case because of the interplay between nanowire shape, stress state, strain relief and interface width. Regarding this correlation, it is necessary to measure the chemical composition with high spatial resolution, independent of the local strain state and in combination with the interface morphology as hereafter illustrated by the (In,Ga)N/GaN example.

Low-loss EELS in TEM has been demonstrated to be a powerful technique for probing the chemical composition on the nanometer scale [KON 12]. Compared to the core-loss EELS, the low-loss EELS method has a slightly decreased spatial resolution due to the relatively large delocalization of the low-loss inelastic scattering. Nevertheless, it offers an improvement in the signal-to-noise ratio.

Further, the increase in signal strength allows a reduced incident probe intensity, limiting the electron beam damage to the sample, which is quite important for (In,Ga)N as it is reported to form In-rich clusters induced by the electron beam irradiation [SME 06].

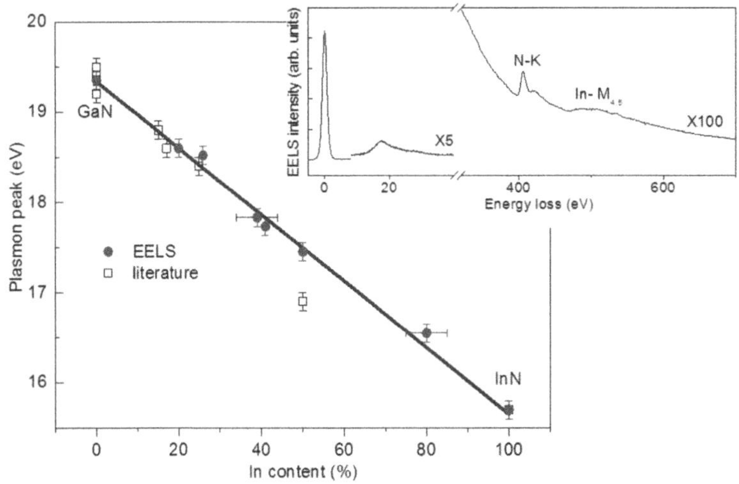

Figure 6.19. *The dependency of plasmon peak on indium concentration in (In,Ga)N alloys, including recently published literature data (the blue line is based on a least squares fit) [KON 12]. Inset: a typical EEL spectrum from (In,Ga)N including the plasmon peak (~19 eV), N K edge (~400 eV) and In $M_{4,5}$ edges (~443 eV)*

Using simultaneous reference measurements of the plasmon peak energy in the low-loss EELS and the indium concentration in the core-loss EELS, a linear relationship between them is established over almost the whole concentration range (see Figure 6.19):

$$E_{\max}(In_xGa_{1-x}N) = 19.35 - 3.6x \ (0 \le x \le 1) \qquad [6.6]$$

Additionally, the local strain effect on the plasmon peak was found to be negligible as demonstrated in reference [KON 12].

Based on this linear relation, the indium concentration profiles along the center of axial (In,Ga)N/GaN nanowires are determined across the interfaces as shown in Figure 6.20. The two presented heterostructures were grown by MBE on patterned GaN substrate with different growth temperatures for (In,Ga)N [BEN 11]. In both cases, the indium concentration amounts to about 35%, a value that is not attainable in planar layers without composition fluctuations or phase separation. On the other hand, the interface width (*W*) – defined as the distance where the concentration varies from 10 to 90% of the maximum value – amounts to 12 nm and 45 nm, respectively. These relatively large values are explainable by the so-called "lattice pulling" effect. In this effect, the lattice misfit strain between (In,Ga)N and GaN impedes the incorporation of indium, thus pulling the composition of the growing (In,Ga)N layer towards that of GaN, i.e. towards lower indium concentrations and lattice matching conditions. This pulling effect acts stronger with increasing final indium concentration. Finally, it is still to be clarified how the pulling effect competes with the other strain relieving mechanisms in nanowires and which are the controlling parameters.

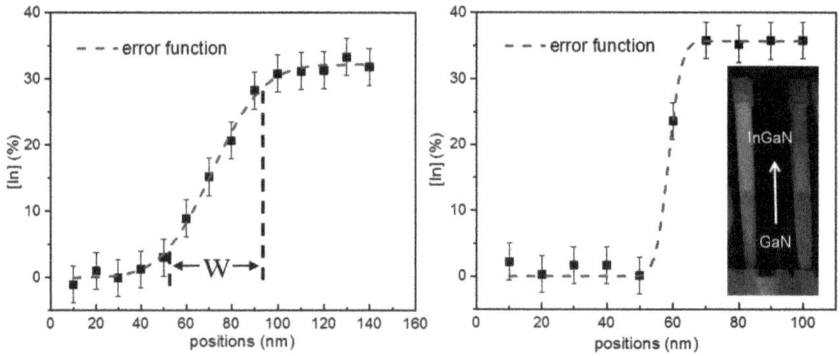

Figure 6.20. *The axial indium concentration profiles measured by line scans of low-loss EELS along two ordered (In,Ga)N/GaN NW heterostructures grown at different temperatures; inset: dark-field scanning TEM image of (In,Ga)N/GaN*

6.6. Conclusions

– The exact knowledge about the microstructure of individual nanowires and nanowire ensembles is an essential condition for the understanding of their physical properties (cf. "structure-property relation").

– Transmission electron microscopy and electron energy-loss spectroscopy provide excellent experimental tools for spatially resolved crystal structure and defect analysis.

– The determination of the atomic structure of extended defects and their relation to the 3D nanowire character is the basis for developing concepts to control or to avoid defects.

– The structural characterization is important to provide feedback to understand the growth mechanisms of III-N nanowires and to improve growth.

6.7. Bibliography

[ALB 13] ALBERT S., BENGOECHEA-ENCABO A., *et al.*, "Selective area growth of In(Ga)N/GaN nanocolumns by molecular beam epitaxy on GaN-buffered Si(111): from ultraviolet to infrared emission", *Nanotechnology,* vol. 24, p. 175303, 2013.

[AMB 02] AMBACHER O., MAJEWSKI J., MISKYS C., *et al.*, "Pyroelectric properties of Al(In)GaN/GaN hetero and quantum well structures", *Journal of Physics: Condensed Matter*, vol. 14, pp. 3399–3434, 2002.

[BEN 11] BENGOECHEA-ENCABO A., BARBAGINI F., FERNANDEZ-GARRIDO S., *et al.*, "Understanding the selective area growth of GaN nanocolumns by MBE using Ti nanomasks", *J. Cryst. Growth*, vol. 325, pp. 89–92, 2011.

[CON 09] CONSONNI V, KNELANGEN M., JAHN U., *et al.*, "Effect of nanowire coalescence on their structural and optical properties on a local scale", *Appl. Phys. Lett.*, vol. 95, p. 241910, 2009.

[CON 10] CONSONNI V., KNELANGEN M., GEELHAAR L., *et al.*, "Nucleation mechanisms of epitaxial GaN nanowires: Origin of their self-induced formation and initial radius", *Phys. Rev. B.,*vol. 81, p. 085310, 2010.

[EGE 96] EGERTON R.F., *Electron-Energy Loss Spectroscopy in the Electron Microscope*, 2nd ed. Plenum, New York, 1996.

[FER 12] FERNANDEZ-GARRIDO S., KONG X., GOTSCHKE T., *et al.*, "Spontaneous nucleation and growth of GaN nanowires: the fundamental role of crystal polarity", *Nano Lett.*, vol. 12, p. 6119, 2012.

[FIT 91] FITZGERALD E.A., "Dislocations in strained-layer epitaxy: theory, experiment, and applications", *Materials Science Reports*, vol. 7, p. 87, 1991.

[GLA 06] GLAS F., "Critical dimensions for the plastic relaxation of strained axial heterostructures in free-standing nanowires", *Phys. Rev. B.*, vol. 74, p. 121302 (R), 2006.

[GRA 07] GRANDAL J., SÁNCHEZ-GARCÍA M.A.,CALLEJA E., *et al.*, "Accommodation mechanism of InN nanocolumns grown on Si(111) substrates by molecular beam epitaxy", *Appl. Phys. Lett.*, vol. 91, p. 021902, 2007.

[HUL 11] HULL D., BACON D.J., (ed.), *Introduction to Dislocations*, 5th ed., Elsevier, p. 109, 2011.

[HIR 77] HIRSCH P., HOWIE A., NICHOLSON R.B., *et al.*, *Electron Microscopy of Thin Crystal*, 2nd ed., Huntington, New York 1977.

[HEL 91] HELLMAN E.S., "The polarity of GaN: a critical review", *MRS Internet J. Nitride Semicond. Res.*, vol. 3, p. 11, 1998.

[JEN 11] JENICHEN B., BRANDT O., PFÜLLER C., *et al.*, "Macro- and micro-strain in GaN nanowires on Si(111)", *Nanotechnology,* vol. 22, p. 295714, 2011.

[JIA 02] JIANG N., EUSTIS T.J., CAI J., *et al.*, "Polarity determination by atomic location by channeling enhanced microanalysis", *Appl. Phys. Lett.*, vol. 80, p. 389, 2002.

[KAV 10] KAVANAGH K.L., "Misfit dislocations in nanowire heterostructures", *Semicond. Sci. Technol*, vol. 25, p. 024006, 2010.

[KON 02] KONG X., HU G.Q., Duan X.F., *et al.*, "Polarity determination for GaN thin films by electron energy-loss spectroscopy", *Appl. Phys. Lett.*, vol. 81, p. 1990, 2002.

[KON 11] KONG X., J. RISTIC M.A., SANCHEZ-GARCIA E., *et al.*, "Polarity determination by electron energy-loss spectroscopy: application to ultra-small III-nitride semiconductor nanocolumns", *Nanotechnol*, vol. 22, p. 415701, 2011.

[KON 12] KONG X., ALBERT S., BENGOECHEA-ENCABO A., *et al.*, "Plasmon excitation in electron energy-loss spectroscopy for determination of indium concentration in (In, Ga)N/GaN nanowires", *Nanotechnol*, vol. 23, p. 415701, 2012.

[LAN 00] LANGFORD J.I., CHUNG F.H., SMITH D.K. (eds.), *Industrial Applications of X-ray Diffraction*, Marcel Deckker Inc., 2000.

[LAW 72] LAWAETZ P., "Stability of the wurtzite structure", *Phys. Rev. B,* vol. 5, p. 4039, 1972.

[LIU 07] LIU F., COLLAZO R., MITA S., *et al.*, "The mechanism for polarity inversion of GaN via a thin AlN layer: Direct experimental evidence", *Appl. Phys. Lett.*, vol. 91, p. 203115, 2007.

[LUN 12] LUNA E., GUZMAN A., TRAMPERT A., "Critical role of two-dimensional island-mediated growth on the formation of semiconductor heterointerfaces", *Phys. Rev. Lett.*, vol. 109, p.126101, 2012.

[LUN 14] LUNA E., GRANDAL J., GALLARDO E., *et al.*, "Microscopy and Mircoanalysis", submitted, 2014.

[MAT 75] MATTHEWS J.W., (ed.), *Epitaxial Growth, Part B*, Academic Press, New York, 1975.

[MKH 06] MKHOYAN K.A., BATSON P.E., CHA J., *et al.*, "Direct determination of local lattice polarity in crystals", *Science,* vol. 312, p. 1354, 2006.

[MOR 09] MORAM M.A., VICKERS M.E., "X-ray diffraction of III-nitrides", *Rep. Prog. Phys.*, vol. 72, p. 036502, 2009.

[PON 96] PONCE F.A., BOUR D.P., YOUNG W.Y., *et al.*, "Determination of lattice polarity for growth of GaN bulk single crystals and epitaxial layers", *Appl. Phys. Lett.*, vol. 69, p. 337, 1996.

[ROB 07] ROBINS L.H., BARTNESS K.A., BARKER J.M., *et al.*, "Optical and structural study of GaN nanowires grown by catalyst-free molecular beam epitaxy. II Sub-band-gap luminescence and electron irradiation effects", *J. Appl. Phys.*, vol. 101, p. 113505, 2007.

[ROU 98] ROUVIERE J.L., WEYHER J.L., SEELMANN-EGGEBERT M., *et al.*, "Polarity determination for GaN films grown on (0001) sapphire and high-pressure-grown GaN single crystals", *Appl. Phys. Lett.*, vol. 73, p. 668, 1998.

[ROU 08] ROUVIERE J.L., BOUGEROL C., AMSTATT B., *et al.*, "Measuring the local lattice polarity in AlN and GaN by high resolution Z-contrast imaging: The case of (0001) and $(000\bar{1})$ GaN quantum dots", *Appl. Phys. Lett.*, vol. 92, p. 201904, 2008.

[SAN 07] SANCHEZ A.M., LOZANO J.G., GARCIA R., *et al.*, "Strain mapping at the atomic scale in highly mismatched heterointerfaces", *Adv. Func. Mater.*, vol. 17, p. 2588, 2007.

[SHA 05] SHAHEDIPOUR-SANDVIK F., GRANDUSKY J., ALIZADEH A., *et al.*, "Strain dependent facet stabilization in selective-area heteroepitaxial growth of GaN nanostructures", *Appl. Phys. Lett.*, vol. 87, p. 233108, 2005.

[SER 73] SERNEELS R., SNYKERS M. DELAVIGNETTE P., *et al.*, "Friedel's law in electron diffraction as applied to the study of domain structures in non-centrosymmetrical crystals", *Phys. stat. sol.* (b), vol. 58, p. 277, 1973.

[SME 06] SMEETON T.M., HUMPHREYS C.J., BARARD J.S., *et al.*, "The impact of electron beam damage on the detection of indium-rich localization centres in InGaN quantum wells using transmission electron microscopy", *J. Mater. Sci.*, vol. 41, p. 2729, 2006.

[SPE 92] SPENCE J.C.H., ZUO J.M., *Electron Microdiffraction,* Plenum, New York, 1992.

[SPI 02] SPIECKER E., "Determination of crystal polarity from bend contours in transmission electron microscope images", *Ultramicroscopy,* vol. 92, p. 111, 2002.

[STA 87] STADELMANN P.A., "EMS–a software package for electron diffraction analysis and HREM image simulation in materials science", *Ultramicroscopy,* vol. 21, p. 131, 1987.

[STA 98] STAMPF C., VAN DE WALLE C., "Energetics and electronic structure of stacking faults in AlN, GaN and InN", *Phys. Rev. B.*, vol. 57, p. R15052, 1998.

[SUM 04] SUMIYA M., FUKE S., "Review of polarity determination and control of GaN", *MRS Internet J. Nitride Semicond. Res.*, vol. 9, p. 1, 2004.

[TAF 83] TAFTØ J., "Structure-factor phase information from two-beam electron diffraction", *Phys. Rev. Lett.,* vol. 51, p. 654, 1983.

[TAF 87] TAFTØ J., "Reciprocity in electron energy-loss spectra from non-centrosymmetric crystals", *Acta Crystallogr., Sect. A: Found. Crystallogr. A*, vol. 43, p. 208, 1987.

[TAK 99] TAKEUCHI, S., SUZUKI, K., "Stacking fault energies of tetrahedrally coordinated crystals", *Phys. stat. sol. (a)*, vol. 171, p. 99, 1999.

[TRA 00] TRAMPERT A., PLOOG K.H., "Heteroepitaxy of large-misfit systems: Role of coincidence lattice", *Cryst. Res. Technol.*, vol. 35, p. 793, 2000.

[TRA 04] TRAMPERT A., RISTIC J., JAHN U., *et al.*, "TEM study of (Ga,Al)N nanocolumns and embedded GaN nanodiscs", *Proc. 13th International Conference on Microscopy of Semiconductor Materials. Journal of Physics: Conference Series,* vol. 209, p. 012033, 2004.

[VAN 73] VAN DER MERWE J.H., HERMAN H. (ed.), *Treatise on Materials Science and Technology*, Academic Press, New York, vol. 2, 1973.

[VUR 03] VURGAFTMAN I., MEYER J.R., "Band parameters for nitrogen-containing semiconductors", *J. Appl. Phys.*, vol. 94, p. 3675, 2003.

[WIL 53] WILLIAMSON G.K., HALL W.H., "X-ray line broadening from filed aluminium and wolfram", *Acta Metal.*, vol. 1, p. 22, 1953.

[WUL 01] WULFF G., "On the question of the rate of growth and dissolution of crystal surfaces", *Z. Kristallogr. Mineral.*, vol. 34, p. 449, 1901.

PART 2

Nucleation and Growth Mechanisms of GaN and ZnO Nanowires

7

Ni Collector-Induced Growth of GaN Nanowires on C-Plane Sapphire by Plasma-Assisted Molecular Beam Epitaxy

7.1. Introduction

Over the past decade, impressive progress has been made to demonstrate and develop applications based on nanowires (NWs) in electronics, optoelectronics, photovoltaics, sensing, and biosciences ([LIE 07], for more details, refer to Chapters 5–10 in Volume 2 [CON 14]).However, these developments were only enabled by the tailoring of NWs with predictable physical properties, which is still nowadays a major concern for real-world device integration. To this aim, the reliable fabrication of NWs with specific dimensions, shape, crystal orientation, composition and position is mandatory. A detailed understanding of the fundamental nucleation and growth phenomena at the atomic scale is thus essential.

In the frame of the "bottom–up" approach where atoms are self-assembled to form increasingly larger structures by mastering the crystallization and growth anisotropy, three different processes have been applied to promote the growth of GaN NWs by molecular beam epitaxy (MBE). The growth of GaN NWs on bare sapphire [GEE 07] can be induced by Ni catalyst seeds or "collectors" [WAC 09], so named because they collect the active species similarly to the well-established vapor–liquid–solid growth process (VLS) [WAG 64]. These species diffuse along the collector surface or through its bulk and crystallize at the lower interface to the substrate, from which the NW elongates further. Alternatively, GaN NWs can also

---

Chapter written by Caroline CHÈZE.

be grown in a self-induced way on Si or on AlN buffers [YOS 97, CAL 07, for more details, refer to Chapter 7] without the need for any foreign material, or by selective area growth where prepatterned masks define the NW location and diameter [SEG 08, see Chapter 10 of this volume]. The growth of GaN NWs by MBE induced by collectors has not been much investigated in comparison to the self-induced one. Mamutin [MAM 99] first reported VLS growth of GaN and InN whiskers on GaAs and Al_2O_3 substrates. In the case of GaN a low temperature GaN buffer was first deposited [POZ 99] and VLS growth was suggested to occur from Ga droplets that were also observable at the tip of the whiskers in reference [MAM 99]. However, no deeper study on this whisker growth mechanism was carried out. More recently, our group successfully promoted the growth of GaN NWs on the C-plane sapphire substrate by employing Ni collector seeds. Remarkably, the required growth window for the collector-induced NWs was very similar to the one that is necessary to the self-induced NW growth on bare Si [CHÈ 10a]. Thus, collector- and self-induced NWs were directly compared. This unique combination offered the possibility to investigate eventual effects of the collectors made of a foreign material on the NW properties.

In this chapter, we present a detailed study of the Ni-induced GaN NW growth by MBE. The first section describes the experimental parameters and growth monitoring methods. In the second section, we investigate the nucleation of the Ni-induced GaN NWs by two complementary *in situ* techniques, line-of-sight quadrupole mass spectrometry (QMS) [KOB 02] and reflection high-energy electron diffraction (RHEED). Thus, we are able to monitor both the incorporation rate of Ga and the crystal structure. The third section deals with GaN NW growth processes by the study of the NW geometry in dependence of the growth parameters. Finally, the fourth section compares the structural and optical properties of these Ni-induced GaN NWs and of self-induced GaN NWs grown on Si under the same conditions.

7.2. Experimental description

The Ni-induced GaN NWs were grown by MBE. Ga was supplied by an effusion cell and N by an RF plasma source. The Ga (N) flux was calibrated by the growth of planar layers in the N (Ga) limited regime [RIE 96] and corresponded to a planar growth rate of 0.4 (2.0) Å/s unless noted otherwise. The substrate temperature for most experiments was 730°C, and was measured by a pyrometer. The growth duration was set to 76 min, and during this time, a 180-nm-thick planar layer would have been grown under suitable conditions (nominal thickness). In addition, samples dedicated to the study of the nucleation stages were grown for 15 s and 30 s only. Prior to the growth of GaN, 4-Å-thin Ni layers were deposited either *ex situ* in an ion-beam coater or *in situ* from an effusion cell in the MBE chamber on C-plane sapphire substrates. Subsequently, the substrates were annealed in the MBE chamber to 780°C for 15 min, which induced the formation of Ni islands, followed by the growth of GaN. For comparison, self-induced GaN NWs were grown on bare Si(111) and Si(001) substrates without any buffer layer. Immediately before insertion into the MBE system, the Si substrates were cleaned by the Radio Corporation of America (RCA) procedure, and before GaN growth, the substrates were annealed at 950°C for 15 min.

During the nucleation of the Ni-induced NWs the amount of Ga desorbing from the substrate was continuously measured by a quadrupole mass spectrometer (QMS) mounted in the line-of-sight to the substrate. This allowed us to acquire a temporal profile of the Ga desorption and to extract the Ga incorporation knowing the Ga flux supplied [KOB 02]. To this aim the QMS response was calibrated from the full desorption of this known Ga flux at 780–800°C on a clean sapphire substrate. Unless indicated otherwise, the growth of GaN was initiated by the opening of both Ga and N shutters simultaneously, which defined the origin of the time scale $t = 0$ s. Two other separate experiments were carried out under identical growth conditions to observe the development of the structure during nucleation by RHEED in the (10-10) and the (1-210) azimuths of sapphire. For the lattice constant measurement, a calibration was carried out by line scans and Gaussian–Lorentzian fitting using the streaky pattern of a sapphire substrate as well as the spotty one of a rough layer of N-rich GaN.

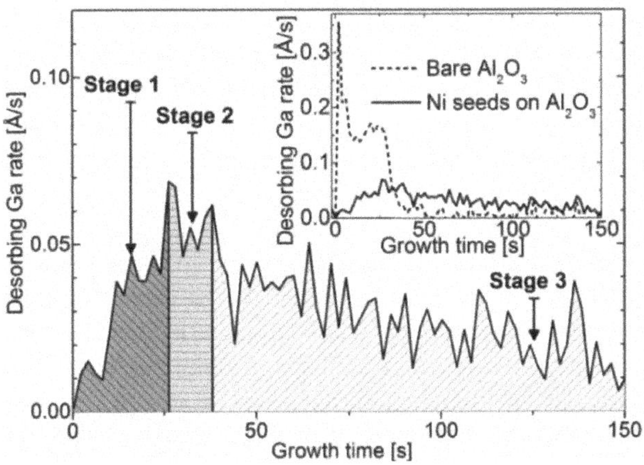

Figure 7.1. *Calibrated QMS profile of Ga desorption during GaN nucleation on a sapphire substrate covered by Ni collectors. The three different stages indicated by the hatching direction correspond to the observation of the three RHEED patterns observed along both the [10-10]$_{Al2O3}$ and the [-12-10]$_{Al2O3}$ azimuths as shown in Figure 7.2. Inset: Calibrated QMS profiles of Ga desorption during GaN nucleation on a sapphire substrate covered by Ni collectors (solid trace, as in Figure 7.1) and on a bare sapphire substrate (dashed trace). (Figure adapted from [CHE 10d]). For a color version of this figure, see www.iste.co.uk/consonni/nanowires1.zip*

The morphology of the samples was investigated by scanning electron microscopy (SEM). Selected samples were prepared for the analysis by cross-sectional transmission electron microscopy (TEM) by the standard procedure involving mechanical thinning and ion milling, and other NWs were simply harvested and transferred to TEM grids. For photoluminescence (PL) measurements, samples were put into a microscope cryostat and excited by the 325-nm line of a He–Cd laser. The laser beam was focused by a 15× objective to a micrometer-sized spot. The PL signal was collected by the same objective, selected by a 0.8 m monochromator, and measured by a cooled Si charge-coupled-device detector.

7.3. Ni-induced GaN nanowire nucleation

Based on the QMS and RHEED *in situ* monitoring of the nucleation, three stages were evidenced [CHÈ 10b]. Figure 7.1 presents the Ga desorption profile acquired by QMS in real-time during the GaN NW nucleation on a sapphire substrate covered by Ni collectors where the three stages are underlined. During stage 1, no Ga desorption occurred upon opening both Ga and N shutters at $t = 0$, i.e., the full Ga flux from the effusion cell is initially incorporated. However, the desorption signal increased gradually during the first 30 s of nucleation. In stage 2, a maximum was attained. Stage 3 started after 35 s with the decrease in the Ga desorption. Finally, after 380 s the nucleation ended, and a steady state was reached, where all the Ga supplied was incorporated again. The initially negligible rate of desorbing Ga atoms is remarkable because for heteroepitaxial nucleation a delay in incorporation of the adsorbate is expected [KOB 02], as we also observed during GaN nucleation under the same conditions of temperature and V/III ratio on bare sapphire (inset in Figure 7.1). In that case the instant Ga-desorption at a rate close to the supplied Ga-rate indicated that the nucleation was indeed initially inhibited on bare sapphire. Complementary experiments (not shown) during which both bare and Ni-covered sapphire substrates were solely exposed to the Ga flux also yielded to the initial Ga incorporation in the presence of the Ni collectors, in contrast to the instant increase of the Ga desorption signal to a saturation level corresponding to the Ga supply rate upon Ga exposure of the bare sapphire surface. These experiments thus proved that the Ga incorporation during stage 1 is not due to GaN nucleation on bare sapphire, and that the incorporation of Ga observed during the first 30 s in Figure 7.1 was only related to the presence of the Ni collectors.

Besides, the RHEED study of the nucleation (Figure 7.2) allowed us to clarify the nature of the change in the Ga incorporation behavior. In both sapphire azimuths, three different RHEED patterns appeared subsequently during the first minute of nucleation, and their appearance coincided, noteworthily, with a change in the Ga incorporation behavior in stages 1–3. Throughout the first 25–30 s that limited nucleation stage 1, the RHEED pattern did not change significantly while

QMS indicated a gradually increasing Ga desorption signal (Figure 7.1). Following stage 1, a different RHEED pattern briefly appeared coinciding with the time interval of the maximum in the Ga desorption signal in stage 2. After about 35 s of nucleation, the pattern characteristic for wurzite (WZ) GaN developed and remained for the rest of the growth. The appearance of this WZ GaN pattern corresponded to the decrease of the Ga desorption signal in Figure 7.1. However, additional spotty reflexes characteristic from stage 2 were still observable even after 3 min of growth.

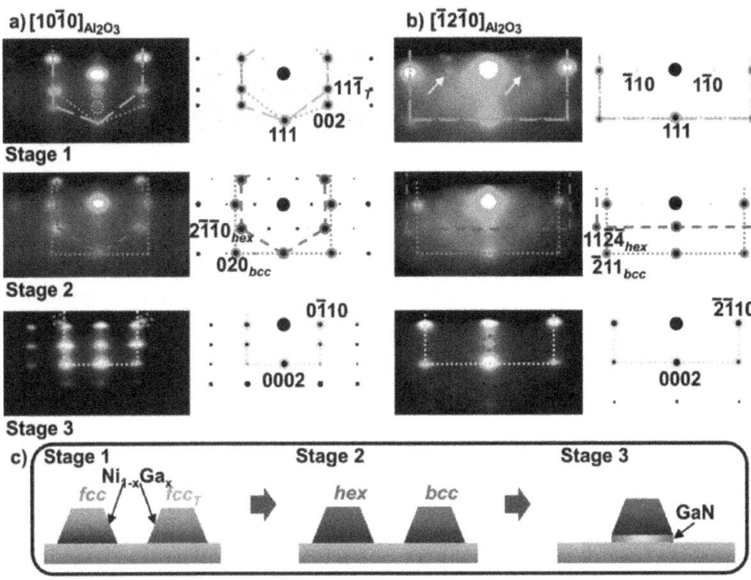

Figure 7.2. *RHEED patterns (left) and corresponding kinematical diffraction patterns simulated with the EMS software [STA 87] (right) along a) the $[10\text{-}10]_{Al2O3}$, and b) the $[-12\text{-}10]_{Al2O3}$ azimuths of the different nucleation stages as indicated in Figure 7.1 and sketched in c). The diameter of the simulated reflexes indicates their intensity, and the weak spots are not seen on the experimental patterns. For stage 1 (top) the dotted and dashed lines stand for the twinned fcc or $L1_2$ $Ni_{1-x}Ga_x$ structures, respectively. In the $[-12\text{-}10]_{Al2O3}$ azimuth, additional (1-10) and (-110) reflexes (arrows) reveal the incorporation of Ga into Ni and the formation of a Ni-Ga solid solution or disordered Ni_3Ga. In the $[10\text{-}10]_{Al2O3}$ azimuth, two additional reflexes were observed in RHEED (dashed circles) that do not exist in the simulation. These reflexes could result from double diffraction. The pattern was simulated using the fcc α'-Ni_3Ga phase. For stage 2 (middle) the dotted and the dashed lines stand for the bcc Ni–Ga structure and for another Ni–Ga compound of hexagonal structure (see text), respectively. The pattern was simulated using the β-NiGa and γ-$Ni_{13}Ga_7$ phases. For stage 3 the dotted lines stand for the wurtzite structure of GaN. Additional reflexes generated by the Ni–Ga compound are also underlined by dotted circles. (Reprinted with permission from [CHÈ 10b]. Copyright 2010 American Chemical Society). For a color version of this figure, see www.iste.co.uk/consonni/nanowires1.zip*

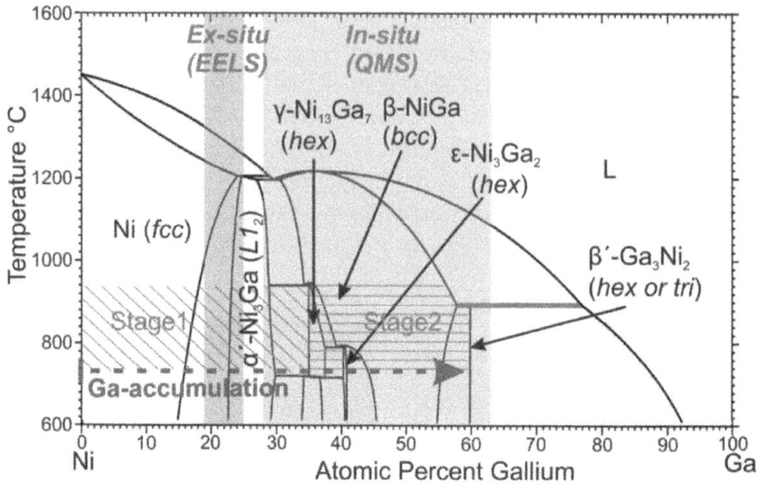

Figure 7.3. *Sketch of the Ni–Ga phase diagram. The shaded area on the left-hand side corresponds to the composition range of the Ni collectors at the tips of GaN NWs measured ex situ by EELS. The shaded area on the right-hand side marks the composition range extracted from the in situ QMS experiments. The dashed arrow indicates the reaction path during nucleation. The crystal structure of the different phases is given in parentheses: hex for hexagonal and tri for trigonal. Unlabeled regions are mixed phases. (Reprinted with permission from [CHÈ 10b]. Copyright 2010 American Chemical Society). For a color version of this figure, see www.iste.co.uk/consonni/nanowires1.zip*

As the sequence of patterns implied structural changes in the collectors, we carefully analyzed and simulated these patterns in connection with the Ni–Ga (Figure 7.3) and Ni–Ga–N phase diagrams [OKA 08, HUL 04, GRÖ 99], and with the estimation of the amount of the incorporated Ga determined by the calibrated QMS signal in order to identify the crystal structure of the collectors throughout nucleation. These studies enabled the precise description and understanding of the collector-induced nucleation of GaN NWs. As the spotty patterns characteristic for electron transmission through the collectors were observed during the whole process, we concluded that a substantial part of the Ni collectors remained solid throughout nucleation and growth, indicating the dominance of the vapor–solid–solid (VSS) mechanism over the VLS one. We identified the three main stages of the nucleation as follows (Figure 7.2(c)): First, Ga incorporation into the Ni collectors, second, transformation of the crystal structure of the Ni–Ga collectors, and third, nucleation of GaN. Prior to growth, the Ni collectors had the fcc crystal structure. In the early stage Ga diffused into the Ni collectors forming a Ni–Ga solid solution or a disordered α′-Ni_3Ga phase with the $L1_2$ structure, as also revealed by the appearance of faint additional (1-10) and (-110) reflexes in the $[-12-10]_{Al2O3}$ azimuth upon shutter opening (Figure 7.2(b) top). Ideally, the $L1_2$ structure of Ni_3Ga can be obtained from the fcc one of Ni by the substitution of Ni atoms with Ga atoms at the corner sites of the fcc lattice, which introduces additional reflexes to the fcc diffraction pattern such as the ones observed. When the solubility limit of Ga into Ni was reached, the α'-Ni_3Ga phase with the $L1_2$ structure formed. Further Ga incorporation lead to a marked modification of the crystal structure of the collectors (stage 2). The β-NiGa bcc structure appeared, and at the same time at least one of the hexagonal phases ε-Ni_3Ga_2, γ-$Ni_{13}Ga_7$, or β'-Ga_3Ni_2 formed. After this intermediate stage and about 35 s after growth was initiated, at last GaN precipitated below the collectors, leading to the formation of the NWs (stage 3).

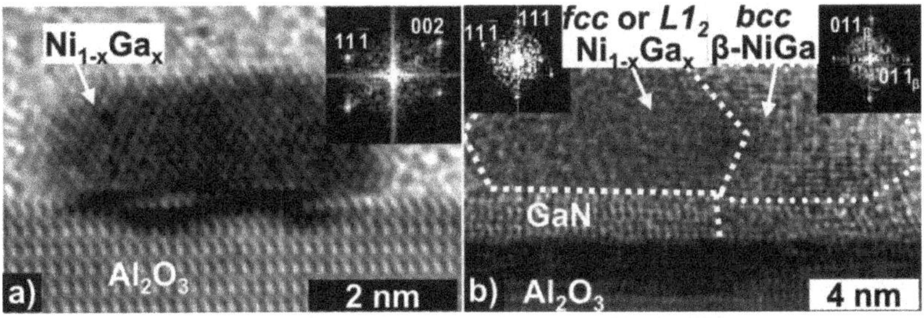

Figure 7.4. *Cross-sectional high-resolution TEM images taken along $[10\text{-}10]_{Al2O3}$ of a) $Ni_{1-x}Ga_x$ collectors after 15 s growth, and b) a sapphire substrate covered by Ni collectors after 900 s Ga exposure. Insets are the corresponding 2D Fourier transforms of the Ni–Ga collectors. In (a) and (b) left, the collectors have an fcc or $L1_2$ structure, while in b) the right one has the bcc structure. (Reprinted with permission from [CHÈ 10b]. Copyright 2010 American Chemical Society)*

Ex situ TEM investigation of dedicated samples for which growth was stopped after 15 s and 30 s was also carried out to evidence the presence of the different phases. After 15 s of growth corresponding to stage 1 (Figure 7.4(a)), a large number of collector seeds with the same structure and orientation determined by RHEED during stage 1 were detected but no GaN nuclei were found. This observation further confirmed that in stage 1 only Ga was incorporated into the Ni collectors while GaN did not grow. However, for the sample grown for 30 s (not shown) or for longer growth duration (inset in Figure 7.8(a)), the Ni collectors were found on top of GaN with the same structure as the sample grown for 15 s. This disagreement with the RHEED data for stage 2 may have resulted from the fairly short duration of this stage but also from the consumption of the Ga excess in the collectors by reactive N still present in the chamber during the cooling of the sample. This assertion was also corroborated by the difference in the Ga concentration range within the Ni collectors measured *in situ* by QMS and *ex situ* post-growth by electron energy loss spectroscopy [LAR 08a]. However, we observed Ni-collectors with β-NiGa bcc phase and the epitaxial orientation detected by RHEED in NW samples whose growth procedure was terminated under Ga excess (Figure 7.4(b) right and reference [KEH 08]). Thus, the β-NiGa bcc phase is effectively observable *ex situ*, but only in the case where a large amount of Ga is supplied just before the end of the growth procedure in order to preserve enough Ga inside the Ni–Ga collectors after the cooling. This finding underlines the importance of *in situ* studies.

Furthermore, we also found a remarkable agreement between the kinetically limited evolution of the collector phase observed in this study and the phase sequence predicted by Ni–Ga–N ternary phase diagrams between 500 and 850°C [HUL 04, GRÖ 99]. Indeed, the solubility of N into Ni is poor and neither ternary Ni–Ga–N, nor nickel-nitrides compounds should be stable at the temperature and pressure conditions used for our NW growth [WRI 91]. Moreover, in all of these ternary diagrams, NiGa is the first thermodynamically stable Ni–Ga alloy to present a tie-line with GaN as the Ga concentration is increased. In other words, only N_2 gas and Ni-rich Ni–Ga alloys can coexist if Ga/Ni ≤ 1. Hence, the phase transition of the collector observed in stage 2 *has* to take place prior to GaN precipitation, and our investigation provides direct evidence for this transition. In addition, the occurrence of the phase transition to high-Ga-content Ni–Ga alloys implies that at least during nucleation Ga diffusion proceeds mainly through the bulk of the collector particles and not along their surface.

The above findings enabled us to interpret the change in the Ga incorporation behavior as follows. Upon opening the Ga and N shutters, Ga readily alloyed with Ni, and all the incoming Ga was incorporated. Much less Ga desorbed from the Ni collectors than from the bare sapphire surface (Inset in Figure 7.1), so at this stage Ga incorporation into the collectors was not the rate-limiting step. As the Ga content in the collector particles rose, their chemical potential changed, and some of the arriving Ga desorbed instead of being incorporated. In stage 2, even more Ga desorbed on the Ni-covered than on the bare surface. Thus, at this point Ga incorporation and/or diffusion in the collectors could have been rate-limiting. Once the collector composition was Ga-rich, GaN nucleation could take place. However, in stage 3 the Ga desorption did not drop instantaneously but gradually and steady-state conditions were reached faster without the Ni collectors. Most likely, this behavior was due to a barrier for GaN nucleation underneath the collectors, similar to the case on bare SiC [KOB 02]. Indeed, as this barrier is the difference in free enthalpy involved in the formation of a stable nucleus, any nucleus of a size smaller than a critical value corresponding to the barrier height would dissociate. Eventually, all of the arriving Ga was incorporated, i.e. growth was limited by the supply, not by any process at the surface. We stress that the transformation of the collector's crystal structure during the VSS nucleation of these Ni-induced NWs may have important effects for the NW crystal structure and shape due to the modification of the particle diameter, contact angle, and bond orientation to the uppermost NW atomic layer [GLA 07, JEP 08].

7.4. Ni-induced GaN nanowire growth mechanism

This section evidences that the nucleation mechanism has direct consequences on the following elongation of the NWs. The careful study of the Ni-induced GaN NW growth on C-plane sapphire in dependence on the growth parameters indeed revealed that, conformal to the nucleation processes, the preferential accumulation of Ga over N atoms inside the Ni collectors leads to Ga-rich conditions and that the NW growth rate is thereby limited by the N-rate [GEE 07, CHÈ 10c].

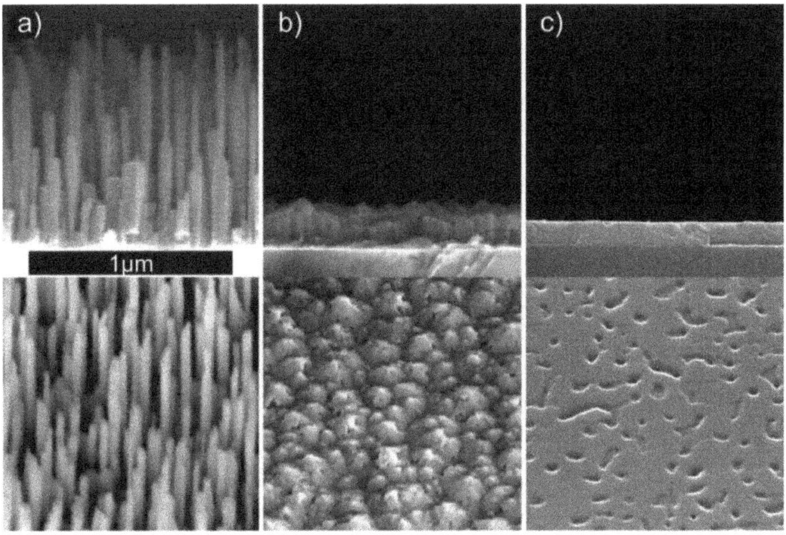

Figure 7.5. *SEM images of a) Ni-induced GaN NWs grown under N excess on C-plane sapphire, GaN layers grown, b) under N excess on bare sapphire, and c) under Ga excess on Ni-covered sapphire. Top and bottom rows correspond to cross-section and bird's eye views, respectively. (Figures (a)–(c) were adapted from [CHÈ 10d])*

SEM images of NW samples grown at 730°C are shown in Figure 7.5(a). The growth conditions were varied over a wide range in order to elucidate the NW elongation mechanism. The growth temperature was set to 730°C, because at this temperature the decomposition of GaN can be neglected. However, the NWs formed in MBE even at the growth temperature as high as 800°C. We found that the parameter essential to the NW growth is the V/III ratio that must be larger than 1, i.e. N-excess conditions must be established to promote the NW growth on sapphire in presence of the Ni collectors. Both growth under N-excess directly on a bare sapphire substrate (Figure 7.5(b)) and growth under Ga-excess in the presence of Ni-collectors (Figure 7.5(c)) yielded planar layers similar to the ones obtained on bare sapphire without Ni. After growth the samples were entirely covered by NWs oriented perpendicularly to the substrate. For a V/III ratio of 5, the diameter of these NWs was 35 ± 20 nm and their length 880 ± 140 nm, which is almost 5 times larger than the nominal thickness. Notice that this length was not completely homogeneous and few shorter NWs also grew between the long ones. The NW density measured from SEM images was 1.1×10^{10} NWs/cm^2.

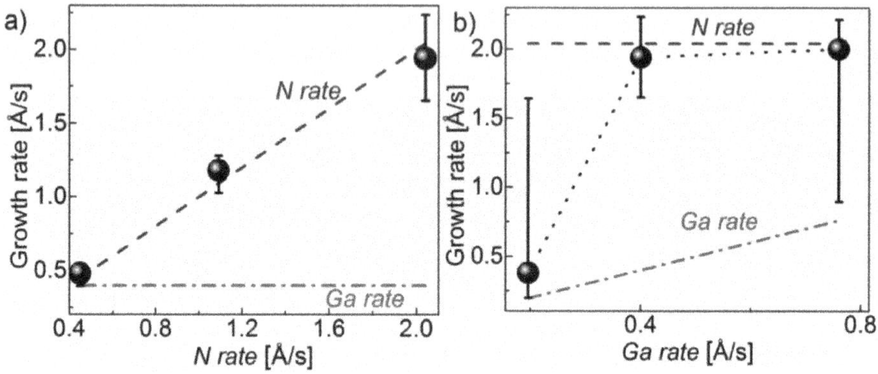

Figure 7.6. *NW growth rate of the Ni-induced GaN NWs a) as a function of the N-rate under constant Ga-flux, and b) as a function of the Ga-rate for NWs grown under constant N-flux. The nominal N(Ga)-rate is indicated by a dashed (dashed-dotted) line and corresponds to the growth rate of a planar layer of GaN on bare sapphire substrate under excess of Ga(N). (Figures (a) and (b) were adapted from [CHÈ 10d]). For a color version of this figure, see www.iste.co.uk/consonni/nanowires1.zip*

Figure 7.6 summarizes the NW growth rates deduced from NW lengths measured by SEM for various combinations of N- and Ga-rates. The error bars indicate the distribution in length of the NWs. The nucleation time was not taken into account in the measurement. However, at the growth temperature of 730°C the nucleation time measured with the QMS was shorter than 10 min for all the samples, which is negligible compared to the growth time of 76 min [CHÈ 10d]. Figure 7.6(a) shows the growth rate of the NWs as a function of the N-rate. The growth rate of the Ni-induced NWs increased with increasing N-rate and closely matched the N-rate, i.e. the growth rate of a planar layer in the N-limited growth regime. Figure 7.6(b) presents the growth rate of the NWs as a function of the Ga-rate. The lowest Ga-rate limits NW growth, presumably because the amount of Ga accumulating in the catalyst seed does not exceed the amount of impinging N. But once the Ga-rate is set over a critical value, the NW growth rate again followed the N-rate.

These results support the conclusion that the catalyst-induced NW growth is governed by the diffusion of Ga along the NW sidewalls and the accumulation of Ga into the Ni-collectors as also observed during the nucleation process [CHÈ 10b]. As the collectors are not enriched in N, their local environment is Ga-rich.

[GEE 07,CHÈ 10c, CHÈ 10d] Thus, as N is locally the species in minority at the NW tip, growth is limited by the impinging N atoms and proceeds at the rate of the supplied N, which is nevertheless still much higher than the impinging Ga rate. Under the globally N-rich conditions that are necessary for NW formation, the Ni-induced NWs thus elongate as if they were grown under Ga-excess.

7.5. Ni-induced GaN nanowire structural and optical properties

For optoelectronic applications it is of crucial interest to know how the material quality affects the optical characteristics of the NWs, and whether there is any difference in this respect between the growth approaches. The present section is therefore devoted to the structural characteristics of the Ni-induced GaN NWs, as well as their optical properties that are compared to the ones of self-induced NWs grown on Si.

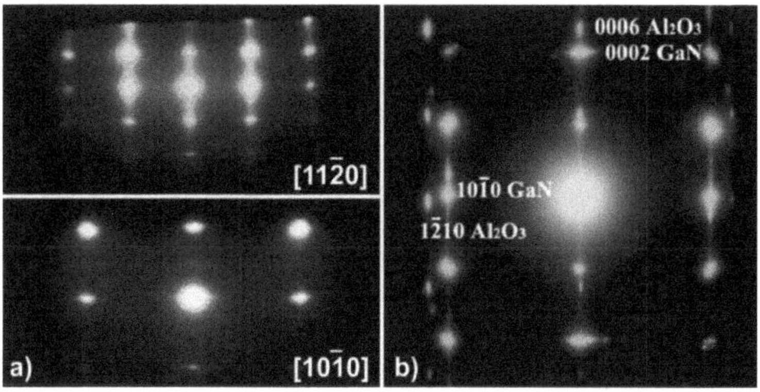

Figure 7.7. *a) RHEED patterns along the [11–20]$_{GaN}$ and [10–10]$_{GaN}$ directions, and b) selected-area electron diffraction pattern along [11-20]$_{GaN}$ of the Ni-induced GaN NWs. (Figures (a)–(c) were adapted from [CHÈ 10d], and c) by courtesy of T. Kehagias, Department of Physics, Aristotle University of Thessaloniki, Greece). For a color version of this figure, see www.iste.co.uk/consonni/nanowires1.zip*

The epitaxial growth of the Ni-induced NWs on C-plane sapphire suggested by their vertical orientation was confirmed by RHEED during the nucleation and at the end of the growth. The RHEED patterns acquired at the end of the growth (Figure 7.7(a)) were spotty as expected for rough surfaces, and the well defined spots corresponded to the WZ GaN structure observed in transmission through the NW crystal [CAL 07, CHÈ 10a, GEE 11]. The data revealed that the NWs grow along the C-axis. The patterns corresponding to the [10-10]$_{GaN}$ and [11-20]$_{GaN}$ azimuths clearly alternated every 30° when the sample was rotated around the [0001] direction, which was a sign for good in-plane orientation. This orientation was as for planar GaN growth on C-plane sapphire substrates: $(0001)_{GaN} \| (0001)_{Al2O3}$ and $[11\text{-}20]_{GaN} \| [10\text{-}10]_{Al2O3}$. This epitaxial orientation was also confirmed by TEM (Figure 7.7(b), [KEH 08]).

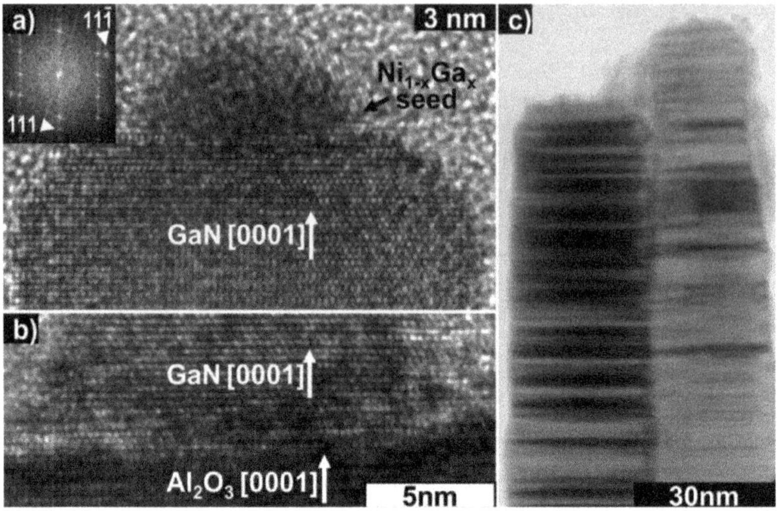

Figure 7.8. *High-resolution cross-sectional TEM images along the (11-20) zone axis of Ni-induced NWs of a) the tip surmounted by the Ni-seed, and b) the sharp interface between the sapphire substrate and the GaN NW. c) Cross-sectional TEM image of harvested Ni-induced GaN NWs showing basal SFs within the NWs. In a) the inset is the 2D Fourier transform of the Ni–Ga seed with the fcc or $L1_2$ structure. (Figures (a)–(c) were adapted from [CHÈ 10d] by courtesy of T. Kehagias, Department of Physics, Aristotle University of Thessaloniki, Greece)*

TEM also revealed the single crystalline nature and WZ structure of the GaN NWs (Figure 7.8(a)). The sharp interface between the GaN NWs and the sapphire substrate without any interfacial layer was evidenced (Figure 7.8(b)). In addition, basal-plane stacking faults (SF) were detected in most of these GaN NWs with the density of 2×10^6/cm but no other extended defects were found (Figure 7.8(c)). SFs of I1 nature [GEE 07, CHÈ 10a, LAR 08b] ran parallel to the (0001) plane through the whole NW diameter, and apparently formed with a random distribution along the NW axis. Very importantly, the observation of the Ni compound particles of 5–20 nm in diameter at the NW tips brought out the confirmation that the NWs grow on sapphire by a VLS or VSS mechanism. Similarly to nucleation stage 1, the fcc or the $L1_2$ structure was revealed by post-growth TEM analysis [CHÈ 10d] corresponding either to Ni or to the phase α'-Ni_3Ga, respectively. This result further supported the reversed phase transformation mechanism from the bcc-NiGa phase to the fcc-like phase by Ga purge after growth completion.

In marked contrast to the Ni-induced NWs, self-induced NWs grown on Si were almost free of any SFs [CAL 07, CHÈ 10d]. The SFs in the Ni-induced NWs were found up to the tip, and according to PL results, both types of NWs were free of strain. Thus, the different types of substrates, Si and sapphire, were not causing the discrepancy in crystal quality provided by the two approaches. Instead, there are several possible explanations for the formation of SFs which are all related to the presence of the Ni-collectors at the NW tip where growth takes place. First, the collector material may have affected the crystal structure of the NW material. As shown in the first section, the solid collector had a cubic NiGa structure during NW growth. Thus, the binding configuration of the GaN nucleus at the NW-collector interface may have adopted the zinc blende (ZB) configuration because it enabled a more favorable bond orientation between the Ga-top layer and the cubic GaNi. Second, Ni may also have been incorporated and stabilized the ZB phase by the reduction of the ZB–WZ structural energy difference, similarly to the incorporation of impurities in GaN, like Si, Mg, In and C [CHI 01], and as has been also experimentally evidenced by the doping with Mg of GaN NWs [CIM 06]. Third, the triple phase line GaN–GaNi–vapor introduced by the Ni-collectors [GLA 07] may have been the location of preferential nucleation for a more favorable ZB phase. Fourth, adatoms of a lower surface mobility at the NW tip could have been trapped at ZB sites [TAR 97, ZYW 98], thereby originating SFs. However, in this case, the absence of SFs in the self-induced NWs also directly pointed at the Ni particles as the origin of the reduced adatom kinetics and the formation of SFs in the Ni-induced NWs.

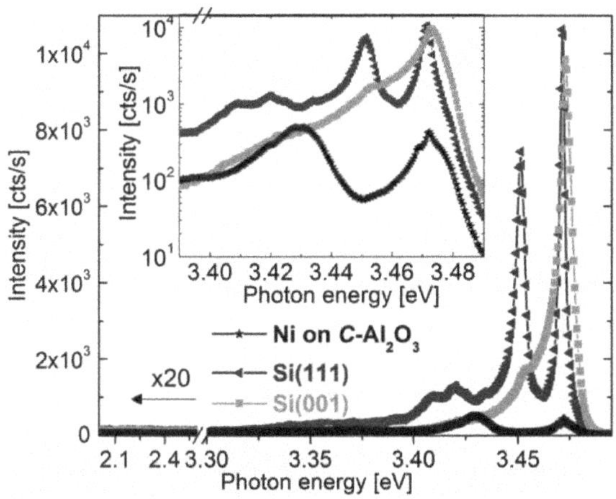

Figure 7.9. *Low temperature (10 K) PL spectra of GaN NWs grown on a sapphire substrate with Ni seeds, on Si(111), and on Si(001). Inset: Detail of the near band edge (semi-log scale). (Reprinted with permission from [CHÈ 10a], Creative Commons license 2.0 CC-BY 2010 Springer Science). For color version of this figure, see www.iste.co.uk/consonni/nanowires1.zip*

Moreover, low temperature (10 K) PL spectra from Ni-induced GaN NWs and from self-induced GaN NWs grown on Si(111) and Si(001) were collected and their intensity was calibrated by comparison with that of a high quality GaN film taken as a reference sample (Figure 7.9). Note that these NWs were grown at 780°C in order to get optimum characteristics. The spectra of all the samples revealed a clear emission line at 3.472 eV related to the donor-bound exciton transition (D^0, X), characteristic of strain-free GaN [ROB 07]. The sharpest peak was obtained for the NWs grown on Si(111), and had a full-width at half-maximum of 3.1 meV. None of the samples exhibited the yellow band, which was an indication of very low densities of the related point defects for both pathways. However, there was a strong difference in the luminescence intensity of the NWs: the overall integrated intensity and the peak intensity of the (D^0,X) line of the NWs obtained by the catalyst-free approach were, respectively, 6 and 25 times higher than the corresponding values for the Ni-induced NWs. Additional emission lines were detected at 3.43 eV for the catalyst-induced NWs, while for the catalyst-free NWs emission features centered at 3.41–3.42 eV and at 3.45 eV were resolved (Figure 7.9, inset). For all these samples there appeared to be a multitude of transition lines in the range of 3.41–3.43 eV. Actually, emission in this energy range was assigned to SFs in bulk [LIU 05], as well as in self-induced NWs [CAL 00, CON 09] where the I_1-type basal-plane SFs were observed to follow boundary dislocations resulting from NW coalescence [CON 09, for more details, refer to Chapter 8 of this book]. In Ni-induced GaN NWs a dominant emission at 3.437 eV was tentatively attributed to neutral acceptor-bound excitons related to Ni contamination [YOO 06]. As discussed above, the main difference in the crystal quality between the two approaches was the occurrence of SFs. Thus, a shift from 3.41–3.42 eV for the self-induced NWs to 3.43 eV for the Ni-induced NWs may have been caused by SFs, and/or Ni contamination. The emission of the peak in the 3.41–3.43 eV region was much stronger compared to that of the (D^0,X) transition for the Ni-induced NWs than for the self-induced NWs. Also, no emission lines related to Ni in GaN have been unambiguously identified so far, and SFs are known to be radiative defects. Therefore, it seems more reasonable to attribute the broad peak at 3.43 eV to SFs. Note that the emission at 3.45 eV was not resolved for the Ni-induced NWs. In agreement with former reports [CAL 00, ROB 07], we typically observed this line for the self-induced GaN NWs. Despite various mechanisms proposed to tentatively explain its origin (for more details, refer to Chapter 3), no firm conclusions were drawn.

The most striking result was that the luminescence of the catalyst-free NWs was about 25 times stronger than that of the catalyst-induced NWs, and this effect was observed in a systematic manner for different samples. A quantitative comparison of PL intensities is particularly difficult for such nanostructures as surface recombination is especially critical (for more details, refer to Chapter 3), but the difference between the self-induced and Ni-induced NWs was so large that it must be a real effect. Thus, in order to establish a valid comparison, we took the following observations into account. First, the morphologies of the samples were very similar. As the GaN NW lateral facet orientation is crucial to the luminescence characteristic from the GaN NWs [TES 11, CHI 07], we asserted that the lateral facet planes which constrict both types of NWs were identical, i.e. M-<10-10>. Second, the surface-to-volume ratio of both types of NWs differed by a factor of only 1.7, i.e. much less than the PL intensities. Thus, based on these observations, we asserted that the different recombination channels in these NWs could not originate from different defects characteristic of particular surfaces [CHI 07, TES 11]. Also, the NW diameters were smaller than the exciton diffusion length, and possible differences in light extraction efficiencies did not suffice as an explanation either. Hence, no significant influence on the emission intensity from the NW surface-to-volume ratio was expected. In addition, the occurrence of SFs in catalyst-induced NWs could explain additional emission lines but not the reduced PL intensity. Last, the position of the (D^0,X) peak indicated that in all cases the NWs were strain-free, so the effect of the different types of substrates was only limited to the base of the NWs. Therefore, these results strongly suggest the presence of an additional non-radiative recombination center in the Ni-induced NWs, and the most plausible explanation is the unintentional incorporation of Ni from the Ni-collectors.

7.6. Conclusion

We have investigated the nucleation and growth processes of GaN NWs induced by Ni-collectors on C-plane sapphire by MBE. We found that the Ni-collectors govern the whole processes by the preferential accumulation of Ga over N atoms. Moreover, the Ni collectors could have further effects on the NW properties. First, Ni contamination and/or the presence of an additional phase, i.e. the Ni-collectors, may originate the formation of SFs. Second, the direct comparison of growth routines in the same MBE chamber under identical conditions strongly suggests that Ni contamination is the cause of the significantly reduced PL intensity of the Ni-collector-induced NWs. The use of collector seeds made of foreign material may offer an alternative way to control the growth of NWs. But both the structural and optical qualities of the self-induced NWs are superior.

7.7. Acknowledgments

I would like to thank my former and current colleagues from the SINOPLE and NODE projects, as well as from Qimonda, Munich, and from the Paul-Drude-Institutfür Festkörperelektronik, Berlin, for their support in collecting the data presented in this chapter. I am also very grateful for the advice and guidance of H. Riechert and L. Geelhaar who critically read this manuscript. The financial support of the European Union through the IST project NODE Grant

No. 015783 and the Marie Curie RTN PARSEM Grant No. MRTN-CT-2004-005583 is also acknowledged.

7.8. Bibliography

[CAL 00] CALLEJA E., SÁNCHEZ-GARCÍA M.A., SÁNCHEZ F.J., *et al.*, "Luminescence properties and defects in GaNnanocolumns grown by molecular beam epitaxy", *Physical Review B*, vol. 62, no. 24, pp. 16826–16834, 2000.

[CAL 07] CALLEJA E., RISTIĆ J., FERNÁNDEZ-GARRIDO S., *et al.*, "Growth, morphology, and structural properties of group-III-nitride nanocolumns and nanodisks", *Physica Status Solidi B*, vol. 244, no. 8, pp. 2816–2837, 2007.

[CHÈ 10a] CHÈZE C., GEELHAAR L., BRANDT O., *et al.*, "Direct comparison of catalyst-free and catalyst-induced GaN nanowires", *Nano Research*, vol. 3, no. 7, pp. 528–536, 2010.

[CHÈ 10b] CHÈZE C., GEELHAAR L., TRAMPERT A., *et al.*, "Collector phase transitions during vapor–solid–solid nucleation of GaN nanowires", *Nano Letters*, vol. 10, no. 9, pp. 3426–3431, 2010.

[CHÈ 10c] CHÈZE C., GEELHAAR L., JENICHEN B., *et al.*, "Different growth rates for catalyst-induced and self-induced GaN nanowires", *Applied Physics Letters*, vol. 97, no. 15, pp. 153105-1–153105-3, 2010.

[CHÈ 10d] CHÈZE C., Investigation and comparison of GaN nanowire nucleation and growth by the catalyst-assisted and self-induced approaches, Doctoral thesis, Humbolt-Universitätzu, Berlin, 2010.

[CHI 01] CHISHOLM J.A., BRISTOWE P.D., "Ab initio study of the effect of doping on stacking faults in GaN", *Journal of Crystal Growth*, vol. 230, nos. 3–4, pp. 432–437, 2001.

[CHI 07] CHIN A.H., AHN T.S., LI H., *et al.*, "Photoluminescence of GaN nanowires of different crystallographic orientations", *Nano Letters*, vol. 7, no. 3, pp. 626–631, 2007.

[CIM 06] CIMPOIASU E., STERN E., KLIE R., *et al.*, "The effect of Mg doping on GaN nanowires", *Nanotechnology*, vol. 17, no. 23, pp. 5735–5739, 2006.

[CON 09] CONSONNI V., KNELANGEN M., JAHN U., *et al.*, "Effects of nanowire coalescence on their structural and optical properties on a local scale", *Applied Physics Letters*, vol. 95, no 24, pp. 241910-1–241910-3, 2009.

[CON 14] CONSONNI V., FEUILLET G., (eds.), *Wide Band Gap Semiconductor Nanowires 2: Heterostructures and Optoelectronic Devices*, ISTE, London, and John Wiley & Sons, New York, 2014.

[GEE 07] GEELHAAR L., CHÈZE C., WEBER W.M., *et al.*, "Axial and radial growth of Ni-induced GaN nanowires", *Applied Physics Letters*, vol. 91, no. 9, pp. 093113-1–093113-3, 2007.

[GEE 11] GEELHAAR L., CHÈZE C., JENICHEN B., *et al.*, "Properties of GaN nanowires grown by molecular beam epitaxy", *IEEE Journal of Selected Topics in Quantum Electronics*, vol. 17, no. 4, pp. 878–888, 2011.

[GLA 07] GLAS F., HARMAND J.C., PATRIARCHE G., "Why does wurtzite form in nanowires of III-V zinc blende semiconductors?", *Physical Review Letters*, vol. 99, no. 14, pp. 146101-1–146101-4, 2007.

[GRÖ 99] GRÖBNER J., WENZEL R., FISCHER G., *et al.*, "Thermodynamic calculation of the binary systems M-Ga and investigation of ternary M-Ga-N phase equilibria (M = Ni, Co, Pd, Cr)", *Journal of Phase Equilibria and Diffusion*, vol. 20, no. 6, pp. 615–625, 1999.

[HES 11] HESTROFFER K., LECLERE C., BOUGEROL C., *et al.*, "Polarity of GaN nanowires grown by plasma-assisted molecular beam epitaxy on Si(111)", *Physical Review B*, vol. 84, no. 24, pp. 245302-1–245302-6, 2011.

[HUL 04] HULL B., MOHNEY S., LIU Z., "Thermodynamic modeling of the Ni–Al–Ga–N system", *Journal of Material Research*, vol. 19, no. 6, pp. 1742–1751, 2004.

[JEP 08] JEPPSON M., DICK K., WAGNER J., *et al.*, "GaAs/GaSb nanowire heterostructures grown by MOVPE", *Journal of Crystal Growth*, vol. 310, no. 18, pp. 4115–4121, 2008.

[KEH 08] KEHAGIAS T., KOMNINOU P., DIMITRAKOPULOS G.P., *et al.*, "Atomic-scale configuration of catalyst particles on GaN nanowires", *Physica Status Solidi A*, vol. 12, no. 5, pp. 3716–3719, 2008.

[KOB 02] KOBLMÜLLER G., PONGRATZ P., AVERBECK R., *et al.*, "Delayed nucleation during molecular-beam epitaxial growth of GaN observed by line-of-sight quadrupole mass spectrometry", *Applied Physics Letters*, vol. 80, no. 13, pp. 2281–2283, 2002.

[LAR 08a] LARI L., MURRAY R.T., BULLOUGH T.J., *et al.*, "Nanoscale compositional analysis of Ni-based seed crystallites associated with GaN nanowire growth", *Physica E*, vol. 40, no. 7, pp. 2457–2461, 2008.

[LAR 08b] LARI L., MURRAY R.T., BULLOUGH T.J., *et al.*, "Defect characterization and analysis of III-V nanowires grown by Ni-promoted MBE", *Physica Status Solidi A*, vol. 205, no. 11, pp. 2589–2592, 2008.

[LIE 07] LIEBER C.M., WANG Z.L., "Functional nanowires", *MRS Bulletin*, vol. 32, no. 2, pp. 99–108, 2007.

[LIU 05] LIU R., BELL A., PONCE F.A., *et al.*, "Luminescence from stacking faults in gallium nitride", *Applied Physics Letters*, vol. 86, no. 2, pp. 021908-1–021908-3, 2005.

[MAM 99] MAMUTIN V.V., "Growth of A(3)-N whiskers and plate-shaped crystals by molecular-beam epitaxy with the participation of the liquid phase", *Technical Physics Letters*, vol. 25, no. 9, p. 741–744, 1999.

[OKA 08] OKAMOTO H., "Ga-Ni (gallium–nickel)", *Journal of Phase Equilibria and Diffusion*, vol. 29, no. 3, p. 296, 2008.

[POZ 99] POZINA G., BERGMAN J.P., MONEMAR B., *et al.*, "Optical and structural characterization of Ga(In)N three-dimensional nanostructures grown by plasma-assisted molecular beam epitaxy", *Physica Status Solidi B*, vol. 216, no. 1, pp. 445–450, 1999.

[RIE 96] RIECHERT H., AVERBECK R., GRABER A., *et al.*, "MBE growth of (In)GaN for LED applications", *Materials Research Society Symposium Proceedings*, vol. 449, p. 149, 1996.

[ROB 07] ROBINS L.H., BERTNESS K.A., BARKER J.M., *et al.*, "Optical and structural study of GaN nanowires grown by catalyst-free molecular beam epitaxy. I. Near-band-edge luminescence and strain effects", *Journal of Applied Physics*, vol. 101, no. 11, pp. 113505-1–113505-8, 2007.

[SEG 08] SEKIGUCHI H., KISHINO K., KIKUCHI A., "Ti-mask selective-area growth of GaN by RF-plasma-assisted molecular-beam epitaxy for fabricating regularly arranged InGaN/GaNnanocolumns", *Applied Physics Express*, vol. 1, no. 1, pp. 124002-1–124002-3, 2008.

[STA 87] STADELMANN P.A., "EMS – A software package for electron diffraction analysis and HREM image simulation in materials science", *Ultramicroscopy*, vol. 21, no. 2, pp. 131–146, 1987.

[TAR 97] TARSA E.J., HEYING B., WU X.H., *et al.*, "Homoepitaxial growth of GaN under Ga-stable and N-stable conditions by plasma-assisted molecular beam epitaxy", *Journal of Applied Physics*, vol. 82, no. 11, pp. 5472–5479, 1997.

[TES 11] TESSAREK C., HEILMANN M., CHRISTIANSEN S., Towards an InGaNnanorod based photovoltaic device: growth of GaNnanorods by metal-organic vapor phase epitaxy, Poster presentation at Next Generation Solar Energy 2011 – Erlangen.

[WAC 09] WACASER B.A., DICK K.A., JOHANSSON J., *et al.*, "Preferential interface nucleation: an expansion of the VLS growth mechanism for nanowires", *Advanced Materials*, vol. 21, no. 2, pp. 153–165, 2009.

[WAG 64] WAGNER R.S., ELLIS W.C., "Vapor-liquid-solid mechanism of single crystal growth", *Applied Physics Letters*, vol. 4, no. 5, pp. 89–90, 1964.

[WRI 91] WRIDT H.A., *Phase Diagrams of Binary Nickel Alloys*, in Nash P., (ed.), ASM International, Materials Park, pp. 213–218, 1991.

[YOO 06] YOO J., HONG Y.J., AN S.J., *et al.*, "Photoluminescent characteristics of Ni-catalyzed GaN nanowires", *Applied Physics Letters*, vol. 89, no. 4, pp. 043124-1–043124-3, 2006.

[YOS 97] YOSHIZAWA M., KIKUCHI A., MORI M., *et al.*, "Growth of self-organized GaN nanostructures on Al_2O_3 (0001) by RF-radical source molecular beam epitaxy", *Japanese Journal of Applied Physics*, vol. 36, Part 2, no. 4B, pp. L459–L462, 1997.

[ZYW 98] ZYWIETZ T., NEUGEBAUER J., SCHEFFLER M., "Adatom diffusion at GaN (0001) and (000-1) surfaces", *Applied Physics Letters*, vol. 73, no. 4, pp. 487–489, 1998.

8

Self-Induced Growth of GaN Nanowires by Plasma-assisted Molecular Beam Epitaxy

8.1. Introduction

GaN nanowires (NWs), also called nanocolumns or nanorods, benefit from their attractive physical properties originating from their high aspect ratio at nanoscale dimensions to form new three-dimensional nanostructures. They could act as building blocks in the next generation of low-cost optoelectronic devices on Si substrates owing to their high crystalline and optical quality [CAL 07, BER 11, LI 12]. Their integration has already been demonstrated in a wide variety of optoelectronic devices such as light-emitting diodes (LEDs) [KIK 04, SEK 08, KIS 12], lasers [JOH 02, GRA 05] or ultra-violet photo-detectors [CHE 07, RIG 10] (for more details, refer to Chapters 5–8 in Vol. 2 [CON 14]).

The growth of GaN NWs can be achieved by several chemical and physical deposition techniques: metal-organic vapor phase epitaxy (MOVPE, [KOE 10], for more details, refer to Chapter 10), hybrid vapor phase epitaxy (HVPE) [KIM 02] and plasma-assisted molecular beam epitaxy (MBE). The (molecular beam epitaxy) MBE growth as a versatile and intermediate temperature process represents a good compromise to form GaN NWs with a very high purity and to assemble them in complex heterostructures. Three different growth modes are usually considered in the MBE growth of GaN NWs: the catalyst-induced approach (for more details, refer to Chapter 7), the selective area growth (for more details, refer to Chapter 9) and the self-induced approach. While the former growth modes are typical for the formation of other III-V semiconductor NWs such as arsenides and phosphides, the

---

Chapter written by Vincent CONSONNI.

self-induced approach is peculiar to GaN NWs: these NWs can spontaneously form without the help of any foreign catalyst, III-element droplets or substrate prepatterning *via* expensive advanced lithography. In that sense, the self-induced term should be distinguished from the self-assisted or self-catalyzed terms used for other III-V semiconductor NWs. The self-induced growth of GaN NWs by MBE was first demonstrated in 1997 by the pioneering works of the groups of Katsumi Kishino [YOS 97, YOS 98] and Enrique Calleja [SAN 98, CAL 99] on Al_2O_3 and Si(111) substrates, respectively. GaN NWs are generally grown in the wurtzite crystalline structure along the polar *c*-axis. As a valuable new MBE growth mode leading to the formation of GaN NWs with remarkable physical properties, the self-induced approach has subsequently been developed by a wide number of groups over the last decade [TU 03, PAR 05, BER 06, CHE 06, MEI 06, SON 07, TCH 07, CHE 08, FUR 08, ARB 09, CON 09, CHE 10c, GOO 11, KUM 12, CHE 13, GRO 13, TAM 13, WIE 13, FAN 14]. However, a detailed fundamental understanding of the self-induced formation mechanisms of GaN NWs by MBE is critical to thoroughly control their structural, optical, electrical, and physical properties, which is a prerequisite for their integration in optoelectronic devices.

In this chapter, the nucleation and growth mechanisms of GaN NWs by MBE are reviewed in detail within the self-induced approach. The aim is to gain a fundamental overview of this new valuable MBE growth mode, which may favor in turn the fabrication of efficient optoelectronic devices made from GaN NWs. A special emphasis is placed on describing the formation mechanisms, from the very onset of the nucleation phase through the elongation phase of NW nuclei to the NW coalescence. Their correlation with the growth conditions is also emphasized as well as the theoretical approaches used for describing the underlying physical processes.

8.2. General principles

8.2.1. *MBE chamber*

The self-induced growth of GaN NWs can typically be achieved in an ultra-high vacuum MBE chamber, which is equipped with standard Knudsen thermal effusion cells and radio frequency (RF) plasma sources. The standard Knudsen thermal effusion cells can produce Ga species as the *III*-element and Si and Mg species as n-and p-type dopants, respectively [CAL 07, BER 11, LI 12]. The Ga species are produced by the direct evaporation of molten Ga metal in the standard Knudsen thermal effusion cell. The Ga flux can be monitored by a beam flux gauge, but its sensitivity varies with time as deposits accumulate on the active components. The formation of AlGaN and InGaN quantum dots or wells can also be considered with Al and In standard Knudsen thermal effusion cells ([KIK 04, SEK 08, KIS 12], Chapters 1 and 2 in Vol. 2). In contrast, the RF plasma sources can supply active N species as the *V*-element. A gas injector for the N_2 gas flow is linked to a RF plasma excitation: an electrical discharge is induced by coupled RF excitations at 13.56 MHz when the plasma source operates. The use of a Si photodiode permits to monitor the plasma discharge by following the spectral peak at about 850 nm. The output voltage from the Si photodiode can further be selected as a reference for the plasma source. In principle, the N species consist of active atomic N as well as excited and ionized molecules despite the use of biased plates to deflect any charge species in the plasma source. The relative amount of N species depends both on the RF power, N_2 gas flow rate, plasma pressure and source geometry. Furthermore, an optical pyrometer can be used for monitoring the growth temperature: it is often calibrated with the measurement of the Ga desorption at the given temperature of 675°C from the appearance and vanishing of 7 × 7 Si spotty reflection high-energy electron diffraction (RHEED) patterns. The MBE chamber is also often equipped with *in situ* characterization tools such as RHEED or line-of-sight quadrupole mass spectrometer (QMS). The RHEED equipment is a very powerful tool to investigate *in situ* the structural morphology and properties of nanostructures grown by PAMBE. The line-of-sight QMS directly gives information about the contaminants in the MBE chamber or more precisely about the Ga desorption rate, for instance, as growth proceeds. Field-emission scanning electron microscopy (FESEM), atomic force microscopy (AFM), transmission electron microscopy (TEM) and X-ray diffraction (XRD) are usually used for determining *ex situ* the structural properties.

8.2.2. *Typical growth conditions*

The self-induced growth of GaN NWs by PAMBE occurs in a fairly broad range of growth conditions as shown in Figure 8.1 [FER 09], which are related to the Ga- and N-rates as well as to the growth temperature [15-56]. The Ga-rate is typically controlled with the Ga effusion cell temperature while the N-rate is varied with the power of the RF plasma source (i.e. 400–500 W) and with the N_2 flux (i.e. 1–2 sccm). The Ga- and N-rates can be defined by the nominal Ga (respectively N) flux corresponding to the growth rate of GaN planar layer deposited on a SiC(111) substrate under N-rich (respectively Ga-rich) conditions. Overall, a highly N-rich vapor phase is required (with effective V/III ratio larger than 1 by taking desorption rates into account) and combined with a high growth temperature in the range of 730°C to 830°C for the self-induced formation of GaN NWs. In contrast, a two-dimensional (2D) compact layer is formed at lower growth temperatures, regardless of the V/III ratio since the diffusion of Ga adatoms is not favored on the surface. Furthermore, no growth takes place at higher growth temperature, regardless of the V/III ratio owing to massive Ga desorption and GaN decomposition.

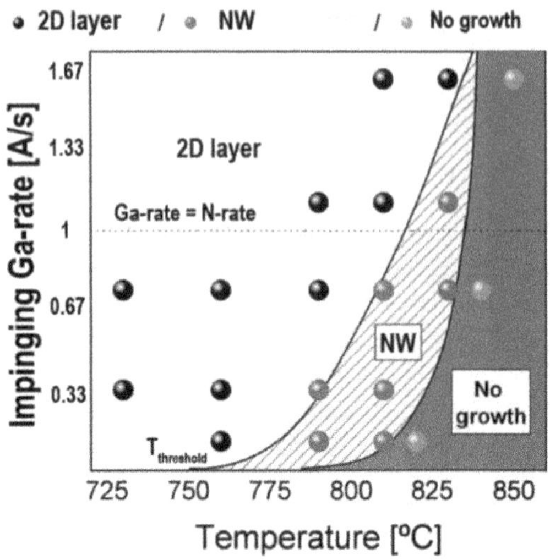

Figure 8.1. *Growth diagram showing the boundary between the growth regimes of 2D layer and NW as a function of impinging Ga-rate and growth temperature for a given N-rate of 1 Å/s (adapted from [FER 09]). For a color version of this figure, see www.iste.co.uk/consonni/nanowires1.zip*

8.2.3. *Nucleation surface effects*

The self-induced growth of GaN NWs has been demonstrated on a wide variety of substrates: *Si(111)* [SAN 98, CAL 99, CAL 00, TU 03, PAR 05, BER 06, CHE 06, MEI 06, VAN 06, BER 07, CAL 07, DEB 07, SON 07, TCH 07, BER 08, FUR 08, RIS 08, STO 08, CON 09, FER 09, LAN 09, CHE 10c, CHE 10b, CON 10, KNE 10, SON 10, BRU 11, CAR 11, CON 11a, CON 11b, CON 11c, GAL 11, GOO 11, MAT 11, CON 12, HES 12, KUM 12, LAR 12, CHE 13, HES 13, SAB 13, WIE 13, FAN 14, SOB 14a, SOB 14b], *Si(001)* [CER 06, CHE 10c, CHE 10b, TAM 13], Al_2O_3 [YOS 97, YOS 98, CAL 00, VAN 06, SEK 07, CHE 08, FOX 09], *diamond* [SCH 12] and *SiC* [FER 12]. Importantly, the formation of *AlN* [YOS 97, YOS 98, SAN 98, CAL 99, CAL 00, BER 06, VAN 06, BER 07, SEK 07, SON 07, BER 08, CHE 08, LAN 09, CON 10, KNE 10, BRU 11, GAL 11, FER 12, LAR 12, CHE 13, HES 13], *GaN* [CAL 00, TU 03, TAM 13] or *amorphous* Al_2O_3 [GOO 11, SOB 14a, SOB 14b] buffer layers as well as Si_xN_y [CAL 00, PAR 05, CHE 06, MEI 06, CAL 07, DEB 07, TCH 07, FUR 08, RIS 08, STO 08, CON 09, FER 09, CHE 10c, CHE 10b, SON 10, CON 11a, CON 11b, CON 11c, MAT 11, CON 12, KUM 12, HES 12, HES 13, SAB 13, WIE 13] or SiO_x [STO 08, FAN 14] interlayers is typically achieved for the growth on top of Si(111) or Si(001) substrates. Basically, the formation mechanisms of GaN NWs strongly depend on the physical properties of the nucleation surface: (1) its chemical nature, (2) its amorphous or crystalline nature driving lattice mismatch, wetting behavior and growth regime, and (3) its structural morphology mainly governing preferential nucleation sites. Most of the literature is dedicated to the self-induced growth on the AlN buffer layer and Si_xN_y interlayer.

The structural properties of the AlN buffer layer are generally driven by the choice of substrate, V/III ratio and growth temperature. At low growth temperature and under Al- or N-rich growth conditions on Si substrate, structural defects and morphological irregularities are formed on the AlN buffer layer such as threading dislocations, hexagonal matrix pits [BER 06, BER 07, BER 08], columnar protrusions [BRU 11] or pedestals [LAR 12, CHE 13], as shown in Figure 8.2.

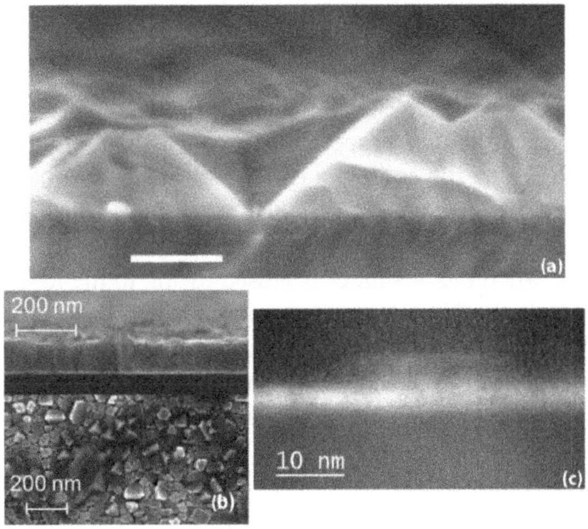

Figure 8.2. *Examples of morphological irregularities on the AlN buffer layer. a) Cross-sectional FESEM image of hexagonal pits (reprinted from [BER 07], copyright 2007, with permission from Elsevier); b) cross-sectional and top-view FESEM images of columnar protrusions(reprinted with permission from [BRU 11], copyright 2011, AIP Publishing LLC); c) high-angle annular dark field scanning TEM image of a pedestal (reprinted with permission from [LAR 12], copyright 2012 American Chemical Society)*

In contrast, at higher growth temperature and under Al-rich conditions on Si or SiC substrates, large 2D islands are formed with a much smaller root-mean square roughness on the AlN buffer layer, as revealed in Figure 8.3 [FER 12]. It is also worth noticing that the structural quality of the AlN buffer layer is dependent upon its thickness: thinner films of several nanometers are generally associated with smoother nucleation surfaces while thicker films of several tens of nanometers are rougher with the formation of morphological irregularities.

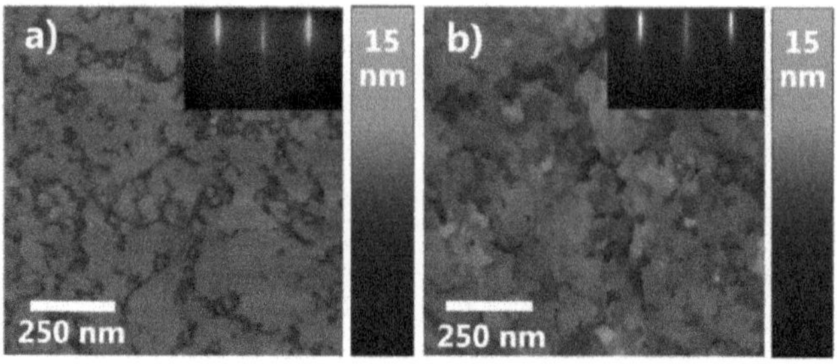

Figure 8.3. *AFM images of the AlN buffer layer grown on a) SiC(0001) and b) SiC(000-1) substrates. The insets show the RHEED pattern along the [11-20] azimuth at the end of the growth (reprinted with permission from [FER 12], copyright 2012 American Chemical Society). For a color version of this figure, see www.iste.co.uk/consonni/nanowires1.zip*

In contrast to the structural properties of the AlN buffer layer, the structural properties of the Si_xN_y interlayer are poorly documented. It should be noted that the self-induced growth of GaN NWs on Si(111) or Si(100) substrates systematically involves the formation of a Si_xN_y interlayer as soon as the RF plasma source is turned on and even when the N shutter is closed owing to the leakage of N species. As mentioned before, the N species are composed of atomic N but also of more complex excited and ionized molecules. In particular, if electrically connecting GaN NWs with the Si substrate is being looked after, the occurrence of the insulating Si_xN_y interlayer should be avoided as much as possible. However, its formation is inevitably favored by (1) the large difference between the Ga-N (2.2 eV) and Si-N (4.5 eV) bond energy and (2) the larger surface density of Si and N adatoms than of Ga adatoms (that more easily desorb), increasing the probability of the Si-N chemical reaction. The structural properties of the Si_xN_y interlayer are generally driven by the nitridation temperature [WIE 13]. An atomically smooth Si_xN_y amorphous interlayer is formed when nitridation is achieved at low temperature (~ 150°C). In contrast, flat islands are grown with a typical size in the range of 100 to 200 nm and separated by about 1 nm deep grooves when nitridation is performed at higher temperature (~ 850°C). Importantly, the chemical nature of the Si_xN_y interlayer has not completely been identified yet and both crystalline and amorphous domains may coexist [CHE 10c].

Over the last decade, it has been found out that two nucleation mechanisms generally proceed, depending on the structural morphology of the nucleation surfaces. Its control is crucial since it governs the structural morphology of NW nuclei and some of the final structural properties of GaN NWs such as their density as well as their smallest radius and length, for instance. When a large number of structural defects and morphological irregularities are formed on the nucleation surfaces, a seeded nucleation mechanism is favored. The morphological irregularities act as preferential nucleation sites and hence as a seed for the self-induced formation of GaN NWs, as shown, for instance, by the nucleation of GaN NWs within the hexagonal matrix pits in Figure 8.4 [BER 07, BER 08].

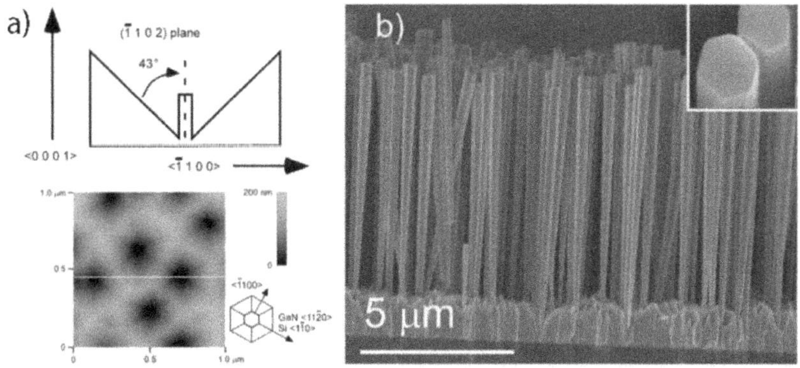

Figure 8.4. *a) Schematic diagram of a hexagonal pit and of GaN NW nucleation within a hexagonal pit. AFM image of hexagonal pits (reprinted from [BER 07], copyright 2007, with permission from Elsevier); b) cross-sectional FESEM images of GaN NWs grown on the AlN buffer layer with hexagonal matrix pits(reprinted from [BER 08], copyright 2008, with permission from Elsevier)*

In contrast, when the nucleation surfaces are much smoother (i.e. as long as a reduced number of structural defects and morphological irregularities are formed on the AlN buffer layer and on the Si_xN_y interlayer), a shape transition related nucleation mechanism is instead predominant and discussed in detail in the following. The preferential nucleation sites are here much less related to structural defects: a nucleation phase driven by thermodynamics results in the self-induced formation of GaN NW nuclei with a critical radius [SON 07, RIS 08, STO 08, LAN 09, CHE 10c, CHE 10b, CON 10, KNE 10, CON 11a, CON 11b, CON 11c, CON 12].

8.3. Nucleation phase

The nucleation phase of GaN NW nuclei is somewhat different from standard crystal growth (in which an initial critical nucleus is formed) in the sense that GaN does not nucleate in the form of NWs. Basically, two successive periods are included in the nucleation phase [SON 07, RIS 08, STO 08, LAN 09, CHE 10c, CHE 10b, CON 10, KNE 10, CON 11a, CON 11b, CON 11c, CON 12]: an initial incubation period corresponds to the nucleation of stable nuclei as clusters, followed by a transition period leading to the formation of NW nuclei after one or several shape transitions, as discussed in detail in this section.

8.3.1. *Incubation period*

8.3.1.1. *Experimental observations*

An incubation period typically occurs on both the AlN buffer layer and Si_xN_y interlayer at the very onset of the nucleation phase, when both N and Ga shutters are open [LAN 09, CHE 10c, CON 10, KNE 10, CON 11a, CON 11c, HES 12, HES 13]. *In situ* RHEED, line-of-sight QMS and grazing incidence X-ray diffraction (GIXRD) measurements with synchrotron radiation have been used to investigate its main characteristics [LAN 09, CHE 10c, CON 10, KNE 10, CON 11a, CON 11c, HES 12, HES 13]. Basically, several physical processes take place on the surface at the very onset of the nucleation phase: Ga and N atoms can adsorb onto, desorb from or diffuse on the surface [RIS 08].The Ga desorption rate roughly corresponds to the Ga impinging flux during the incubation period, indicating that only very few Ga adatoms are incorporated [CHE 10c]. Metastable 2D GaN nuclei are instead expected to form since the probability of creating a stable nucleus with a critical radius is low at growth temperatures higher than 750°C. The incubation period is, therefore, a metastable phase, during which no stable nucleus is formed. It can be characterized by an incubation time, which can vary in a typical range of several tens of seconds to several tens of minutes. Importantly, the incubation time strongly depends on growth conditions [LAN 09, CON 11c, LIM 12] such as growth temperature or Ga-rate, as shown in Figure 8.5 [CON 11c]. The supply of Mg atoms for p-type doping also shortens drastically the incubation time for identical growth conditions, increasing the nucleation rate of stable nuclei [LIM 12]. The incubation period typically results in a delay in the nucleation phase and eventually ends when stable nuclei are formed.

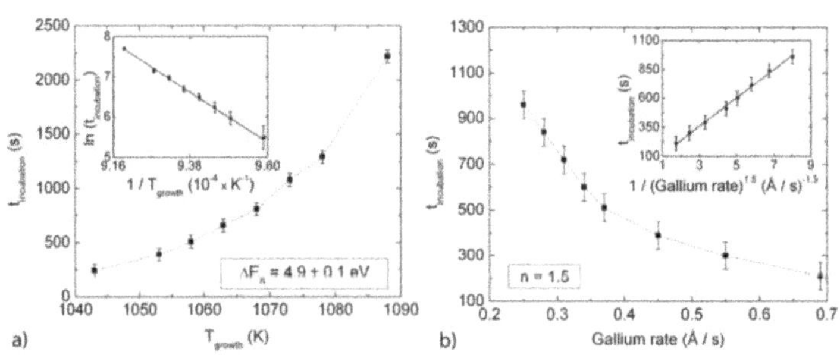

Figure 8.5. *a) Evolution of the incubation time with the growth temperature. The Ga-rate is 0.45 Å/s, which corresponds to a V/III ratio of 6.2. The inset represents the variation of* $\ln(t_{incubation})$ *versus* $1/T_{growth}$ *desired from equation [8.1] in which a nucleation energy of 4.9 ± 0.1 eV is inferred from the fit (i.e. solid line); b) evolution of the incubation time with the Ga-rate. The growth temperature is 780°C. The inset represents the variation of* $t_{incubation}$ *vs* $1/R^{1.5}$*, deduced from equation [8.1] in which a nucleus critical size-dependent exponent p(n) of 1.5 is inferred from the fit (i.e. solid line) (reprinted with permission from [CON 11c], Copyright 2011, AIP Publishing LLC)*

If the growth is directly achieved on the AlN buffer layer, stable nuclei are not NWs but instead develop as clusters, which can be approximated by spherical cap-shaped islands, as shown in the high-resolution transmission electron microscopy (HRTEM) images of Figure 8.6 [CON 10]. The preferential nucleation sites may or not be related to structural defects such as crystallite boundaries [SEK 07, LAN 09] [FER 12]. Furthermore, the formation of a 2D wetting layer is still a matter of debate [SON 07, LAN 09, CON 10, KNE 10] and may depend on the Ga-rate and growth temperature.

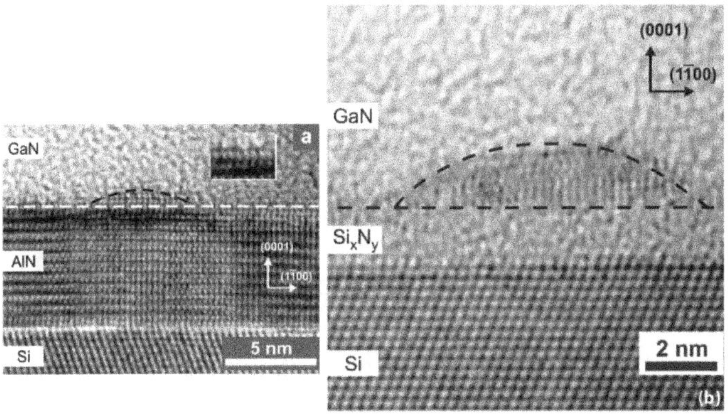

Figure 8.6. *HRTEM images of spherical cap-shaped islands formed at the end of the incubation period during the nucleation phase on a) the AlN buffer layer (reprinted with permission from [CON 10], copyright 2010 by the American Physical Society) and b) the Si_xN_y interlayer (reprinted with permission from [CON 11a], copyright 2011 by the American Physical Society)*

Importantly, for the growth involving the Si_xN_y interlayer, the incubation time also depends on the experimental procedure used in the MBE chamber in addition to growth conditions. When nitridation is initially performed with active N species and followed by the supply of Ga atoms to start the growth, stable nuclei develop as clusters [STO 08, CON 11a], which can again be approximated by spherical cap-shaped islands as revealed in the HRTEM image of Figure 8.6 [CON 11a]. The preferential nucleation sites may be surface defects such as surface steps or roughness on the Si_xN_y interlayer. In contrast, when both nitridation with active N species and start of the growth with the supply of Ga species are achieved simultaneously, the incubation period is somehow more complicated [CHE 10c]. In this case, a β-Si_3N_4 crystalline layer is initially formed, similarly to other nitridation methods such as electron-dissociated N_2 [TAB 97], NH_3 [WU 02], NO [WAN 01], and subsequently amorphized [HES 12]. The incubation period includes both the formation and amorphization phases of the β-Si_3N_4 crystalline layer as well as the metastable phase for the nucleation of stable nuclei. The formation and amorphization phases last about 1 min in a conventional MBE chamber and do not depend on growth temperature [HES 13]. The amorphization of the *β*-Si_3N_4 crystalline layer is thus required for the nucleation of stable nuclei, as shown by the increase in the GaN XRD signal as the *β*-Si_3N_4 XRD signal vanishes in Figure 8.7 [HES 12]. The amorphization phase has tentatively been assigned to the superficial implantation of residual, low-energy N ions in the N plasma [HES 12, HES 13]. Interestingly, the need for the amorphization of the *β*-Si_3N_4 crystalline layer can be correlated with the self-induced growth of GaN NWs on Al_2O_3 surfaces: while GaN NWs can nucleate on the amorphous Al_2O_3 buffer layer, their formation is hampered on crystalline Al_2O_3 substrates [SOB 14a].

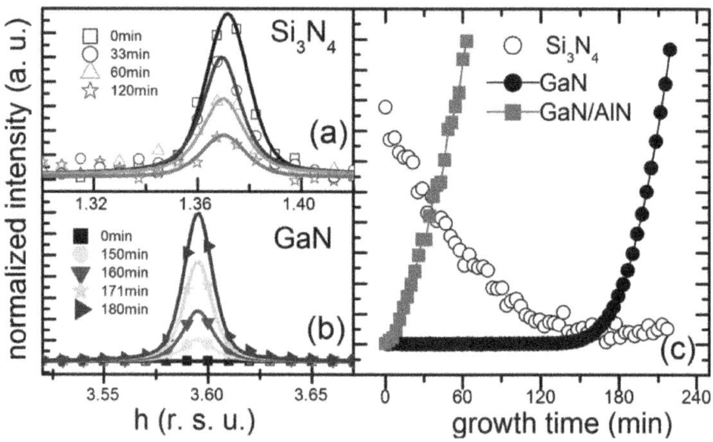

Figure 8.7. *GIXRD scans around a) 10-10 and b) 30-30 reflections during the nucleation phase, showing the evolution of β-Si_3N_4 and GaN peaks, respectively. The peak integrated intensity is reported versus growth time in c). Open and full dots represent the evolution of the β-Si_3N_4 and GaN peak integrated intensity, respectively. For comparison, full squares represent the evolution of the GaN peak integrated intensity when GaN NWs are grown on the AlN buffer layer (reprinted with permission from [HES 12], copyright 2012, AIP Publishing LLC). For a color version of this figure, see www.iste.co.uk/consonni/nanowires1.zip*

Furthermore, the presence of clusters on top of or even embedded in the Si_xN_y interlayer is shown in the HRTEM images of Figure 8.8 [CHE 10c], revealing that the incubation period is fairly complicated in the present case.

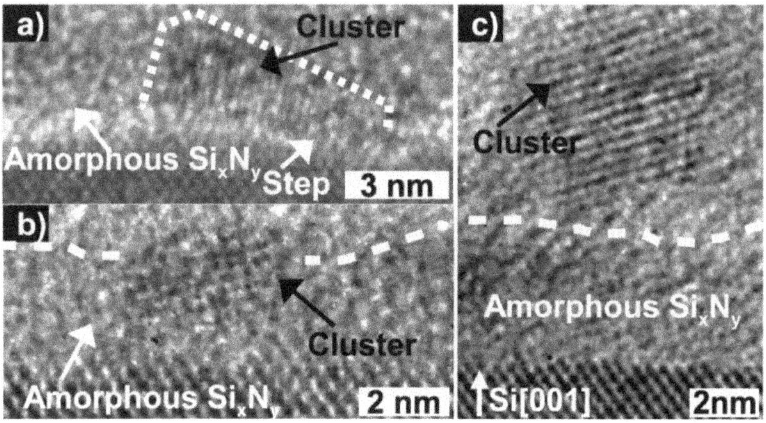

Figure 8.8. *Cross-section HRTEM images of a dedicated sample grown on Si(001) at the very onset of the nucleation phase. The sample is covered with GaN clusters a) in contact with the underlying Si substrate at a step, b) embedded in the amorphous Si_xN_y interlayer and c) on top of the amorphous Si_xN_y interlayer (reprinted with permission from [CHE 10c] Copyright 2010, AIP Publishing LLC)*

8.3.1.2. *Theoretical modeling*

The incubation period can be described via the incubation time $t_{\text{incubation}}$ in the framework of standard island nucleation theory [CON 11c]:

$$t = 1/(IS) \text{ where } I = I_0 R^{p(n)} \exp\left(\frac{\Delta E_n}{k_B T_{growth}}\right) \qquad [8.1]$$

where I is the nucleation rate, S is the surface area, R is the Ga atom arrival rate per unit surface area and unit time, T_{growth} is the growth temperature and I_0 is a R- and T_{growth}-independent constant. Also, $\Delta E_n = (n+1)E_D - E_{SD} + E_n$ is the nucleation energy barrier where E_D is the desorption activation energy, E_{SD} is the surface diffusion activation energy, E_n and n are the dissociation energy and the number of atoms in a critical-sized nucleus. p(n) is a nucleus critical size-dependent exponent. The plots of $\ln(t_{\text{incubation}})$ *vs.* $1/T_{\text{growth}}$ and $t_{\text{incubation}}$ *vs.* $1/R^{p(n)}$ allows the determination of important parameters: both a nucleation energy of 4.9 ± 0.1 eV and a very small nucleus critical size has been deduced, as shown in Figure 8.5 [CON 11c]. The orders of magnitude are quite usual for MBE growth.

8.3.2. *Transition period*

8.3.2.1. *Experimental observations*

Following the incubation period corresponding to the time needed for the formation of stable nuclei, a transition period is subsequently characterized by the coarsening of stable nuclei and by one or several shape transitions eventually leading to the self-induced formation of NW nuclei [LAN 09, CON 10, KNE 10, CON 11a]. The main characteristics of the transition period are strongly affected by the physical properties of the nucleation surface, such as its chemical nature and its amorphous or crystalline nature.

For the growth on the AlN buffer layer, the transition period is deeply related to the relaxation of the strong epitaxial constraint, the AlN/GaN lattice mismatch being 2.5% [LAN 09, CON 10, KNE 10]. The epitaxial strain is actually relieved in two consecutive stages, as shown in Figure 8.9(e) [KNE 10]. An initial, partial elastic relaxation takes place through successive first-order shape transitions: spherical cap-shaped islands coarsen and transform into truncated pyramid-shaped islands and then into full pyramid-shaped islands at critical sizes, as revealed in Figure 8.9(a)–(c) [CON 10]. The residual epitaxial strain is reduced through each island shape transition, as revealed by the successive plateau in Figure 8.9(e) [KNE 10]. Subsequently, plastic relaxation occurs when the stored elastic strain energy is large enough: a misfit dislocation nucleates at the edges of full pyramid-shaped islands, where the stress field is maximal and moves towards a quasi-central position to completely relieve the residual epitaxial strain [CON 10]. A final shape transition is eventually induced, leading to the self-induced formation of NW nuclei. Interestingly, the theoretical and experimental critical radius of 13 nm for the occurrence of plastic relaxation within full pyramid-shaped islands is equal to the initial radius of NW nuclei [CON 10]. This indicates that NW nuclei do contain at least one misfit dislocation at the interface with the AlN buffer layer and cannot form below a critical radius of 13 nm driven by the AlN/GaN lattice mismatch [CON 10, KNE 10]. This emphasizes the critical role of plastic relaxation as a necessary condition for the self-induced formation of NW nuclei [LAN 09, CON 10, KNE 10].

For the growth on the Si_xN_y interlayer, the epitaxial constraint is not expected or is at least much weaker. The transition period is not affected by strain relaxation effects, although the Laplace pressure-induced strain is significant for nanosized islands [CON 11a]. In contrast to the growth on the AlN buffer layer, spherical cap-shaped islands coarsen by retaining their shape and hence undergo abruptly one single first-order shape transition at the smaller critical radius of 5 nm, as shown in Figure 8.10 [CON 11a].

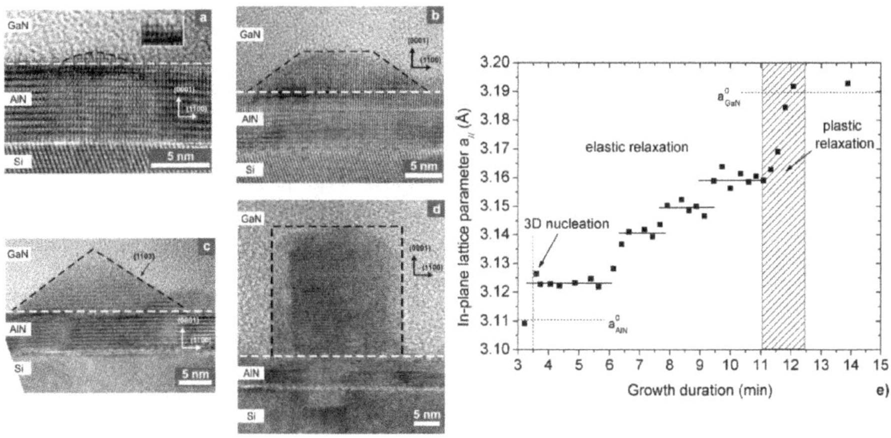

Figure 8.9. *HRTEM images collected on dedicated samples grown on the AlN buffer layer with different durations in the transition period during the nucleation phase, revealing the following respective GaN island shapes; a) a spherical cap-shaped island with an inset representing a high magnification of the first AlN monolayers at the interface; b) a truncated pyramid-shaped island; c) a full pyramid-shaped island; d) a NW nucleus. The pyramid-shaped island and NW are hexahedral: the full pyramid-shaped island presents a facet angle of 32° with respect to the (0001) plane, suggesting that the (1-103) side facet is involved (reprinted with permission from [CON 10], copyright 2010 by the American Physical Society). e) Evolution of the GaN in-plane lattice parameter from RHEED patterns as a function of growth duration. The bulk lattice constants of GaN and AlN are indicated by dashed lines (from [KNE 10] Copyright 2010, 10P Publishing, Reproduced by permission of 10P Publishing All rights reserved)*

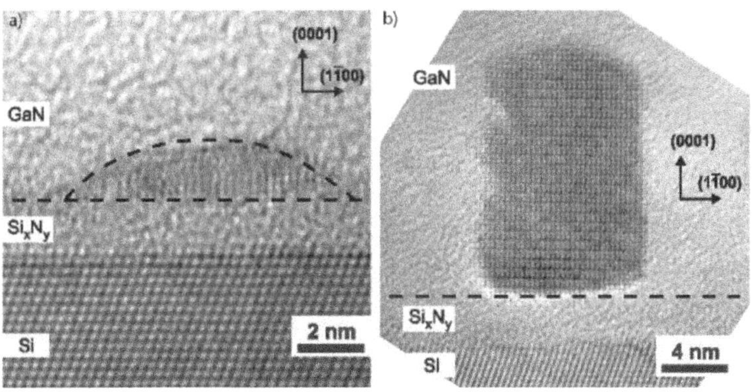

Figure 8.10. *HRTEM images collected on dedicated samples grown on the amorphous Si_xN_y interlayer with different durations in the transition period during the nucleation phase, revealing the following respective GaN island shapes: a) a spherical cap-shaped islands and b) a NW nucleus. The NW is hexahedral (reprinted with permission from [CON 11a], copyright 2011 by the American Physical Society)*

On the one hand, the absence of faceting process somehow excludes the role of a significant epitaxial constraint, which is in correlation with the HRTEM images showing that the Si_xN_y interlayer would be amorphous [FUR 08, STO 08, CON 09, CHE 10c, CON 11a, CON 11b, CON 11c]. On the other hand, GaN NWs grown on the Si_xN_y interlayer exhibit a strong in-plane orientation with a typical twist angle of several degrees. The in-plane orientation of GaN NWs is strongly dependent upon nitridation temperature, as shown by X-ray pole figures in Figure 8.11 [WIE 13]. In addition, GaN NWs grown on the amorphous Al_2O_3 buffer layer do not have any in-plane orientation either [SOB 14b]. The apparent discrepancy on the Si_xN_y interlayer is still a matter of debate but may be reconciled as follows. The epitaxial information from Si substrate may be transferred across the thin Si_xN_y interlayer [CAL 07], but the exact underlying physical processes have not been identified yet [RIS 08]. More likely, both amorphous and crystalline domains may be mixed in the Si_xN_y interlayer [CHE 10c, HES 12, WIE 13]. Furthermore, the presence of grooves in the amorphous Si_xN_y interlayer may also give rise to the direct nucleation on crystalline Si substrate.

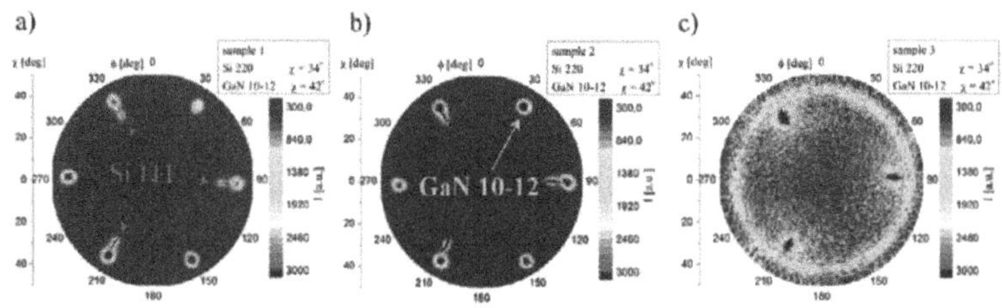

Figure 8.11. *XRD 10-12 pole figures of a) sample 1 nitridized at 850°C, b) sample 2 nitridized at 450°C and (a) sample 3 nitridized at 150°C, showing the Ga NW orientations relative to Si(111) substrate surface. The three peaks at κ = 34° originate from Si 220 reflection while the six peaks at κ = 42° come from GaN 10-12 reflection (from [WIE 13] © 2013 IOP Publishing. Reproduced by permission of IOP Publishing. All rights reserved). For a color version of this figure, see www.iste.co.uk/consonni/nanowires1.zip*

8.3.2.2. *Theoretical modeling*

Surface energy anisotropy has been suggested as being the main driving force for the final shape transition towards NW nuclei for both the growth on the AlN buffer layer [CON 10] and on the Si_xN_y interlayer [CON 11a, DUB 12a]. The NW nuclei are composed of polar *c*-plane top facets and non-polar *m*-plane vertical sidewalls, as revealed by X-ray pole figures and HRTEM images [BER 08, LAR 08, CHE 10c, CON 12]. In the wurtzite crystalline structure, the calculations of surface energy from *ab-initio* simulations are not straightforward: for instance, when the chemical bonding is broken along the *c*-axis, the *c*-plane surface energy cannot be precisely determined due to the occurrence of distinct Ga and N polarity. The average of the −*c*- and +*c*-plane surface energies can, however, be calculated and is about

123 meV/Å^2 in the Ga-rich limit [NOR 99]. In contrast, it is well known that the *m*-plane surface energy of 118 meV/Å^2 is typically one of the lowest surface energies together with the *a*-plane surface energy of 121 meV/Å^2 [NOR 96]. In the framework of thermodynamic considerations, a scaling thermodynamic model has been developed to account for the nucleation mechanisms on the amorphous Si_xN_y interlayer [DUB 12a]. It is based on the comparison of the formation enthalpy of spherical cap-shaped islands and NWs with the same volume. By inserting the scaling law as revealed in Figure 8.12(a) in the formation enthalpy difference, the driving force for the final shape transition has been found to be surface energy anisotropy, provided that the *m*-plane surface energy is low enough, as shown in Figure 8.12(b) [DUB 12a]. Also, edge energy has been proposed to act as an energy barrier since it usually increases with the inclination angle of island facets.

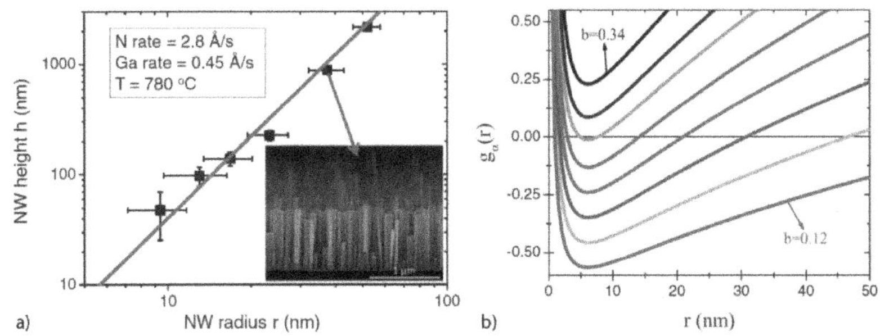

Figure 8.12. *a) Super-linear dependence $h = 0.14r^{2.46}$ of GaN NW height h on radius r from statistical analyses of FESEM and TEM images. The inset is a typical cross-sectional FESEM image of GaN NWs; b) normalized difference g_α in surface energy between spherical cap-shaped islands and NWs with the same volume as a function of NW radius r for several values of NW sidewall surface energy-dependent factor b (reprinted with permission from [DUB 12a], copyright 2012 by the American Physical Society). For a color version of this figure, see www.iste.co.uk/consonni/nanowires1.zip*

8.4. Growth phase

Following the nucleation phase leading to the self-induced formation of NW nuclei, their elongation subsequently takes place during the growth phase and is governed by kinetics, as discussed in detail in the present section [CAL 07, DEB 07, BER 08, RIS 08, SON 10, MAT 11, CON 12].

8.4.1. *Elongation period*

The elongation of GaN NW nuclei during the growth phase mainly originates from the adsorption and adatom surface diffusion on the substrate surface and on the NW vertical sidewalls, as well as from geometrical effects. In addition, although the former physical processes on the surfaces can be described in one single isolated GaN NW, it turns out that shadowing and collective effects in the GaN NW ensemble can also play a significant role on the elongation period.

8.4.1.1. *Role of adatom surface diffusion and incorporation rates*

Many physical processes are involved in the elongation of GaN NWs, both on the substrate surface and on the NW top facet and vertical sidewalls. The adsorption of Ga and active N species from the direct impinging fluxes, their desorption or the surface diffusion of Ga adatoms are all critical processes in the self-induced approach just like in the catalyst-induced approach [RIS 08]. These processes also depend strongly on growth conditions such Ga-rate and growth temperature, for instance. The NW elongation is generally limited by Ga species, and the axial growth rate is accordingly smaller than N-rate [SAN 98, SON 07, FER 09, CHE 10b, SON 10, CON 12]. More rarely, the NW elongation has also been reported as limited by N species [YOS 97, TCH 07, SON 07, LAN 08, SON 10, FER 13]. When limited by Ga species, the axial growth rate is systematically larger than the Ga-rate, showing that the NW elongation is assisted by the surface diffusion of Ga adatoms on the NW vertical sidewalls. Therefore, at a given time, the NW length is inversely proportional to its diameter [CAL 07, DEB 07, STO 08, DUB 12b, SIB 12, DUB 13], as revealed in Figure 8.13 [DEB 07]. This indicates that the axial growth rate is increased as NWs are thinner.

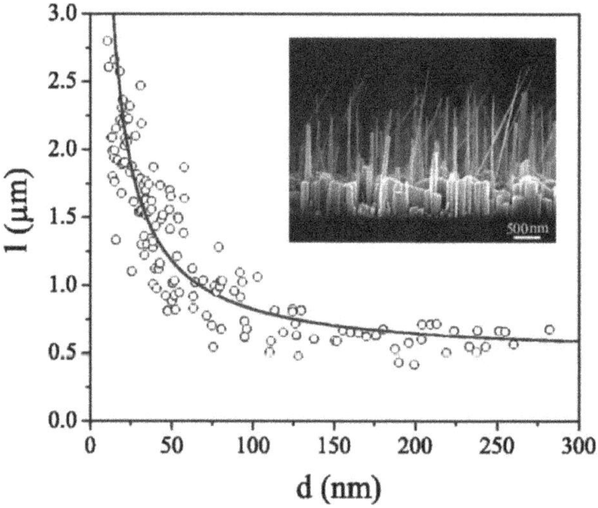

Figure 8.13. *NW length L and diameter d dependence for a given growth time of 360 min: open dots represent experimental data while blue line yields the theoretical fit from the relation: $L = C_1(1+C_2/d)$. C_1 and C_2 are two constants related to the thickness of the compact layer and to the diffusion-induced growth, respectively (reprinted with permission from [DEB 07]. Copyright 2007, AIP Publishing LLC)*

Importantly, the NW vertical sidewalls are composed of non-polar *m*-planes (with a surface state energy of 0.6 ± 0.2 eV below the conduction band), which are intrinsically pinned [LYM 13]. The surface diffusion processes have been investigated from density-functional theory (DFT) calculations on these non-polar *m*-planes [LYM 09]. The diffusion barriers for Ga adatoms on non-polar *m*-planes are strongly anisotropic in the plane: the diffusion barrier is smaller along the *a*-axis (i.e. 0.21 eV) than along the *c*-axis (i.e. 0.93 eV), since the separation distance between two neighboring adsorption sites is shorter. In contrast to the polar *c*-planes, the non-polar *m*-planes are unstable against atomic N, favoring the formation of desorbing N_2 molecules [LYM 09]. Therefore, the surface diffusion of N adatoms on the NW vertical sidewalls is unlikely. The polar *c*-planes can easily be stabilized thermodynamically and kinetically: 2 × 2 N adatom reconstructions are formed under extreme N-rich conditions, while the diffusion barriers for N adatoms are huge. Thus, it is expected that the incorporation of Ga adatoms is much more likely on the polar *c*-planes than on the non-polar *m*-planes [LYM 09]. This accounts for the higher nucleation rate on the polar *c*-planes and hence for the preferential axial growth of GaN NWs.

8.4.1.2. *Role of MBE chamber geometry*

In the MBE chamber, the angles of the effusion cells and RF plasma source typically lie in the range of 20° to 40° with respect to the normal to the substrate surface. Accordingly, two different growth regimes are expected to occur as the growth phase of GaN NW nuclei proceeds. The NW top facet is simultaneously impinged by the fluxes of Ga and active N species, such that a classical MBE process takes place. In contrast, the NW vertical sidewalls are impinged by the fluxes of Ga and active N species at different time and a migration-enhanced MBE process occurs. Even in the absence of surface diffusion, different species arrival rates are thus imposed on the NW top facet and vertical sidewalls from single geometrical considerations: the axial growth rate is expected to be 5–6 (i.e. $\pi/\tan\alpha$ where α is the angle of the effusion cell to the normal to the substrate surface) times larger than the radial growth rate, taking into account both the substrate rotation and the angles of the effusion cell and RF plasma source [FOX 09]. Furthermore, the substrate rotation may also enhance Ga adatom surface diffusion under N-rich conditions since almost no N species arrive on the NW vertical sidewalls during the deposition of Ga adatoms. Geometrical effects have experimentally been revealed by investigating both the role of substrate rotation and the angles of effusion cells and RF plasma source [FOX 09, GAL 11]. Under identical growth conditions, well-aligned hexagonal-shaped NWs are formed with substrate rotation [FOX 09]. In contrast, differently aligned tadpole-shaped NWs with the occurrence of very long inclined NWs are grown without substrate rotation [FOX 09]. The structural morphology of GaN NWs is also strongly affected by the angle of the RF plasma source with respect to the normal to the substrate surface, as shown in Figure 8.14

[GAL 11]. Significant tapering effects are induced by increasing the angle of the RF plasma source to almost 67°: the larger amount of active N species impinging the NW vertical sidewalls is in favor of the radial growth by limiting Ga desorption [GAL 11].

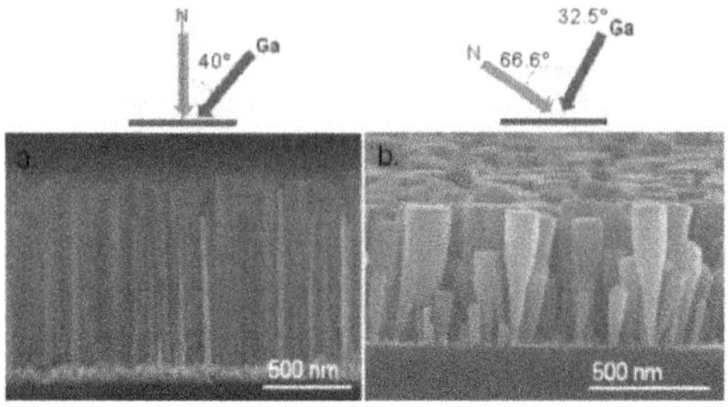

Figure 8.14. *Cross-sectional FESEM images of GaN NWs with the RF plasma source either a) normal to the substrate surface or b) at a large angle of about 67° relative to the substrate surface (from [GAL 11] © 2011 IOP Publishing. Reproduced by permission of IOP Publishing. All rights reserved). For a color version of this figure, see www.iste.co.uk/consonni/nanowires1.zip*

8.4.1.3. *Experimental observations and theoretical modeling*

The axial growth rate of GaN NWs can experimentally be determined from two different approaches. The NW length as deduced by SEM imaging can be multiplied by the growth time, as determined by RHEED. The determination of growth time with a high precision is however required, especially the actual growth time from which GaN NW nuclei start elongating. Otherwise, AlN markers can be used periodically but the inherent formation of an AlN shell on the NW vertical sidewalls could strongly affect the surface diffusion of Ga adatoms [GAL 11]. As a consequence, the axial growth rate should be determined carefully and significant error bars usually occur.

The NW axial growth rate is dependent upon growth conditions such as growth time [PAR 05, CAL 07, SON 07, TCH 07, FUR 08, FOX 09, CAR 11, GAL 11, CON 12, DUB 12b, SCH 12, DUB 13], Ga-rate [CAL 99, CAL 00, PAR 05, SON 07, FER 09, CHE 10b, SON 10, CAR 11, CON 12], growth temperature [SON 07, FER 09, LIM 10, CAR 11, MAT 11, CON 12] and MBE chamber geometry within the angle of the effusion cell with respect to the normal to the substrate surface [FOX 09, GAL 11]. As growth time gets longer, the NW length also increases [PAR 05, CAL 07, SON 07, TCH 07, FUR 08, FOX 09, CAR 11, GAL 11, CON 12, DUB 12b, SCH 12, DUB 13]. However, the NW axial growth rate initially decreases strongly and then saturates, as shown in Figure 8.15(b) [CON 12]: as the NW length gets larger than the diffusion length of Ga adatoms on the NW vertical sidewalls, the contribution from the substrate surface is expected to vanish and hence the axial growth rate saturates [CON 12]. In contrast, when limited by Ga species, the NW axial growth rate almost linearly increases initially with Ga-rate [CAL 99, CAL 00, PAR 05, SON 07, FER 09, CHE 10b, SON 10, CAR 11, CON 12], as revealed in Figure 8.15(c) [CON 12]. However, a more efficient nucleation on the NW vertical sidewalls can account for a slight deviation from the perfect linearity [CON 12]. Subsequently, the NW axial growth rate is expected to saturate with Ga-rate, when limited by active N species [CAL 99, CAL 00, PAR 05, SON 07, FER 09, CHE 10b, SON 10, CAR 11, FER 13]. Thus, the NW axial growth rate can be limited by either Ga species or active N species, depending on the fluxes used and on the V/III ratio. The evolution of the NW axial growth rate with growth temperature [SON 07, FER 09, LIM 10, CAR 11, MAT 11, CON 12] involves two different regimes, as presented in Figure 8.15(d) [CON 12]. For increasing low growth temperature, the axial growth rate is controlled by the surface diffusion of Ga adatoms and hence increases. In contrast, for increasing higher growth temperature, the axial growth rate is driven by the desorption of Ga adatoms and hence decreases [MAT 11, CON 12]. The optimal growth temperature corresponding to the largest axial growth rate is about 780°C.

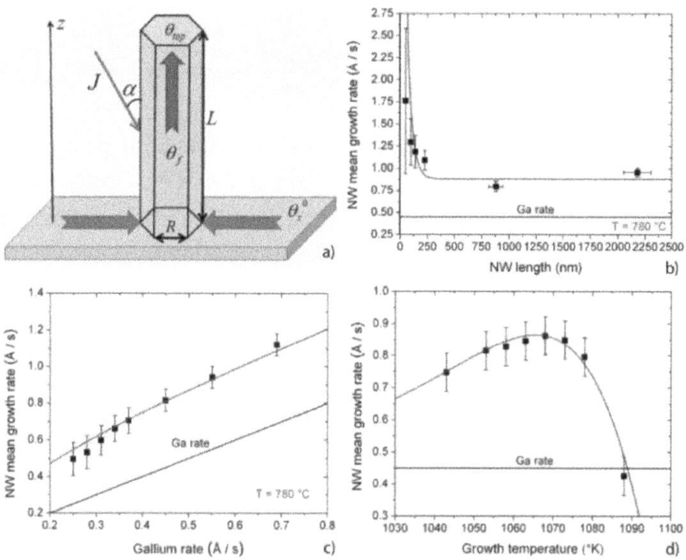

Figure 8.15. *a) Schematic diagram of the theoretical modeling for the surface diffusion-induced NW growth within the self-induced approach. Evolution of the mean NW axial growth rate as a function of b) NW length (i.e., growth time), c) Ga-rate and d) growth temperature: full dots represent experimental data while the medium gray line yields the theoretical fit obtained by deriving equation [8.2]. The dark gray line indicates the value of Ga-rate. The N-rate is 2.8 Å/s.(Reprinted with permission from [CON 12], Copyright 2012 by the American Physical Society). For a color version of this figure, see www.iste.co.uk/consonni/nanowires1.zip*

In light of the theoretical modeling of the axial growth rate as developed for the catalyst-induced approach of arsenides and phosphides III-V semiconductor NWs [JOH 05, DUB 08], a surface diffusion-induced growth model has been used by Debnath *et al.* [DEB 07] and Galopin *et al.* [GAL 11]. A small Ga adatom diffusion length in the range of 40 to 50 nm has been reported on the NW vertical sidewalls [DEB 07, GAL 11, CON 12], pointing out that only Ga atoms adsorbing on the NW vertical sidewalls close to the NW top facet take part in the NW elongation. Furthermore, the difference in chemical potentials between the NW vertical sidewalls and top facet is positive (i.e. 39 meV/atom), indicating that a net flow of Ga adatoms propagates from the NW vertical sidewalls to the NW top facet [GAL 11].

More recently, Consonni *et al.* have reported a quantitative description for the growth rate of GaN NWs by specifically developing a surface diffusion-induced growth model for the self-induced approach [CON 12]. By considering materials balance and by solving diffusion equations, the axial growth rate d*L*/dt of an isolated hexagonal-shaped GaN NW is given by [CON 12]:

$$\frac{3\sqrt{3}R^2}{2\Omega_s}\frac{dL}{dt}=\frac{3\sqrt{3}R^2}{2}(J\cos\alpha-J_{des})-6RD_f\left(\frac{dn_f}{dz}\right)_{z=L} \qquad [8.2]$$

in which R is the radius of the NW cross-section assumed as being constant in a first approximation, Ω_s is the elementary volume in the NW, J is the Ga impinging rate, α is the angle with respect to the normal to the substrate surface, J_{des} is the desorption rate from the NW top facet, D_f and n_f are the diffusion coefficient and concentration of Ga adatoms on the NW vertical sidewalls, respectively.

The net Ga flux contributing to the NW elongation is given on the left-hand side of equation [8.2]. The adsorption–desorption growth rate on the NW top facet is given by the first term on the right-hand side. The diffusion flux of Ga adatoms from the substrate surface and NW vertical sidewalls to the NW top facet is supplied by the second term on the right-hand side. Interestingly, the present theoretical modeling is in very good agreement with the experimental results: the evolution of the axial growth rate with growth conditions can be fitted by one single set of identical fitting parameters, as shown in Figure 8.15, and results in the determination of a large number of key kinetic parameters [CON 12]. The effective diffusion lengths of Ga adatoms on the NW vertical sidewalls and on the amorphous Si_xN_y interlayer are only 45 and 100 nm, respectively. Also, the nucleation energy of 5.44 eV is larger on the NW vertical sidewalls than on their top facet: this is somehow in agreement with DFT calculations [LYM 09] predicting that the incorporation rates are larger on the NW top facet than on the NW vertical and that the NW elongation is preferred along the *c*-axis.

8.4.1.4. *Shadowing and collective effects in the NW ensemble*

The limitations of the previous theoretical modeling described by Consonni *et al.* [CON 12] are related to two major hypotheses: the hexagonal-shaped GaN NW is considered as isolated while its radius is taken as constant. However, the NW length at a given growth time is quite significantly dispersed, as presented in Figure 8.16(b) [SAB 13]. This can be associated with two additional physical phenomena. First, each isolated hexagonal-shaped GaN NW undergoes the growth phase with a different actual growth time owing to the local fluctuations in the nucleation phase. For instance, the NW length dispersion is more pronounced by raising growth temperature, which is assigned to the different incubation and transition times for the formation of isolated GaN NW nuclei in the NW ensemble [CON 11c, MAT 11]. Second, in addition to the local fluctuations in the nucleation phase, shadowing and collective effects should be taken into account when a NW ensemble is considered with a high density of the order of several hundreds of NWs per μm^2 [CON 12]. Basically, the surface physical processes involved in the growth of neighboring NWs are expected to strongly interact with each other: Ga atoms can initially be desorbed from an isolated NW and subsequently be readsorbed at the vertical sidewalls of an isolated neighboring NW, resulting in strong collective effects in the NW ensemble. The self-induced growth of GaN NWs is thus mediated by the exchange of Ga atoms in between neighboring NWs. Sabelfeld *et al.* have shown that shadowing and collective effects in the NW dense ensemble should be considered to precisely account for the NW length dispersion [SAB 13]. Despite shadowing effects, collective effects favor the growth of shorter NWs, which are fed by longer NWs. They also give rise to a narrow NW length distribution instead of a broad length distribution for longer NWs, as shown in Figure 8.16(c) [SAB 13]. The development of stochastic mean field differential equations can describe the evolution of the bimodal length distribution for short and long NWs as growth proceeds, as revealed in Figure 8.16(d) [SAB 13].

8.4.2. *Radial growth*

Although the elongation of GaN NWs is preferentially achieved along the polar *c*-axis, the radial growth along the non-polar *m*-axis is not completely annihilated [PAR 05, CAL 07, SON 07, FUR 08, FOX 09, SON 10, CAR 11, DUB 12b, DUB 13]. Basically, the radial growth takes place as soon as the NW length exceeds the Ga adatom diffusion length of about 40 to 50 nm. The NW radial growth is improved by increasing Ga-rate (i.e. decreasing V/III ratio) [YOS 98, FER 09, SON 10], decreasing growth temperature [RIS 08, MAT 11] and/or using Mg and Be doping [FUR 08, LIM 10, STO 11]. It is expected that the distinct Ga and N incorporation rates may be balanced on the non-polar *m*- and polar *c*-planes by varying growth conditions.

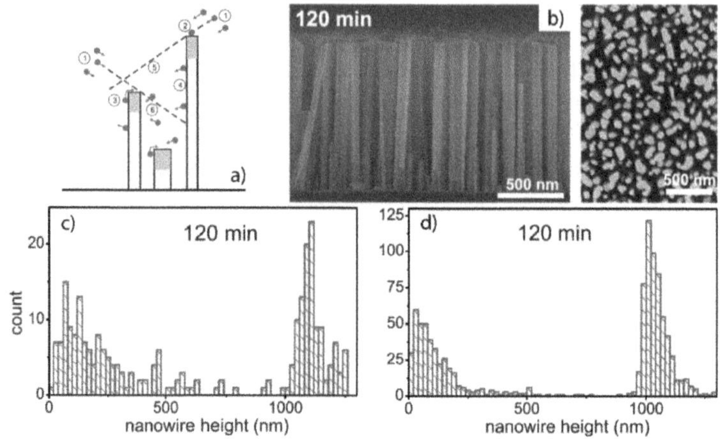

Figure 8.16. *a) Schematic diagram of the growth of a NW ensemble: 1- the incoming atom flux, 2- the impingement on the NW top facet, 3- the diffusion along the NW vertical sidewalls with further contribution to the growth at the NW top facet, 4- the desorption from the NW vertical sidewalls, 5- the shadowing of short NWs by longer neighboring NWs and 6- the adsorption of the desorbed atoms on a neighboring NW; b) cross-sectional and top-view FESEM images of GaN NWs; c) experimental NW length distribution from FESEM image analysis; d) calculated NW length distribution from stochastic mean field differential equations involving collective effects in the NW ensemble (reprinted with permission from [SAB 13]. Copyright 2013, AIP Publishing LLC). For a color version of this figure, see www.iste.co.uk/consonni/nanowires1.zip*

The physical processes responsible for the radial growth on the local scale are still open questions. The radial growth may proceed via the successive nucleation and flow of GaN single steps (i.e. monolayers) on the NW vertical sidewalls. These GaN single steps may nucleate at the bottom of GaN NWs and, subsequently, their ascending growth may occur toward the NW top [GAL 11]. An alternative scenario considers the nucleation of these GaN single steps on top of GaN NWs and, then, their descending growth toward the NW bottom [FER 13]. Importantly and as presented in Figure 8.12, the NW length L scales with its radius R according to the scaling power law $L \sim L_0 (R/R_0)^{\alpha}$, α being a growth index depending on growth conditions and L_0 and R_0 being the initial NW length and radius [DUB 12b, DUB 13]. This shows that the self-induced growth of GaN NWs follows scaling growth kinetics [DUB 12b]. The evolution of the NW length L and radius R with growth time is given by the scaling laws, as shown in Figure 8.17 [DUB 12b]. For low NW density, the nucleation and ascending flow of one single step on the vertical sidewalls toward the NW top account for the scaling law [DUB 12b]. For higher NW density, the scaling law is also explained by shadowing effects [DUB 12b]. As the NW density is increased with growth time, a transition regime is expected for intermediate NW density.

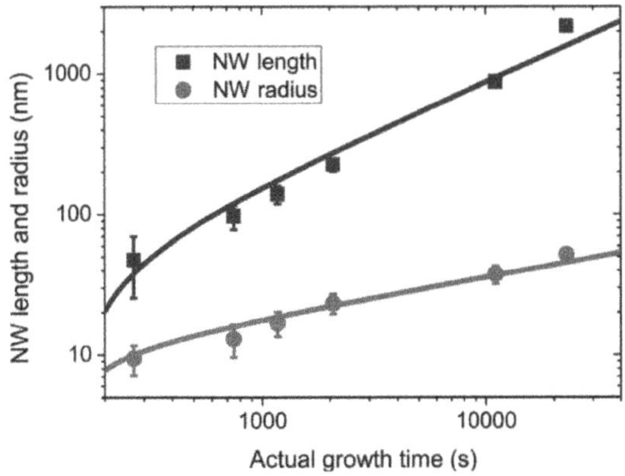

Figure 8.17. *Evolution of the NW length and radius as a function of growth time: full dots represent experimental data while the upper and lower lines yield the theoretical fit from the relations:* $L = [1+C(t-t_0)]^{\alpha/(\alpha+1)}$ *and* $R = [1+C(t-t_0)]^{1/(\alpha+1)}$ *where* t_0 *is the nucleation time. C is a constant and* $\alpha = 2.46$ *is the growth index (reprinted with permission from [DUB 12b]. Copyright 2012, AIP Publishing LLC). For a color version of this figure, see www.iste.co.uk/consonni/nanowires1.zip*

More recently, the radial growth rate has been manipulated by varying growth conditions such as Ga-rate [FER 13] and growth temperature [TAM 13] as the elongation period proceeds.

By using the initial template consisting of predeposited GaN NW arrays acting as substrates, the effects of Ga-rate (i.e. V/III ratio) on the radial growth during the elongation period have been shown in the N-limited regime [FER 13]. One critical point for the radial growth is related to the finite amount of active N species on the NW top facet when the effective Ga-rate (from the direct impinging flux and adatom surface diffusion) is sufficiently large [FER 13]. Importantly, the effective Ga-rate directly depends on the NW length and radius. For short NWs, the radial growth is active and the GaN NW radius increases, as shown in Figure 8.18 [FER 13]: the high surface density of Ga adatoms is favored by their surface diffusion along the vertical sidewalls besides the direct impinging flux; therefore, the formation of islands is achieved on the NW side facet and the radial growth is induced. In contrast, for longer NWs, the radial growth is inactive and the GaN NW radius gets constant, as revealed in Figure 8.18 [FER 13]: the surface density of Ga adatoms is decreased as the NW radius increases; as a result, the formation of islands on the NW side facet is annihilated and, thus, the radial growth is suppressed [FER 13]. Therefore, the GaN NW radius can be self-regulated in the N-limited regime.

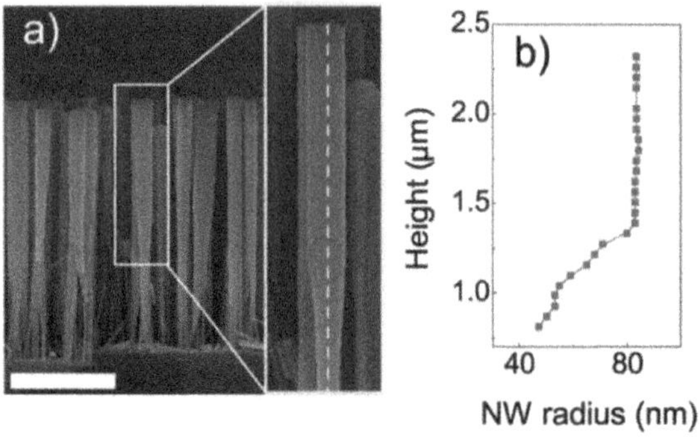

Figure 8.18. *a) Cross-sectional FESEM image of the overgrown GaN NW templates using Ga- and N-rates of 2 and 1.8 Å/s, respectively. The FESEM image includes both the GaN NW template as a substrate and the second step of the growth in the yellow rectangular area; the scale bar has a length of 1 µm; b) evolution of the GaN NW radius measured along the dashed line shown in (a) during the second step of the growth (reprinted with permission from [FER 13]. Copyright 2013 American Chemical Society)*

An alternative approach to significantly control the NW radius consists of switching growth temperature in dedicated cycles from high through low to high values as growth proceeds [TAM 13]. The variation of growth temperature, as the elongation period proceeds, directly affects the surface diffusion of adatoms as well as their desorption. This can lead to the formation of GaN NWs with a given radius that is variable along the NW length, depending on when the growth temperature has been changed [TAM 13].

8.4.3. *Coalescence period*

In the self-induced approach, the density of GaN NWs is high and of the order of 10^{10} cm^{-2}, which is favorable for establishing a coalescence period. Indeed, the differential twist and tilt of nearby GaN NWs result in their coalescence owing to geometrical considerations. For instance, GaN NWs grown on the Si_xN_y interlayer are slightly tilted by an angle of a few degrees with respect to the normal to the substrate surface, which is most likely due to the surface roughness: the differential tilt is expected to induce the coalescence of these NWs. Also, GaN NWs are slightly twisted with respect to one another on the Si_xN_y interlayer. In this case, the radial growth is expected to induce their coalescence. From geometrical considerations, the coalescence of two closed nearby GaN NWs usually occurs at a mean NW length of several hundreds of nanometers and is associated with the formation of structural defects and with the generation of local microstrain [CON 09, FAN 14]. A network of boundary dislocations is, for instance, formed to accommodate both differential tilt and twist between two closed nearby GaN NWs, above which I_1-type basal plane stacking faults are nucleated, as shown in Figure 8.19 [CON 09]. Triple twin domains are also formed as coalescence proceeds by increasing Mg doping [ARB 09]. Importantly, when the differential tilt and twist of two nearby GaN NWs are increased, the formation of more pronounced structural defects takes place such as zipper dislocation arrays for instance [GRO 13]. Furthermore, a fluctuating and highly inhomogeneous microstrain from ±0.015 to ±0.03% is generated in two coalesced GaN NWs, while single GaN NWs are free of strain [JEN 11]. This can account for the increase in the photoluminescence linewidth above 1 meV as well as for the specific XRD peak profiles in coalesced GaN NWs [KAG 12].

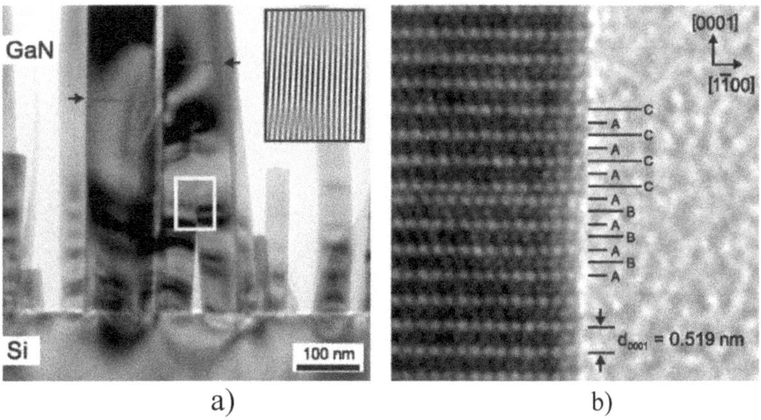

Figure 8.19. *a) Cross-section TEM image of two coalesced NWs. The black arrows indicate the location of stacking faults. The white box indicates the location of boundary dislocations just above the contact point. The inset reveals a Fourier-filtered enhancement showing only the (1-100) planes to emphasize the edge character of the boundary dislocations. b) HRTEM image of a basal plane SF as marked in (a). The SF is of I_1-type and starts on the Aa layer (reprinted with permission from [CON 09]. Copyright 2009, AIP Publishing LLC)*

The optical properties of two coalesced NWs are also strongly affected: while the NW top retains a very high optical quality, radiative transitions at 3.42 eV (Y_2 line) are assigned to excitons bound to I_1-type basal plane stacking faults in the NW center. The occurrence of radiative transitions at 3.21 (Y_7 line) and 3.35 eV (Y_4 line) is also typical and may originate from point defects at edge threading dislocations in coalesced GaN NWs [CHE 06, LEF 11], for more details refer to Chapter 3.

In a NW ensemble, the magnitude of coalescence can quantitatively be assessed by distinguishing between single NWs and NW clusters through the determination of the coalescence degree $\alpha_{\text{coalescence}}$ [CON 11b]: $\alpha_{\text{coalescence}} = 1 - (\rho_{\text{cluster}}/\rho_{\text{NW}})$ where ρ_{cluster} and ρ_{NW} are the cluster and NW density, respectively. A free-standing NW is considered a single cluster. A coalescence degree of 0 (i.e. $\rho_{\text{cluster}} = \rho_{\text{NW}}$) means that coalescence does not occur at all. In contrast, a coalescence degree of 1 (i.e. $\rho_{cluster} << \rho_{NW}$) yields that all GaN NWs are coalesced. The coalescence degree typically lies in the range of 0.3 to 0.8 for standard Ga-rates and growth temperatures, as presented in Figure 8.20 [CON 11b]. Interestingly, the coalescence degree is strongly reduced by raising growth temperature, as shown in Figure 8.20(b) [CON 11b]. Also, the modification of Ga-rate (i.e. V/III ratio) as the elongation proceeds may somehow reduce the coalescence degree significantly by annihilating the radial growth in the N-limited regime. Furthermore, it is worth mentioning that coalescence can be helpful for the top contact planarization in the fabrication of GaN NW-based LEDs (Chapter 2 in Vol. 2).

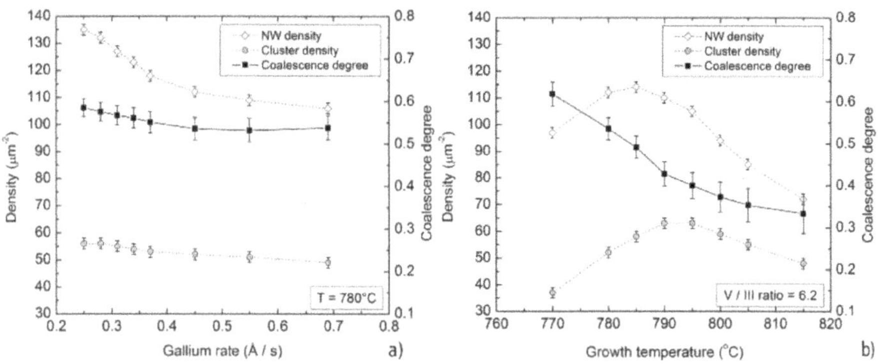

Figure 8.20. *a) Evolution of the NW density, cluster density and coalescence degree as a function of Ga-rate. The growth temperature is 780°C. b) Evolution of the NW density, cluster density and coalescence degree as a function of growth temperature. The Ga-rate is 0.45 Å/s, which corresponds to a V/III ratio of 6.2 (reprinted with permission from [CON 11b]. Copyright 2011, AIP Publishing LLC)*

8.4.4. *Density effects*

Both nucleation phase and coalescence period play a significant role on the density of GaN NWs, which strongly depends on growth conditions, such as growth time [CAL 07, SON 07, FUR 08, LAN 09, CON 11b, SCH 12], Ga-rate [FER 09, CAR 11, CON 11b] and growth temperature [SON 07, LIM 10, CAR 11, CON 11b]. Nucleation and coalescence effects on the density of GaN NWs can be disentangled by distinguishing again between single NWs and NW clusters. A good example is given by the evolution of the density of GaN NWs with growth time. While not considering coalescence and single NW counting, the density of GaN NWs initially increases significantly with growth time until reaching a maximum [CAL 07, FUR 08, SCH 12]. It eventually decreases owing to coalescence, as revealed in Figure 8.21 [CAL 07]. However, while considering coalescence and single NW counting, the density of GaN NWs actually increases continuously with growth time until reaching a plateau [SON 07, LAN 09, CON 11b].

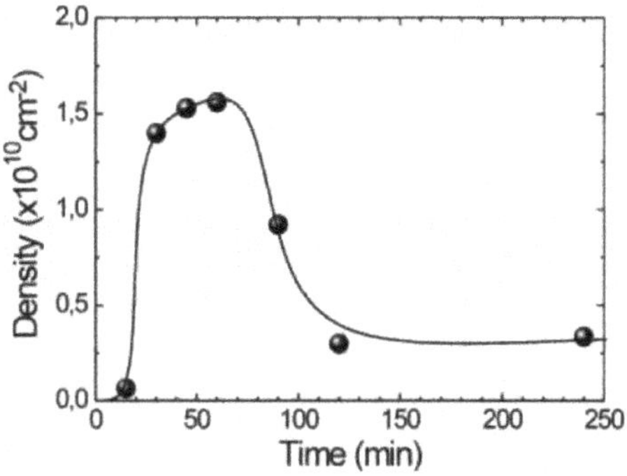

Figure 8.21. *Evolution of the NW density as a function of growth time (line guides the eye) (reprinted with permission from [CAL 07]. Copyright 2007 American Chemical Society)*

Interestingly, the evolution of the density of GaN NWs does not follow standard island nucleation theory. The density of GaN NWs is typically increased by decreasing Ga-rate [FER 09, CAR 11, CON 11b], as shown in Figure 8.20(a) [CON 11b]. In contrast, it initially increases up to a maximum by raising growth temperature and eventually decreases drastically. It should be noted that the final density of GaN NWs is directly related to the initial density of spherical cap-shaped islands, which are formed during the incubation period of the nucleation phase. Still, the growth conditions are supposed to favor or hamper their shape transitions toward GaN NW nuclei during the transition period of the nucleation phase. For instance, raising growth temperature is expected to thermally activate the shape transitions from spherical cap-shaped islands, initially increasing the density of GaN NWs [CON 11b]. However, for higher growth temperature, the subsequent decrease in the density of GaN NWs can be accounted for by massive desorption of Ga adatoms during the incubation period of the nucleation phase and hence by the very low initial density of spherical cap-shaped islands [SON 07, LIM 10, CAR 11, CON 11b], as shown in Figure 8.20(b) [CON 11b].

8.4.5. *Polarity effects*

Negatively charged N planes are stacked alternately with positively charged Ga planes along the growth polar *c*-axis of the wurtzite crystalline structure of GaN. Since no center of inversion occurs, the [0001] and [000-1] directions are polar and not equivalent. By convention, the +c [0001] (respectively, –c [000-1]) direction corresponds to the Ga (respectively, N) polarity: the vector starts with a Ga (respectively, N) atom, points to a N (respectively, Ga) atom and is aligned along the polar *c*-axis. The issue of how polarity affects the nucleation phase of GaN NWs has recently been an intense matter of debate [CHE 08, FUR 08, CHE 10a, BRU 11, HES 11, KON 11, DEN 12, FER 12, LAR 12, MAT 12,]. The identification of the polarity of GaN NWs can be performed by convergent beam electron diffraction (CBED) [CHE 08, FUR 08, CHE 10a, FER 12, LAR 12], electron energy loss spectroscopy (EELS) [KON 11, FER 12], resonant XRD in synchrotron around the Ga-K edge [HES 11], probe Cs corrected TEM measurements [MAT 12], sensitive chemical etching [LAR 12] and piezoresponse force microscopy [BRU 11]. Among all these characterization techniques, the resonant XRD in synchrotron is particularly attractive while offering the polarity measurement over a NW ensemble by benefiting from the breakdown of the Friedel's law, as shown in Figure 8.22 [HES 11].

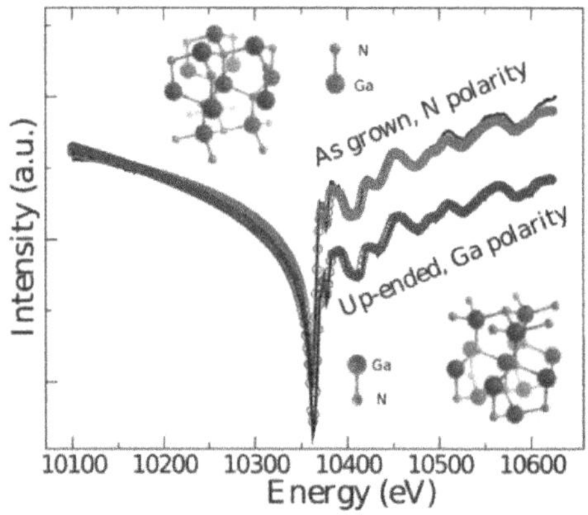

Figure 8.22. *Evolution of the intensity of the (11-22) and (-1-12-2) pair with the energy around the Ga K-edge. Because of the non-centrosymmetrical characteristic of the GaN wurtzite crystalline structure, a discrepancy in intensity is clearly observed. In this way, the N-polar and Ga-polar GaN NWs can be distinguished (reprinted with permission from [HES 11], Copyright 2011 by the American Physical Society). For a color version of this figure, see www.iste.co.uk/consonni/nanowires1.zip*

Generally speaking, GaN NWs can be either Ga-polar [CHE 08, FUR 08, CHE 10a, BRU 11] or N-polar [CHE 08, BRU 11, HES 11, KON 11, DEN 12, LAR 12, MAT 12, SOB 14a]. However, a general consensus is being established in favor of the N-polarity of GaN NWs grown on the Si_xN_y interlayer [HES 11, KON 11, DEN 12, LAR 12, MAT 12]. Similarly, GaN NWs grown on the amorphous Al_2O_3 buffer layer has been found N-polar [SOB 14a]. However, the polarity of GaN NWs on the AlN buffer layer is more complex [CHE 08, BRU 11, FER 12, LAR 12,]. When structural defects and morphological irregularities are formed on the AlN buffer layer, the formation of GaN NWs with either Ga- or

N-polarity may be due to the mixed polarity of the underlying seed layer [CHE 08, BRU 11]. When the AlN buffer layer is much smoother, with a reduced number of structural defects, only N-polar GaN NWs are formed [FER 12]. A high density of N-polar GaN NWs is grown on the N-polar AlN buffer layer while only very few N-polar GaN NWs develop on the Al-polar AlN buffer layer. In the latter case, inversion domain boundaries are nucleated to reverse the polarity of GaN NWs and their formation may be assisted by surfactant effects involving Si atoms on the surface, for instance [FER 12].

8.5. Conclusion

The nucleation and growth mechanisms of GaN NWs by PAMBE within the self-induced approach have been a matter of intense debate over the last decade. Their understanding has made significant progress: the self-induced approach is still attractive as a low-cost and versatile technique to assemble GaN NW arrays with a very high purity.

The formation mechanisms have been reviewed here from the nucleation phase through the growth phase to the coalescence period. These mechanisms are strongly dependent upon the structural morphology of the nucleation surface. When the nucleation surface is highly smooth with a reduced number of structural defects, the nucleation phase basically results in the formation of NW nuclei and is driven by thermodynamics: spherical cap-shaped islands are initially formed following an incubation period and subsequently undergo one or several shape transitions toward GaN NW nuclei, depending on the nature of the nucleation surface. The GaN NW nuclei are composed of polar *c*-plane top facet and non-polar *m*-plane vertical sidewalls. The driving force for the shape transition is expected to be surface energy anisotropy. In contrast, the subsequent growth mechanisms lead to the elongation of NW nuclei and are governed by kinetics. The underlying physical processes involve surface phenomena both in individual NWs and in NW ensembles *via* collective effects. The MBE chamber geometry (i.e. angle of effusion cells and plasma source as well as substrate rotation) also plays a significant role. In the light of the nucleation and growth mechanisms described here, the necessary ingredients (i.e. large effective V/III ratio and high growth temperature) for the self-induced formation of GaN NWs can be more deeply understood. The need for a growth temperature above 750°C is in favor of (1) the surface diffusion of Ga adatoms to induce localized nucleation and three-dimensional (3D) growth modes; (2) the low density of spherical cap-shaped islands to hamper further coalescence process; (3) the thermal activation of the shape transitions toward GaN NW nuclei; (4) the surface diffusion of Ga adatoms on the NW vertical sidewalls to promote the axial growth over the radial growth. The high V/III ratio is required to (1) ensure surface energy anisotropy as the main driving force for the shape transitions toward GaN NW nuclei, (2) the different incorporation rates on the NW vertical sidewalls and top such that the elongation phase is favored.

Overall, the self-induced growth of GaN NWs is still highly attractive. However, despite the significant progress made in the understanding of the formation mechanisms, several critical aspects are still challenging and should be investigated further. Among these, the polarity effects, the precise role of the Si_xN_y interlayer, the nature of nucleation sites and correlated density and positioning control or the doping effects should carefully be considered in order to make sure that the fabrication of efficient optoelectronic devices based on GaN NWs is achievable.

8.6. Acknowledgments

The author would like to thank all the coauthors of his own papers reviewed in this work, especially M. Knelangen, L. Geelhaar, A. Trampert and H. Riechert from Paul-Drude-Institut für Festkörperelektronik, Berlin, Germany.

8.7. Bibliography

[ARB 09] ARBIOL J., ESTRADE S., PRADES J.D., *et al.*, "Triple-twin domains in Mg doped GaN wurtzite nanowires: structural and electronic properties of this zinc-blende-like stacking", *Nanotechnology*, vol. 20, p. 145704, 2009.

[BER 06] BERTNESS K.A., ROSHKO A., SANFORD N.A., *et al.*, "Spontaneously grown GaN and AlGaN nanowires", *Journal of Crystal Growth*, vol. 287, pp. 522–527, 2006.

[BER 07] BERTNESS K.A., ROSHKO A., MANSFIELD L.M., *et al.*, "Nucleation conditions for catalyst-free GaN nanowires", *Journal of Crystal Growth*, vol. 300, pp. 94–99, 2007.

[BER 08] BERTNESS K.A., ROSHKO A., MANSFIELD L.M., *et al.*, "Mechanism for spontaneous growth of GaN nanowires with molecular beam epitaxy", *Journal of Crystal Growth*, vol. 310, pp. 3154–3158, 2008.

[BER 11] BERTNESS K.A., SANFORD N.A., DAVYDOV A.V., "GaN nanowires grown by molecular beam epitaxy", *IEEE Journal of Selected Topics in Quantum Electronics*, vol. 17, no. 4, pp. 847–858, 2011.

[BRU 11] BRUBAKER M.D., LEVIN I., DAVYDOV A.V., *et al.*, "Effect of AlN buffer layer properties on the morphology and polarity of GaN nanowires grown by molecular beam epitaxy", *Journal of Applied Physics*, vol. 110, p. 053506, 2011.

[CAL 99] CALLEJA E., SANCHEZ-GARCIA M.A., SANCHEZ F.J., *et al.*, "Growth of III-nitrides on Si(111) by molecular beam epitaxy: doping, optical, and electrical properties", *Journal of Crystal Growth*, vol. 201–202, pp. 296–317, 1999.

[CAL 00] CALLEJA E., SANCHEZ-GARCIA M.A., SANCHEZ F.J., *et al.*, "Luminescence properties and defects in GaN nanocolumns grown by molecular beam epitaxy", *Physical Review B*, vol. 62, no. 24, pp. 16826–16834, 2000.

[CAL 07] CALARCO R., MEIJERS R.J., DEBNATH R.K., *et al.*, "Nucleation and growth of GaN nanowires on Si(111) performed by molecular beam epitaxy", *Nano Letters*, vol. 7, no. 8, pp. 2248–2251, 2007.

[CAL 07] CALLEJA E., RISTIC J., FERNANDEZ-GARRIDO S., *et al.*, "Growth, morphology, and structural properties of group-III-nitride nanocolumns and nanodisks", *Physica Status Olidi B*, vol. 244, no. 8, pp. 2816–2837, 2007.

[CAR 11] CARNEVALE S.D., YANG J., PHILLIPS P.J., *et al.*, "Three-dimensional GaN/AlN nanowire heterostructures by separating nucleation and growth processes", *Nano Letters*, vol. 11, pp. 866–871, 2011.

[CER 06] CERUTTI L., RISTIC J., FERNANDEZ-GARRIDO S., *et al.*, "Wurtzite GaN nancolumns grown on Si(001) by molecular beam epitaxy", *Applied Physics Letters*, vol. 88, p. 213114, 2006.

[CHE 06] CHEN H.Y., LIN H.W., SHEN C.H., *et al.*, "Structure and photoluminescence properties of epitaxially oriented GaN nanorods grown on Si(111) by plasma-assisted molecular beam epitaxy", *Applied Physics Letters*, vol. 89, p. 243105, 2006.

[CHE 07] CHEN R.S., CHEN H.Y., LU C.Y., *et al.*, "Ultrahigh photocurrent grain in m-axial GaN nanowires", *Applied Physics Letters*, vol. 91, p. 223106, 2007.

[CHE 08] CHERNS D., MESHI L., GRIFFITHS I., *et al.*, "Defect reduction in GaN/(0001) sapphire films grown by molecular beam epitaxy using nanocolumn intermediate layers", *Applied Physics Letters*, vol. 92, p. 121902, 2008.

[CHE 10a] CHEZE C., GEELHAAR L., BRANDT O., *et al.*, "Direct comparison of catalyst-free and catalyst-induced GaN nanowires", *Nano Research*, vol. 3, pp. 528–536, 2010.

[CHE 10b] CHEZE C., GEELHAAR L., JENICHEN B., *et al.*, "Different growth rates for catalyst-induced and self-induced GaN nanowires", *Applied Physics Letters*, vol. 97, p. 153105, 2010.

[CHE 10c] CHEZE C., GEELHAAR L., TRAMPERT A., *et al.*, "In situ investigation of self-induced GaN nanowire nucleation on Si", *Applied Physics Letters*, vol. 97, p. 043101, 2010.

[CHE 13] CHEN Y.T., ARAKI T., PALISAITIS J., *et al.*, "Nucleation of single GaN nanorods with diameters smaller than 35 nm by molecular beam epitaxy", *Applied Physics Letters*, vol. 103, p. 203108, 2013.

[CON 09] CONSONNI V., KNELANGEN M., JAHN U., *et al.*, "Effects of nanowire coalescence on their structural and optical properties on a local scale", *Applied Physics Letters*, vol. 95, p. 241910, 2009.

[CON 10] CONSONNI V., KNELANGEN M., GEELHAAR L., *et al.*, "Nucleation mechanisms of epitaxial GaN nanowires: origin of their self-induced formation and initial radius", *Physical Review B*, vol. 81, p. 085310, 2010.

[CON 11a] CONSONNI V., HANKE M., KNELANGEN M., *et al.*, "Nucleation mechanisms of self-induced GaN nanowires grown on an amorphous interlayer", *Physical Review B*, vol. 83, p. 035310, 2011.

[CON 11b] CONSONNI V., KNELANGEN M., TRAMPERT A., *et al.*, "Nucleation and coalescence effects on the density of self-induced GaN nanowires grown by molecular beam epitaxy", *Applied Physics Letters*, vol. 98, p. 071913, 2011.

[CON 11c] CONSONNI V., TRAMPERT A., GEELHAAR L., *et al.*, "Physical origin of the incubation time of self-induced GaN nanowires", *Applied Physics Letters*, vol. 99, p. 033102, 2011.

[CON 12] CONSONNI V., DUBROVSKII V.G., TRAMPERT A., *et al.*, "Quantitative description for the growth rate of self-induced GaN nanowires", *Physical Review B*, vol. 85, p. 155313, 2012.

[CON 13] CONSONNI V., "Self-induced growth of GaN nanowires by molecular beam epitaxy: a critical review of the formation mechanisms", *Physica Status Solidi RRL*, vol. 7, no. 10, pp. 699–712, 2013.

[CON 14] CONSONNI V., FEUILLET G., (eds.), *Wide Band Gap Semiconductor Nanowires 2: Heterostructures and Optoelectronic Devices*, ISTE, London, and John Wiley & Sons, New York, 2014.

[DEB 07] DEBNATH R.K., MEIJERS R., RICHTER T., *et al.*, "Mechanism of molecular beam epitaxy growth of GaN nanowires on Si(111)", *Applied Physics Letters*, vol. 90, p. 123117, 2007.

[DEN 12] DEN HERTOG M.I., GONZALEZ-POSADA F., SONGMUANG R., *et al.*, "Correlation of polarity and crystal structure with optoelectronic and transport properties of GaN/AlN/GaN nanowire sensors", *Nano Letters*, vol. 12, pp. 5691–5696, 2012.

[DUB 08] DUBROVSKII V.G., SIBIREV N.V., HARMAND J.C., *et al.*, "Growth kinetics and crystal structure of semiconductor nanowires", *Physical Review B*, vol. 78, p. 235301, 2008.

[DUB 12a] DUBROVSKII V.G., CONSONNI V., TRAMPERT A., *et al.*, "Scaling thermodynamic model for the self-induced nucleation of GaN nanowires", *Physical Review B*, vol. 85, p. 165317, 2012.

[DUB 12b] DUBROVSKII V.G., CONSONNI V., GEELHAAR L., *et al.*, "Scaling growth kinetics of self-induced GaN nanowires", *Applied Physics Letters*, vol. 100, p. 153101, 2012.

[DUB 13] DUBROVSKII V.G., TIMOFEEVA M.A., "Modeling GaN nanowire growth on silicon", *Technical Physics Letters*, vol. 39, no. 1, pp. 127–129, 2013.

[FAL 11] FALTA J., SCHMIDT T., GANGOPADHYAY S., *et al.*, "Ultra-thin high-quality silicon nitride films on Si(111)", *Euro Physics Letters*, vol. 94, p. 16003, 2011.

[FAN 14] FAN S., ZHAO S., LIU X., *et al.*, "Study on the coalescence of dislocation-free GaN nanowires on Si and SiO_x", *Journal of Vacuum Science and Technology B*, vol. 32, no. 2, p. 02C114, 2014.

[FER 09] FERNANDEZ-GARRIDO S., GRANDAL J., CALLEJA E., *et al.*, "A growth diagram for plasma-assisted molecular beam epitaxy of GaN nanocolumns on Si(111)", *Journal of Applied Physics*, vol. 106, p. 126102, 2009.

[FER 12] FERNANDEZ-GARRIDO S., KONG X., GOTSCHKE T., *et al.*, "Spontaneous nucleation and growth of GaN nanowires: the fundamental role of crystal polarity", *Nano Letters*, vol. 12, pp. 6119–6125, 2012.

[FER 13] FERNANDEZ-GARRIDO F., KAGANER V.M., SABELFELD K.K., *et al.*, "Self-regulated radius of spontaneously formed GaN nanowires in molecular beam epitaxy", *Nano Letters*, vol. 13, pp. 3274–3280, 2013.

[FOX 09] FOXON C.T., NOVIKOV S.V., HALL J.L., *et al.*, "A complementary geometric model for the growth of GaN nanocolumns prepared by plasma-assisted molecular beam epitaxy", *Journal of Crystal Growth*, vol. 311, p. 34233427, 2009.

[FUR 08] FURTMAYR F., VIELEMEYER M., STUTZMANN M., *et al.*, "Nucleation and growth of GaN nanorods on Si(111) surfaces by plasma-assisted beam epitaxy – the influence of Si- and Mg-doping", *Journal of Applied Physics*, vol. 104, p. 034309, 2008.

[GAL 11] GALOPIN E., LARGEAU L., PATRIARCHE G., *et al.*, "Morphology of self-catalyzed GaN nanowires and chronology of their formation by molecular beam epitaxy", *Nanotechnology*, vol. 22, p. 245606, 2011.

[GEE 11] GEELHAAR L., CHEZE C., JENICHEN B., *et al.*, "Properties of GaN nanowires grown by molecular beam epitaxy", *IEEE Journal of Selected Topics in Quantum Electronics*, vol. 17, p. 878, 2011.

[GOO 11] GOODMAN K., PROTASENKO V., VERMA J., *et al.*, "Moleular beam epitaxial growth of gallium nitride nanowires on atomic layer deposited aluminum oxide", *Journal of Crystal Growth*, vol. 334, pp. 113–117, 2011.

[GRA 05] GRADECAK S., QIAN F., LI Y., *et al.*, "GaN nanowire lasers with low lasing thresholds", *Applied Physics Letters*, vol. 87, p. 173111, 2005.

[GRO 13] GROSSKLAUS K.A., BANERJEE A., JAHANGIR S., *et al.*, "Misorientation defects in coalesced self-catalyzed GaN nanowires", *Journal of Crystal Growth*, vol. 371, pp. 142–147, 2013.

[HES 11] HESTROFFER K., LECLERE C., BOUGEROL C., *et al.*, "Polarity of GaN nanowires grown by plasma-assisted molecular beam epitaxy on Si(111)", *Physical Review B*, vol. 84, p. 245302, 2011.

[HES 12] HESTROFFER K., LECLERE C., CANTELLI V., *et al.*, "In situ study of self-assembled GaN nanowire nucleation on Si(111) by plasma-assisted molecular beam epitaxy", *Applied Physics Letters*, vol. 100, p. 212107, 2012.

[HES 13] HESTROFFER K., DAUDIN B., "A RHEED investigation of self-assembled GaN nanowire nucleation dynamics on bare Si and on Si covered with a thin AlN buffer layer", *Physica Status Solidi RRL*, vol. 7 no. 10, pp. 835–839, 2013.

[JEN 11] JENICHEN B., BRANDT O., PFULLER C., *et al.*, "Marco- and micro-strain in GaN nanowires on Si(111)", *Nanotechnology*, vol. 22, p. 295714, 2011.

[JOH 02] JOHNSON J.C., CHOI H.J., KNUTSEN K.P., *et al.*, "Single gallium nitride nanowire lasers", *Nature Materials*, vol. 1, p. 106, 2002.

[JOH 05] JOHANSSON J., SVENSSON C.P.T., MARTENSSON T., *et al.*, "Mass transport model for semiconductor nanowire growth", *The Journal of Physical Chemistry B*, vol. 109, pp. 13567–13571, 2005.

[KAG 12] KAGANER V.M., JENICHEN B., BRANDT O., *et al.*, "Inhomogeneous strain in GaN nanowires determined from x-ray diffraction peak profiles", *Physical Review B*, vol. 86, p. 115325, 2012.

[KIK 04] KIKUCHI A., KAWAI M., TADA M., *et al.*, "InGaN/GaN multiple quantum disk nanocolumn light-emitting diodes grown on (111) Si substrates", *Japanese Journal of Applied Physics*, vol. 43, no. 12A, pp. L1524–L1526, 2004.

[KIM 02] KIM H.M., KIM D.S., KIM D.Y., *et al.*, "Growth and characterization of single-crystal GaN nanorods by hydride vapor phase epitaxy", *Applied Physics Letters*, vol. 81, p. 2193, 2002.

[KIS 12] KISHINO K., KAMINURA J., KAMIYAMA K., "Near-infrared InGaN nanocolumn light-emitting diodes operated at 1.46 μm", *Applied Physics Express*, vol. 5, p. 031001, 2012.

[KNE 10] KNELANGEN M., CONSONNI V., TRAMPERT A., *et al.*, "In situ analysis of strain relaxation during catalyst-free nucleation and growth of GaN nanowires", *Nanotechnology*, vol. 21, p. 245705, 2010.

[KOE 10] KOESTER R., HWANG J.S., DURAND C., *et al.*, "Self-assembled growth of catalyst-free GaN wires by metal-organic vapour phase epitaxy", *Nanotechnology*, vol. 21, p. 015602, 2010.

[KON 11] KONG X., RISTIC J., SANCHEZ-GARCIA M.A., *et al.*, "Polarity determination by electron energy-loss spectroscopy: application to ultra-small III-nitride semiconductor nanocolumns", *Nanotechnology*, vol. 22, p. 415701, 2011.

[KUM 12] KUMAR P., TUTEJA M., KESARIA M., *et al.*, "Superstructure of self-aligned hexagonal GaN nanorods formed on nitrided Si(111) surface", *Applied Physics Letters*, vol. 101, p. 131605, 2012.

[LAN 08] LANDRE O., SONGMUANG R., RENARD J., *et al.*, "Plasma-assisted molecular beam epitaxy growth of GaN nanowires using indium-enhanced diffusion", *Applied Physics Letters*, vol. 93, p. 183109, 2008.

[LAN 09] LANDRE O., BOUGEROL C., RENEVIER H., *et al.*, "Nucleation mechanism of GaN nanowires grown on (111)Si by molecular beam epitaxy", *Nanotechnology*, vol. 20, p. 415602, 2009.

[LAR 08] LARGEAU L., DHEERAJ D.L., TCHERNYCHEVA M., *et al.*, "Facet and in-plane crystallographic orientations of GaN nanowires grown on Si(111)", *Nanotechnology*, vol. 19, p. 155704, 2008.

[LAR 12] LARGEAU L., GALOPIN E., GOGNEAU N., *et al.*, "N-polar GaN nanowires seeded by Al droplets on Si(111)", *Crystal Growth & Design*, vol. 12, p. 2724, 2012.

[LEF 11] LEFEBVRE P., FERNANDEZ-GARRIDO S., GRANDAL J., *et al.*, "Radiative defects in GaN nanocolumns: correlation with growth conditions and sample morphology", *Applied Physics Letters*, vol. 98, p. 083104, 2011.

[LI 12] LI S., WAAG A., "GaN based nanorods for solid state lighting", *Journal of Applied Physics*, vol. 111, p. 071101, 2012.

[LIM 10] LIMBACH F., SCHAFER-NOLTE E.O., CATERINO R., *et al.*, "Morphology and optical properties of Mg doped GaN nanowires in dependence of growth temperature", *Journal of Optoelectronics and Advanced Materials*, vol. 12, no. 6, pp. 1433–1441, 2010.

[LIM 12] LIMBACH F., CATERINO R., GOTSCHKE T., *et al.*, "The influence of Mg doping on the nucleation of self-induced GaN nanowires", *AIP Advances*, vol. 2, p. 012157, 2012.

[LYM 09] LYMPERAKIS L., NEUGEBAUER J., "Large anisotropic adatom kinetics on nonpolar GaN surfaces: consequences for surface morphologies and nanowire growth", *Physical Review B*, vol. 79, p. 241308(R), 2009.

[LYM 13] LYMPERAKIS L., WEIDLICH P.H., EISELE H., *et al.*, "Hidden surface states at non-polar GaN (10-10) facets: intrinsic pinning of nanowires", *Applied Physics Letters*, vol. 103, p. 152101, 2013.

[MAT 11] MATA R., HESTROFFER K., BUDAGOSKY J., *et al.*, "Nucleation of GaN nanowires grown by plasma-assisted molecular beam epitaxy: the effect of temperature", *Journal of Crystal Growth*, vol. 334, pp. 177–180, 2011.

[MAT 12] DELA MATA M., MAGEN C., GAZQUEZ J., *et al.*, "Polarity assignment in ZnTe, GaAs, ZnO, and GaN-AlN nanowires from direct dumbbell analysis", *Nano Letters*, vol. 12, pp. 2579–2586, 2012.

[MEI 06] MEIJERS R., RICHTER T., CALARCO R., *et al.*, "GaN-nanowhiskers: MBE-growth conditions and optical properties", *Journal of Crystal Growth*, vol. 289, pp. 381–386, 2006.

[NOR 96] NORTHRUP J.E., NEUGEBAUER J., "Theory of GaN(10-10) and (11-20) surfaces", *Physical Review B*, vol. 53, pp. R10477–R10480, 1996.

[NOR 99] NORTHRUP J.E., NEUGEBAUER J., "Indium-induced changes in GaN (0001) surface morphology", *Physical Review B*, vol. 60, pp. R8473–R8476, 1999.

[PAR 05] PARK Y.S., LEE S.H., OH J.E., *et al.*, "Self-assembled GaN nano-rods grown directly on (111)Si substrates: dependence on growth conditions", *Journal of Crystal Growth*, vol. 282, pp. 313–319, 2005.

[RIG 10] RIGUTTI L., TCHERNYCHEVA M., BUGALLO A.D., *et al.*, "Ultra violet photodetector based on GaN/AlN quantum disks in a single nanowire", *Nano Letters*, vol. 10, pp. 2939–2943, 2010.

[RIS 08] RISTIC J., CALLEJA E., FERNANDEZ-GARRIDO S., *et al.*, "On the mechanisms of spontaneous growth of III-nitride nanocolumns by plasma-assisted molecular beam epitaxy", *Journal of Crystal Growth*, vol. 310, pp. 4035–4045, 2008.

[SAB 13] SABELFELD K.K., KAGANER V.M., LIMBACH F., *et al.*, "Height self-equilibration during the growth of dense nanowire ensembles: order emerging from disorder", *Applied Physics Letters*, vol. 103, p. 133105, 2013.

[SAN 98] SANCHEZ-GARCIA M.A., CALLEJA E., MONROY E., *et al.*, "The effect of the V/III ratio and substrate temperature on the morphology and properties of GaN- and AlN-layers grown by molecular beam epitaxy on Si(111)", *Journal of Crystal Growth*, vol. 183, pp. 23–30, 1998.

[SCH 12] SCHUSTER F., FURTMAYR F., ZAMANI R., *et al.*, "Self-assembled GaN nanowires on diamond", *Nano Letters*, vol. 12, pp. 2199–2204, 2012.

[SEK 07] SEKIGUCHI H., NAZAKATO T., KIKUCHI A., *et al.*, "Structural and optical properties of GaN nanocolumns grown on (0001) sapphire substrates by rf-plasma-assisted molecular-beam epitaxy", *Journal of Crystal Growth*, vol. 300, pp. 259–262, 2007.

[SEK 08] SEKIGUCHI H., KISHINO K., KIKUCHI A., "GaN / AlGaN nanocolumn ultraviolet light-emitting diodes grown on n-(111) Si by RF-plasma-assisted molecular beam epitaxy", *Electronics Letters*, vol. 44, no. 2, pp. 151–152, 2008.

[SIB 12] SIBIREV N.V., TCHERNYCHEVA M., CIRLIN G.E., *et al.*, "Effect of diffusion from a lateral surface on the rate of GaN nanowire growth", *Semiconductors*, vol. 46, no. 6, pp. 838–841, 2012.

[SOB 14a] SOBANSKA M., KLOSEK K., BORYSIUK J., *et al.*, "Enhanced catalyst-free nucleation of GaN nanowires on amorphous Al_2O_3 by plasma-assisted molecular beam epitaxy", *Journal of Applied Physics*, vol. 115, p. 043517, 2014.

[SOB 14b] SOBANSKA M., WIERZBICKA A., KLOSEK K., *et al.*, "Arrangement of GaN nanowires grown by plasma-assisted molecular beam epitaxy on silicon substrates with amorphous Al_2O_3 buffers", *Journal of Crystal Growth*, 2014.

[SON 07] SONGMUANG R., LANDRE O., DAUDIN B., "From nucleation to growth of catalyst-free GaN nanowires on thin AlN buffer layer", *Applied Physics Letters*, vol. 91, p. 251902, 2007.

[SON 10] SONGMUANG R., BEN T., DAUDIN B., *et al.*, "Identification of III-N nanowire growth kinetics via a marker technique", *Nanotechnology*, vol. 21, p. 295605, 2010.

[STO 08] STOICA T., SUTTER E., MEIJERS R.J., *et al.*, "Interface and wetting layer effect on the catalyst-free nucleation and growth of GaN nanowires", *Small*, vol. 4, no. 6, pp. 751–754, 2008.

[STO 11] STOICA T., CALARCO R., "Doping of III-nitride nanowires grown by molecular beam epitaxy", *IEEE Journal of Selected Topics in Quantum Electronics*, vol. 17, no. 4, pp. 859–868, 2011.

[TAB 97] TABE M., YAMAMOTO T., "Initial stages of nitridation of Si(111) surfaces: X-ray photoelectron spectroscopy and scanning tunneling microscopy studies", *Surface Science*, vol. 376, pp. 99–112, 1997.

[TAM 13] TAMBO H., ASAHI H., "Control of GaN nanorod diameter by changing growth temperature during molecular-beam epitaxy", *Journal of Crystal Growth*, vol. 383, pp. 57–62, 2013.

[TCH 07] TCHERNYCHEVA M., SARTEL C., CIRLIN G., *et al.*, "Growth of GaN free-standing nanowires by plasma-assisted molecular beam epitaxy: structural and optical characterization", *Nanotechnology*, vol. 18, p. 385306, 2007.

[TU 03] TU L.W., HSIAO C.L., CHI T.W., *et al.*, "Self-assembled vertical GaN nanorods grown by molecular beam epitaxy", *Applied Physics Letters*, vol. 82, no. 10, pp. 1601–1603, 2003.

[VAN 06] VAN NOSTRAND J.E., AVERETT K.L., CORTEZ R., *et al.*, "Molecular beam epitaxial growth of high-quality GaN nanocolumns", *Journal of Crystal Growth*, vol. 287, pp. 500–503, 2006.

[WAN 01] WANG X.S., ZHAI G., YANG J., *et al.*, "Nitridation of Si(111)", *Surface Science*, vol. 494, pp. 83–94, 2001.

[WIE 13] WIERZBICKA A., ZYTKIEWICZ Z.R., KRET S., *et al.*, "Influence of substrate nitridation temperature on epitaxial alignment of GaN nanowires to Si(111) substrate", *Nanotechnology*, vol. 24, p. 035703, 2013.

[WU 02] WU C.L., HSIEH J.L., HSUEH H.D., *et al.*, "Thermal nitridation of the Si(111)-(7×7) surface studied by scanning tunneling microscopy and spectroscopy", *Physical Review B*, vol. 65, p. 045309, 2002.

[YOS 97] YOSHIZAWA M., KIKUCHI A., MORI M., *et al.*, "Growth of self-organized GaN nanostructures on Al_2O_3(0001) by RF-radical source molecular beam epitaxy", *Japanese Journal of Applied Physics*, vol. 36, pp. L459–L462, 1997.

[YOS 98] YOSHIZAWA M., KIKUCHI A., FUJITA N., *et al.*, "Self-organization of GaN/$Al_{0.18}Ga_{0.82}N$ multi-layer nano-columns on (0001)Al_2O_3 by RF molecular beam epitaxy for fabricating GaN quantum disks", *Journal of Crystal Growth*, vol. 189–190, pp. 138–141, 1998.

9

Selective Area Growth of GaN Nanowires by Plasma-Assisted Molecular Beam Epitaxy

9.1. Introduction

Group III-nitrides semiconductors are widely used nowadays in optoelectronic devices, in particular, InGaN alloys for light emitting diodes (LED) working in the whole visible spectral region [NAK 00, YAM 08, KUY 07], as well as for white light generation. Most of the efforts devoted to these materials focused on *c*-plane oriented, two-dimensional quantum well (QW) structures. However, one of the main factors limiting the performance of these planar structures is the high density of non-radiative defects that are generated during the growth of the material due to the increasing lattice mismatch between GaN and InGaN alloys with higher In content. In addition, the growth of high quality InGaN alloys with In content above 25% poses big challenges derived from the quite different optimal growth temperatures for GaN and InN. Heteroepitaxy of InGaN/GaN QWs on sapphire, Si, or SiC is particularly prone to suffer from the above-mentioned problems. However, it has been shown that dislocation- and strain free III-nitrides can be grown on Si(100) and (111), as well as on amorphous SiO_2 substrates [CAL 00, MEI 06, THI 06, YOS 97, PAR 04] by forming quasi one-dimensional structures such as self-assembled nanocolumns (NCs). Additionally, NC-based optical devices, and in particular LEDs, could benefit from a better light extraction efficiency as compared to standard planar LEDs, because of guiding effects along the NC axis.

Chapter written by Miguel A. SANCHEZ-GARCIA, Steven ALBERT,
Ana M. BENGOECHEA-ENCABO, Francesca BARBAGINI and Enrique CALLEJA.

During the last decade, much effort has been dedicated to the growth of III-nitrides-based self-assembled NCs by Plasma-Assisted Molecular Beam Epitaxy (PAMBE). This has allowed a much better knowledge of some fundamental physical properties of this material system, like the band gap value of InN [DAV 02, WU 02, GRA 05], as well as on the mechanisms involved in light emission from almost defect-free nanocrystals and the growth mechanisms [YOS 97, SAN 98, FER 09, RIS 08] (for more details, refer to Chapter 8). Though self-assembled NCs are easy to grow, efficient devices based on self-assembled NCs, like LEDs (for more details, refer to Chapter 5 in Volume 2 [CON 14]), suffer severe limitations derived from a strong dispersion (morphological, electrical) inherent to the self-assembled process. Problems of dark spots generated at defective NCs (crystal defects upon NCs merging) and/or an inhomogeneous current injection may derive from a poor electroluminescence (EL) yield [BAV 11]. White LEDs based on InGaN/GaN self-assembled nanocolumnar structures have been demonstrated, showing multicolor emission derived from an inhomogeneous distribution of In due to, either changes in NCs diameter (that affect the axial growth rate); axial and radial strain variations, or the inherent tendency of InGaN alloys to have composition fluctuations [KIS 07, LIN 10].

These problems can be overcome using selective area growth (SAG) techniques. The SAG aims to achieve growth only on selected regions, which are defined by openings in a thin mask material layer deposited previously on top of the substrate. Frequently used mask materials are Ti, SiO_2 and Si_3N_4. After the deposition of the mask material, a patterning of the nanoholes is needed. Various methods can be used for this purpose, such as laser interferometric lithography [HER 06], e-beam lithography (EBL) [KIS 09, BEN 11], photolithography [LI 11] or colloidal lithography (COL) [BEN 12]. Nanoimprint lithography can also be applied for this purpose, especially for the fabrication of a large number/area of patterned substrates and at a lower cost (industrial application) [ERI 11]. In any case, the need of a previous patterning of the mask layers for the SAG makes this approach more demanding from a technological point of view, in comparison to the simpler self-assembled method.

9.2. Mask preparation

The first requirement to achieve selectivity deals with the fabrication of an appropriate mask layer. The most common materials used as a mask layer are Ti, SiO_2 and Si_3N_4. This analysis will focus on the use of Ti, although the conclusions can also be applied to the other materials. The mask preparation includes two different processes: (1) the deposition of the mask material (Ti) on the substrate and (2) the opening (patterning) of a nanohole array. Both processes are crucial to achieve selective growth and high crystal quality of the ordered NCs, and they need to be optimized separately. The opening of the nanoholes in the mask layer can be achieved by different lithography methods. In this work, three techniques are analyzed: e-beam (EBL), focused ion beam (FIB) and colloidal lithography (COL).

For the deposition of the mask material, there are some critical steps that, if not taken into consideration, might lead to a total failure of the selective growth. Apart from the need of a material that stands the high temperatures used for SAG, two main factors must be considered: the surface roughness and the adhesion of the mask material to the substrate. Each of them is treated separately in the following sections.

The presence of surface roughness of the mask material around, or between the nanoholes, even lower than 1 nm, may lead to a complete failure of the SAG of GaN NCs. Figure 9.1 shows the ideal case of a smooth mask (a) (prepared by EBL), the case of a mask (prepared also by EBL under non-ideal conditions) with local increase in roughness around the holes (b), and the case of a FIB mask with bumps around the holes (c). In the ideal case of Figure 9.1(a), the surface roughness of the mask measured by AFM is 0.5 ± 0.1 nm RMS, very similar to the values of both the bare GaN template and of the deposited Ti layer. With these roughness values, the MBE growth leads to a proper selective nucleation of NCs inside the holes, with no traces of nucleation on the mask (Figure 9.1(d)).

In both cases (b) and (c), an accurate AFM analysis reveals the presence of material around the holes that locally increases the surface roughness to 1 nm or higher. For the mask prepared by EBL (Figure 9.1(b)) this morphology is attributed to the re-deposition of material around the holes during the dry etch (RIE) process, while in the case of FIB masks, surface bumps around the holes are probably formed when bombarding the Ti with a high dose of Ga^+ ions. In either case, the mask roughness leads to a typical donut-shaped morphology, as shown in Figures 9.1(e) and (f), where many thin NCs nucleate all around the hole, which remains empty. In some cases, scanning electron microscopy (SEM) pictures taken after growth revealed a single hollow column as shown in Figure 9.1(f). This peculiar growth layout is likely due to the difficulty for the Ga and N adatoms to diffuse from the mask surface into the holes due to the roughness that acts as a barrier. For this reason, the NCs nucleate at the perimeter of these donut-shaped bumps. The presence of accumulated material around the holes can only be detected by accurate AFM analysis, while SEM pictures show smooth surfaces even when there is a small local roughness increase. To minimize this material re-deposition and achieve optimal roughness values, the plasma etch conditions used in EBL technique (mainly plasma power and time) or the dose when using FIB need to be optimized. The surface roughness is not an issue when preparing the masks by COL, since there is no etching involved, as it is explained below. In this case, surfaces look like those in Figure 9.1(a).

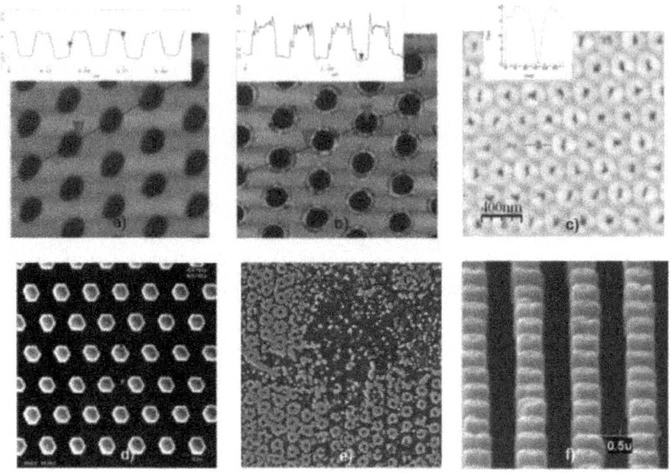

Figure 9.1. *Top row: smooth mask obtained by EBL a), and masks with increased roughness in the area around the holes obtained by EBL b), and FIB c). Insets: cross-section AFM analysis of the mask. Bottom row: d), e) and f) growth results using the respective masks in the upper row. (Adapted from [BAR 11]). For a color version of this figure, see www.iste.co.uk/consonni/nanowires1.zip*

A too thick (well above 10 nm) Ti layer barely adheres to the GaN substrate, resulting in delamination of the mask during the MBE growth, as shown in Figure 9.2(a). This delamination is observed when cooling down after the PA-MBE growth (from approximately 675°C to room temperature). This issue is solved by optimizing both the adhesion of the Ti film to the substrate and the thickness of the Ti film. To improve adhesion, the initial GaN template (as received) is cleaned with n-methyl-pyrrolidone (NMP) at 90°C by sweeping the surface several times with a soft stick, then rinsing with iso-propanol (IPA) and de-ionized water (DIW). Prior to Ti deposition, the substrate is heated in an oven up to 300°C for 30 min to desorb water molecules from the surface. The optimal thickness of the Ti layer can be found empirically. If mask adhesion is optimized, but delamination still occurs, it could be due to strain generated by thermal shock when cooling down. To minimize strain, the thickness of the Ti mask was reduced from 10 nm to 7 nm. A thinner mask resulted in most cases in the complete degradation of the Ti material at high temperature during the growth, as shown in Figure 9.2(b). This point can be overcome if the mask is previously nitridated as it will be described in the next section. Mask adhesion to the substrate is a factor to consider in all of the proposed methods, meaning EBL, FIB and COL. Both optimal substrate cleaning and mask thickness are of fundamental importance for a good adhesion of the Ti layer to the substrate. The thickness of the mask is also related to the aspect ratio of the nanoholes to be opened. A very high aspect ratio (small diameter, large thickness) may hinder the nucleation (clogging). In this work, typical sizes of the nanoholes are between 50 and 250 nm.

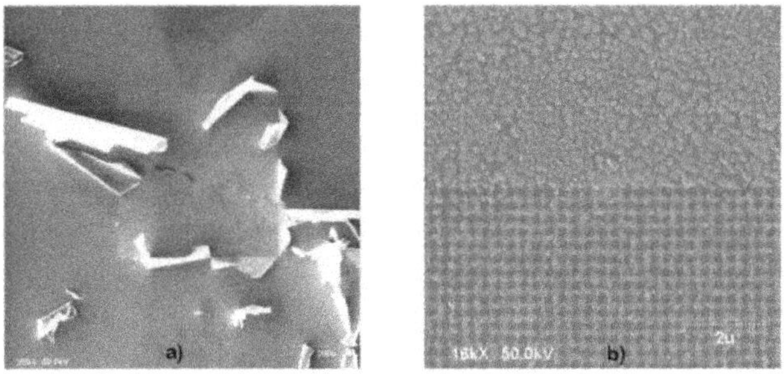

Figure 9.2. *a) Peeling-off of the Ti mask after growth in case of a 10 nm Ti layer thickness. b) Degradation of a 5 nm Ti mask during growth at high temperature. (Adapted from [BAR 11])*

Once the mask has been deposited and the surface roughness/adhesion issues have been solved, nanopatterning is performed. The geometry of the mask (i.e. diameter of the nanoholes and distance between them, *pitch*) also needs to be taken into consideration to achieve perfect SAG. If the pitch is too large, all the Ga impinging directly on the Ti mask may desorb from it before reaching the closest nanoholes by diffusion, or be trapped by an N atom and nucleate on the mask. In both cases, proper SAG is not achieved. Therefore, as it will be shown in the next section, the Ga and N fluxes, together with the substrate temperature need to be adjusted properly for a particular diameter-pitch geometry of the mask.

As mentioned before, the opening of the nanoholes in the mask layer is analyzed in this work by three techniques: EBL, FIB and COL.

The EBL process uses a 350 nm layer of positive resist (ZEP520A). After baking at 190°C for 2 min, the sample is transferred to the EBL system. The nanoholes were opened in the resist using a current of 0.5×10^{-9} A and an exposure time ranging from 15 to 45 μs. The process conditions are optimized to obtain arrays of nanoholes with diameters ranging from 50 to 200 nm and pitch (center-to-center distance) varying from 80 nm to 300 nm. After developing the resist, the sample is dry-etched in O_2/CF_4 plasma at 10^{-2} mbar and 110 W (Plasmalab) for 2 min and 10 s. The post-etch residue is removed by dipping in NMP at 80°C for 30 min and abundantly rinsing in IPA and DIW. AFM analysis revealed an etching depth between 7 and 10 nm, thus, in some cases the GaN material underneath the titanium mask was also etched around 3 nm. In this process, the final cleaning step is crucial, since residues after the cleaning process can be detrimental to achieve SAG.

With FIB technique, nanoholes are opened in a one-step process using a liquid–metal Ga^+ ion source at 30 KeV (ionLine Raith GmbH, Germany). The process conditions are optimized to obtain arrays of nanoholes with 100 nm diameter and pitch 250 nm (30 pA, ion dose 10^{17} cm^{-2}). No extra cleaning steps were applied after the ion etching. AFM analysis revealed an etching depth between 10 and 15 nm. Re-deposition of Ti (1–2 nm) appears in some cases at the edges of the holes.

The colloidal lithography method is based on the initial formation of a colloidal solution, made of polystyrene (PS) nanobeads, on top of the substrate. The PS nanobeads show a close-packaged arrangement in large areas (domains up to 400 × 400 μm^2). Once the nanobeads are arranged on the surface, reactive ion etching (RIE) oxygen plasma (40 W; oxygen pressure of 80 mTorr, 1 nm/s) is used in order to reduce the diameter of the beads to the desired diameter and, consequently, the pitch. After the plasma etching, 7 nm of Ti are evaporated on the surface and a lift-off process results in an array of nanoholes on the Ti layer (150 nm diameter, typically) covering the GaN surface.

To summarize this section, the structure and morphology of the nanohole-patterned Ti mask become the driving factors that determine whether selective area growth is achieved or not. Main features of the mask that might negatively influence the ordered growth are surface roughness, adhesion of the Ti to the GaN substrate and surface cleaning. EBL, FIB and COL processes can be used to fabricate good quality masks patterned with nanoholes, which serve as nucleation sites for the selective area growth of GaN NCs. The first two processes (EBL and FIB), while providing perfect uniformity (ordered arrangement of the holes), require special equipment (expensive and time consuming) and critical cleaning stages following the etching steps. Possible residues will hamper the achievement of selectivity during the MBE growth. The COL process is a much faster and cheaper process that neither requires sophisticated systems (like EBL or FIB) nor special cleaning steps, since there is no etching process involved (only lift-off). However, the arrangement of the nanoholes is not perfectly uniform.

9.3. Selectivity, nucleation mechanism and morphology control of the nanocolumns

Following the optimization of the mask, the MBE growth conditions must be carefully tuned to achieve selectivity, meaning growth only within the holes while avoiding it on the mask (Ti for this particular work). Selectivity depends critically on the sticking coefficients, diffusion lengths and desorption rates of the metal adatoms, as well as on the geometry of the nanoholes array in the mask (pitch). This growth process differs considerably from the self-assembled process, where the main nucleation driving force is the lattice mismatch between the substrate and the NCs material [FER 09, RIS 08]. This section addresses the effects of the MBE growth parameters (substrate temperature and III/V flux ratio) and nanohole mask geometry on the GaN NCs SAG on GaN templates (on sapphire), covered with a Ti nanohole mask. The main objective is to understand the growth process and to establish optimal conditions to achieve and to control selectivity and homogeneity. This section also highlights the key parameters to produce the transition from the initial nucleation stage within the nanoholes to a vertical growth regime.

Since the temperature values for SAG are quite high (⩾900°C), nitridation of the Ti mask, prior to the growth, is needed to prevent its degradation. The nitridation process is similar to that proposed by Sekiguchi *et al.* [SEK 08], following two steps: a first step at low temperature (460°C during 10 min) and a second step at high temperature (880°C during 3 min). The plasma source during nitridation is set to 580 W at nitrogen flux of 1.2 sccm that corresponds to an equivalent stoichiometric GaN growth rate higher than 30 nm/min. Electron energy loss spectroscopy measurements (not shown) prove the formation of TiN, more stable than Ti at high temperatures. This nitridation process (avoiding degradation at high temperature) together with a low Ti mask surface roughness (below 0.5 nm, RMS value, which favors the Ga diffusion) are the key starting points needed to achieve SAG.

GaN NCs SAG was studied in a set of samples using masks with nanoholes of 150 nm diameter and 300 nm pitch. The duration of all growths was 30 min, while growth temperatures varied from 900 to 960°C, and the nominal III/V flux ratios from 0.7 to 2.7. These nominal flux ratios are quite high when compared to the typical ones used for self-assembled NCs (III/V << 1) [FER 09]. However, considering the high temperatures needed for SAG, the actual III/V ratios at the growth front are much smaller than the nominal ones because of the very strong Ga desorption and GaN decomposition rates. These high temperatures are needed for SAG in order to enhance both Ga surface diffusion (higher probability for Ga adatoms to reach the nanoholes) and Ga surface desorption to avoid or minimize the formation of GaN NCs on the mask [SEK 08]. This is indeed a very delicate trade-off. Figure 9.3 shows the corresponding plan-view SEM images.

For a nominal III/V flux ratio of 0.7 (Ga = 10 nm/min; N = 15 nm/min) and 940°C (Figure 9.3(a)), GaN NCs grow both on the Ti mask and within the nanoholes (no selectivity). A substrate temperature increment of 10°C is not enough to achieve full selectivity (Figure 9.3(b)). Finally, at 960°C perfect selectivity was achieved (Figure 9.3(c)). At even higher temperature, growth is no longer possible (under these conditions) because of the very strong Ga desorption and GaN decomposition rates. From the point of view of selectivity, the growth temperature and nitrogen flux (or reciprocal III/V ratio) act in opposite ways [FER 09], since a higher growth temperature promotes the diffusion and desorption of Ga adatoms (better selectivity), while a higher nitrogen flux increases the probability of GaN formation on the mask before Ga desorption occurs (worse selectivity) [FER 08]. Thus growth conditions to obtain selectivity are not unique, which allows us to use a lower substrate temperature by increasing the III/V ratio. This is shown in Figures 9.3(d)–(f), which correspond to III/V nominal ratios of 3.2 (Ga = 16 nm/min; N = 5 nm/min); 0.9 (Ga = 10 nm/min; N = 11 nm/min); and 0.7 (Ga = 10 nm/min; N = 15 nm/min), respectively.

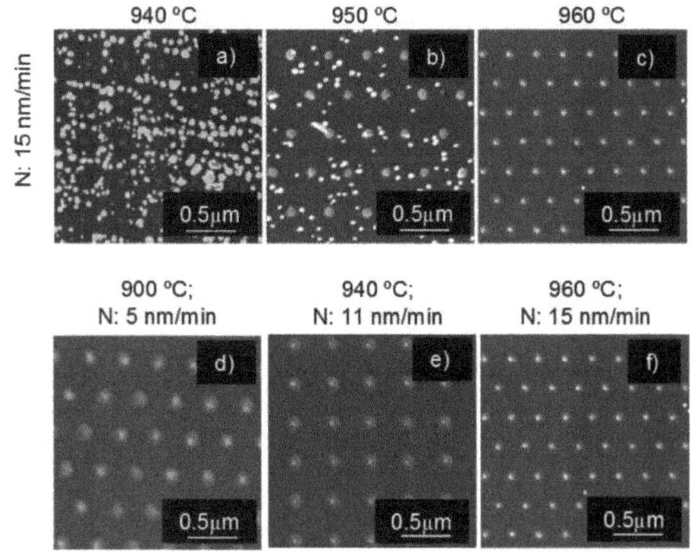

Figure 9.3. *Plan-view SEM images of a series of samples grown for 30 min on Ti masks with nanoholes of around 150 nm diameter and 300 nm pitch. Growth temperature and nominal III/V ratio are varied (different N fluxes for a given Ga flux) in order to check their effect on the NC nucleation selectivity. (Adapted from [BEN 11])*

In addition to selectivity, the substrate temperature also critically affects the growth inside the nanoholes. Considering that an increase in the growth temperature reduces exponentially (by desorption) the Ga amount that incorporates into the nanoholes by diffusion or direct impinging, a high enough growth temperature may suppress or inhibit the vertical growth. Such a case is shown in Figure 9.4(a), where ordered GaN (aimed to form NCs) grown with Φ_{Ga} = 15 nm/min, Φ_N = 5 nm/min at 940°C reveal a perfect selectivity but, instead of NCs, the growth results in GaN pyramids. Once this geometry is reached, the size and shape of the GaN pyramids does not change when growth goes on [BEN 11]. This inhibition of the vertical growth after the initial nucleation stage seems to be related to a lack of sufficient Ga within the nanoholes and a lower growth rate on *r*-planes. In this particular case (Figure 9.4(a)) the nominal Φ_{Ga} was not high enough to grow NCs. Nucleation took place because, upon starting growth, a strong Φ_{Ga} transient (shutter opened) yields an initial value higher (20 nm/min) than the nominal value (15 nm/min). This leads to a higher local III/V ratio within the nanohole for a short time, until the transient is over. This mechanism will dominate growth whenever the chosen nominal Φ_{Ga} is too small to keep the vertical growth.

Keeping the previous nominal Φ_{Ga}, but lowering the temperature from 940 to 900°C, i.e. significantly reducing the Ga desorption rate, leads to vertical growth of ordered NCs, as shown in Figure 9.4(b), that still have pyramidal top facets (*r*-planes) [BEN 11]. In the first case of suppressed vertical growth and even in the second case (Figures 9.4(a) and (b)) the growth conditions are most likely N-rich, i.e. Φ_{Ga} limits the growth rate. A further reduction in the growth temperature down to 860°C leads to the formation of ordered GaN NCs with a flat top (*c*-plane), as shown in Figure 9.4(c). Based on the above arguments this morphology change from pyramidal to flat top can be attributed to an increase of the local III/V ratio within the nanoholes due to a decrease in the Ga desorption rate. Since this morphology transition basically depends on the increased amount of Ga reaching the nanoholes, either by reducing Ga desorption (temperature) or by increasing the nominal impinging Φ_{Ga}, this suggests that a change from N-rich to Ga-rich conditions is involved. It is known that for homoepitaxial growth of GaN, the layer morphology is determined by the growth regime either Ga-rich or N-rich [TAR 97, ADE 02] and that N-rich conditions yield a faceted surface.

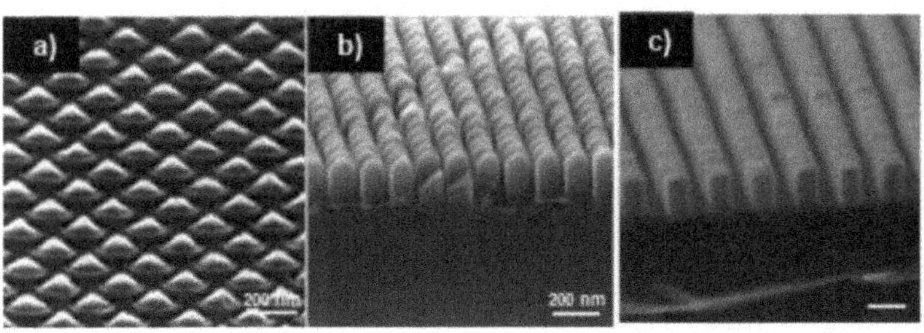

Figure 9.4. *SEM images of different ordered GaN NC morphologies, grown under the same nominal III/V ratio (= 3) but different temperatures; a) 940°C, b) 900°C and c) 860°C. (Adapted from [ALB 12])*

The control of the morphology of the top part of the GaN NC is critical to grow nanocolumnar heterostructures including embedded InGaN nanodisks, because the actual geometrical shape of the nanodisks will strongly vary, as shown by the HAADF-STEM images in Figures 9.5(a) and (b). For flat top GaN NCs the InGaN nanodisk has parallel (*c*-plane) interfaces to the GaN beneath and above it (cap layer) showing a true disk geometry. For pyramidal GaN NCs, the InGaN inclusion has parallel interfaces to the existing *r*-planes of both the GaN beneath and above it (cap). The optical properties of the two cases, with nominally identical nanodisks (same growth time and In flux) are quite different. The PL spectra shown in Figure 9.5(c) reveal a peak centered at 2.75 eV in the case of the "pyramidal-shaped" InGaN nanodisks, while the flat top nanodisk shows PL emission peaking at 2.32 eV, that is, ~0.4 eV red-shifted. Besides this energy shift, a reduction in the emission intensity by a factor higher than 40 can be observed for the flat top InGaN nanodisks. All these features can be understood as mainly due to the internal electric field (spontaneous and/or piezo origin) and the associated electron–hole wavelength overlap, that strongly depend on the crystal orientation.

Considering the "nominal" thickness of 25 nm for the InGaN nanodisks in both samples, pure quantum confinement effects can be neglected. However, the different polar character of the NCs planes; the different strain associated; and even possible differences in In incorporation on different crystal planes may be considered responsible for the observed differences. The reduced intensity observed in the case of InGaN nanodisks with *c*-plane top geometry might as well be partially caused by the presence of non-radiative defects, as reported for the case of InGaN/GaN QWs [CHE 01].

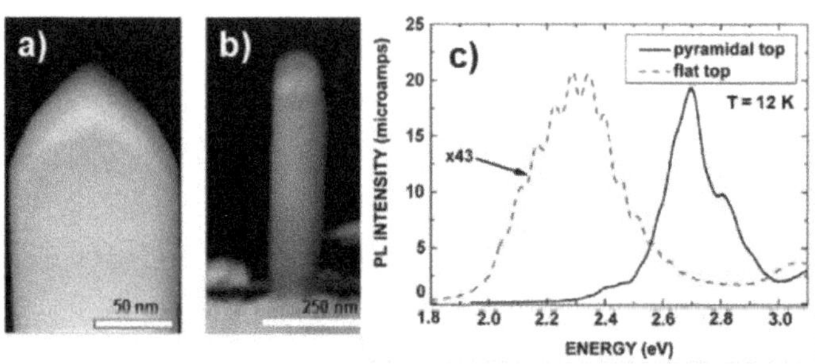

Figure 9.5. *a) STEM bright-field image of InGaN QDisk embedded in ordered GaN NCs with pencil like top, and b) with flat to; c) low temperature PL measurements of samples shown in Figures 9.5(a) and (b). (Adapted from [ALB 12])*

To summarize this section, ordered GaN NCs can be grown by PAMBE on GaN templates at different substrate temperatures and impinging Φ_{Ga} fluxes. Depending on the growth temperature, three growth regimes can be distinguished: (1) GaN pyramids with no further vertical growth, (2) GaN NCs with a top formed by *r*-planes, and (3) GaN NCs with *c*-plane top surfaces. The effects of the growth temperature and ϕ_{Ga}on the NCs morphology have been described in terms of changes in the local III/V ratio within the nanoholes. The impact of morphology changes, in particular, the polar character of the planes acting as growth front, on the optical properties of InGaN nanodisks has been assessed by low temperature PL measurements that show a clear red-shift and reduced PL emission intensity for the InGaN nanodisks grown on flat top (*c*-plane) GaN NCs.

9.4. Growth of ordered NCs for LEDs applications

This section is dedicated to the SAG of InGaN/GaN NC structures on Ti-masked GaN/sapphire templates by PAMBE, as opposed to self-assembled (not ordered) InGaN heterostructures (for more details, refer to Chapter 2 in Volume 2). Ordered InGaN NCs emitting in the red, green and blue (RGB) spectral range are described. Once the control of emission wavelength is achieved, white light emission is obtained at room temperature by either using composition-graded InGaN active regions or a monolithic integration of RGB emitting InGaN sections in each NC, making these structures attractive building blocks for new-generation RGB nano-LEDs.

When growing InGaN/GaN nanocolumnar heterostructures, selectivity may be lost due to the much lower temperatures needed to incorporate In. One of the requirements for the selective growth of InGaN NCs on top of GaN NCs is a sufficient length of the latter in order to prevent (by shadowing effect) the nucleation of InGaN on the mask. To reach this critical GaN NC length in a reasonable time, a high growth rate is needed. For this purpose, in all samples used the InGaN region started after growing 600 nm high GaN NCs, following a two step process: (1) 1 h growth of GaN NCs at 880°C with an impinging Ga-flux (Φ_{Ga}) of 16 nm/min and a N-flux (Φ_N) of 5 nm/min, (2) 1 h of GaN NC grown at 840°C with Φ_{Ga}= 8 nm/min and Φ_N= 14 nm/min. The first step at higher temperature and lower active nitrogen is needed to achieve selectivity [BEN 11]. The growth rate under these conditions is about 3 nm/min. Step two, with a lower temperature and higher active nitrogen, provides a growth rate of 7 nm/min while keeping selectivity and nanocolumnar shape.

9.4.1. *InGaN with single color emission*

The first objective of this section is to study the impact of the growth temperature, In/Ga ratio, and III/V ratio on the SAG of InGaN/GaN NCs with the goal to achieve emission in the blue, green and red spectral range. For this purpose, a series of four InGaN samples (T1 to T4) was grown for 1 h under the conditions described in Table 9.1.

Sample number	Impinging fluxes in nm/min (x10[14] atoms/s cm[2])			T_{sample} in °C
	Ga	In	N	
T1	4.3 (3.17)	4.3 (2.27)	10 (7.37)	750
T2	4.3 (3.17)	4.3 (2.27)	10 (7.37)	700
T3	4.3 (3.17)	4.3 (2.27)	10 (7.37)	650
T4	2.3 (1.7)	6.3 (3.33)	14 (10.3)	650

Table 9.1. *Growth conditions for the InGaN NCs grown on top of GaN NCs*

Figure 9.6 shows cross-sectional SEM pictures of the samples described in Table 9.1 (T1 to T4). Considering the height of the initial GaN part (600 nm) the growth rates of the InGaN region of InGaN/GaN samples can be directly estimated from these images, being 2, 4 and 2.2 nm/min for T1, T2 and T3 respectively. These results indicate a growth rate increase from 750°C to 700°C followed by a decrease from 700°C to 650°C.

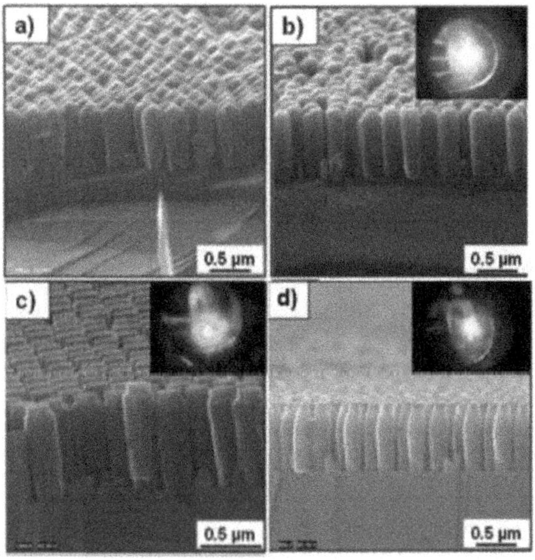

Figure 9.6. *SEM images of samples: a) T1, b) T2, c) T3 and d) T4 (insets show photographs of the emission colour taken during the PL experiments). For a color version of this figure, see www.iste.co.uk/consonni/nanowires1.zip*

This behavior can be interpreted in terms of a higher metal (mostly Ga) availability from 750°C to 700°C due to a reduced metal desorption rate [GAL 07], while from 700°C to 650°C, in addition to a further decrease of metal desorption, a significant reduction of the Ga-adatom diffusion length along the NC sidewalls will occur, leading to a shortage in Ga supply to the NC top side and a subsequent reduction of the InGaN growth rate. This implies that the shortage of Ga supply must dominate over the lower In desorption rate in the 700°C to 650°C range. Aside from the effects on growth rate, both mechanisms (lower In desorption and Ga shortage) would determine a strong increase of the In% upon reduction of the growth temperature, as it will be shown later by PL measurements.

The InGaN growth rate in sample T4 (3.8 nm/min) is 1.6 nm/min higher than that of T3 grown at the same temperature (650°C). Although the total amount of group III atoms supplied in sample T4 is lower than in sample T3 (around 8% less), it has higher In/Ga ratio and higher nitrogen flux that enhances the probability of In incorporation, compared to sample T3. Based on this and the increased growth rate, a higher In% is expected in sample T4.

Low temperature (LT) PL spectra of samples T1 to T4 are shown in Figure 9.7(a). The PL spectrum of sample T1 shows the characteristic strain-free GaN D^0X line at 3.47 eV with full width half maximum (FWHM) of 6 meV revealing a very high crystal quality, and the absence of any InGaN peak. This is not surprising, considering that 750°C is too high to allow for any measurable In incorporation.

However, PL spectra from samples T2 and T3 show a continuous red-shift that scales inversely with the growth temperature, corresponding to an increase of In%. Sample T4, as it was expected, shows a further increase in the In content indicated by the PL peak located at 2.04 eV. The In content estimated from the PL peak position [WU 03] is ~18% for T2, ~30% for T3, and 38% for T4.

Figure 9.7(b) shows the LT-CL spectra of a single NC in sample T4, where several In-related peaks are observed. Spatially resolved CL measurements performed in the same NC from sample T4 (Figure 9.7(c)) reveal a red-shift toward the NC apex. This behavior is assumed to be present in all samples T2, T3 and T4, as indicated by the high energy broad PL shoulder observed in PL (Figure 9.7(a)). Changes of In composition may be due to a combination of the following mechanisms: (1) at growth temperatures between 650°C and 700°C InN decomposition still plays a significant role leading to a partial decomposition of the InGaN NC section, which in turn will supply more In to the growth front yielding a higher In composition towards the top; (2) shadowing effects due to an NC diameter increase during the InGaN growth are more effective on the Ga atoms diffusing up the sidewalls, compared to In atoms, due to their different diffusion lengths [SEK 10]; and (3) the lattice pulling effect, i.e. the effect of strain, present in InGaN

NCs [TOU 11], on the In incorporation in the InGaN lattice. Indium atoms tend to segregate to more relaxed regions in order to minimize strain energy, leading to a lower In% region close to the InGaN/GaN interface [PER 01]. The appearance of a distinct CL peak at 2.84 eV, seemingly originating from the InGaN/GaN interface (Figures 9.7(b) and (c)), may indicate an abrupt partial strain relaxation when reaching a certain critical thickness.

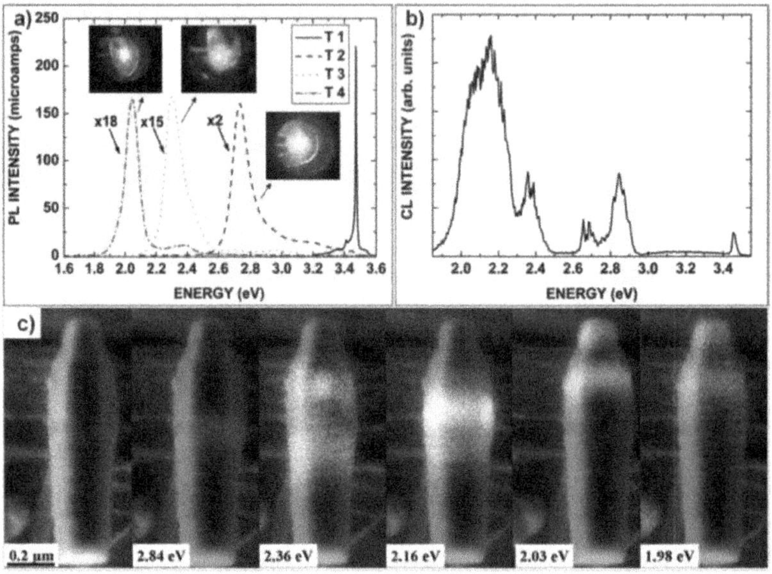

Figure 9.7. *a) 12 K PL spectra of InGaN/GaN NCs samples (T1 to T4) showing blue, green and red emission (inset) , b) 10 K CL spectra from a single InGaN/GaN NC of sample T4, c) spatially resolved CL measurements at 10 K in a single InGaN/GaN NC of sample T4 (corresponds to the single NC spectra shown in Figure 9.7(b)). (Adapted from [ALB 13a]). For a color version of this figure, see www.iste.co.uk/consonni/nanowires1.zip*

All the results presented so far were obtained using GaN templates. However, there is also a great interest in growing ordered nanostructures on Si(111) substrates. SAG of III nitride NCs on Si substrates would result in cheap, easy to process devices. A few attempts at MBE SAG of GaN NCs directly (no buffer) on masked Si substrates showed that multiple very thin NCs nucleate within each mask nanohole (Figure 9.8) [KIS 08, CAL 07]. It seems that the wetting of GaN on bare Si surfaces is quite difficult. These tiny NCs may eventually merge into a single one, but plenty of defects generated upon NCs coalescence. To overcome this problem, GaN-buffered Si substrates can be used [ALB 13c].

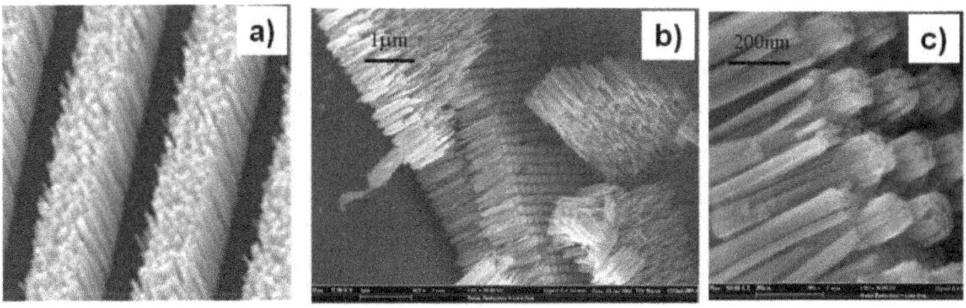

Figure 9.8. *a) SAG of GaN NCs on Si(111) using a mask with stripes, (adapted from [KIS 08]), b) and c) SAG of GaN NCs on Si(111) using a mask with nanoholes, (adapted from [CAL 07]). In both cases, multiple GaN NCs nucleate within the openings (either stripes or holes), not obtaining a single NC*

Carrying the same study as the one presented above on GaN templates, perfect selectivity can also be achieved and ordered InGaN NCs with a given nominal In composition are grown by SAG [ALB 13c]. Figure 9.9 shows the SEM picture and PL spectra of ordered InGaN/GaN NCs grown on GaN-buffered Si(111) substrates. For this particular sample, the PL emission is centered on 2.2 eV (green). A PL-related internal quantum efficiency (IQE) at room temperature can be estimated from the ratio between the integrated PL intensities at room temperature and low temperature (T = 12 K), assuming an IQE of 1 at T = 12 K, and obtaining a value around 40% for this sample, which is very high.

The possibility of tuning the In composition in the InGaN section by modifying the growth parameters is also confirmed. For this, a series of samples were grown (S1 to S4 in Figure 9.10) covering a wide range of In compositions all the way to 100% In content (pure InN). More details can be found in reference [ALB 13c]. Normally, a decrease of the growth temperature leads to a pronounced red-shift of the PL emission, indicating an increase in the In incorporation. However, besides the growth temperature, the III/N ratio can also be used to change the In content. In order to achieve the SAG of InN NCs (100% In) on top of GaN, sample S4 was grown at much lower temperature (300°C) to drastically reduce the InN decomposition. In all cases, the nanocolumnar shape, as well as the selectivity, is retained (not shown here). The PL spectrum of sample S4 (Figure 9.10) shows a peak at 0.78 eV, which is a bit higher than the energy corresponding to fully relaxed high quality (low residual doping) InN layers, which is known to be at 0.63 eV [GAL 06]. Most likely the origin of such a blue-shift in the PL emission from InN NCs must be the presence of a high electron concentration. It should be noted also that the growth temperature used for this sample (300°C) is well below the optimal one needed for InN (around 475°C) [GRA 07].

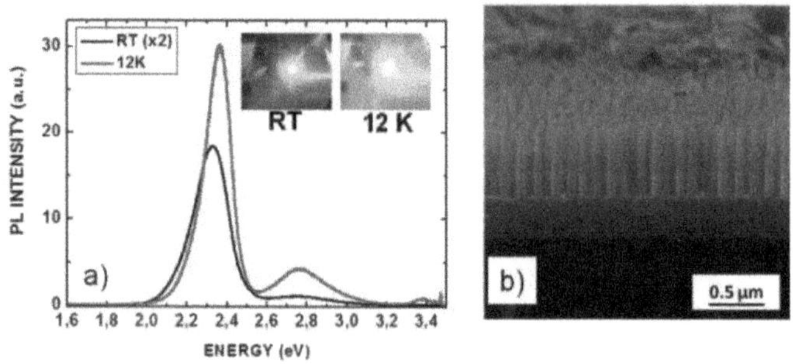

Figure 9.9. *InGaN/GaN NCs grown on GaN-buffered Si(111) substrates: a) RT and 12 K PL spectra, b) SEM image. For a color version of this figure, see www.iste.co.uk/consonni/nanowires1.zip*

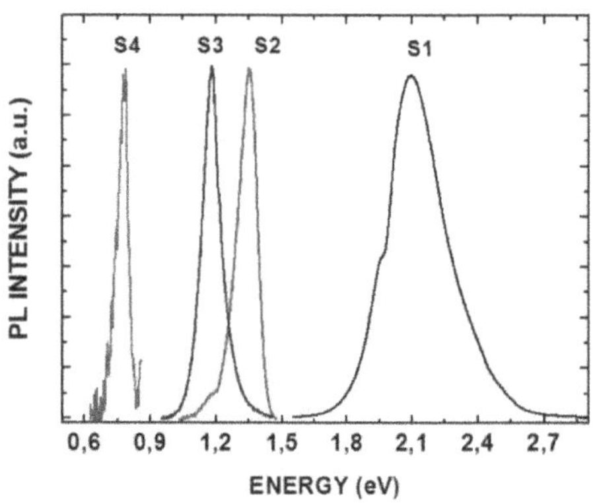

Figure 9.10. *12 K PL spectra of InGaN/GaN NCs samples (S1 to S4) grown on GaN-buffered Si(111) substrates, covering the whole composition range. For a color version of this figure, see www.iste.co.uk/consonni/nanowires1.zip*

To summarize, ordered InGaN NCs are obtained over a wide and the whole composition range, on GaN templates and on GaN-buffered Si(111) substrates respectively. Following the control of wavelength, two different approaches are presented in the next two sections to generate white light emission: (1) composition-graded InGaN active regions, and (2) by monolithic integration of RGB emitting InGaN sections in each NC.

9.4.2. *InGaN with a gradient In composition for white light emission*

This section deals with the SAG of InGaN/GaN nanostructures for white light LEDs by using composition graded InGaN active regions. For this purpose, two samples (with the growth conditions described in Table 9.2) are analyzed. In sample T5 the InGaN region was grown for 1 h using a temperature gradient from 725°C to 625°C. Sample T6 has the same InGaN graded region as T5 but with the addition of three thin GaN barriers (grown for 5 min; Φ_{Ga}= 2.3 nm/min; Φ_N= 14 nm/min) within the graded InGaN region at specific temperature values (700°C, 675°C, and 650°C).

Sample number	Impinging fluxes in nm/min (x10^{14} atoms/s cm^2)			T_{sample} in °C
	Ga	In	N	
T5	2.3 (1.7)	6.3 (3.33)	14 (10.3)	725→625
T6	2.3 (1.7)	6.3 (3.33)	14 (10.3)	725→625

Table 9.2. *Growth conditions for the InGaN NCs with gradient composition*

Figure 9.11(a) shows RT and LT PL spectra of sample T5 where a broad band (FWHM of 0.51 eV) yields white light emission at LT. Spatially resolved CL data taken at 10 K from a single NC (Figure 9.11(c)) show a monotonic decrease of energy from 2.98 eV in the middle-low NC region down to 2.12 eV at the NC apex, as expected from the growth temperature gradient.

The temperature gradient, from 725°C to 625°C, implies an increasing In content toward the NC apex caused by: (1) a reduced In desorption and InN decomposition, and (2) a reduction of the Ga-atoms diffusion length along the NC sidewalls leading to an effective Ga shortage at the growth front (NC apex). In addition, the mechanisms discussed before as the origin of In% changes in nominally composition fixed InGaN regions, may also be present, though their effects may be more complicated to estimate due to the temperature gradient used.

RT-PL intensity from sample T5 shows a significant quenching, particularly strong for the medium–high energy range of the spectrum (Figure 9.11(a)), that yields a greenish-yellow emission. This behavior, reported before in self-assembled InGaN/GaN NCs [ALB 11, GUO 10], was attributed to carrier diffusion to higher In% regions (higher localization) upon thermal excitation. In order to check this effect, in sample T6, several GaN barriers were incorporated during the InGaN region growth (at 700°C, 675°C and 650°C) trying to block or minimize carrier diffusion. Figure 9.11(b) shows a broad PL spectrum at 12 K (FWHM of 0.34 eV) peaking at 2.4 eV, similar to that observed in sample T5. However, the RT spectrum, though still showing a significant quenching, still retains emission peaks at 1.95 eV, 2.16 eV, and 2.5 eV, leading to yellow-white emission (inset in Figure 9.11(b)). The RT emission improvement is likely due to the specific structure of this sample that includes thin GaN barriers within the active InGaN graded region. Together with a reduced intensity quenching, sample T6 also shows improved internal quantum efficiency (IQE) estimated from PL integrated intensity ratio (from 1.7 eV to 3.3 eV) measurements at 12 K and RT, being 4.7% for sample T5 and 7.8% for sample T6.

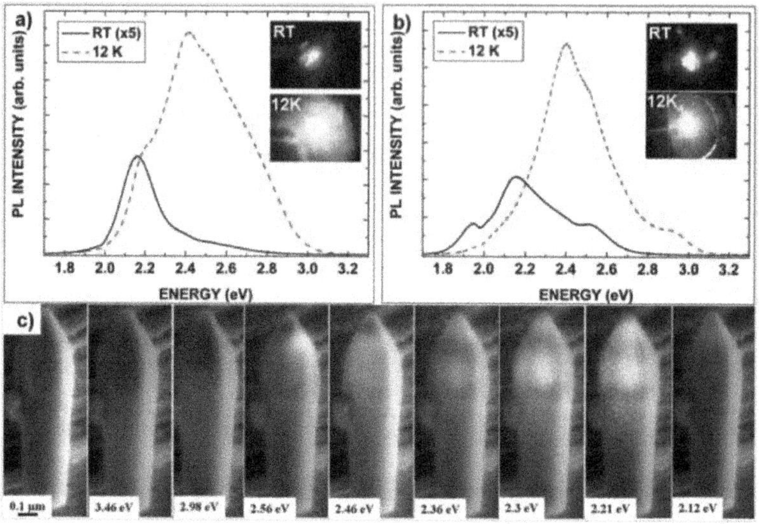

Figure 9.11. *PL spectra of InGaN/GaN NCs of samples T5 and T6 with graded In composition due to a temperature gradient of 1.67 °C/min: a) without GaN barriers, b) with GaN barriers inserted at 700°C, 675°C and 650°C (insets show photographs of the emission color taken during the PL experiments) and c) spatially resolved CL measurements at 10 K in a single InGaN/GaN NC of sample T5. (Adapted from [ALB 13a]). For a color version of this figure, see www.iste.co.uk/consonni/nanowires1.zip*

To summarize this section, ordered InGaN/GaN NCs with a composition graded InGaN active regions showed white light emission. Thermal quenching of the high energy emission PL intensity in these graded samples, likely due to carrier diffusion to higher In% regions, is significantly reduced by insertion of thin GaN barriers. Optimized nanostructure design produces yellow-white emission at RT.

9.4.3. *RGB structures for white light emission*

In this case two samples (T7, T8) are analyzed with a InGaN region formed by stacking InGaN portions emitting in the RGB range (samples T4, T3 and T2 of previous section), respectively. In both samples the blue emitting part is grown first, followed by the green emitting part and the red part. This stacking order is chosen in order to minimize the decomposition of the high In-content part, i.e. red emitting part. The overall growth time of the three InGaN sections was 1 h in both samples, while in sample T7, total time was shared equally by the three regions for 20 min (20/20/20), in sample T8 the growth times were modified to 8/10/41 min to demonstrate the emission tuneability by changing the thickness of the respective InGaN parts and also to enhance the low energy emission part.

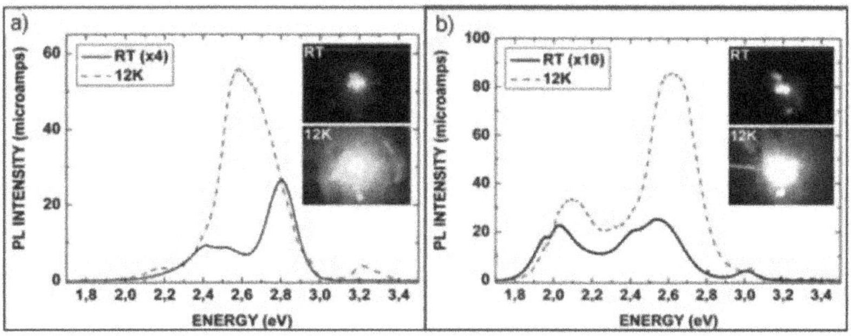

Figure 9.12. *Room and low temperature (12 K) PL spectra of a) samples T7 and b) T8. (Adapted from [ALB 13b]). For a color version of this figure, see www.iste.co.uk/consonni/nanowires1.zip*

Figure 9.12(a) shows LT and RT PL spectra of sample T7 in which each segment of the RGB structure was grown for 20 min. A blue-white emission is observed both at LT and RT, though a significant intensity quenching occurs at RT, particularly strong for the green-blue range. In addition, the spectrum does not show a significant contribution in the red region. In order to tune the RGB structure towards real white light emission, sample T8 was grown using different growth times for each color section, namely, 8 min for the blue, 11 min for the green and 41 min for the red, aiming to enhance the low-energy (red) emission. The growth times of the respective InGaN regions were calculated based on the integrated RT-PL intensities of the single color emissions of samples T2–T4 with the goal to achieve a more balanced spectral contribution of the three regions. Figure 9.12(b) shows LT and RT-PL spectra of sample T8 covering the energy range from 1.8 to 2.8 eV (almost the whole visible spectrum) yielding white colour. Still a significant intensity quenching at RT is observed at the green-blue region of the spectrum. This behavior, reported before for self-assembled InGaN/GaN [ALB 11, GUO 10] is suggested to be related to carrier diffusion to regions of higher In% (higher localization) upon thermal excitation.

Therefore, it is shown that the monolithic integration of three InGaN portions emitting in the red-green-blue colors in a single nanocolumnar heterostructure leads to white light emission. It is also proven that the emission color can be tuned by playing with the thickness variation of the respective InGaN regions.

9.5. Growth of ordered GaN nanocolumns on non-polar and semi-polar directions

An important issue concerning optoelectronic devices based on III-nitrides is the presence of strong internal polarization fields that may be detrimental to the efficiency. This is particularly relevant for the case of heterostructures grown along the *c*-axis, which is the most common growth direction. To overcome this problem, the growth on non-polar or semi-polar directions (where polarization fields are zero or reduced) has been proposed. There is currently an intense activity dedicated to the growth of high quality non-polar and semi-polar material [MIE 10], with a particular emphasis on non-polar LEDs [KIM 07, FAR 12]. However, the growth of high crystal quality non-polar or semi-polar GaN thick layers is still very challenging.

In the case of SAG of GaN NCs on *c*-plane GaN templates, it was shown in previous sections that the growth front is generally formed by semi-polar (*r*-planes) that yield a "pencil-like" shape [BEN 11]. This profile is then transferred to the InGaN QDisks embedded within the GaN NC. Though in this structure, the effects of internal fields in the active region of the device (i.e. nanoLED) may be reduced compared with polar planes, the most effective solution is to grow along non-polar directions, either by growing core–shell (coaxial) heterostructures on the lateral non-polar facets (*m*-planes) of axial polar NCs (for more details, refer to Chapter 6 in Volume 2), or by growing axial heterostructures on non-polar planes (substrates) such as [10-10] (*m*-planes) or [11-20] (*a*-planes). In both approaches, the internal fields are nominally zero. This section analyzes the SAG of GaN NCs on non-polar and semi-polar GaN templates using colloidal lithography.

Figure 9.13 shows SEM images of SAG *a*-plane [11-20] GaN NCs keeping a perfect selectivity. However morphology differences are clearly observed when compared with previous results on *c*-plane GaN. While *c*-plane GaN NCs had a typical hexagonal structure with pyramidal top surfaces (pencil-like), nanostructures grown on *a*-plane substrates show elongated geometries with several lateral facets, as observed by SEM and AFM (inset in Figure 9.13.(a)).

The quality of these NCs was analyzed by low temperature (8 K) PL measurements. A clear improvement is observed when comparing the PL spectra from the initial *a*-plane GaN template and the one from the SAG material (Figure 9.14). Besides the much higher intensity emission of the NCs (a factor of 500), the initial template shows a broad yellow band emission centered around 2.25 eV, related to defects, which is not present in the SAG sample.

Figure 9.13. *Plane view and cross section SEM pictures of a-plane SAG GaN NCs. The inset in a) corresponds to an AFM measurement. (Adapted from [BEN 12]). For a color version of this figure, see www.iste.co.uk/consonni/nanowires1.zip*

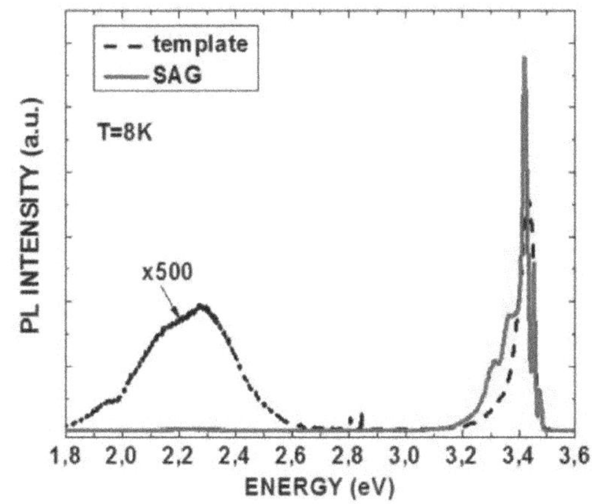

Figure 9.14. *Low temperature (8 K) PL spectra of the bare a-plane (11-20) GaN template and after SAG of GaN NCs. For a color version of this figure, see www.iste.co.uk/consonni/nanowires1.zip*

In order to check the morphology evolution of the *a*-plane nanostructures with time, a second sample, similar to the first sample, was grown during a longer time (Figure 9.13(c)). As growth proceeds, the nanostructures' shape evolves toward elongated and arrow-headed one, in a similar way as those grown by MOVPE [JIN 09]. Figures 9.15(a)–(c) show magnified images of the evolution sequence. In some cases the onset of the nanostructures' coalescence along the *c*-axis can also be observed (Figure 9.15(c)). It is quite significant to notice that the preferential growth direction is the *c*-axis (in plane) leading to a strong growth anisotropy between *a*-and *c*-planes.

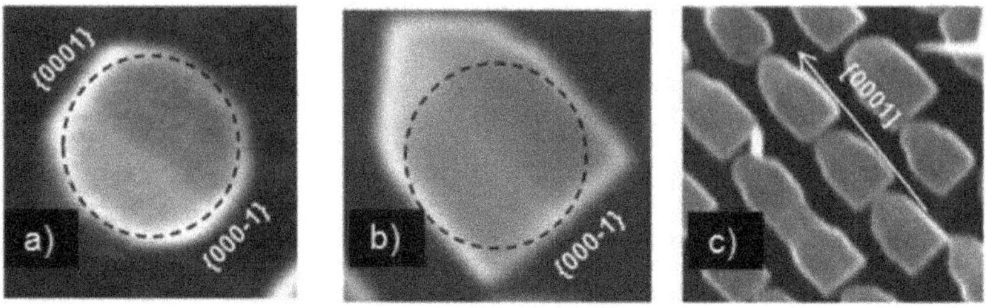

Figure 9.15. *Magnified images of the typical morphology evolution during the growth of sample in Figure 9.13(c). The dashed circle indicates the initial nanohole in the Ti mask. (Adapted from [BEN 12])*

Figure 9.16 shows the results of a longer growth time, where almost full coalescence is achieved (as stripes along the c-direction). This may result in very interesting to obtain high quality, low dislocation non-polar GaN films on non-polar GaN templates of much lower quality.

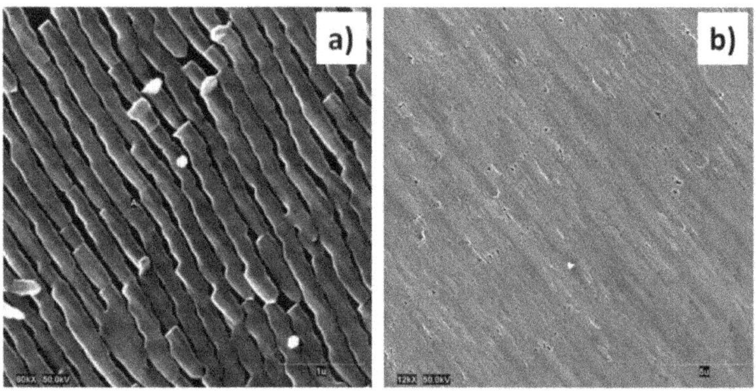

Figure 9.16. *SEM images of non-polar GaN NCs, as those in Figure 9.13, but with a longer growth time. Coalescence along stripes in the (0001) direction is clearly observed*

SAG of GaN NCs was also performed on semi-polar [11-22] GaN templates. Under similar growth conditions as for non-polar templates, perfect selectivity is achieved. The morphology of the nanostructures selectively grown was, again, different to that of non-polar substrates. Figure 9.17 shows SEM pictures of samples grown during 1 h (Figure 9.17(a) and (b)), and 3 h (Figures 9.17(c) and (d)). For a long growth time, the nanostructures no longer grow perpendicular to the substrate, but tilted around 32° relative to the substrate surface. TEM analysis (not shown here) proves that the GaN nanostructures' elongation (preferential growth axis) occurs parallel to the [0001] Ga-polar direction and the perfect epitaxial alignment with the substrate. Two features can be mentioned: (1) three facets different in nature (polar, semi-polar, and non-polar) are exposed to the molecular beams; and (2) from TEM measurements, most of the GaN nanostructures volume seems to be free of defects, because stacking faults from the substrate are filtered reaching only the nanostructures bottom area. This last result is also very interesting when trying to apply this type of nanostructure for the generation of defect-free semi-polar GaN templates by the coalescence of the NCs.

Figure 9.17. *SEM images of GaN ordered growth by MBE on semi-polar [11-22] templates. a) and b) correspond to 1 h growth and c) and d) to 3 h growth. (Adapted from [BEN 13])*

The crystal quality of these nanostructures, as compared to that of the substrate improved quantitatively (by several orders of magnitude) and qualitatively (excitonic features dominate over stacking faults-related luminescence), as confirmed by low temperature (8 K) PL measurements (Figure 9.18). The broad emission from the (11–22) GaN template at around 3.43 eV, is due to luminescence from boundary stacking faults (BSFs) [LIU 05]. After SAG, a dominant band edge emission related to the donor bound exciton DX_0 at 3.473 eV, not even observed in the substrate, with a line width of only 2.7 meV, is observed. This is indicative of a high quality and strain-free material. In addition, a weak emission at 3.42 eV (due to BSFs) can be still observed, which is assumed to originate from either the underlying GaN template or from the bottom part of the NCs.

To summarize, colloidal lithography is successfully applied for the SAG by MBE of non-polar [11-20] and semi-polar [11-22] GaN NCs. In both cases, PL analysis shows a strong improvement in crystal quality after SAG, due to an efficient filtering of stacking faults and dislocations coming from the initial template. In both cases the preferential growth direction corresponds to the *c*-plane. These achievements may represent the first steps for future optoelectronic applications, where, in addition to an optimal crystal quality, internal electric fields can be drastically reduced or avoided.

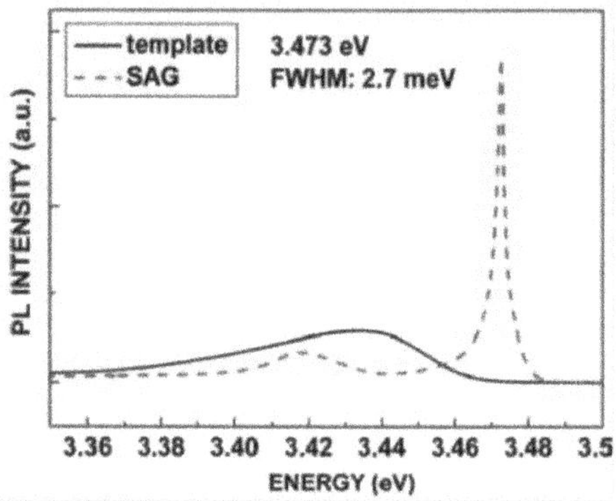

Figure 9.18. *Low temperature (8 K) PL spectra of the bare (11–22) GaN template and after SAG of GaN NCs. Adapted from [BEN 13]. For a color version of this figure, see www.iste.co.uk/consonni/nanowires1.zip*

9.6. Summary

The structure and morphology of the nanohole-patterned Ti mask becomes the driving factor that determines whether SAG is achieved or not. Main features of the mask that might negatively influence the ordered growth are surface roughness, adhesion of the Ti to the GaN substrate and surface cleaning. EBL, FIB and COL processes can be used to fabricate good quality masks patterned with nanoholes, which serve as nucleation sites for SAG of GaN NCs. The first two processes (EBL and FIB), while providing perfect uniformity (ordered arrangement of the holes), require special equipments (expensive and time-consuming) and critical cleaning stages following the etching steps. The COL process is a much faster and cheaper process that requires neither sophisticated systems (like EBL or FIB) nor special cleaning steps, since there is no etching process involved (only lift-off). However, the arrangement of the nanoholes is not perfectly uniform.

Ordered GaN NCs are grown by PAMBE on GaN templates. Depending on the growth conditions, three growth regimes can be distinguished: (1) GaN pyramids with no further vertical growth, (2) GaN NCs with a top formed by *r*-planes, and (3) GaN NCs with *c*-plane top surfaces. The impact of morphology changes, in particular, the polar character of the planes acting as a growth front, on the optical properties of InGaN nanodisks has been assessed by low temperature PL measurements that show a clear red-shift and reduced PL emission intensity for the InGaN nanodisks grown on flat top (*c*-plane) GaN NCs.

Ordered InGaN NCs are obtained over a wide and the whole composition range, on GaN templates and on GaN-buffered Si(111) substrates respectively. Following the control of the In composition (i.e. wavelength emission), two different approaches are presented to generate white light emission: (1) composition-graded InGaN active regions, and (2) by monolithic integration of RGB emitting InGaN sections in each NC.

The composition graded InGaN active regions emitted white light, though a strong thermal quenching of the PL intensity, likely due to carrier diffusion to higher In% regions is observed. This effect is significantly reduced by insertion of thin GaN barriers. An optimized nanostructure design produces yellow-white emission at RT.

However, the monolithic integration of three InGaN portions emitting in the red-green-blue colors in a single nanocolumnar heterostructure also leads to a strong white light emission. In this case, it is also proven that the emission color can be tuned by playing with the thickness variation of the respective InGaN regions.

Finally, SAG by MBE of non-polar [11-20] and semi-polar [11-22] GaN NCs is presented. In both cases, PL analysis shows a strong improvement in crystal quality after SAG, due to an efficient filtering of stacking faults and dislocations coming from the initial template. In both cases the preferential growth direction corresponds to the *c*-plane. These achievements may represent the first step for future optoelectronic applications, where, in addition to an optimal crystal quality, internal electric fields can be strongly reduced or avoided.

9.7. Bibliography

[ADE 02] ADELMANN C., BRAULT J., JALABERT D., *et al.*, "Dynamically stable gallium surface coverages during plasma-assisted molecular-beam epitaxy of (0001) GaN", *Journal of Applied Physics*, vol. 91, p. 9638, 2002.

[ALB 11] ALBERT S., BENGOECHEA-ENCABO A., LEFEBVRE P., *et al.*, "Emission control of InGaN nanocolumns grown by molecular-beam epitaxy on Si(111) substrates", *Applied Physics Letters*, vol. 99, p. 131108, 2011.

[ALB 12] ALBERT S., BENGOECHEA-ENCABO A., LEFEBVRE P., *et al.*, "Selective area growth and characterization of InGaN nano-disks implemented in GaN nanocolumns with different top morphologies", *Applied Physics Letters*, vol. 100, p. 231906, 2012.

[ALB 13a] ALBERT S., BENGOECHEA-ENCABO A., SANCHEZ-GARCIA M.A., *et al.*, "Selective area growth and characterization of InGaN nanocolumns for phosphor-free white light emission", *Journal of Applied Physics*, vol. 113, p. 114306, 2013.

[ALB 13b] ALBERT S., BENGOECHEA-ENCABO A., SANCHEZ-GARCIA M.A., *et al.*, "Monolithic integration of blue, green, and red light emitting nanocolumnar structures for phosphor free white light emission", *Applied Physics Letters*, vol. 102, p. 181103, 2013.

[ALB 13c] ALBERT S., BENGOECHEA-ENCABO A., SÁNCHEZ-GARCÍA M.A., *et al.*, "Selective area growth of In(Ga)N/GaN nanocolumns by molecular beam epitaxy on GaN-buffered Si(111): from ultraviolet to infrared emission", *Nanotechnology*, vol. 24, pp. 175303–175310, 2013.

[BAR 11] BARBAGINI F., BENGOECHEA-ENCABO A., ALBERT S., *et al.*, "Critical aspects of substrate nanopatterning for the ordered growth of GaN nanocolumns", *Nanoscale Research Letters*, vol. 6, p. 632, 2011.

[BAV 11] BAVENCOVE A.L., TOURBOT G., GARCIA J., *et al.*, "Submicrometre resolved optical characterization of green nanowire-based light emitting diodes", *Nanotechnology*, vol. 22, p. 345705, 2011.

[BEN 11] BENGOECHEA-ENCABO A., BARBAGINI F., FERNANDEZ-GARRIDO S., *et al.*, "Understanding the selective area growth of GaN nanocolumns by MBE using Ti nanomasks", *Journal of Crystal Growth*, vol. 325, p. 89, 2011.

[BEN 12] BENGOECHEA-ENCABO A., ALBERT S., SANCHEZ GARCIA M.A., *et al.*, "Selective area growth of a- and c-plane GaN nanocolumns by molecular beam epitaxy using colloidal nanolithography", *Journal of Crystal Growth*, vol. 353, pp. 1–4, 2012.

[BEN 13] BENGOECHEA-ENCABO A., ALBERT S., ZUÑIGA-PEREZ J., *et al.*, "Selective area growth and characterization of InGaN/GaN nanocolumns on semi-polar (11-22) GaN templates", *Applied Physics Letters*, vol. 103, p. 241905, 2013.

[CAL 00] CALLEJA E., SANCHEZ-GARCIA M.A., SANCHEZ F.J., *et al.*, "Luminescence properties of GaN nanocolumns grown by molecular beam epitaxy", *Physical Review B*, vol. 62, pp. 16826–16834, 2000.

[CAL 07] CALLEJA E., RISTIĆ J., FERNÁNDEZ-GARRIDO S., *et al.*, "Growth, morphology, and structural properties of group-III-nitride nanocolumns and nanodisks", *Physica Status Solidi (b)*, vol. 244, pp. 2816–2837, 2007.

[CHE 01] CHERNS D., HENLEY S.J., PONCE F.A., "Edge and screw dislocations as nonradiative centers in InGaN/GaN quantum well luminescence", *Applied Physics Letters*, vol. 78, p. 2691, 2001.

[CON 14] CONSONNI V., FEUILLET G., (eds.), *Wide Band Gap Semiconductor Nanowires 2: Heterostructures and Optoelectronic Devices*, ISTE, London, and John Wiley & Sons, New York, 2014.

[DAV 02] DAVYDOV V.Y., KLOCHIKHIN A.A., SEISYAN R.P., *et al.*, "Absorption and emission of hexagonal InN. Evidence of narrow fundamental band gap", *Physica Status Solidi (b)*, vol. 229, p. R1, 2002.

[ERI 11] ERIKSSON T., YAMADA S., KRISHNAN P., *et al.*, "High volume nanoimprint lithography on III/V substrates: Imprint fidelity and stamp lifetime", *Microelectronic Engineering*, vol. 88, p. 293, 2011.

[FAR 12] FARRELL R.M., YOUNG E.C., WU F., *et al.*, "Materials and growth issues for high-performance nonpolar and semipolar light-emitting devices", *Semiconductor Science & Technology*, vol. 27, pp. 024001–15, 2012.

[FER 08] FERNANDEZ-GARRIDO S., KOBLMULLER G., CALLEJA E., *et al.*, "*In situ* GaN decomposition analysis by quadrupole mass spectrometry and reflection high-energy electron diffraction", *Journal of Applied Physics*, vol. 104, p. 033541, 2008.

[FER 09] FERNANDEZ-GARRIDO S., GRANDAL J., CALLEJA E., *et al.*, "A growth diagram for plasma-assisted molecular beam epitaxy of GaN nanocolumns on Si(111)", *Journal of Applied Physics*, vol. 106, p. 126102, 2009.

[GAL 06] GALLINAT C.S., KOBLMÜLLER G., BROWN J.S., *et al.*, "In-polar InN grown by plasma-assisted molecular beam epitaxy", *Applied Physics Letters*, vol. 89, p. 032109, 2006.

[GAL 07] GALLINAT C.S., KOBLMÜLLER G., BROWN J.S., *et al.*, "A growth diagram for plasma-assisted molecular beam epitaxy of In-face InN", *Journal of Applied Physics*, vol. 102, p. 064907, 2007.

[GRA 05] GRANDAL J., SANCHEZ-GARCÍA M.A., CALLE F., *et al.*, "Morphology and optical properties of InN layers grown by molecular beam epitaxy on Silicon substrates", *Physica Status Solidi (c)*, vol. 2, no. 7, pp. 2289–2292, 2005.

[GRA 07] GRANDAL J., SÁNCHEZ-GARCÍA M.A., CALLEJA E., *et al.*, "Accommodation mechanism of InN nanocolumns grown on Si(111) substrates by molecular beam epitaxy", *Applied Physics Letters*, vol. 91, p. 021902, 2007.

[GUO 10] GUO W., ZHANG M., BANERJEE A., *et al.*, "Catalyst-free InGaN/GaN nanowire light emitting diodes grown on (001) silicon by molecular beam epitaxy", *NanoLetters*, vol. 10, p. 3355, 2010.

[HER 06] HERSEE S., SUN X., WANG X., "The controlled growth of GaN nanowires", *NanoLetters*, vol. 6, p. 1808, 2006.

[JIN 09] JINDAL V., SHAHEDIPOUR-SANDVIK F., "Theoretical prediction of GaN nanostructure equilibrium and nonequilibrium shapes", *Journal of Applied Physics*, vol. 106, p. 083115, 2009.

[KIM 07] KIM K., SCHMIDT MATHEW C., SATO H., *et al.*, "Study of nonpolar m-plane InGaN/GaN multiquantum well light emitting diodes grown by homoepitaxial metal-organic chemical vapor deposition", *Applied Physics Letters*, vol. 91, p. 181120, 2007.

[KIS 07] KISHINO K., KIKUCHI A., SEKIGUCHI H., *et al.*, "InGaN/GaN nanocolumn LED emitting from blue to red", *Proceeding of the SPIE*, vol. 6473, p. 64730T, 2007.

[KIS 08] KISHINO K., HOSHINO T., ISHIZAWA S., *et al.*, "Selective-area growth of GaN nanocolumns on titanium-mask-patterned silicon (111) substrates by RF-plasma-assisted molecular-beam epitaxy", *Electronic Letters*, vol. 44, p. 819, 2008.

[KIS 09] KISHINO K., SEKIGUCHI S., KIKUCHI A., "Improved Ti-mask selective-area growth (SAG) by rf-plasma-assisted molecular beam epitaxy demonstrating extremely uniform GaN nanocolumn arrays", *Journal of Crystal Growth*, vol. 311, p. 2063, 2009.

[KUY 07] KUYKENDALL T., ULRICH P., ALONI S., *et al.*, "Complete composition tunability of InGaN nanowires using a combinatorial", *Nature Materials*, vol. 6, pp. 951–956, 2007.

[LEF 01] LEFEBVRE P., MOREL A., GALLART M., *et al.*, "High internal electric field in a graded-width InGaN/GaN quantum well: accurate determination by time-resolved photoluminescence spectroscopy", *Applied Physics Letters*, vol. 78, p. 1252, 2001.

[LI 11] LI S., FUENDLING S., WANG X., *et al.*, "Polarity and its influence on growth mechanism during MOVPE growth of GaN sub-micrometer rods", *Crystal Growth & Design*, vol. 11, p. 1573, 2011.

[LIN 10] LIN H.-W., LU Y.-J., CHEN H.-Y., *et al.*, "InGaN/GaN nanorod array white light-emitting diode ", Applied Physics Letters, vol. 97, p. 073101, 2010.

[LIU 05] LIU R., BELL A., PONCE F.A., *et al.*, "Luminescence from stacking faults in gallium nitride", *Applied Physics Letters*, vol. 86, p. 021908, 2005.

[MEI 06] MEIJERS R., RICHTER T., CALARCO R., *et al.*, "GaN-nanowhiskers: MBE-growth conditions and optical properties", *Journal of Crystal Growth*, vol. 289, pp. 381–386, 2006.

[MIE 10] DE MIERRY P., KRIOUCHE N., NEMOZ M., *et al.*, "Semipolar GaN films on patterned r-plane sapphire obtained by wet chemical etching", *Applied Physics Letters*, vol. 96, p. 231918, 2010.

[NAK 00] NAKAMURA S., FASOL G., PEARTON S.J., *The Blue Laser Diode: The Complete Story*, Springer, Berlin, 2000.

[PAR 04] PARK Y.S., PARK C.M., FU D.J., *et al.*, "Photoluminescence studies of GaN nanorods on Si (111) substrates grown by molecular-beam epitaxy", *Applied Physics Letters*, vol. 85, p. 5718, 2004.

[PER 01] PEREIRA S., CORREIA M.R., PEREIRA E., *et al.*, "Compositional pulling effects in InxGa1-xN/GaN layers:a combined depth-resolved cathodoluminescence and Rutherford backscattering/channeling study", *Physical Review B*, vol. 64, p. 205311, 2001.

[RIS 08] RISTIC J., CALLEJA E., FERNANDEZ-GARRIDO S., *et al.*, "On the mechanisms of spontaneous growth of III-nitride nanocolumns by plasma-assisted molecular beam epitaxy", *Journal of Crystal Growth*, vol. 310, pp. 4035–4045, 2008.

[ROM 06] ROMANOV A.E., BAKER T.J., NAKAMURA S., *et al.*, "Strain-induced polarization in wurtzite III-nitride semipolar layers", *Journal of Applied Physics*, vol. 100, p. 023522, 2006.

[SAN 98] SANCHEZ-GARCIA M.A., CALLEJA E., MONROY E., *et al.*, "The effect of the III/V ratio and substrate temperature on the morphology and properties of GaN- and AlN-layers grown by molecular beam epitaxy on Si(1 1 1)", *Journal of Crystal Growth*, vol. 183, p. 23, 1998.

[SEK 08] SEKIGUCHI H., KISHINO K., KIKUCHI A., "Ti-mask selective-area growth of GaN by RF-plasma-assisted molecular-beam epitaxy for fabricating regularly arranged InGaN/GaN nanocolumns", *Applied Physics Express*, vol. 1, p. 124002, 2008.

[SEK 10] SEKIGUCHI H., KISHINO K., KIKUCHI A., "Emission color control from blue to red with nanocolumn diameter of InGaN/GaN nanocolumn arrays grown on same substrate", *Applied Physics Letters*, vol. 96, p. 231104, 2010.

[TAR 97] TARSA E.J., HEYING B., HU X.H., *et al.*, "Homoepitaxial growth of GaN under Ga-stable and N-stable conditions by plasma-assisted molecular beam epitaxy", *Journal of Applied Physics*, vol. 82, p. 5472, 1997.

[THI 06] THILLOSEN N., SEBALD K., HARDTDEGEN H., *et al.*, "The state of strain in single GaN nanocolumns as derived from micro-photoluminescence measurements", *NanoLetters*, vol. 6, pp. 704–708, 2006.

[TOU 11] TOURBOT G., BOUGERO C., GRENIER A., *et al.*, "Structural and optical properties of InGaN/GaN nanowire heterostructures grown by PA-MBE", *Nanotechnology*, vol. 22, p. 075601, 2011.

[WU 02] WU J., WALUKIEWICZ W., YU K.M., *et al.*, "Unusual properties of the fundamental band gap of InN", *Applied Physics Letters*, vol. 80, p. 3967, 2002.

[WU 03] WU J., WALUKIEWICZ W., "Band gaps on InN and group III nitride alloys" *Superlattices & Microstructures*, vol. 34, pp. 63–75, 2003.

[YAM 08] YAM F.K., HASSAN Z., "InGaN: an overview of the growth kinetics, physical properties and emission mechanisms", *Superlattices & Microstructures*, vol. 43, pp. 1–23, 2008.

[YOS 97] YOSHIZAWA M., KIKUCHI A., MORI M., *et al.*, "Growth of self-organized GaN nanostructures on Al_2O_3(0001) by RF-radical source molecular beam epitaxy", *Japanese Journal of Applied Physics*, vol. 36, pp. L459–L462, 1997.

10

Metal-Organic Vapor Phase Epitaxy Growth of GaN Nanorods

10.1. Introduction

The growth of GaN wires[1] by metal-organic vapor phase epitaxy (MOVPE) can be seen in its historical context to understand the present challenges of these materials. The GaN compound was first synthesized by Johnson *et al.* in 1932 [JOH 32] by passing ammonia over hot Ga, and its wurtzite structure was determined by Jusa and Hahn in 1938 [JUS 38]. The first epitaxial growth of high-quality thin films occurred in 1969 by halide vapor phase epitaxy with the technological development of appropriate substrate materials (i.e. sapphire) [MAR 69]. These materials were n-type doped with electron concentrations ranging from 10^{18} to 10^{20} cm^{-3}. Such high n-type background concentrations made it difficult to achieve p-type doping. After this initial progress made during the 1970s, the pace of research on GaN became slow, because of the lack of high-quality epilayers inducing a very large density of extended defects, such as dislocations, and also because of the lack of success in making p-type GaN to realize p-n junctions. The origin of high background doping was attributed at this time to oxygen incorporation during the growth instead of nitrogen vacancies. Important progress has then been made since the 1980s with the development of several techniques: molecular beam epitaxy (MBE), hydride vapor phase epitaxy (HVPE) and metal-organic chemical vapor deposition (MOCVD). This last method, that will be detailed in this chapter, has been used for the realization of many semiconductor devices and complex heterostructures [STR 99, WAT 13] (for more details, refer to Volume 2 [CON 14]). Strenuous efforts were made to optimize growth conditions and introduce more

Chapter written by Joël EYMERY.

1 Also called *rods* or *pillars* in the literature as a function of their length and aspect ratio with the *nano* prefix when it applies.

suitable buffer layers to reduce the n-type background doping levels and to reduce defect density. Remarkable progress was made by insertions of low-temperature AlN [AMA 88, AKA 89] or low-temperature GaN buffer layers before GaN growth [NAK 92]. Concerning the p-type doping, although the group II element Mg has long been expected to be a good acceptor dopant and significant amounts of Mg could be incorporated into GaN during MOVPE growth, it was impossible to detect positive charge carriers at room temperature. The films turned out to be highly resistive. A rather fortuitous discovery allowed Amano *et al.* [AMA 89] to demonstrate that post-growth low-energy electron beam irradiation activates Mg-doped GaN atoms and converts them from the as-grown highly resistive state to a p-type conductive state. Nakamura *et al.* [NAK 92] have improved the activation of the Mg acceptors in MOVPE grown GaN by thermal annealing in N_2. A hole concentration as high as 3×10^{17} cm^{-3} was achieved with a resistivity of only 2 Ωcm. These advances led us to realize the first blue GaN based p-n junction light emitting diode (LED) in 1993 [NAK 94] and later to green and yellow LEDs. This field is still under strong development with the achievement of lighting devices that may also contain wires and rods (see a review in [LI 12]), but also with the realization of efficient lasers for several visible wavelengths.

Another fast development of the GaN materials lies in the power application. Nitrides provide efficient alternative technologies to surpass the silicon power Metal oxide semiconductor field effect transistors (MOSFET) that asymptotically approached its theoretical bounds. They take advantage of the high "ON-resistance" and breakdown voltage due to the high critical electric field strength of these materials [DOR 06] and also of the high electron mobility and saturation velocity. Basic GaN transistor structure [KHA 93] involves source and drain electrodes pierced through a top AlGaN layer to form ohmic contacts with an underlying dimensional electron gas (2DEG). This creates a short-circuit between the source and the drain until the 2DEG "pool" of electrons is depleted and the semi-insulating GaN crystal can block the flow of current. In order to deplete the 2DEG, a gate electrode is placed on top of the AlGaN layer. In many of these early depletion-mode GaN transistors, this gate electrode was formed as a Schottky contact to the top surface. By applying negative voltage to this contact, the Schottky barrier becomes reverse biased, and the electrons underneath are depleted. Due to the absence of inversion symmetry in wurtzite III-nitride materials, the polarization-induced electric fields in N-polar heterostructures are opposite to those of the Ga-polar counterpart, inducing a 2DEG above instead of below the wide-bandgap barrier layer. These considerations allow us to play with polarity and crystallographic plane orientations to optimize the electronic structure designs [WON 13].

These two famous nitride application examples (light emitters and power transistors) enlighten some requirements to make full use of GaN materials in wire-based devices (for more details, refer to Volume 2) or simply to obtain relevant objects to perform fundamental physics (for more details, refer to Chapters 1–6). They stressed the importance of the managements of the extended and point defect density, of the n and p doping and of the orientation/polarity control to use GaN wires as a template for heterostructure growth, and more generally for devices.

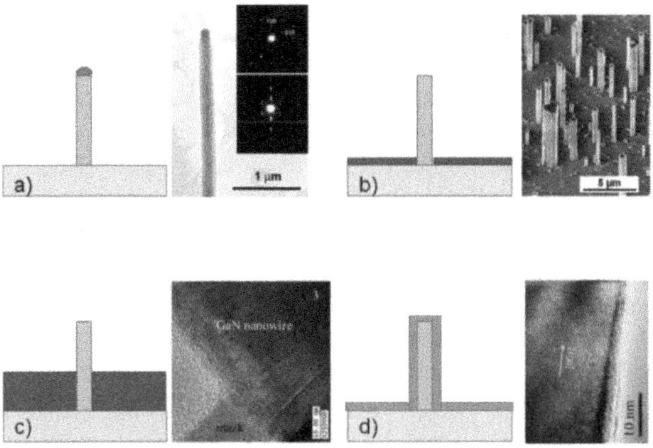

Figure 10.1. *Schematics of the principal methods developed to obtain GaN wires by MOVPE: a) Au catalyst-assisted (Reprinted with permission from Macmillan [Nature Materials] ([KUY 07]), Copyright © 2007). (see also Ga self-catalyst [TES 13]), b) catalyst-free self-organized growth (From [KOE 10] © 2010 IOP Publishing. Reproduced by permission of IOP Publishing. All rights reserved), c) selective area method (Reprinted with permission from [HER 06]. Copyright 2006 American Chemical Society) (see also guided growth [LI 11]) and d) oxide-assisted growth (Reprinted from [SHI 01], Copyright 2001, with permission from Elsevier). For simplicity, wires are drawn perpendicular to the substrate, but disorientations can occur as a function of the technique used. For a color version of this figure, see www.iste.co.uk/consonni/nanowires1.zip*

The following will review some solutions that have been explored in MOCVD wire growths (see the schematics shown in Figure 10.1) in order to keep the GaN material advantages and also to add the new ones specific to the wire geometry or size effect. The emphasis will be placed on catalyst-assisted, catalyst-free/self-organized and selected-area growths of GaN without going into the details of the realization of InAlGaN heterostructures that deserve longer discussions.

10.2. Catalyst-assisted growth

The vapor–liquid–solid growth mechanism has been employed with many metal catalysts (for example Fe [CHE 00, DUA 00], Ni [CHE 00, DUA 00, QIA 04], Au [SUN 05, AHL 11] and Co [CHE 00]) to synthesize GaN wires by chemical vapor deposition (CVD) and laser-assisted methods. More details concerning the kinetics and thermodynamics of this catalyst-assisted growth can be found in [GIV 75, MOH 09, MOH 10]. Several kinetics pathways can control the overall mechanism: mass transport in the gas phase, chemical reaction at the vapor–liquid interface, diffusion in the liquid phase and incorporation of atoms in a crystal lattice. Because the mixing and precipitation in the liquid phase are thermodynamic processes that work toward equilibrium, a phase diagram is very useful to predict how the catalyst will work in terms of alloys and compounds formation, eutectic points, etc. Most of the published processes were using Ga metal as the high-temperature vapor source, but it turned out that this method was quite difficult to implement in the large area of continuous-flow reactors due to heterogeneous vapor pressure effects. To overcome these drawbacks, the MOVPE growth of GaN wires with triangular cross-section has been demonstrated with Fe, Ni and Au catalysts in 2003 [KUY 03] and with In catalyst in 2005 [SU 05a]. Standard trimethylgallium (TMG) and ammonia sources were used as Ga and N precursors, and 2–10 nm thin films of Ni, Fe or Au were thermally evaporated onto a silicon substrate or *c*- and *a*-plane sapphire substrates. Electron diffraction indicated that the growth direction was not always along the usual GaN *c*-axis, but may depend on the catalysts to be along the [110] direction for Fe and [210] direction for Au and Ni catalysts. The smooth non-equivalent facet surface planes of these triangular cross-section structures have different growth rates, but they can be used as efficient templates for further heterostructure growths. For example, nanowires grown with Ni-catalysts along the [110] direction with {1 -1 0 1} and {0 0 0 1} facets have been used to perform optical cavities [SEO 08], to grow InGaN/GaN multiple quantum wells and to demonstrate wavelength-controlled lasers [QIA 08].

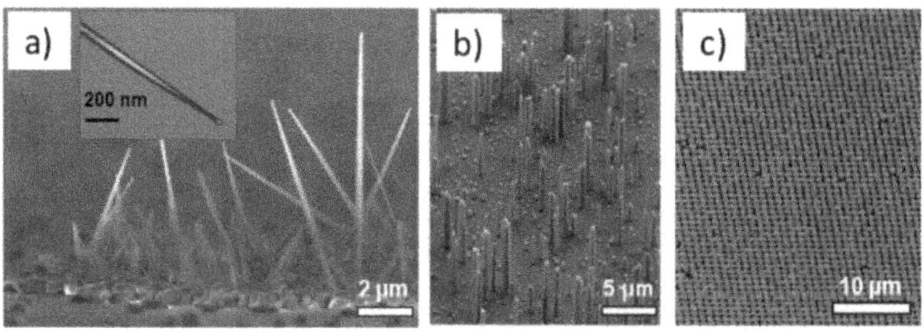

Figure 10.2. *Examples of MOVPE growths of GaN wires obtained at CEA/INAC with standard precursors using a) Ni-nitrate catalysts, b) self-catalyzed/self-organized way and c) selective area method. For a color version of this figure, see www.iste.co.uk/consonni/nanowires1.zip*

Catalysts can be provided by layer deposition and dewetting [AHL 11], evaporating salt solutions [SU 05b] or directly from colloids or cluster deposition.

The surface preparation (annealing procedure in a given atmosphere) has to be carefully managed in terms of catalyst-positioning and chemical state before the wire growth to get a good catalytic efficiency and a control of the wire diameter. Quite small nanowire diameters can be easily obtained from this method in the 10–200 nm range, i.e. much smaller than what is generally obtained with self-catalyzed MOCVD techniques. They provide convenient systems to explore size-dependent properties, such as impurity activation energy [YOO 09]). The main difficulty to overcome for this catalyst growth consists of obtaining a regularly oriented growth, or better, an epitaxial relationship with the substrate (see Figure 10.2(a)). Guided growth of millimetre-long horizontal nanowire has been obtained [TSI 11], but a growth perpendicular to the surface (or with well-defined epitaxial angles) is generally preferred to simplify the integration of wires inside devices. Up until now, tri-dimensional integration, such as branched heterostructures [SU 05] using secondary catalysts on the sidewalls of initial wires, has not been much explored for nitride wires. The impact of the process parameters has been studied in detail for most of these systems, and the general conclusions have pointed out the important parameters to consider: i.e. the V/III ratio, the pressure and the temperature. The optimization of these parameters (using for example a very low V/III ratio compared to planar growth) can lead to non-tapered GaN wires without threading dislocations, but with some stacking faults. For the chemical analysis of the wires and their catalyst droplets, energy-dispersive X-ray and electron energy loss spectroscopy were employed confirming the nature of the catalyst material. In a general way, the droplet contamination inside the wire has to be limited to its maximum to avoid impurity-related defects [RES 05, TSO 09] affecting both transport [OH 10] and luminescence [KUB 07] properties.

10.3. Catalyst-free and self-organized growth

Although, catalyst-assisted MOVPE growth of III-nitride nanowires has turned out to be an efficient and easy approach, the drawbacks in terms of degrading the optical and structural properties due to the metal contamination and the difficulty to get well aligned wires in epitaxy with the substrate lead to a huge development of epitaxial catalyst-free self-organized III-nitride wires with several growth methods. To date, the catalyst-free III-nitride wire growth along the *c*-axis has been intensely achieved by MBE (for more details, refer to Chapters 7–9). However, this technique presents quite a low growth rate (generally less than 200 nm/h), and the uniformity on large wafer scale has not been optimized yet making the realization of a commercial production difficult. It has been shown that oxide-assisted growth [SHI 01] can provide disoriented GaN wire assemblies and that HVPE allows us to obtain vertically aligned GaN wires on c-sapphire in the catalyst-free approach [KIM 02, KIM 04]. The greater achievement of the CVD method with chloride precursors ($InCl_3$ and $GaCl_3$) is the complete composition tunability for the catalyst-free $In_xGa_{1-x}N$ ternary alloy nanowires synthesis with a visible emission from 400 to 750 nm (for x = 0–0.6) [KUY 07] that have been utilized to fabricate visible LEDs [HAH 11]. However, this approach also seems to be rather difficult to implement and has been little reported in the literature. At present, the MOCVD catalyst-free GaN wire growth has been achieved by several groups, who demonstrate fast and convenient methods to get randomly positioned ±c-oriented wires in epitaxial relationship with the substrate and with good photoluminescence properties (i.e. without a significant optical defect band). The possibility to realize rod structures has first been demonstrated using standard nitride precursors (TMG) and ammonia by decreasing the V/III ratio and adding silane [HAF 98]. This study was mainly focused on the changes of growth morphologies by adding Si [HAF 98] (and then Mg [BEA 98]) elements into GaN, not to the realization of one-dimensional nanostructures. A few years later, complex single molecule precursors bisazido (dimethylaminopropyl) gallium and its ethyl derivative have been used to study in detail the influence of temperature and carrier gas ratio on the morphology of GaN nanostructures, i.e. self-organized nanopillars, nanorods and randomly oriented nanowires [WHO 02, KHA 03]. The main idea of this approach was to avoid using ammonia or any other nitrogen precursors to circumvent the low-efficiency cracking of these molecules. This advantage was probably not strong enough to motivate a new process development by the industrial growth community. The self-catalyzed and self-organized MOCVD growth of GaN wires gained new developments more recently [KOE 10, LI 11a, ALL 12, TES 13] with its use as a template for device applications using standard commercial reactors and precursors (trimethyl- or triethyl-based combined with ammonia, i.e. what is generally used for MOVPE planar growth). It has been shown that several basic conditions were required to get a one-dimensional growth in a self-catalyzed and self-organized way. They can be illustrated by the growth of GaN wires on sapphire along the –c direction. Such wires exhibit N-polarity with a flat -*c*-plane surface at their top and lateral $\{1\bar{1}00\}$ *m*-plane facets forming irregular hexagons (see Figure 10.2(b)). They are in epitaxy with sapphire, and their unit cells are rotated by 30° around the *c*-axis: $[1\bar{1}00]_{GaN} // [1\bar{2}10]_{Sap.}$ and $[0001]_{GaN} // [0001]_{Sap}$. Their usual diameters are in the 350 nm^{-2} µm range[2] and densities less than 10^8 cm^{-2}. MOVPE has a fast average longitudinal growth rate that can reach 130 µm/h with adapted flow and pressure conditions. It allows us to grow very long wires (> 500 µm) of enlarged diameter (~2 µm) on the full wafer in less than 4 h [EYM 13]. With such diameters, the argument of free-surface elastic relaxation for the growth of heterostructures does not apply, and these wires can be considered as quasi-bulk materials (for more

2 Small diameter wires in the 50–100 nm range have also been demonstrated in the literature with few standard conditions: high density for single molecule precursors [KHA 03] and low density for very low flow of standard precursors [CHE 10].

details, refer to Chapter 2). Nevertheless, the threading dislocations are still eliminated by bending toward the sidewall facet [HER 06, COL 10, CHE 11, TES 13], and the interfacial strain resulting from the large lattice mismatch between GaN and sapphire has an exponential decay of the same order as the wire diameter [LI 09]. Inversion domain boundaries (IDBs) separating –c and +c oriented crystals can be observed inside the wires [NOR 96]. They nucleate at the bottom of the wires on surface defects [LIU 07] and propagate along the full height [CHE 11]. The probability to observe IDB increases with the wire diameter, and several configurations may occur as shown by transmission electron microscopy or X-ray coherent diffraction imaging. As an example, it has been observed on some rods inner cylinder ("core") with Ga polarity and outer "shell" with N polarity [TES 13].

Several process chart flows have been proposed in the literature to grow these wires. From a general point of view, they can be summarized by three main steps illustrated in Figure 10.3 [EYM 13, ALL 12, KO 12].

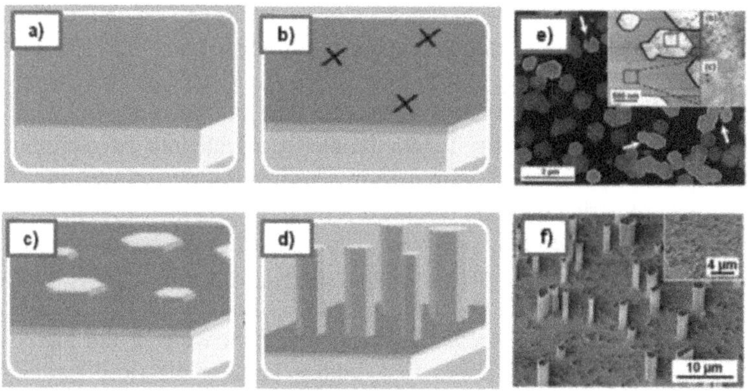

Figure 10.3. *Important steps driving the MOVPE growth of catalyst-free self-organized GaN wires. a) Choice of the substrate that impacts directly epitaxial relationships, polarity, cleaning, dopant diffusion. b) Creation/deposition of intermediate layers to have growth selectivity or to control the polarity. Eventual surface preparation (for example annealing, defect creation) that allows introducing preferential locations for the next step (crosses in the figure). c) Control of the nucleation seeds to initiate the wire growth. d) One-dimensional wire growth. e) Illustration of (c) and (d) steps (From [KOE 10] © 2010 IOP Publishing. Reproduced by permission of IOP Publishing. All rights reserved). For a color version of this figure, see www.iste.co.uk/consonni/nanowires1.zip*

1) The *surface preparation* that allows the chemical cleaning of the surface and the polarity selection, mandatory to drive the epitaxy of polar GaN (see Figure 10.3(a) and (b)). It has been shown for example that an annealing under NH_3 favors N-polar wires on c-sapphire surface, and a thin AlN layer deposition protects the silicon surface from TMG etching [CHE 10, LI 11b, SAL 13].

2) The *seed nucleation step* that will define the size and density of the wires (see Figure 10.3(c)). Two main methods have been proposed up to now to manage this point. The first uses a *selective-growth layer* that prevents the GaN nucleation on the whole substrate surface except in spontaneously formed holes [KOE 10, TES 13]. Silane gas (SiH_4) and ammonia are injected *in situ* in the reactor before starting GaN growth to get a SiN_x thin layer, well-known for its selective-growth properties for bi-dimensional growths. The second avoids the deposition of this peculiar layer, but still opens silane also during the initial growth stages [ALL 11]. In general, the nucleation step has different growth conditions (materials quantity, temperature, carrier gas, etc.) than the wire themselves to control the size/shape and density of the seeds. All these process parameters have to be adjusted for each reactor's geometry and size.

3) The *one-dimensional growth* that is favored by the choice of specific growth conditions (see Figure 10.3(d)). They are also optimized for each reactor, and the general tendency is to largely reduce the ammonia flux with respect to planar layers (in order to get a lower V/III molar ratio) and to add silane gas during the GaN wire growth. The role of silane and SiN_x layer (also called anti-surfactant layer in [MAR 13]) is not completely understood, it certainly changes the kinetics and equilibrium surface properties with an increase of the sidewall energies with respect to the top surface energy. Density-functional theory calculations have predicted that a Ga-polar +*c*-plane surface is formed by a Ga bilayer with Si incorporated underneath under Ga-rich condition silane injection [ROS 06], and a catalytic role of this bi-layer can be suspected, but the –c N-polar surface energy calculations have not been compared to m-plane energies for different gas atmospheres. Under a large amount of silane addition, the SiN_x can also be formed on the wire facets. This effect has been proposed to position core-shell multi-quantum wells at the top of a wire [EYM 12] with a two-step process: first, with a large silane flow leading to the selective-growth layer formation and secondly with a decrease/stop of the silane flow allowing the shell to grow on the unprotected part.

It is convenient to grow the wires in the mass-transport regime of the MOVPE. For a variation of the growth temperature (for example between 850 and 1000°C), the total amount of deposited material for a given growth time (t) must be constant [STR 99, MIT 01]. In this case, chemical reactions are not limited by kinetic or thermodynamic considerations, which can be very complex due to the precursor gas phase reactivity and to reactor's geometry [THO 96, COL 03, PAR 06, CRE 07, MOS 07]. Vapor phase diffusion is generally considered to determine the concentration above the wires, and a gas diffusivity parameter can be used to model the mass transport in the stagnant layer immediately above the surface [STR 99, SO 13]. The MOCVD incorporation pathways and their differences with respect to the MBE case [DUB 05, SIB 12] can be illustrated by the analysis of the increase of the wire length (l) as a function of the growth time.

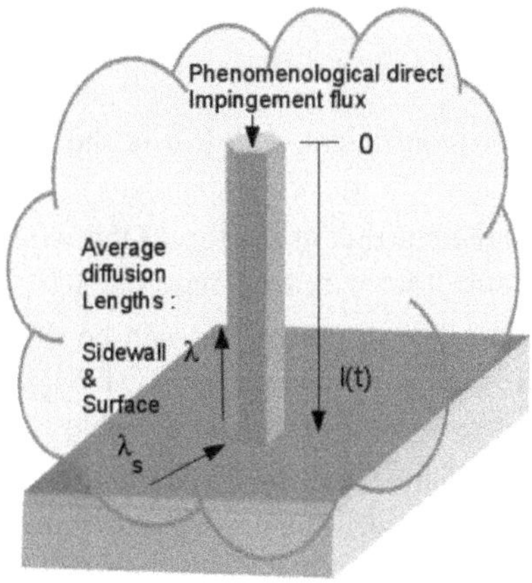

Figure 10.4. *Schematics of the MOVPE incorporation pathways for the growth of GaN wires corresponding to the model of Ruth and Hirth [RUT 64]. For simplicity, only one wire is considered. The gas phase provides adatoms on the surface, on the sidewalls and at the top of the wire. Average diffusion lengths and direct impingement have to be considered. This model can be generalized to selective growth arrays leading to a redistribution/concurrence between the incorporation sites*

For simplicity, it is assumed that the wire diameter is constant and equal to 2 R (i.e. we have a perfect one-dimensional growth). The first two contributions to the growth come from the gas phase collection. Indeed, as illustrated in Figure 10.4, we can consider: (1) the direct impingement from the top wire facet; (2) upward flow coming from the sidewall facet diffusion. Note that the direct impingement is a phenomenological parameter that can be related to self-catalyst effect and/or to tip-enhanced collection depending on environment and geometry. Moreover, the facet diffusion plays an important role in MOCVD due to the very long diffusion length associated with this technique [STR 99, EYM 13]. The third pathway corresponds to (3) the flow coming from the neighborhood substrate surface of the wires that may also fill facet surface diffusion and tip incorporation. It can, therefore, contribute differently to the growth as a function of the wire length: for short wires, the tip can be directly supplied by the surrounding surface. These three terms were considered by Ruth and Hirth [RUT 64] quite a long time ago and have been incorporated in a model relating the time of growth t and the length l in the following equation:

$$t = \frac{V\lambda}{\sqrt{2}(V^2 - I^2)} Log\left[\frac{\left(\beta + \frac{I}{V}\right)\cosh\left(\frac{\sqrt{2}\,l}{\lambda}\right) + \left(1 + \frac{\beta I}{V}\right)\sinh\left(\frac{\sqrt{2}\,l}{\lambda}\right)}{\beta + \frac{I}{V}}\right] - \frac{I\,l}{V^2 - I^2} \qquad [2.11]$$

where $\lambda = \sqrt{2\frac{D}{\omega}}$ is the average facet diffusion length given by the Einstein relation (D is the surface diffusion coefficient and $1/\omega$ is the adatom stay time prior to evaporation), V is a linear-growth rate corresponding to the limiting case $\lambda \gg R$, i.e. when the direct impingement flux at the top of the wire I can be neglected. The parameter β defines the substrate-wire exchange rate ($\beta < 1$ corresponds to a low exchange rate). The value of these parameters can be directly estimated from the shape of the $l(t)$ curve at small t (see details in [RUT 64]). As shown in [EYM 13] for GaN wires surrounded by a SiN_x layer, the β value is small (~ 0.1) confirming the small substrate-wire exchange rate imposed by the selective layer preventing species sticking and also by the surface sidewall properties. It also demonstrates the predominance of the gas phase incorporation mechanisms and the major role played by the facet diffusion term (V = 18.5 nm/s and I = 4 nm/s), but as said already, the mechanism driving the direct incorporation at the top of the wire (related to I) is still not explained at this time, and the presence of a Ga droplet at the top of the wire – during the growth – cannot be directly checked because it may be removed by the cooling procedure[3]. Finally, the value of the average diffusion length along the facets ($\lambda = 7.5\ \mu m$ for the previous growth conditions) agrees with the limit between the early stage of the growth and the linear regime where the diffusion flow has reached a steady state regime, but the calculations of surface energies and diffusion barriers are difficult to perform due to the influence of silicon making a quantitative comparison impossible.

10.4. Selected-area growth

The selected-area growth (SAG) of a "microprism" (5–16 μm in diameter) has been developed originally for the realization of laser microcavities to take advantage of the facet smoothness [AKA 97]. Selective growth of GaN under H_2 was obtained by low-pressure MOVPE using a mask-patterned GaN epitaxial layer on a (0001) sapphire substrate. The orientation of 50 nm SiO_2 mask-patterning was found to be important because of the 30° rotation of the crystallographic orientation between the GaN layer and the sapphire substrate already described. A few years later, the SAG growth of GaN nanowires on patterned substrates using intermediate porous anodic-aluminium-oxide template (AAO) was demonstrated [DEB 05]. Using the same type of substrates (GaN/c-sapphire with patterned SiO_x), the MOVPE SAG of pyramid capped GaN nanowires with 60–80 nm diameter and 100 nm length was also

3 Moreover, a decrease of temperature when ammonia is stopped can spontaneously create some gallium droplets that can be misinterpreted as self-catalyst signature.

managed [COL 10]. Unfortunately, due to the non-regular formation of the nanoporous within the AAO layer, the position and the diameter of the wires were not precisely controlled. To overcome this problem, the patterning of the mask has been developed with different techniques, for example by using laser interference lithography and dry etching of SiN_x layers [WAN 06]. By this way, the precise location and diameter of the objects can be well-controlled by the growth mask patterning, and the resulting array has the regularity for further growth. The long range order is for example compatible with some applications, such as photonic crystals. Actually, the methods cited above mainly correspond to *confined* SAG growth, and a constant diameter is not guaranteed for nanostructures emerging from the opening. Two main techniques based on *pulsed* and *continuous* growths have been developed to decrease the lateral growth (see a review in [BAE 13] and Figure 10.5).

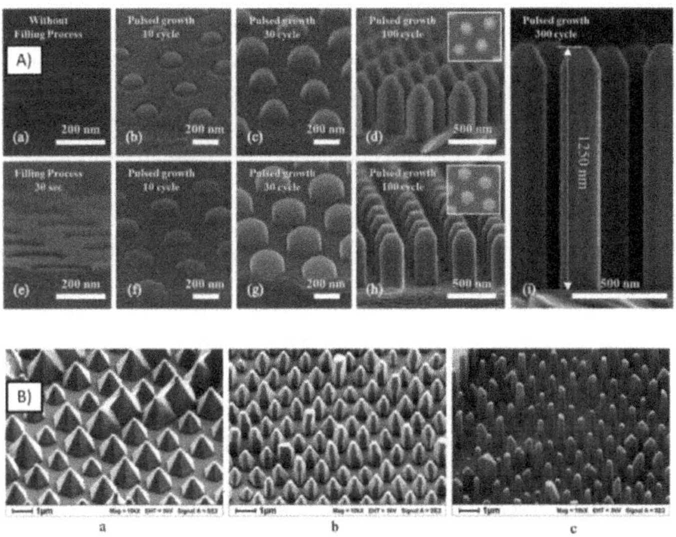

Figure 10.5. *Illustration of (a) pulsed growth (Reproduced from [JUN 04] with permission of The Royal Society of Chemistry) and(b) continuous growth (From [BER 10] © 2010 IOP Publishing. Reproduced by permission of IOP Publishing. All rights reserved) of GaN pillars obtained by MOVPE. Part A) shows the influence of the filling of hole patterns on a GaN template with SiO_2 growth mask as a function of pulsed growth cycles. Part B) shows the influence of the H_2/N_2 carrier gas mixture on the nanostructure morphology. C-sapphire substrates with SiO_2 growth mask are used with (a) pure N_2, (b) $H_2/N_2 = 1/2$, (c) $H_2/N_2 = 2/1$ ratio*

In the *pulsed-growth* mode [HER 06], the III and V precursors are provided alternately after a proper filling of the mask holes. As previously stated, the growth mechanisms are strongly dependent on the growth conditions and on the substrate polarity. In these systems, the growth is usually performed on a Ga-polar layer, and the one-dimensional growth has been explained [LIN 12] by a higher sticking coefficient of Ga adatoms on the *c*-plane than on the *m*-plane and to a longer residual time on the *c*-plane (see also the calculation of the atom diffusion on ±c-surfaces [ZYW 98]). This influence of the Ga adatom kinetic behavior has been analyzed with respect to the nanostructure morphology evolution by changing the Ga interruption duration at a fixed ammonia flow rate and growth temperature [LIN 12] and also by combining this process with NH_3 time interruptions [JUN 14]. The durations of the Ga and NH_3 interruption times have, therefore, to be carefully optimized, and convenient metrics have to be defined to describe the kinetics, such as the pulsed-mode growth ratio (N/Ga injection time)/(N/Ga interruption time) [JUN 14]. The aim is to control the filling of the mask holes at the beginning of the growth and then to prevent the lateral growth with convenient facet (and shape) morphologies (see Figure 10.5). It is done in conventional setups at the price of quite a low growth rate, but still competitive with respect to MBE. Similarly, MOVPE-pulsed mode growth has been further developed to demonstrate the ability of using wire arrays for an improvement of blue and green LED performance [TAN 10] or to reduce the dislocation density in the 10^7 cm^{-2} range by performing the wire coalescence [TAN 09]. In this last work, the geometry of the pattern on the SiN_x mask layer is varied with a nanoimprint technique to get different diameters and pitch distances. It uses a Si mold design realized by e-beam lithography, and this method can be transposed to wafer scale mass production (for more details, refer to Chapters 5–8 in Volume 2).

In the *continuous-growth* method, the fabrication of N-face GaN nanorods (diameters down to 220 nm and 8:1 aspect ratio) has been demonstrated by using a well-adjusted carrier gas mixture between hydrogen and nitrogen without dopant addition to avoid any "surfactant influence" (especially Si) [BER 10, BER 11]. It managed to grow well-aligned N-face GaN sub-micrometer wire array directly on c-sapphire substrate with patterned SiO_x mask layer by nanoimprint lithography. The statistical variation of the rod diameter was about 100 nm, with an average rod diameter of 400 nm, and the verticality of the side facets can be realized only by using an optimized H_2/N_2 carrier gas mixture (Figure 10.5). It has been proposed that hydrogen has a passivation effect on the nitrogen atoms' incorporation depending on the facet orientations [LI 11b]. Additionally, it was shown that the aspect ratio of the wires was successfully influenced by increasing the pitch of the pattern, while maintaining the hole-diameter constant. Note that very long wires (or a very high aspect ratio) cannot nevertheless be achieved without silane addition. Finally, a similar continuous-growth method has combined the recipes of the self-organized way (i.e. silane addition) and carrier gas adjustment to get selected-area growth on sapphire [CHE 11] evidencing another possibility to tailor these materials (see Figure 10.2(c)).

10.5. Discussion and conclusion

This chapter has reported the wealth of solutions to perform the growth of GaN wires by MOVPE. Varying the growth conditions, most of these techniques have been extended to nitride alloys, such as $In_xGa_{1-x}N$ [QIA 05] and $Al_xGa_{1-x}N$ [SU 05b], to build new heterostructures based on quantum wells and quantum dots. It allowed us to satisfy the variety of applications that can be addressed by the wire geometry (for more details, refer to Chapters 5–8 in Volume 2). Catalyst-assisted growth was widely adopted to get "disordered" materials, but catalyst-free solutions seem now to be prevalent in current GaN growth technology for its low contamination level and the possibility to obtain epitaxial materials. Moreover, selective-area growth provides an efficient solution to precisely position the wires and to have a very narrow distribution of the geometrical features. N and P dopings necessary for optoelectronic applications have been reached (in a core-shell geometry for p-doping) [QIA 05, KOE 10], but further studies are still necessary to understand incorporation mechanisms, distributions and limits [PAR 13].

The growth of GaN wires is still a very active, ongoing and interesting topic. The concept of kinetic Wulff plots for polar and non-polar orientations [LEU 12] seems to be interesting to understand the role of the large number of parameters involved in this complex MOVPE environment [JIN 09, DU 05]. It allows correlating growth dynamics with commonly observed morphologies, including anisotropic nucleation of the islands and wire facet expansions, and points out the necessity to understand the dependence of facet energies as the function of the atmosphere (including dopant elements that can create surface compounds), materials' quantity, temperature, pressure. Finally, the mechanism of selective-area growth of GaN wires needs to be further understood, especially to understand the interplay between the different incorporation mechanisms as the function of the pattern geometry.

10.6. Bibliography

[AHL 11] AHL J.-P., BEHMENBURG H., GIESEN C., *et al.*, "Gold catalyst initiated growth of GaN nanowires by MOCVD", *Physica Status Solidi C*, vol. 8, no. 7–8, pp. 2315–2317, 2011.

[AKA 89] AKASAKI I., AMANO H., KOIDE Y., *et al.* "Effects of AlN buffer layer on crystallographic structure and on electrical and optical properties of GaN and $Ga_{1-x}Al_xN$ ($0 < x \leqq 0.4$) films grown on sapphire substrate by MOVPE", *Journal of Crystal Growth*, vol. 98, nos. 1–2, pp. 209–219, 1989.

[AKA 97] AKASAKA T., KOBAYASHI Y., ANDO S., *et al.*, "GaN hexagonal microprisms with smooth vertical facets fabricated by selective metalorganic vapor phase epitaxy", *Applied Physics Letters*, vol. 71, no. 15, pp. 2196–2198, 1997.

[ALL 11] ALLOING B., VEZIAN S., TOTTEREAU O., *et al.*, "On the polarity of GaN micro- and nanowires epitaxially grown on sapphire 0001 and Si 111 substrates by metal organic vapor phase epitaxy and ammonia-molecular beam epitaxy", *Applied Physics Letters*, vol. 98, pp. 011914–3, 2011.

[ALL 12] ALLOING B., BERAUDO E., CORDIER Y., *et al.*, "Fabrication and growth of GaN-based micro and nanostructures", *International Journal of Nanotechnology*, vol. 9, pp. 412–427, 2012.

[AMA 88] AMANO H., AKASAKI I., HIRAMATSU K., *et al.*, "Effect of the buffer layer in metalorganic vapour phase epitaxy of Ga Non sapphire substrate", *Thin Solid Films*, vol. 163, pp. 415–420, 1988.

[AMA 89] AMANO H., KITO M., HIRAMATSU K., *et al.*, "p-type conduction in Mg-doped GaN treated with low-energy electron beam irradiation (LEEBI)", *Japanese Journal of Applied Physics*, vol. 28, pp. l2112–l2114, 1989.

[BAE 13] BAE S.-Y., LEE J.-Y., MIN J.-H., *et al.*, "Morphology evolution of pulsed-flux Ga-polar GaN nanorod growth by metal organic vapor phase epitaxy and its nucleation dependence", *Applied Physics Express*, vol. 6, pp. 075501–075504, 2013.

[BEA 98] BEAUMONT B., VAILLE M., NATAF G., *et al.*, "Mg-enhanced lateral overgrowth of GaN on patterned GaN/sapphire substrate by selective Metal Organic Vapor Phase Epitaxy", *MRS Internet Journal Nitride Semiconductor Research,* vol. 3, pp. 20, 1998.

[BER 10] BERGBAUER W., STRASSBURG M., KOLPER M., *et al.*, "Continuous-flux MOVPE growth of position-controlled N-face GaN nanorods and embedded InGaN quantum wells", *Nanotechnology*, vol. 21, no. 30, pp. 305201–305205, 2011.

[BER 11] BERGBAUER W., STRASSBURG M., KOLPER M., *et al.*, "N-face GaN nanorods: Continuous-flux MOVPE growth and morphological properties", *Journal of Crystal Growth*, vol. 315, no. 1, pp. 164–167, 2011.

[CHE 00] CHEN C.-C., YEH C.-C., "Large-scale catalytic synthesis of crystalline gallium nitride nanowires", *Advanced Materials*, vol. 12, pp. 738–741, 2000.

[CHE 10] CHEN X.J., PERILLAT-MERCEROZ G., SAM-GIAO D., *et al.*, "Homoepitaxial growth of catalyst-free GaN wires on N-polar substrate", *Applied Physics Letters*, vol. 97, no. 15, pp. 151909–3, 2010.

[CHE 11] CHEN X.J., HWANG J.S., PERILLAT-MERCEROZ G., *et al.*, "Wafer-scale selective area growth of GaN hexagonal prismatic nanostructures on c-sapphire substrate", *Journal of Crystal Growth*, vol. 322, pp. 15–22, 2011.

[COL 03] COLTRIN M.E., MITCHELL C.C., "Mass transport and kinetic limitations in MOCVD selective-area growth", *Journal of Crystal Growth*, vol. 254, no. 1–2, pp. 35–45, 2003.

[COL 10] COLBY R., LIANG Z.W., WILDESON I.H., *et al.*, "Dislocation Filtering in GaN Nanostructures", *NanoLetters*, vol. 10, no. 5, pp. 1568–1573, 2010.

[CON 14] CONSONNI V., FEUILLET G., (eds.), *Wide Band Gap Semiconductor Nanowires 2: Heterostructures and Optoelectronic Devices*, ISTE, London, and John Wiley & Sons, New York, 2014.

[CRE 07] CREIGHTON J.R., WANG G.T., COLTRIN M.E., "Fundamental chemistry and modeling of group-III nitride MOVPE", *Journal of Crystal Growth*, vol. 298, pp. 2–7, 2007.

[DEB 05] DEB P., KIM H.Y., RAWAT V., *et al.*, "Faceted and vertically aligned GaN nanorod arrays fabricated without catalysts or lithography", *NanoLetters*, vol. 5, no. 9, pp. 1847–1851, 2005.

[DOR 06] DORA Y., CHAKRABORTY A., MCCARTHY L., *et al.*, "High breakdown voltage achieved on AlGaN/GaN HEMTs with integrated slant field plates", *IEEE Electron Device Letters,* vol. 27, no. 9, pp. 713–715, 2006.

[DU 05] DU D.X., SROLOVITZ D.J., COLTRIN M.E., *et al.*, "Systematic prediction of kinetically limited crystal growth morphologies", *Physical Review Letters*, vol. 95, pp. 155503–155504, 2005.

[DUA 00] DUAN X., LEBER C.M., "Laser-assisted catalytic growth of single crystal GaN nanowires", *Journal of the American Chemical Society*, vol. 122, pp. 188–189, 2000.

[DUB 05] DUBROVSKII V.G., CIRLIN G.E., SOSHNIKOV I. P., *et al.*, "Diffusion-induced growth of GaAs nanowhiskers during molecular beam epitaxy: theory and experiment", *Physical Review B*, vol. 71, pp. 205325–205326, 2005.

[EYM 12] EYMERY J., SALOMON D., CHEN X.J., *et al.*, Process for catalyst-free selective growth on a semiconductor structure, patent number WO2012136665, 2012.

[EYM 13] EYMERY J., CHEN X.J., DURAND C., *et al.*, "Self-organized and self-catalyst growth of semiconductor and metal wires by vapour phase epitaxy: GaN rods versus Cu whiskers", *Comptes Rendus Physique,* vol. 14, pp. 221–227, 2013.

[GIV 75] GIVARGIZOV E.I., "Fundamental aspects of VLS growth", *Journal of Crystal Growth*, vol. 31, pp. 20–30, 1975.

[HAH 11] HAHN C., ZHANG Z.Y., FU A., *et al.*, "Epitaxial growth of InGaN nanowire arrays for light emitting diodes", *ACS Nano*, vol. 5, no. 5, pp. 3970–3976, 2011.

[HAF 98] HAFFOUZ S., BEAUMONT B., GIBART P., "Effect of magnesium and silicon on the lateral overgrowth of GaN patterned substrates by metal organic vapor phase epitaxy", *Journal of Nitride Semiconductor Research*, vol. 3, p. 8, 1998.

[HER 06] HERSEE S.D., SUN X.Y., WANG X., "The controlled growth of GaN nanowires", *NanoLetters*, vol. 6, pp. 1808–181, 2006.

[HUA 02] HUANG Y., DUAN X., CUI Y., *et al.*, "Gallium nitride nanowire nanodevices", *NanoLetters*, vol. 2, no. 2, pp. 101–104, 2002.

[JIN 09] JINDAL V., SHAHEDIPOUR-SANDVIK F., "Theoretical prediction of GaN nanostructure equilibrium and nonequilibrium shapes", *Journal of Applied Physics*, vol. 106, no. 8, pp. 083115–083117, 2009.

[JOH 32] JOHNSON W.C., PARSON J.B., CREW M.C., "Nitrogen compounds of Gallium III", *Journal of Physical Chemistry*, vol. 36, no. 10, pp. 2651–2654, 1932.

[JUN 14] JUNG B.O., BAE S.-Y., KATO Y., *et al.*, "Morphology development of GaN nanowires using a pulsed-mode MOCVD growth technique", *CrystEngComm*, 2014, (Advance Article).

[JUS 38] JUSA R., HAHN H., "Über die Kristallstrukturen von Cu_3N, GaN und InN metallamide und metallnitride", *Zeitschrift für anorganische und allgemeine Chemie*, vol. 239, pp. 282–287, 1938.

[KHA 03] KHANDERI J., WOHLFART A., PARALA H., *et al.*, "MOCVD of gallium nitride nanostructures using $(N_3)_2Ga\{(CH_2)_3NR_2\}$, R = Me, Et, as a single molecule precursor: morphology control and materials characterization", *Journal of Materials Chemistry*, vol. 13, pp. 1438–1446, 2003.

[KHA 93] KHAN M.A., BHATTARAI A.R., KUZNIA J.N., *et al.* , "High-electron-mobility transistor based on a GaN-$Al_xGa_{1-x}N$ heterojunction", *Applied Physics Letters*, vol. 63, no. 9, pp. 1214–1215, 1993.

[KIM 02] KIM H.-M., KIM D.S., PARK Y.S., *et al.*, "Growth of GaN nanorods by a hydride vapor phase epitaxy method", *Advanced Materials*, vol. 14, pp. 991–993, 2002.

[KIM 04] KIM H.-M., CHO Y.-H., LEE H., *et al.*, "High-brightness light emitting diodes using dislocation-free indium gallium nitride/gallium nitride multiquantum-well nanorod arrays", *NanoLetters*, vol. 4, no. 6, pp. 1059–1062, 2004.

[KO 12] KO S.-M., KIM J.-H., KO Y.-H., *et al.* "Growth mechanism of catalyst-free and mask-free heteroepitaxial GaN submicrometer- and micrometer-sized rods under biaxial strain: variation of surface energy and Adatom kinetics", *Crystal Growth & Design*, vol. 12, pp. 3838–3844, 2012.

[KOE 10] KOESTER R., HWANG J.-S., DURAND C., *et al.*, "Self-assembled growth of catalyst-free GaN wires by MOVPE", *Nanotechnology*, vol. 21, pp. 015602–015609, 2010.

[KUB 07] JI H., KUBALL M., BURKE R.A., *et al.*, "Vibrational and optical properties of GaN nanowires synthesized by Ni-assisted catalytic growth", *Nanotechnology*, vol. 18, pp. 445704–445707, 2007.

[KUY 03] KUYKENDALL T., PAUZAUSKIE P., LEE S., *et al.*, "Metalorganic chemical vapor deposition route to GaN nanowires with triangular cross section", *NanoLetters*, vol. 3, no. 3, pp. 1063–1066, 2003.

[KUY 07] KUYKENDALL T., ULRICH P., ALONI S., *et al.*, "Complete composition tunability of InGaN nanowires using a combinatorial approach", *Nature Materials*, vol. 6, no. 12, pp. 951–956, 2007.

[LEU 12] LEUNG B., SUN Q., YERINO C.D., *et al.*, "Using the kinetic Wulff plot to design and control nonpolar and semipolar GaN heteroepitaxy", *Semiconductor Science and Technology*, vol. 27, pp. 024005–024012, 2012.

[LI 09] LI Q., LIN Y., CREIGHTON J.R., *et al.*, "Nanowires-templated lateral epitaxial growth of low-dislocation density nonpolar a-plane GaN on r-plane sapphire", *Advanced Materials*, vol. 21, no. 23, pp. 2416–2420, 2009.

[LI 11a] LI S.F., FUNDLING S., WANG X., *et al.*, "Selective area growth of GaN rod structures by MOVPE: dependence on growth conditions", *Physica Status Solidi C*, vol. 8, nos. 7–8, pp. 2318–2320, 2011.

[LI 11b] LI S.F., FUNDLING S., WANG X., *et al.*, "Polarity and its influence on growth mechanism during MOVPE growth of GaN sub-micrometer rods", *Cryst. Growth & Design*, vol. 11, pp. 1573–1577, 2011.

[LI 12] LI S.F., WAAG A., "GaN based nanorods for solid state lighting", *Journal of Applied Physics*, vol. 111, no. 7, pp. 071101–071123, 2012.

[LIN 12] LIN Y.T., YEH T.W., DAPKUS P.D, "Mechanism of selective area growth of GaN nanorods by pulsed mode metalorganic chemical vapor deposition", *Nanotechnology*, vol. 23, pp. 465601–465606, 2012.

[LIU 07] LIU F., COLLAZO R., MITA S., *et al.*, "The mechanism for polarity inversion of GaN via a thin AlN layer: direct experimental evidence", *Applied Physics Letters*, vol. 91, pp. 203115–3, 2007.

[MAR 13] MARKURT T., LYMPERAKIS L., NEUGEBAUER J., *et al.*, "Blocking growth by an electrically active subsurface layer: the effect of Si as an antisurfactant in the growth of GaN", *Physical Review Letters*, vol. 110, pp. 036103–036105, 2013.

[MAR 69] MARUSKA H.P., TIETJEN J.J., "The preparation and properties of vapor-deposited single-crystal-line GaN", *Applied Physics Letters*, vol. 15, no. 10, pp. 327–329, 1969.

[MIT 01] MITCHELL C.C., COLTRIN M.E., HAN J., "Mass transport in the epitaxial lateral overgrowth of gallium nitride", *Journal of Crystal Growth*, vol. 222, nos. 1–2, pp. 144–153, 2001.

[MOH 09] MOHAMMAD S.N., "For nanowire growth, vapor-solid-solid (vapor-solid) mechanism is actually vapor-quasisolid-solid (vapor-quasiliquid-solid) mechanism", *Journal of Chemical Physics*, vol. 131, no. 22, pp. 224702–224713, 2009.

[MOH 10] MOHAMMAD S.N., "General theoretical model for the vapor-phase growth and growth rate of semiconductor nanowires", *Journal of Vacuum Science and Technology B*, vol. 28, pp. 329–352, 2010.

[MOS 07] MOSCATELLI D., CAVALLOTTI C., "Theoretical investigation of the gas-phase kinetics active during the GaN MOVPE", *Journal of Physical Chemistry A*, vol. 111, no. 21, pp. 4620–4631, 2007.

[NAK 92] NAKAMURA S., MUKAI T., SENOH M., *et al.*, "Thermal annealing effects on P-type Mg-doped GaN films", *Japanese Journal of Applied Physics*, vol. 28, pp. L139–L142, 1992.

[NAK 94] NAKAMURA S, MUKAI T., SENOH M., "Candela-class high-brightness InGaN/AlGaN double-heterostructure blue-light-emitting diodes", *Applied Physics Letters*, vol. 64, pp. 1687–1669, 1994.

[NOR 96] NORTHRUP J.E., NEUGEBAUER J., ROMANO L.T., "Inversion domain and stacking mismatch boundaries in GaN", *Physical Review Letters*, vol. 77, no. 1, pp. 103–106, 1996.

[OH 10] OH E., LEE B.W., SHIM S., *et al.*, "Platinum-assisted vapor-liquid-solid growth of GaN nanowires and their properties", *Journal of the Korean Physical Society*, vol. 56, no. 1, pp. 100–103, 2010.

[PAR 06] PARIKH R.P., ADOMAITIS R.A., "An overview of gallium nitride growth chemistry and its effect on reactor design: application to a planetary radial-flow CVD system", *Journal of Crystal Growth*, vol. 286, pp. 259–278, 2006.

[PAR 13] PARK J.-H., KIM M.-H., KISSINGER S., *et al.*, "Vertically p–n-junctioned GaN nanowire array diode fabricated on Si(111) using MOCVD", *Nanoscale*, vol. 5, pp. 2959–2966, 2013.

[QIA 04] QIAN F., LI Y., GRADECAK S., *et al.*, "Gallium nitride-based nanowire radial heterostructures for nanophotonics", *NanoLetters*, vol. 4, no. 10, pp. 1975–1979, 2004.

[QIA 05] QIAN F., GRADECAK S., LI Y., *et al.*, "Core/multishell nanowire heterostructures as multicolor, high-efficiency light-emitting diodes", *NanoLetters*, vol. 5, no. 11, pp. 2287–2291, 2005.

[QIA 08] QIAN F., LI Y., GRADECAK S., *et al.*, "Multi-quantum-well nanowire heterostructures for wavelength-controlled lasers", *Nature Materials*, vol. 7, pp. 701–706, 2008.

[RES 05] RESHCHIKOV M.A., MORKOC H., "Luminescence properties of defects in GaN", *Journal of Applied Physics*, vol. 97, pp. 061301–061303, 2005.

[ROS 06] ROSA A.L., NEUGEBAUER J., "Understanding Si adsorption on GaN(0001) surfaces using first-principles calculations", *Physical Review B*, vol. 73, pp. 205314–12, 2006.

[RUT 64] RUTH V., HIRTH J.P., "Kinetics of diffusion-controlled whisker growth", *Journal of Chemical Physics*, vol. 41, no. 10, pp. 3139–3149, 1964.

[SAL 13] SALOMON D., DUSSAIGNE A., LAFOSSAS M., *et al.*, " Metal organic vapour-phase epitaxy growth of GaN wires on Si (111) for light-emitting diode applications", *Nanoscale Research Letters*, vol. 8, no. 1, pp. 61–65, 2013.

[SIB 12] SIBIREV N.G., TCHERNYCHEVA M., CIRLIN G.E., *et al.*, "Effect of diffusion from a lateral surface on the rate of GaN nanowire growth", *Semiconductors*, vol. 46, no. 6, pp. 838–841, 2012.

[SEO 08] SEO M.-K., YANG J.-K., JEONG K.-Y., *et al.*, "Modal characteristics in a single-nanowire cavity with a triangular cross section", *NanoLetters*, vol. 8, no. 12, pp. 4534–4538, 2008.

[SHI 01] SHI W.S., ZHENG Y.F., WANG N., *et al.*, "Microstructures of gallium nitride nanowires synthetized by oxide-assisted method", *Chemical Physics Letters*, vol. 345, pp. 377–380, 2001.

[SO 13] SO B.M., YOUN S.B., IM I.-T., "Growth rate and composition of InGaN during InGaN/GaN quantum wells selective area metal-organic vapor phase epitaxy considering surface diffusion", *Journal of Industrial and Engineering Chemistry*, vol. 19, no. 5, pp. 1747–1751, 2013.

[SU 05a] SU J., CUI G., GHERASIMOVA M., *et al.*, "Catalytic growth of group III-nitride nanowires and nanostructures by metalorganic chemical vapor deposition", *Applied Physics Letters*, vol. 86, no. 1, pp. 013105–3, 2005.

[SU 05b] SU J., GHERASIMOVA M., CUI G., TSUKAMOTO H., *et al.*, "Growth of AlGaN nanowires by metalorganic chemical vapor deposition", *Applied Physics Letters*, vol. 87, pp. 183108–3, 2005.

[SUN 05] SUN X.S., YU Z., JIN Y., *et al.*, "Growth of GaN nanorods via Au catalyst-assisted CVD", *Chemical Vapor Deposition*, vol. 11, pp. 433–436, 2005.

[SUN 08] SUN Q., YERINO C.D., KO T.S., *et al.*, "Understanding nonpolar GaN growth through kinetic Wulff plots", *Journal of Applied Physics*, vol. 104, no. 9, pp. 093523–5, 2008.

[STR 99] STRINGFELLOW G.B., *Organometallic Vapor-Phase Epitaxy: Theory and Practice*, Academic Press, London, 1999.

[TAN 09] TANG T.-Y., SHIAO W.-Y., LIN C.-H., *et al.*, "Coalescence overgrowth of GaN nanocolumns on sapphire with patterned metal organic vapor phase epitaxy", *Journal of Applied Physics*, vol. 105, no. 2, pp. 023501–023505, 2009.

[TAN 10] TANG T.-Y., LIN C.-H., YUNG-SHENG CHEN Y.-S., *et al.*, "Nitride nanocolumns for the development of light-emitting diode", *IEEE Transactions on Electron Devices*, vol. 57, no. 1, pp. 71–78, 2010.

[TES 13] TESSAREK C., BASHOUTI M., HEILMANN M., *et al.*, "Controlling morphology and optical properties of self-catalyzed, mask-free GaN rods and nanorods by metal-organic vapor phase epitaxy", *Journal of Applied Physics*, vol. 114, pp. 144304–144310, 2013.

[THO 96] THON A., KUECH T.F., "High temperature adduct formation of trimethylgallium and ammonia", *Applied Physics Letters*, vol. 69, no. 1, pp. 55–57, 1996.

[TSI 11] TSIVION D., SCHVARTZMAN M., POPOVITZ-BIRO R., *et al.*, "Guided growth of millimeter-long horizontal nanowires with controlled orientations", *Science*, vol. 333, no. 6045, pp. 1003–1007, 2011.

[TSO 09] TSOKKOU D., OTHONOS A., ZERVOS M., "Defect states of chemical vapor deposition grown GaN nanowires: effects and mechanisms in the relaxation of carriers", *Journal of Applied Physics*, vol. 106, pp. 054311–054316, 2009.

[WAN 06a] WANG G.T., TALIN A.A., WERDER D.J., *et al.*, "Highly aligned, template-free growth and characterization of vertical GaN nanowires on sapphire by metal–organic chemical vapour deposition", *Nanotechnology*, vol. 17, pp. 5773–5780, 2006.

[WAN 06b] WANG X., SUN X.Y., FAIRCHILD M., *et al.*, "Fabrication of GaN nanowire arrays by confined epitaxy", *Applied Physics Letters*, vol. 89, no. 23, pp. 233115–3, 2006.

[WAN 13] WANG X., JAHN U., LEDIG J., *et al.*, "The MOVPE growth mechanism of catalyst-free self-organized GaN columns in H_2 and N_2 carrier gases", *Journal of Crystal Growth*, vol. 384, pp. 61–65, 2013.

[WAT 13] WATSON I., "Metal organic vapour phase epitaxy of AlN, GaN, InN and their alloys: a key chemical technology for advanced device applications", *Coordination Chemistry Reviews,* vol. 257, pp. 2120–2141, 2013.

[WHO 02] WOHLFART A., DEVI A., MAILE E., *et al.*, *Chemical Communications*, vol. 9, pp. 998–999, 2002.

[WON 13] WONG M.H., KELLER S., NIDHI D.S., *et al..*, "N-polar GaN epitaxy and high electron mobility transistors", *Semiconductor Science and Technology*, vol. 28, no. 7, pp. 074009–074022, 2013.

[YOO 09] YOON J., GIRGIS A.M., SHALISH I., *et al.*, "Size-dependent impurity activation energy in GaN nanowires", *Applied Physics Letters*, vol. 94, pp. 142102–142103, 2009.

[ZHO 03] ZHONG Z., QIAN F., WANG D., *et al.*, "Synthesis of p-type gallium nitride nanowires for electronic and photonic nanodevices", *NanoLetters,* vol. 3, no. 3, pp. 343–346, 2003.

[ZYW 98] ZYWIETZ T., NEUGEBAUER J., SCHEFFLER, "Adatom diffusion at GaN (0001) and (000^1) surfaces", *Applied Physics Letters*, vol. 73, pp. 487–489, 1998.

11

Metal-Organic Chemical Vapor Deposition Growth of ZnO Nanowires

11.1. Introduction

Metal-organic chemical vapor deposition (MOCVD) (also called metal-organic chemical vapor phase epitaxy (MOVPE) to emphasize the crystallographic relationship) has been, for a long time, one of the most commonly employed techniques in the microelectronics industry for the realization of high-performance electronic and optoelectronic devices [RAZ 89]. Besides the III–V and silicon-based semiconductors, a wide variety of functional materials can be achieved by using MOCVD, including II–VI semiconductors, oxides, magnetic compounds, etc. A large number of elements (from columns I–VII, as well as transitions metals) can be commercially supplied under several organometallic forms with a high purity, thus enlarging the field of potential future applications. That makes MOCVD a technique of choice for developing original materials and/or architectures, from the basic research to the industrial fabrication. The ability to transport chemical species from the bubbler to the reactor, the accurate control of the partial pressures (reactants and dopants), and the high uniformity and reproducibility that are obtained over large areas constitute the main advantages of MOCVD.

II–VI semiconductors including selenides, tellurides and sulfides have been widely grown in the past few decades for a large number of applications in IR, visible and UV emitting devices, single photon sources, high energy detectors and solar cells, etc. More recently, the MOCVD growth of oxide compounds in the (Zn, Mg, Cd, Mn)-O system has attracted much research attention, motivated by a unique combination of physical properties in zinc oxide (ZnO)-based materials, which paves the way for original realizations in optoelectronics, sensors, energy harvesting

---

Chapter written by Vincent SALLET.

or biomedical science. Approximately now for 15 years, crystal growth studies have been carried out using MOCVD with the aim of achieving ZnO and its related alloys under their best crystalline structure. This has been guided by the strong motivation for the realization of high-quality two-dimensional (2D) films, and, later, for the growth of one-dimensional (1D) nanostructures. In most cases, MOCVD reactors have been modified or redesigned to facilitate the uniform deposition over large areas at relatively high temperatures, and under an oxidizing atmosphere. SiC-made substrate holders are commonly heated to more than 1,000°C under nitrous oxide without drastic degradation. Separate inlets for group II and oxygen precursors avoid the premature reactions (e.g. showerhead developed by Thomas Swan). In fact, the MOCVD system and the sources that are used to grow ZnO nanostructures are indeed quite similar to those used for the realization of 2D films. Only the growth parameters, such as O/Zn partial pressure ratio ($R_{VI/II}$), reactor pressure or temperature, must be adjusted to allow the transition from 2D to 1D growth. Readers should note that in the following text dealing with MOCVD, we will be mainly interested in vertical arrays of ZnO nanowires (NWs). This issue is of great importance for their future applications, for example in optoelectronics or light-emitting diodes. In the case of uniform, well-ordered, and vertical arrays of ZnO NWs, the technology for top and bottom contacts would be strongly facilitated.

To grow ZnO nanostructures, several organometallic zinc sources have been employed, including dimethylzinc (DMZn) [LIA 10], dimethylzinc-triethylamine (DMZn-TEN) [SAL 08, MON 12], dimthylzinc-bidentate ether adducts [ASH 11], and diethylzinc (DEZn) [PAR 02, KAR 10, KWA 09, SON 08, KIT 08, PAR 09, BAN 03, JEO 04], this latter precursor being the most commonly used. Regarding the oxygen source, two main molecules have been reported, O_2 and N_2O (nitrous oxide). O_2 has been widely used between 300 and 700°C, whereas N_2O has concerned growth at higher temperatures, in the range 600–1,000°C. Most zinc organometallics are pyrophoric liquids that react with air at room temperature. Strong parasitic reactions in the lines and in the reactor inlet are consequently expected when using O_2 as the oxygen source. To overcome this problem, special MOCVD reactors have been designed to limit the distance between the injectors and the substrate holder, and the working pressure has been reduced to several torr in order to limit the interactions between the gas species.

11.2. Thermodynamics

As stated by Stringfellow [STR 99], "thermodynamics underlies the epitaxial growth for all techniques, including MOVPE". In the following section, we will emphasize some of the thermodynamics issues that will give us a better understanding of the MOCVD growth of ZnO nanostructures.

Because of their extensive use in microelectronics, the thermal decomposition of Zn-organometallics has been studied by several authors [DUM 92, THI 09]. Under a hydrogen atmosphere, DEZn starts to decompose at 300°C, with an activation energy calculated around 51 kcal/mol, and leads to the formation of ethylene and ethane following a two-step homolytic fission:

$$(C_2H_5)_2Zn \Rightarrow C_2H_5^*Zn + C_2H_5^*$$

$$C_2H_5^*Zn \Rightarrow Zn + C_2H_5^*$$

$$2C_2H_5^* \Rightarrow C_4H_{10} \quad \text{or} \quad 2C_2H_5^* \Rightarrow C_2H_6 + C_2H_4 \qquad [11.1.]$$

The recombination of ethyl radicals gives the hydrocarbons by-products, which have to be eliminated and sent to the reactor exhaust. An incomplete decomposition process may induce the incorporation of carbon radicals in the deposited material [MAR 13, LIU 10]. In the presence of oxidizing molecules in the gas phase, the decomposition onset of DEZn is shifted to lower temperatures [THI 09]. Gas phase and surface reactions occur and facilitate the growth of ZnO even at temperatures as low as 100°C (with water) when using the atomic layer deposition (ALD) process [GUZ 12].

Arrhenius plots of the growth rate as a function of the reciprocal temperature are given in two references [BAN 03, YE 07a]. DEZn and O_2 were used as precursors. Similar dependences were observed (see Figure 11.1(a)), i.e. three distinct growth regimes: (1) below 400°C, a kinetic regime governed by surface reactions; (2) a plateau between 400 and ~600–650°C, attributed to a mass-transport regime limited by the species diffusion across the boundary layer; and (3) at high temperature, a desorption regime of the incoming molecules that leads to a decrease of the growth rate. Consequently, for 2D growth as for 1D nanostructures, most of the works using DEZn and O_2 as precursors were reported in the temperature range 400–700°C. ZnO deposition has also extensively studied at higher temperature, up to 1,000°C [KLI 04] by employing another molecule, N_2O, as the oxygen source, benefitting from its higher decomposition temperature. The complete dissociation of N_2O thermally occurs at temperatures above ~800°C or at lower temperature (300–500°C) in the presence of a catalyst [BAM 72, KAP 96] following the reaction:

$$2N_2O \Rightarrow 2N_2 + O_2 \qquad [11.2.]$$

Nevertheless, the decomposition of N_2O as a function of the temperature has been investigated by FT-IR spectroscopy [MAE 06] and mass spectroscopy [LIU 10], and tends to show that a slight proportion (a few percent) of N_2O could dissociate at lower temperature, down to 400°C. No chemical reaction was observed between ZnO and N_2O at room temperature [LIU 10], thus confirming that N_2O is a suitable precursor to avoid premature reactions in the line and/or the reactor inlet.

In most MOCVD processes, Zn and O species are transported and injected separately into the reactor chamber, just above the growing surface. Thus, it is believed that heterogeneous reactions occur on the substrate: (1) adsorption of the reactant species, (2) decomposition, (3) transport to incorporation sites and (4) elimination of the hydrocarbon by-products to the exhaust. Nevertheless, gas phase reaction between the precursors may also occur before reaching the substrate surface, for example within the boundary layer where the molecules must diffuse. Maejima *et al.* [MAE 06] suggest that, in the range 300–600°, $(C_2H_5)_2OZnOH$ adducts are formed and transferred to the growing surface, and, at higher temperature (600–900°C), $Zn(OH_2)$ could be the dominant molecule impinging the substrate. This point is important since the mobility of adatoms or admolecules plays a major role in the final crystal morphology. The issue (which species are mobile: Zn and O, ZnO, or other adducts?) is still to be fully elucidated and would deserve more accurate investigations.

Arrhenius plots of the growth rate as a function of the reciprocal temperature were also given by Ribeaud [RIB 11] when using DEZn and N_2O as precursors (Figure 11.1(b)). Two growth regimes have been observed: a transport regime between 630 and 930°C, and a desorption regime at $T_g > 930$°C.

11.3. Growth of ZnO nanowires

Recently, a wide variety of 1D semiconductor and oxide nanostructures, including nanowires, nanorods, nanoneedles, nanocombs, nanoribbons, nanosprings and nanoflowers, etc., has been produced by using vapor transport or chemical vapor deposition (CVD) techniques. However, the topic is not so new, if we consider the previous works published in the 1950s and 1960s. For example, the growth of whiskers has been achieved [SEA 55, ARN 56] and studied within Frank's dislocation model, where an emerging screw dislocation acts as a step and a sink for the incoming atoms. The crystal defect continuously provides a surface step for the whisker vertical raise. Later, the imperfect capture sites have been extended to monoatomic steps or impurities without being a prerequisite for the start of "filamentary growth" [SCH 67]. In many experiments, it was explained that the mobility dependence of atoms on the crystal plane orientation will play a major role, and so do the mechanisms of capture at surface steps. We will see that these considerations are still the most relevant.

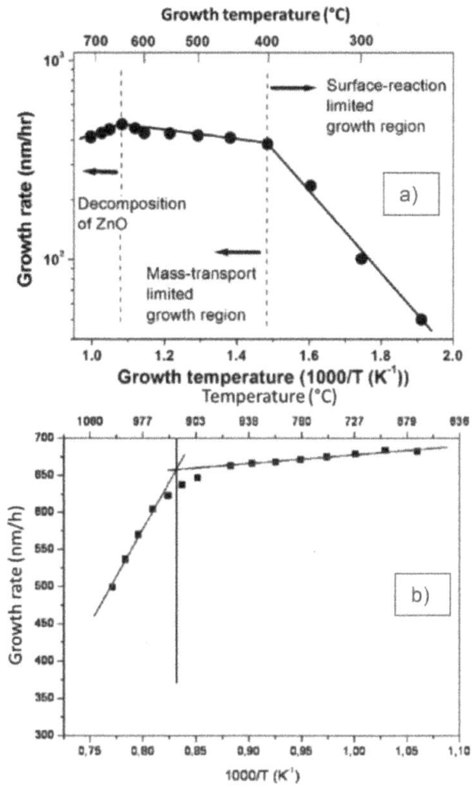

Figure 11.1. *Arrhenius plots of the growth rate versus reciprocal temperature: a) using O_2 as precursor [BAN 03] (reprinted with permission from Elsevier, copyright 2003); b) using N_2O [RIB 11]*

Regarding the realization and applications of ZnO nanostructures, details can be found in several reviews [WAN 09, DJU 10]. Nanostructures' characteristics, properties, length, shape, density, vertical alignment, physical properties, etc., and also the location on which they grow (selective growth) are of extreme importance for the aimed applications in photonics and optoelectronics. In this section, we focus on the *spontaneous growth* mechanisms of vertical arrays ZnO NWs, when using the MOCVD process. But the readers will find an important body of literature related to the particular catalyst-assisted growth, also called vapor–liquid–solid (VLS) growth, which has been also extensively studied, however, mainly by using vapor phase transport, so that only a few experiments have been conducted using MOCVD technique. The gold-assisted growth of ZnO NWs will be detailed in section 11.6, focusing on the specific mechanisms responsible for the emergence of a 1D structure below the catalyst droplet.

Today, the issue of the growth mechanisms that lead to the spontaneous formation of ZnO NWs is progressing, but still far from being fully understood. In near-equilibrium growth, e.g. in wet chemical deposition, the difference in chemical potential between the liquid and the solid phases is small, and the shape of the growing crystal may be determined by the minimization of the total surface energy and the Wulff theorem. But the difference in surface energies of ZnO planes, i.e. the degree of anisotropy, is not so large and is not sufficient to account for the high aspect ratios measured in ZnO NWs. In ZnO, surface energies of 2.25 J/m^2, 1.12 J/m^2 and 1.06 J/m2 have been calculated for C-(0001), M-(10-10) and A-(11-20)-plane, respectively [NA 09]. Consequently, other surface aspects must be considered too, such as the polar nature of the C-plane. Overall, kinetics is believed to control the nanowire growth. In MOCVD process, the chemical potential of the gas phase is much higher than that of the solid phase, thus making this method very non-equilibrium. In the following, we will review the parameters that govern the MOCVD growth of ZnO NWs: VI/II ratio, temperature, supersaturation and reactor pressure. Another dominant factor will be the substrate or buffer layer on which the NWs epitaxially grow. Experimentally, ZnO nanostructures exhibit three directions of fast growth: <0001>, <11-20> and <10-10>. Growth conditions or the choice of substrate surface allow us to preferentially favor one fast direction over the two others. In addition to the polar nature of the (0001) surface, this explains the wide variety of ZnO nano-objects already achieved. Beyond the surface energy considerations, each facet of the ZnO crystal has its specific adsorption/incorporation sites for the incoming gas species/atoms. Besides, due to the difference in dangling bounds, defects, or impurities on the growth front, the mobility of adatoms after precursor decomposition is believed to depend on the orientation and polarity of the growing facet.

Another driving force at the origin of 1D growth could be the occurrence of Zn or ZnO_x ($x<1$) clusters. Their formation under zinc-rich conditions could induce the self-catalyzed growth. Like in the VLS process, liquid Zn/ZnO_x droplets would act as preferential adsorption surfaces for the gas species, and consequently initiate the NW growth. Nevertheless, on the contrary to the GaAs case where a Ga droplet remaining at the tip of the nanowire has been observed [RAM 13], Zn or ZnO relics have not commonly been observed by the crystal growers. Rather, flat (0001) top surfaces or well-faceted tips are observed on MOCVD grown ZnO NWs.

Mastering of the nanowire growth paves the way for more complex 1D architectures, such as core-shell ZnO/ZnMgO heterostructures (for more details, refer to Chapter 3 of Volume 2 [FEI 14]) and type II heterojunctions (for more details, refer to Chapter 4 of Volume 2). It also opens the field of applications, with original devices for lightning and solar cells that will be presented further in Chapters 9 and 10 in Volume 2.

11.4. Spontaneous growth of ZnO nanowires: growth condition effects

From a macroscopic point of view, a growing crystal under given conditions is defined by different facets, which have specific kinetic parameters and surface activities that will determine the final morphology. This is particularly relevant in the MOCVD process, where a number of key parameters can be varied separately. Selecting and reading carefully the important literature related to the MOCVD growth of ZnO NWs, and taking into account our own experience on the topic, we will shed light on the relationship between the growth conditions and ZnO nanostructures morphologies in order to better understand the mechanisms involved.

References	Precursors	Temperature (°C)	Pressure (torr)	Substrate
[CHE 09]	$DMZn/O_2$	350	200	$MgAl_2O_4$
[LIU 04]	$DEZn/O_2$	700	3.10^{-3}–0,2	Sapphire, $Pt/SiO_2/Si$, Si
[SON 08]	$DEZn/O_2$	700		GaN/sapphire
[KIT 08]	$DEZn/O_2$	450 and 750	5	Sapphire
[JEO 04]	$DEZn/O_2$	360–500	5	Sapphire
[PAR 05a]	$DEZn/O_2$	<200 to >380	5	Sapphire
[PAR 09]	$DEZn/O_2$	180–330	1	Sapphire
[CHI 06]	$DEZn/O_2$	200 and 450	6	Sapphire
[PAR 07a]	$DEZn/O_2$	300	0.64	(111) Si
[ZHA 04]	$DEZn/O_2$	350–500	0.06–10	Sapphire
[PAR 05b]	$DEZn/O_2$	700	5	Sapphire
[YUA 04]	$Zn(C_5H_7O_2)_2$	500–600	760	Glass
[CHE 06]	$DEZn/O_2$	480		(100) Si
[KHR 11]	$Zn(C_5H_7O_2)_2$	500	760	(100) Si
[OGA 03]	$DEZn/N_2O$	500–800		Sapphire
[MAE 03]	$DEZn/N_2O$	400–900	200	A-sapphire
[KLI 04]	$DEZn/N_2O$	700–1,050	150	A-sapphire
[BEH 09]	$DEZn/N_2O$	800–1,000	75–300	A-sapphire
[MON 12]	$DMZn\text{-}TEN/N_2O$	800	760	Sapphire
[YE 07]	$DEZn/N_2O$	650	60–80	(111) Si
[ROS 09]	$DEZn/N_2O$	750–800	75–115	Sapphire
[PER10, PER 12]	$DEZn/N_2O$	750–850	75	Sapphire
This chapter (Figures 11.7 and 11.8)	$DEZn/O_2$ $DEZn/N_2O$	480 850	50	A-sapphire

Table 11.1. *Growth parameters used for the realization of ZnO NWs by MOCVD and related to the references presented in this chapter*

11.4.1. *O/Zn ($R_{VI/II}$) ratio*

One dominant factor appearing to cause a transition from 2D to 1D growth is the O/Zn ratio ($R_{VI/II}$) defined by the partial pressures of the precursors sent in the reactor:

$$R_{\frac{VI}{II}} = \frac{O\ gas\ species\ partial\ pressure}{Zn\ gas\ species\ partial\ pressure}$$

First, we will focus on the MOCVD processes using O_2 as the oxygen source.

Chen *et al.* [CHE 09] have employed DMZn and O_2 at 350°C, with a reactor pressure of 200 torr. Keeping O_2 flow constant at 1,000 cc/min, the DMZn flow was varied from 0.19 to 0.61 cc/min, i.e. $R_{VI/II}$ divided by ~6. The morphology turns from a vertical array of separate nanorods to a dense film. A high nucleus density and an increased lateral growth, which favor the grains coalescence, explain the evolution. This morphology transition was also reported by Jeong *et al.* [JEO 04] at 500°C, when reducing O_2/DEZn ratio from 407 to 82 (reactor pressure of 5 torr). The reduction was achieved by increasing the DEZn flow, and therefore, rising the Zn supersaturation, which is suggested to promote the 2D growth.

Regarding the effect of $R_{VI/II}$ increase, an evolution toward thin-film growth was reported in [CHE 09], keeping the DMZn flow constant at 0.19 cc/min and varying the O_2 flow from 300 to 2,000 cc/min ($R_{VI/II}$ multiplied by ~6). Using plasma-enhanced CVD at 700°C, it was observed that, in a plasma with 90% oxygen in helium, high density nucleation occurs and continuous thin films are deposited on sapphire [LIU 04]. The oxygen content must be reduced to 20% to promote the growth of ZnO NWs. Under O-rich conditions (high $R_{VI/II}$) at 800 °C, Song *et al.* [SON 08] have produced corrugated ZnO nanorods, where the common (10-10) facets disappear to the detriment of (10-1-1) and (10-11) facets. A growth model is proposed involving a reduction of the Zn atoms diffusion along the nanorod.

Therefore, from these results, some preliminary tendencies can be noticed: (1) high Zn supersaturation favors the nucleation and the lateral growth, to the benefit of the formation of a packed layer, and (2) high oxygen contents tend to limit the diffusion of Zn atom and also favor the 2D growth

Now, let us consider the case when using N_2O as the oxygen source.

Ogata's group was one of the pioneers of ZnO MOCVD growth using DEZn and N_2O as precursors. In particular, they have emphasized the role of high O/Zn ratio (up to 100,000) in the achievement of flat 2D films on sapphire [OGA 03]. On the contrary, they showed a morphology transition to three-dimensional (3D) and columnar deposition mode when setting low $R_{VI/II}$ (~500). Growth temperature had to be decreased from 800 to 500°C to promote the growth of nanorods, 10–50 nm in diameter and 0.5–1 μm in length. Similar results have been reported by Maejima *et al.* [MAE 03], where nanorods were deposited at 400°C and using $R_{VI/II}$ = 62,500. However, we must note that at such low temperatures the decomposition rate of N_2O is poor, and the effective $R_{VI/II}$ must be much lower. Indeed low growth temperature (Tg) is not a necessary condition, but low O/Zn ratio. Kling *et al.* [KLI 04] have grown vertical nanopillars with diameters of 50 nm and lengths of several micrometers using high temperatures, in the range of 700–1,050°C, and very low $R_{VI/II}$, between 30 and 120. This behavior has also been investigated by Behrends *et al.* [BEH 09] in a Thomas Swan close-coupled showerhead, also designed for the growth of GaN-based materials. $R_{VI/II}$ has been varied from 220 to 7,000 at Tg > 800°C, and results confirm what is commonly observed in the MOCVD growth of ZnO using N_2O, i.e. low ratios are preferred for the realization of NWs, and, at the opposite, high ratios lead to the formation of 2D layers.

Going further, we have systematically studied and varied the O/Zn ratio by two means [MON 12]: increasing the oxygen flow while keeping the zinc flux constant, and *vice versa*. This was important since it has allowed us to separate two contributions: (1) the effect of the excess of oxygen precursor and (2) the effect of the growth rate or supersaturation. Growth experiments were carried out at 800°C and atmospheric pressure on sapphire, using DMZn-TEN and N_2O as precursors. In a first set of samples, N_2O flow was changed from 300 to 900 cc/min, while maintaining DMZn-TEN at 75 cc/min. Morphological transitions were observed by scanning electron microscopy (SEM) from nanorods ($R_{VI/II}$~100), to nanotubes ($R_{VI/II}$~200), and finally to a dense film ($R_{VI/II}$~300), as displayed in Figure 11.2. Growth rate was found to be equivalent for all the samples, indicating that the growth was governed by zinc flux.

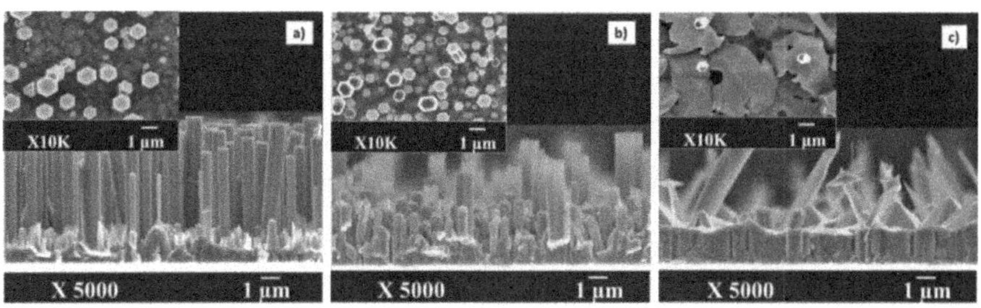

Figure 11.2. *SEM image of ZnO nanorods: a) nanotubes, b) rough layer and c) grown as O/Zn ratio increases*

In a second series of samples, $R_{VI/II}$ has been varied the opposite way, i.e. by increasing DMZn-TEN flow from 15 to 95 cc/min while keeping constant N_2O flow at 300 cc/min. Several morphologies were obtained from a disordered assembly of nanowalls and very thin NWs (at high $R_{VI/II}$~500) to large hexagonal microrods with well vertical alignment (at low $R_{VI/II}$~100). In this case, the morphology evolution results in the combination of O/Zn effect and supersaturation effect. To analyze the specific effect of the growth rate/supersaturation, a third set of samples has been realized at fixed $R_{VI/II}$ and increasing simultaneously N_2O and DMZn-TEN flows. The results will be presented further in section 11.4.5.

For the discussion, let us focus on the effect of the increase of the oxygen flux (toward high $R_{VI/II}$). Interestingly, Chen *et al.* [CHE 02] have used plasma-assisted molecular beam epitaxy (MBE) and reflection high-energy electron diffraction (RHEED) to study the morphology evolution of a growing ZnO surface. If the nature of the mobile species on the surface can still be discussed in the MOCVD process, in the case of MBE, we can assume that Zn and O atoms are the diffusing species. It is shown that under O-rich condition, i.e. high $R_{VI/II}$, oxygen tends to stabilize the incoming Zn atoms. RHEED oscillation damping indicates that the surface mobility of adatoms is strongly affected by $R_{VI/II}$, thus resulting in a change in the epilayer morphology. Murphy *et al.* [MUR 04] have used a stochastic growth model where the effect of lateral diffusion length (small under O-rich conditions, and large under Zn-rich) explains the surface morpholy changes, in good agreement with experimental observations.

In this framework, the different features observed in our samples can be understood and explained by the evolution of Zn adatoms mobility. A schema is given in Figure 11.3.

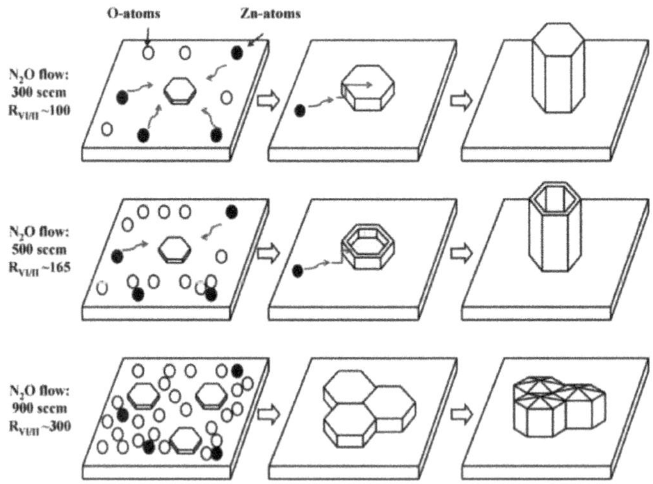

Figure 11.3. *Proposed model for the morphology changes as O/Zn ratio increases [MON 12]. (Reprinted with permission from Elsevier, copyright 2012)*

At low $R_{VI/II}$ values, the mobility of Zn atoms is high, and the diffusion length is sufficient enough to allow the surface anisotropies govern the growth process. The formation of C-oriented nanorods with M-plane facets tends to minimize the total surface energy. Hexagonal nanorods with well-defined facets and a high aspect ratio are synthetized. On the contrary, using conditions of high $R_{VI/II}$ values, high concentrations of oxygen atoms adsorbed on the surface tend to stabilize the mobile Zn, and favor the lateral growth and the formation of a 2D layer. At intermediate $R_{VI/II}$, the migration of Zn adatoms will be also difficult, but it could be more affected on the C-plane growing surface than on M-plane facets, due to their different number of specific incorporation sites (where Zn atom can be stabilized by oxygen). As a result, Zn atoms would tend to diffuse along lateral M-facets, reach the tip of the nanorod, and then would be incorporated at the edge of the (0001) top surface, to finally form a nanotube.

11.4.2. *Substrates*

The nature of the substrate, its orientation and even its polarity, as well as the buffer/nucleation layer on which the NWs grow constitute another major factor. Like in any epitaxy, the crystalline structure of the substrate will determine the emergence and the morphology of the nanostructures. It is interesting to note that it will influence their quality in a lesser degree, as it is possible to achieve high crystalline quality ZnO NWs on cheap and poor-quality materials. The relationship between the substrate (or buffer) and the grown NWs will allow the vertical alignment.

The most commonly used substrate has been sapphire, like for GaN growth. On the contrary to GaN, the fabrication of large area and high-quality ZnO substrates is possible [OHS 04]. Nevertheless, ZnO substrates are expensive and the market is too small and not enough developing to lead to price reduction. Furthermore, one motivation in growing ZnO NWs instead of 2D films is to get rid of expensive substrates and achieve high-quality crystals on a low-cost material. Only a few papers deal with ZnO NWs on ZnO substrates, mainly to study the growth mechanisms [PER 12]. C-axis sapphire has been much employed but it seems A-plane sapphire gives the best results. This is due to the very small lattice mismatch between 4×[a-ZnO lattice parameter] (4×3.25Å=13Å), and c-Al_2O_3 (12.99Å). ZnO NWs grow on A-plane sapphire with the epitaxial relationship: $[0001]_{ZnO}$ // $[11\text{-}20]_{sapphire}$ and $[11\text{-}20]_{ZnO}$ // $[0001]_{sapphire}$. Using R-sapphire, we can realize A-plane ZnO 2D films with C-axis parallel to the surface. The epitaxial relationship was found to be $[11\text{-}20]_{ZnO}$ // $[1\text{-}102]_{sapphire}$ and $[0001]_{ZnO}$ // $[1\text{-}101]_{sapphire}$. As a result, it is not surprising that Kar *et al.* [KAR 10] have realized nanopencil-like structures randomly oriented on the R-plane substrate surface. Sapphire substrates can be fabricated over large areas with high crystal quality. The drawbacks are its insulating character and its bad thermal conductivity. Most of the time, a thermal treatment above 1,000°C under inert gas allows the reconstruction of the surface and the generation of atomically flat terraces.

Vertical arrays of ZnO NWs have also been achieved on glass [YUA 04], and silicon [YE 07b, CHE 06, KHR 11]. The use of amorphous substrates, or the occurrence of native SiO_2 on Si impedes the epitaxial relationship with the growing layer. On these substrates, a thin ZnO buffer layer is often grown at low temperature, typically in the range of 200–400°C. After annealing, the aim is to realize a highly textured seeding layer, with preferential grain orientation along (0001), that will favor the subsequent growth of vertically aligned NWs [KHR 11]. All the processes (deposition of the nucleation layer, annealing and NWs growth) can be carried out in the same run. The effect of ZnO buffer/nucleation layers (orientation, roughness, etc.) has also been investigated, with regard to the growth on sapphire [PAR 06, KWA 09]. In particular, depending on the growth temperature used for this buffer, (0001) vertical NWs or [10-10] 61°-inclined NWs were fabricated [PAR 06].

To reduce the effect of the mismatch, ZnO NWs have been grown on GaN/sapphire templates. Song *et al.* [SON 08] observed that with the introduction of a rough GaN layer, ZnO 1D nanostructures selectively grow on the top of faceted GaN hillocks. Sputtered AlN interfacial layer have also been used on silicon substrates [KAR 11].

11.4.3. *Growth temperature*

In the MOCVD process, temperature considerations are crucial for several reasons. As mentioned before, temperature controls the decomposition of the precursors as well as the regime in which the material grows. An accurate control of the temperature of the different parts of the reactor (susceptor, inlet, walls that can be eventually cooled, etc.) allows us to limit the premature reactions so as to make possible a good uniformity and reproducibility of the process. In 2D film growth, an adequate growth temperature has a major influence on the optical, electrical and structural properties, as well as impurity diffusion from the substrate or between the two grown layers. In 1D nanostructures, temperature also strongly affects the final morphology.

Using O_2 as a precursor, Kitamura *et al.* [KIT 08] have grown, in the first step, vertical nanorods with diameter approximately 100 nm. Then, in the second step, ultrafine NWs (diameter 18 nm) have been introduced on the top of the previous thick NWs by increasing the growth temperature to 750°C. The vertical growth rate of the NWs is measured to rise from 37 to 130 nm/min in steps 1 and 2, respectively.

The proposed mechanism involves an enhancement of the migration of adatoms that would rapidly diffuse from low energy (10-10) M-surfaces (lateral facets) to high energy (0001) top plane. Jeong *et al.* [JEO 04] have observed a morphology transition from a rough film (at 360°C) to ZnO NWs with aspect ratio of 12 (at 500°C). Similarly, Park *et al.* [PAR 05a] reported morphological changes from columnar layer (200–260°C), to nanorods (260–320°C), to nanoneedles (320–380°C), and NWs (Tg > 380°C). The faster diffusion of Zn and O species that would enhance the anisotropic growth is assumed to account for the shape evolution when Tg increases. A detailed analysis of the surface morphology, with regard to the growth temperature, is given in [PAR 09]. The investigated range was 180–330°C, therefore rather low Tg if one compares with those employed in MOCVD process using O_2. The results are illustrated in Figure 11.4. At 330°C, well-aligned nanorods are formed and show pyramidal tips with an average angle of 80° between the inclined facet and the horizontal plane. Decreasing Tg to 280°C results in a shape evolution of the nanorods that exhibit (01-10) steps and (01-12) terraces at the side facets, significant of an enhancement of the lateral growth. At 240°C, the formation of additional (01-13) planes at the tip occurs to the detriment of (01-10) facets. Moreover, an asymmetric morphology of the tip is observed, made of higher index (01-12) and (01-13) facets, and is explained by spiral growth behavior. A further decrease of the growth temperature to 180 °C leads to the formation of a rough layer with (01-13) facets. Interestingly, it is also observed at the interface that the 30° twist of the ZnO lattice with respect to the Al_2O_3 lattice, which induces the $[01\text{-}10]_{ZnO}$ // $[11\text{-}20]_{sapphire}$ epitaxial relationship, only occurs at temperature above 280°C (which was also reported by Chin *et al.* [CHI 06]).

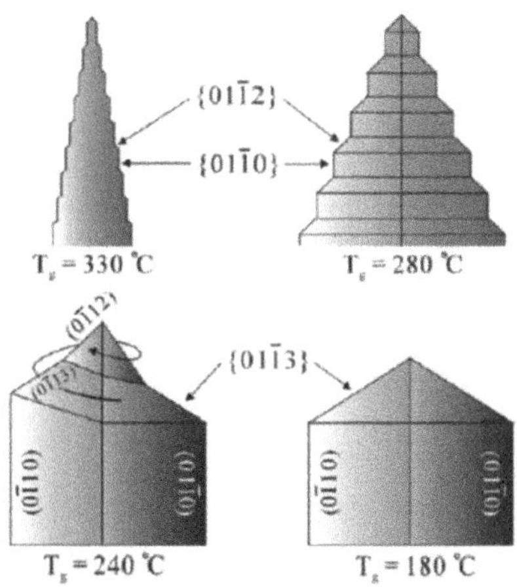

Figure 11.4. *Schematic diagram of facet morphology for different growth temperatures [PAR 09]. (Reproduced with permission from IOP Publishing. © 2009 IOP Publishing. All rights reserved)*

Using N_2O, ZnO growth is performed at a higher temperature to ensure the decomposition of the oxidizing molecule. The advantage is to achieve high-quality material, and, in the case of 1D growth, to give high surface mobility that will enhance the surface anisotropy of the growing facets. The drawbacks are the energy-consuming process and the possible diffusion of impurities from the substrate. Behrends *et al.* [BEH 09] have varied the growth temperature between 800°C and 1,000°C. At 800°C, mostly nanowalls are deposited whereas at 900°C and 1,000°C ZnO NWs emerge at the top of 3D islands.

11.4.4. *Pressure*

MOCVD reactors can be operated at low pressure. Depending on the pumping system capacity and the employed gas flow rates, the deposition may be carried out in a wide range of pressures, typically from a few torr at atmospheric pressure. We must note that the so-called metalorganic molecular beam epitaxy (MOMBE) system operates at much pressure, approximately 10^{-3} torr in [LIU 04]. Concerning the growth of ZnO NWs, it appears that the reactor pressure has a significant influence on the final morphology. Low-pressure conditions limit the premature reactions in the gas phase between Zn- and O-species, and influence the thickness of the boundary layer through which these gas species will diffuse.

Using O_2 a as precursor, reactor pressures between 0.1 and 10 torr were typically set. Park et al. [PAR 07a] observed that increasing the working pressure from 1 to 5 torr enhance the lateral growth and the formation of 2D layer. Zhang et al. [ZHA 04] investigated the influence of the pressure in the range [0.06 – 10 torr] and showed morphology transitions from nanowalls to nanotube and to nanorods.

Using N_2O as the oxygen source, commonly set reactor pressures are higher, in the range of 50–150 torr, as compared with the case of O_2. Behrends *et al.* [BEH 09] have varied the working pressure between 75 and 300 torr. They reported an increase of the diameter of the NWs. Like in the O_2 case, increasing the working pressure favors the lateral growth of the ZnO nanostructures. We also observed the same behavior carrying out experiments at 50 and 70 torr in a horizontal reactor, as shown in Figure 11.5.

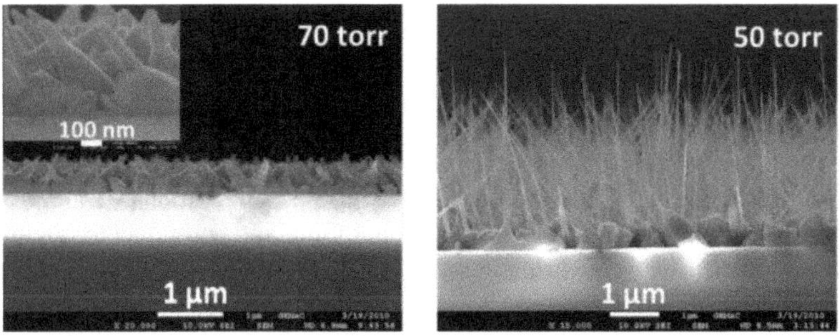

Figure 11.5. *SEM images of two ZnO NWs samples grown with different reactor pressures, 70 and 50 torr*

11.4.5. *Supersaturation and growth rate*

We have specifically investigated the effect of the precursors flow rates keeping constant O/Zn ratio and all other conditions [MON 12]. DMZn-TEN and N_2O partial pressures were simultaneously increased to enrich the gas phase, thus leading to an increase of the growth rate. It was not surprising to observe that ZnO NWs grow bigger using high deposition rates, but at least two features of the nanostructures were changed: (1) their vertical alignment and (2) their aspect ratio. Indeed, a better vertical alignment of the nanorod is clearly achieved with the increase of the precursor flows, as illustrated in Figures 11.6(a) and (b). The aspect ratio of nanorods as a function of the precursor flow rates is shown in Figure 11.6(c). It decreases from 60 to 10 when the growth rate increases. This indicates that high precursor flow rates, which produce a high quantity of grown material (see the inset in Figure 11.6(c)), favor the lateral growth and consequently a low aspect ratio.

In crystal growth, supersaturation is a determining factor. Here, it is defined by:

$$S=(p-p_0)/p_0 \qquad [11.3.]$$

where p and p_0 are the real gas phase and the equilibrium pressures. Increasing the precursor partial pressures increases the chemical potential of the gas phase, and thus the driving force for growth. According to the BCF model [BUR 51], supersaturation facilitates the growth from the gas phase induced by the advance of atomic steps where adatoms migrate and incorporate at kink sites. Interestingly, Ruth *et al.* [RUT 64] have also shown that high supersaturation favors the lateral growth as it enhances the nucleation on whisker sides. More recently, Ye *et al.* [YE 05] have studied the influence of the supersaturation on the ZnO nanostructures. Growths were performed by vapor phase transport in a horizontal tube, and using ZnO and graphite powders as sources. They proposed that, under a certain supersaturation, the final morphology is determined by the competition of capture of incoming atoms by facets of different surface energies.

In our work, the growth rate is controlled by the zinc partial pressure, and increasing values of supersaturation were set through the control of the precursors flows in the gas phase. Thus, in agreement with the contributions of the previous authors, we conclude that precursors' supersaturation, and consequently the growth rate, has a significant influence on the ZnO nanostructures. As a consequence, growing ZnO NWs with a fast growth rate for a short time, or slowly for a long time, all other conditions being kept, does not lead to the same morphology.

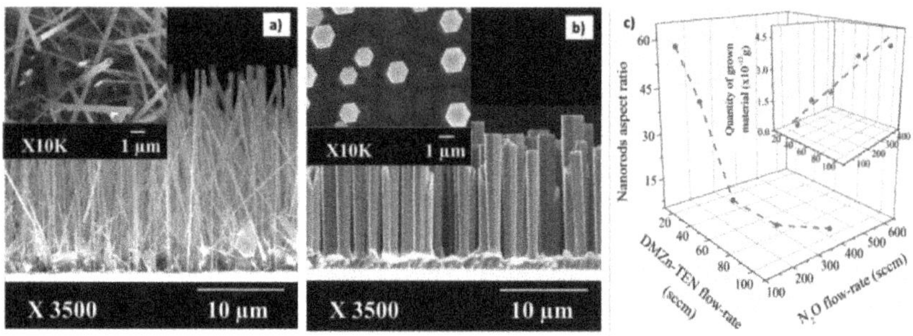

Figure 11.6. *SEM images of two ZnO NWs samples grown: a) at low and b) at high growth rate; c) aspect ratio and estimation of the deposited mass as a function of the precursor flow rates. For a color version of this figure, see www.iste.co.uk/consonni/nanowires1.zip*

11.4.6. *Structural characterization and growth mechanisms*

To study the mechanisms involved in the growth of ZnO NWs, scanning electron and transmission electron microscopies have been extensively employed to assess the morphology and the structural properties at different stages of the MOCVD process. Nucleation has also been studied to better understand how ZnO NWs can emerge from the substrate, under specific growth conditions. Whatever the growth technique, germination steps and strain relaxation could play a major role in the achievement of ZnO NWs. The formation of ZnO nuclei depends on the substrate nature, roughness and preparation, and it is worth emphasizing that ZnO NWs have been realized on surfaces presenting strongly different physical properties. In the case of sapphire, due to the large lattice mismatch, ZnO nucleus is highly strained, and dislocations quickly appear as the island or wetting layer further develops.

Park *et al.* [PAR 05b] reported on the early stage of ZnO nanoneedles; DEZn and O_2 were used as precursors. They observed that 3D ZnO islands start to nucleate and then cover the sapphire surface to form a continuous rough layer. Only after that, nanoneedles develop on this in-plane strained layer, associating a relaxation step of the compressive strain. It is inferred that the 1D growth can be explained by a strain-induced mechanism.

We have also grown ZnO NWs with similar growth conditions; high density arrays are obtained. The morphology shown in Figure 11.7 is typical of what can be found in the literature. The interfacial rough ZnO layer is measured approximately 100 nm.

Figure 11.7. *SEM images (top and cross views) of ZnO NWs on sapphire, grown using O_2. Using N_2O and DEZn, several authors have also investigated the onset of the ZnO NWs growth on sapphire. Even for the shortest deposition time, vertical nanowires are always observed on the top of 3D islands [BEH 09, MON 12, ROS 09, PER 10, PER 12], as shown for a 20 sec growth in Figure 11.8(a)*

It is interesting to note that in these articles, an initially thick/rough strained interfacial layer could not be evidenced, on which the NWs would develop as the growth proceeds further, as is the case using O_2 as precursor. The growth mechanism indeed seems different using N_2O, with NWs forming at the first stage, possibly due to higher growth temperatures. Nevertheless, a very thin wetting layer has been observed by Perillat *et al.* [PER 10], with a thickness of 8 monolayers. In their work, these authors have accurately studied the initial growth using TEM. After 20 s, nanorods and pyramidal island are yet formed. Some vertical nanorods grow directly on the ZnO wetting layer/sapphire, while others grow on the top of 3D islands. Some grains do not exhibit NWs on their summit. After 1 min deposition, the situation is very similar. Islands have evolved into well-faceted pyramids, showing or not a nanowire of their top. No ZnO NW standing directly on the surface is observed, indicating that the pyramids have also developed around the bottom of the previously formed NWs. TEM analysis clearly confirms that the NWs can originate not only from de ZnO/sapphire interface (at the onset of the growth), but also from the summit of the pyramids as the deposition continues. We have grown samples at 30 min using very similar conditions. Our results are in agreement with

Perillat's work. The morphology shows large pyramids with micrometer size, and hexagonal NWs grown along (0001) C-plane direction, with (10-10) M-plane lateral facets (see Figure 11.8(b)). Diameters and lengths are measured approximately 100 nm and 5 µm, respectively, which is typical for ZnO NWs fabricated by MOCVD using DEZn and N_2O precursors. We can note that the density is much lower in comparison to ZnO NWs grown using O_2.

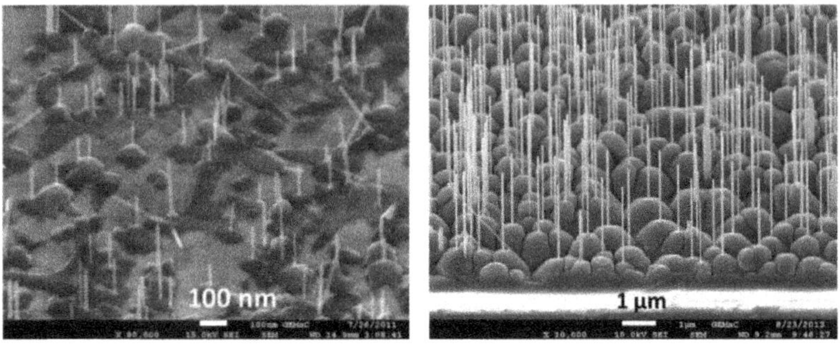

Figure 11.8. *SEM images of ZnO NWs samples grown using N_2O: a) at 20 s and b) 30 min*

Perillat *et al.* have further carried out TEM characterization [PER 12]. In particular, CBED analysis has allowed us to determine the polarity of the different domains of the deposited material. This work has highlighted opposite polarities between NWs and pyramids, and informs us largely on the growth mechanisms. Indeed, if we try to better understand the process responsible for the formation of the ZnO NWs, it is crucial to note that whatever the substrate nature, NWs grow along C-axis with Zn-polarity whereas pyramids are found to be O-polar. As a result, we can infer that the emergence of the ZnO nanowire is caused by a polarity inversion in the growing crystal, at a very small scale. Inverted domain boundaries (IDBs), which mark the frontier between Zn-polar nanowire base and O-polar pyramid/underlayer, have been observed at several locations (see Figure 11.9): (1) at the top the pyramid, (2) inside the pyramid, and (3) at the interface with the substrate/wetting layer. It is worth noting that, once created, IDBs can move up, or stay fixed, during the simultaneous growth of the NW and pyramid that will proceed further. At the interface with the substrate, it is difficult to distinguish whether the IDB starts from the wetting layer or directly from the sapphire.

The formation of IDBs could be induced by the segregation of metallic impurities, such as aluminum, which would come from the sapphire or the hydrothermally grown ZnO substrate. Atomic steps on the sapphire surface may also account for the nucleation of IDBs that would, in this case, help to accommodate the two growing crystals on each side of the step. In the literature, the control of the polarity of ZnO has been reported by treating thermally sapphire substrates, or by using specific buffer layers, such as CrN and CrO_3 [PAR 07b]. In that sense, we can also imagine that local heterogeneities on the sapphire substrate may be responsible for the formation of Zn-polar nuclei (which would further give rise to ZnO NWs) surrounded by O-polar wetting layer.

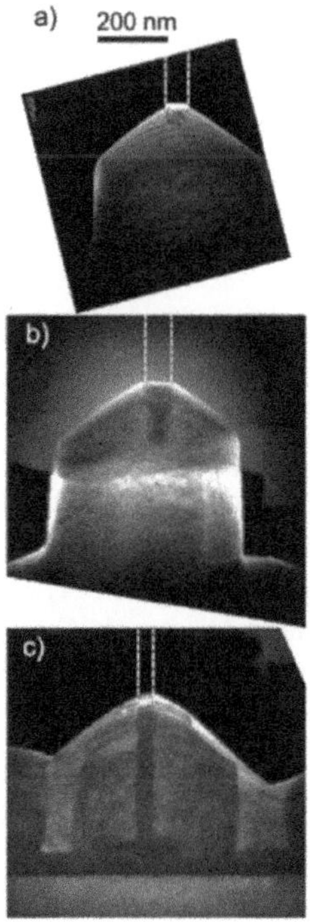

Figure 11.9. *Two-beam TEM images of ZnO pyramids and NWs: a) IDB at the top of the pyramid, b) IDB inside and c) IDB at the ZnO/sapphire interface [PER 12]. (Reproduced with permission from IOP Publishing, © 2012 IOP Publishing. All rights reserved)*

We have mentioned the possible effect of aluminum impurities on polarity inversion. Let us linger over the role of metal atoms in ZnO growth and consider some of the reported works. Adachi *et al.* [ADA 11] have shown that heavily doping with aluminum turns the polarity of the ZnO layer from O-polar to Zn-polar. Using angle-varied X-ray photoelectron spectroscopy, Kabisch *et al.* [KAB 13] confirms that Al-doping causes an inversion of the polarity at the interface between the ZnO layer and the sapphire substrate. Wang *et al.* [WAN 09] have synthesized nanoring structures, by using vapor-solid growth process in a horizontal tube, and introducing indium impurities. TEM analysis shows IDBs initiated by the occurrence of an indium plane on the C-growing surface. Yan *et al.* [YAN 98] similarly reported that in annealed In_2O_3-ZnO films, a single In-O layer induces a polarity inversion between two adjacent ZnO layers. We can also carry the discussion forward by considering what happens in metal-catalyzed growth (VLS process). In the following section, we will show that the VLS NWs and the catalyst-free NWs grow following opposite crystal polarities: +C (Zn-polar) for the spontaneous NWs, and –C (O-polar) for the catalyzed NWs, independently of the nature of the substrate or of the growth conditions. This result emphasizes the particular influence of the metallic droplet on the polarity of ZnO growth. Consequently, from all these observations, we are tempted to infer that aggregates of metal impurities could induce, the same way, the emergence on Zn-polar NWs. These aspects would deserve further investigations in order to fully understand the formation of IDBs in ZnO growth.

To conclude this section, let us summarize and compare the two proposed mechanisms involved in the MOCVD process using O_2 or N_2O as oxygen source.

Using O_2 and a temperature approximately 500°C, the growth starts with the nucleation of ZnO seeds, with a high density. Depending on the substrate (or buffer layer if a template is used), the epitaxial relationship allows the orientation of the growing seeds along the [0001] direction. The grain covering of the surface, associated with coalescence, leads to rough C-axis-oriented layer (on amorphous surface, highly C-axis-textured seed layer can also be grown). Subsequently, due to specific growth conditions (temperature, O/Zn ratio, etc.) as well as a possible strain relaxation mechanism, NWs emerge from the grains and a dense vertical array of ZnO NWs is achieved. The development of the surface induced by 1D shape would help to release the compressive strain in the nucleation layer. The whole ZnO structure is Zn-polar.

Using N_2O, growth is performed at a higher temperature, approximately 850°C. On sapphire, it starts with the covering of a thin ZnO wetting layer, quickly followed by the formation of ZnO pyramids. Again, growth conditions induce these pyramids to be O-polar with {1-10x} facets. The growth rate along –C direction is low. This should only result in the deposition of a rough 3D layer, but polarity inversion phenomena occur simultaneously, and since the onset of the growth, leading to the emergence of Zn polar NWs. Once initiated, specific conditions make the growth of NWs very fast, with a high aspect ratio. Impurity aggregates could be at the origin of the polarity inversion. The morphology results in the simultaneous deposition of slow O-polar pyramids and fast Zn-polar NWs. It is worth noting that the same morphologies are obtained on O-polar ZnO substrates (thus lattice matched) indicating that strain probably does not play any role in the emergence of 1D structures [PER 12]. Interestingly on sapphire, the formation of the very thin ZnO wetting layer returns to the case of homoepitaxial growth. We believe that the strong anisotropy between the growths along +C-axis (Zn-polar) and –C-axis (O-polar) accounts the best for the experimental observations.

Finally, we can conclude that MOCVD is a very suitable technique for the realization of ZnO nanostructures. By individually adjusting the growth parameters (O/Zn ratio, temperature, pressure, supersaturation) and/or selecting the precursors (O_2, N_2O), it allows the variation of shape, density, aspect ratio and alignment of the ZnO NWs on a wide variety of substrates. Like for GaN (for more details, see Chapter 10), MOCVD deposition of ZnO NWs can be potentially achieved with good uniformity over large areas.

11.5. Selective area growth of ZnO nanowires

Beyond the achievement of low-cost devices for which the developed surface induced by the nanostructuration of the material constitutes the major advantage, many applications would require an accurate control of the NW properties, i.e. alignment, morphology, conductivity and location. This is especially critical for light emitters in optoelectronics. To obtain such devices based on large and position-controlled arrays of ZnO NWs with a perfect uniformity, selective area growth is needed to overcome the uncertainty of the nucleation process.

Using MOCVD, Hong *et al.* [HON 07] have selectively grown ZnO NWs on the top of GaN micro-patterns (pyramids-like, with highly anisotropic surface energies), which were previously formed on silicon substrates. A two-step process requiring the sequence of two growths is, however, needed. Thierry *et al.* [THI 11] have developed the selective growth of ZnO NWs directly from a patterned Si3N4 layer. The dielectric mask is deposited on a ZnO/sapphire template substrate, and features a network of holes with diameter ranging from 50 to 200 nm, with a pitch set at 300 or 500 nm. The growth is carried out at 880°C using DEZn and N2O as precursors, with a reactor pressure of 90 torr. The whole process, requiring one step of electronic lithography (or nanoimprint) and one step of growth, is illustrated in Figure 11.10.

The results reveal that, with an accurate set of parameters, the selective area growth of 1D nanostructures is promoted. Figure 11.11 shows very homogeneous arrays, and comparison can be done with a reference sapphire sample without mask (Figure 11.11(left)). The high mobility of adatoms on the dielectric layer allows their diffusion toward the holes where nucleation takes place. The nanostructures are remarkably located on the openings of the mask. The shape of the nanostructures depends on the hole diameter and pitch, but also strongly depends on the initial polarity of the ZnO buffer layer constituting the template; zinc-polar surface is preferred to allow a perfect uniformity.

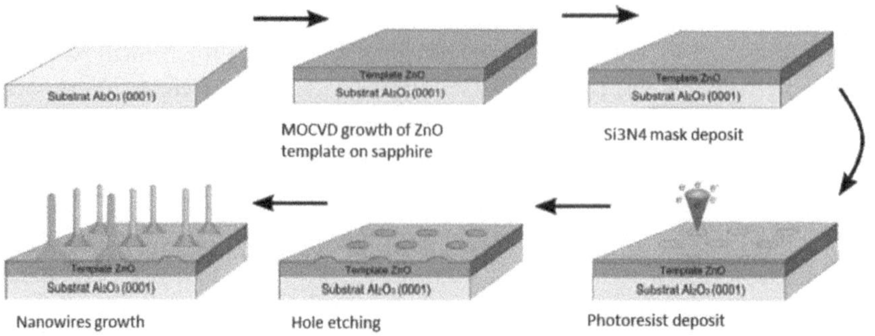

Figure 11.10. *Detailed process for the realization of a patterned substrate using electron beam lithography, and subsequent selective area growth of ZnO NWs [THI 11]. For a color version of this figure, see www.iste.co.uk/consonni/nanowires1.zip*

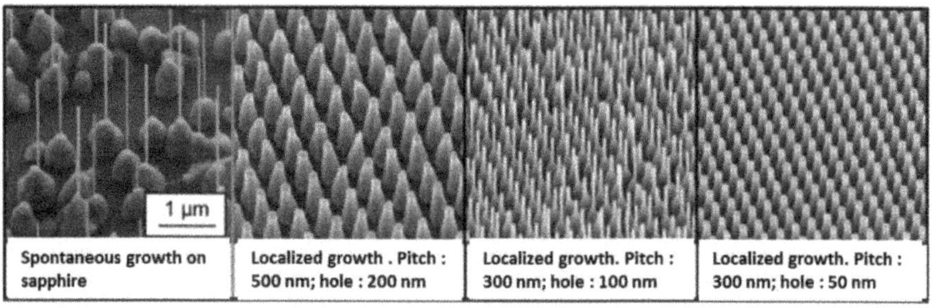

Figure 11.11. *SEM images of selectively grown ZnO NWs on patterned substrates with different holes, diameters and pitch [THI 11]. For comparison, the spontaneous growth on sapphire without mask is presented in the left image*

11.6. Catalyst-assisted growth of ZnO NWs

If zinc oxide and gallium nitride reveal themselves well-adapted materials for the spontaneous growth of 1D and hierarchical structures at the nanoscale, we must recognize that this is not the case for most of the other semiconductors, belonging to the Si-Ge, III–V, or II–VI families. Actually, within the so-called "bottom-up" approach, the growth method leading to the realization of 1D semiconductor nanostructures often involves a catalyzed mechanism. In 1964, Wagner and Ellis [WAG 64] grew silicon wires using gold droplets, previously deposited on a silicon substrate. In this process, named vapor–liquid–solid (VLS), the liquid alloyed droplet (Au-Si eutectic) shows a preferential surface for the gas species adsorption; the wire grows catalytically after diffusion and precipitation of silicon atoms at the liquid–solid interface. The concept has been further applied to (Ga, In)As [BAU 08], GaN [GOT 08], ZnSe [CHA 06], etc.

In a "pure VLS growth", the temperature is chosen accurately after consideration of the metal-semiconductor phase diagram (see Figure 11.12.) so that:

– the catalyst or alloyed droplet is liquid, and acts as a sink where vapor reactants preferentially adsorb;

– at least one component of the semiconductor is soluble in the metal catalyst, and diffuse through the droplet;

– a solubility limit exists so that the droplet becomes supersaturated;

– semiconductor atoms are precipitated at the growing liquid–solid interface.

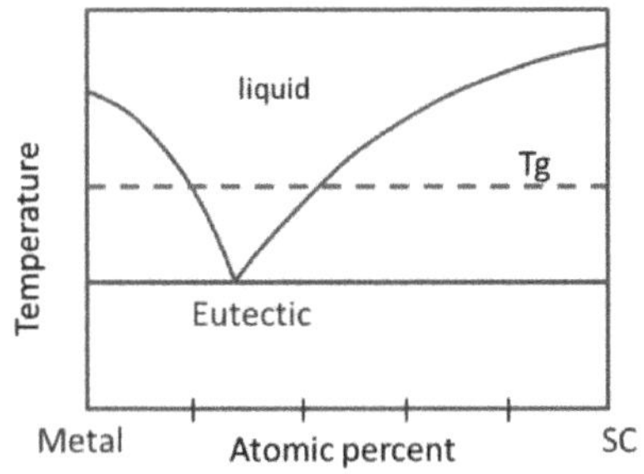

Figure 11.12. *Schematic metal-SC phase diagram for VLS growth. For a color version of this figure, see www.iste.co.uk/consonni/nanowires1.zip*

In reality, for many materials, the growth mechanism often differs from the study case of silicon NWs, and we will see that this is also true for ZnO. Many times in the literature, the process will be called "VLS" but the real mechanism, still debated, could be more vapor-solid-solid (VSS), the metal catalyst remaining solid. Potentially, catalyzed methods offer great possibilities to control the size, the diameter (ideally given by one of the metal particles), the shape and especially the location of the nanostructures. Axial heterojunctions [BAU 08] are formed and even quantum dots can be inserted efficiently [PAN 03] by changing the precursors in the vapor; if need be, the NW heterojunctions may also show high lattice mismatch since strain can be relaxed via lateral surfaces.

Catalyst-assisted ZnO NWs (as opposed to catalyst-free) have been mostly synthesized in a vapor-phase-transport reactor due to the simplicity and versatility of the equipment [WAN 09, LEV 05, RES 07, RAM 10]. Typically, an alumina boat containing a mixture of high-purity ZnO powder and graphite in a 1:1 molar ratio is placed in a horizontal tube at 1,000°C. The substrate (e.g. sapphire, silicon or fused glass), is located downstream at a lower temperature (~700–900°C). Catalyzed growth has also been carried out in MBE [ISA 13] and MOCVD [SAL 10] chambers, but to a much lower extent.

11.6.1. *Evidence of the catalytic effect*

To evidence the catalytic effect of the gold particle on the growth of ZnO, we have introduced two kinds of substrates in a horizontal MOCVD reactor, namely bare sapphire and Au-coated sapphire substrates. Metal droplets with a diameter of a few tens of nanometers had been formed after evaporation of a 6 nm Au layer and subsequent annealing at 800°C. The two samples were placed close to the exhaust of the reactor, in a zone where the zinc precursor concentration is very low, due to its consumption upstream. In these conditions, the zinc supersaturation is too low to initiate the onset of the growth on the bare sapphire, as the incoming Zn molecules desorb from the surface before having time to gather and form nuclei. No deposition occurs as shown in Figure 11.13(a). However, when the sapphire is coated with Au particles, a significant deposit is observed (Figure 11.13(b)). It is clear that the gold surface has acted as a preferential site for the adsorption of Zn molecules. Hence, the ZnO growth is catalyzed and the morphology seems to result from the "crawling" behavior of the gold droplets on the surface. Crawling-like NWs emerge from the metal droplet following the VLS process, but further develop horizontally, without any preferential direction. Their formation has been reported by several authors [LEV 05, RES 07, RAM 10] and indicates that the onset of the growth is critical.

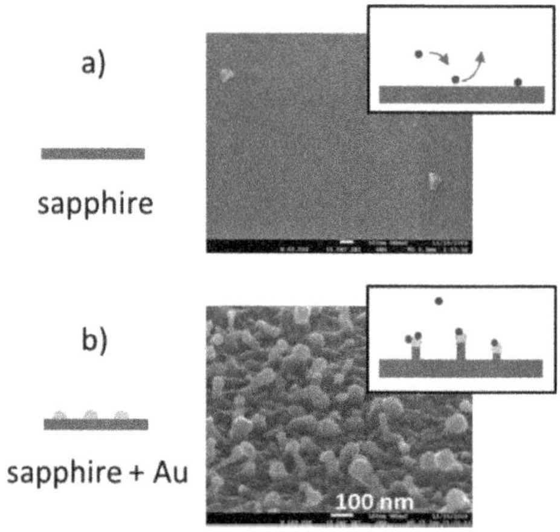

Figure 11.13. *ZnO growth by MOCVD under very low zinc supersaturation conditions: a) on bare sapphire and b) on Au-particles coated sapphire. For a color version of this figure, see www.iste.co.uk/consonni/nanowires1.zip*

11.6.2. *MOCVD growth of catalyst-assisted ZnO nanowires*

We have extensively studied the MOCVD growth of ZnO 1D nanostructures by using the gold-catalyzed approach. DEZn and N_2O were used as zinc and oxygen sources, and helium as carrier gas, in a reactor operating at reduced temperature, between 30 and 70 torr. Several sets of samples were grown varying key parameters such as temperature, O/Zn ratio ($R_{VI/II}$), pressure reactor and substrate nature. A large variety of ZnO nanostructures has been realized and shown in Figure 11.14 (on ZnO substrate) and Figure 11.15 (on sapphire).

On (0001) ZnO substrates, vertical NWs are achieved and are measured to be of approximately 1–2 μm long. Their diameter, measured between 60 and 100 nm, is larger than the one of the corresponding droplet. It is more or less constant along the smallest NWs, but slightly larger at the bottom for the biggest. All NWs exhibit a gold droplet on their top, indicating the successful catalyzed process, and besides indicating that no spontaneous growth occurred on Au-coated substrates (let us note that bare sapphire sample placed side by side exhibits catalyst-free NWs).

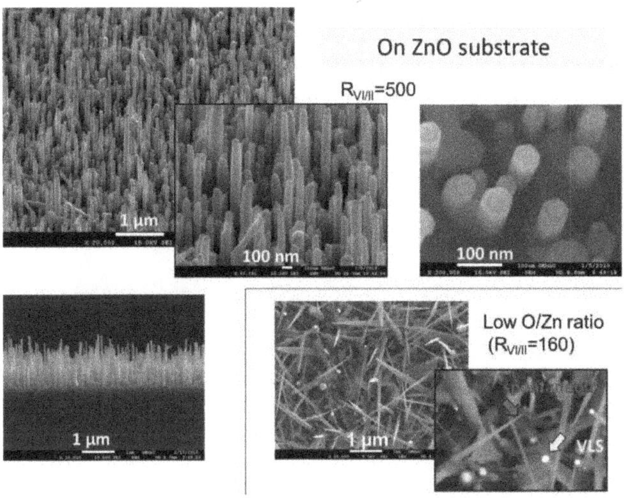

Figure 11.14. *SEM images of catalyst-assisted ZnO NWs on ZnO substrates. For a color version of this figure, see www.iste.co.uk/consonni/nanowires1.zip*

In an ideal VLS process, the metal particle would control the diameter of the growing nanowire (i.e. equal diameters should be found between those two entities). Our observations suggest that the growth mechanism may be mixed, including a VLS process below the droplets responsible for the longitudinal growth, and a vapor-solid process (VS, without catalyst) that allows the lateral enlarging and faceting of the NWs.

As for the spontaneous growth, the variation of $R_{VI/II}$ has been studied. When lowering this ratio (divided by ~3), quite different morphologies are achieved, and the important feature is the occurrence of a significant number of needles that have grown without catalyst. Their diameter shrinks from 100 nm at the bottom to 20 nm at the top. We can understand this result by considering what has been exposed in section 11.4.1, and the discussion related to the role of O/Zn ratio on the morphology transition from 2D to 1D growth. Using $R_{VI/II}$=500, conditions are adequate for the emergence of catalyst-free NWs, but the presence of the gold droplets inhibits the spontaneous growth. Due to the preferential adsorption of zinc precursor (and only zinc) by the metal surface, the number of Zn molecules impinging the ZnO substrate surface is much decreased, and consequently, the effective O/Zn ratio is much higher and thus not suitable for 1D growth. So we must drastically reduce $R_{VI/II}$ in the gas phase to 160 in order to keep an effective ratio sufficiently low to promote the spontaneous growth of NWs. The needle shape indicates that lateral growth on the facets is modified by $R_{VI/II}$ conditions.

Compared with the growth on ZnO substrate, the use of sapphire leads to different shapes of nanostructures inclined along various directions, and the density has been enhanced. Contrary to the previous observations, only a few NWs emerge vertically, and a lot of flat ZnO structures, similar to ribbons, developed on sapphire.

Figure 11.15. *SEM images of catalyst-assisted ZnO NWs on sapphire substrates*

Reducing the reactor pressure from 50 to 30 torr favors the occurrence of sharp ZnO NWs with diameter approximately 30 nm, as well as large flat triangles (Figure 11.15, upper-left). We can also notice a thick interfacial layer, rough and discontinuous, between the sapphire and the nanostructures. Within the investigated pressure range [30–70 torr], the catalyzed process is the unique growth mechanism since all the nanostructures that have been observed feature a gold droplet on their top.

11.6.3. *Formation of ZnO nanoribbons*

In the important literature related to this topic, three directions of fast growth can be found. As said previously, [0001] corresponds to C-axis and it concerns the major part of the NWs and nanoneedles. [10-10] and [11-20] directions are perpendicular to M and A planes, and are often observed as fast directions in nanobelts and nanoribbons growth [CHE 05, FAN 06, WAN 09].

To figure out the structural properties of the ZnO nanoribbons grown in this chapter, especially on sapphire substrates, transmission electron microscopy has been carried out. Figure 11.16 shows a TEM image of two ZnO nanoribbons. The first nanoribbon (NRib1) is a ribbon that features an elongated/flat Au droplet on its tip. This particular shape of the metal catalyst could be responsible for the development of nanoribbons instead of NWs, but only a few were observed in the samples. The second nanoribbon (NRib2) is a more typical one. The length of such nanoribbons can reach up to 4 µm and the width was measured approximately 500 nm at the bottom, close to the base. The upper part exhibits an elongated triangular shape. No structural defects such as dislocations or stacking faults could be observed in the body of the ribbons. We clearly see the spherical gold particle on the top of NRib2 so that the shape of this catalyst droplet should not lead to a flat nanostructure. One feature attracts our attention: the base of the ribbon seems to have developed from a nanowire perpendicularly grown, i.e. along the substrate surface (so called "crawling nanowire").

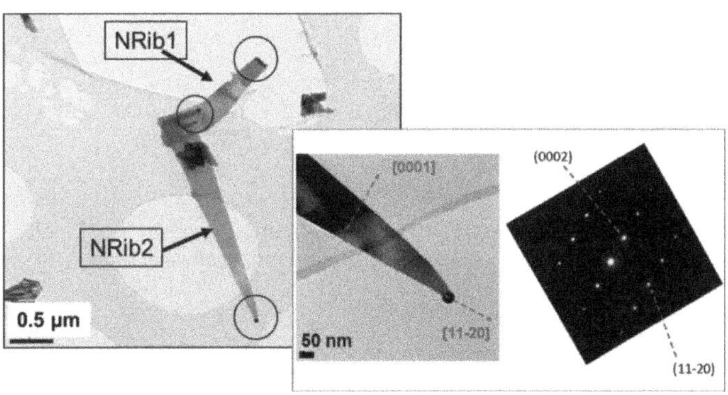

Figure 11.16. *TEM images and electron diffraction pattern of catalyst-assisted ZnO nanoribbons on sapphire substrates. For a color version of this figure, see www.iste.co.uk/consonni/nanowires1.zip*

Figure 11.16 also shows a bright-field image with its electron diffraction pattern of the top of ribbon NRib2. Electron diffraction clearly indicates that the main growth direction is [11-20], and is governed by the VLS mechanism induced by the droplet. However, the ribbon also extends along the [0001] direction probably in close connection with the "crawling nanowire" that has grown at its base. This lateral growth follows a VS process and leads to flat structures. Thus, we propose in Figure 11.17 the following four-step mechanism to illustrate the nanoribbon synthesis:

1) At the onset of the growth, a lot of crawling NWs grow and push the gold droplets along the sapphire surface.

2) NWs develop, following various paths, the most vertical NWs growing faster than the "crawling NWs".

3) A NW starts to grow in the [11-20] direction from a gold droplet located on the side of a crawling NW.

4) A nanoribbon develops from the two structures growing along [11-20] and [0001] directions, i.e. longitudinal and lateral directions, respectively.

This mechanism is similar to the growth mechanism of nanocombs in the sense that two preferential growth directions are observed. However, the longitudinal growth along [11-20] differs from the [10-10] direction, which has been observed in the case of nanocombs [CHE 05].

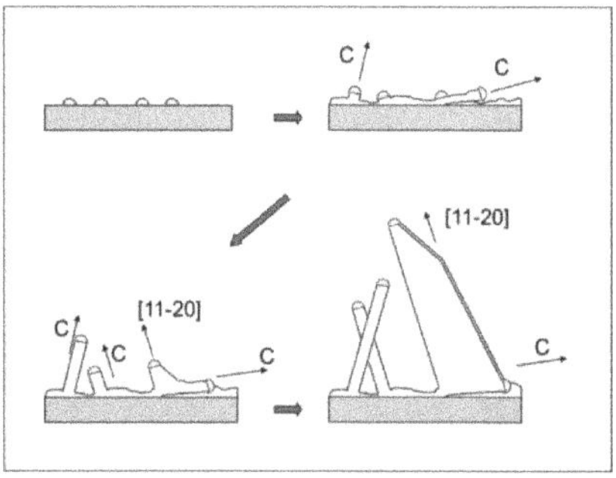

Figure 11.17. *Schematic process of the nanoribbon catalyzed growth. For a color version of this figure, see www.iste.co.uk/consonni/nanowires1.zip*

11.6.4. *Discussion of the catalyzed-growth mechanisms: VLS or not?*

We have already seen that 1D ZnO nanostructures can be achieved by using the catalyzed-assisted method. But what is the exact mechanism? And how can it be compared to the VLS mechanism established for silicon NWs?

First, we must recall that there is no solubility of oxygen in gold. Besides, it is believed that neither the oxygen nor CO adsorbs on the gold surface [ZHA 97]. Hence, considering the ideal VLS process, only Zn molecules will stick onto the catalyst surface, and only Zn atoms will potentially diffuse through the liquid droplet after alloying. Nevertheless, several works strongly suggest that the gold particle remains solid, or, at the most, only a thin surface layer is in the liquid state. In particular, Simon *et al.* [SIM 07] could not detect any zinc by TEM+EDX analysis in gold droplets after growth at 900°C. Using X-ray diffraction, the gold lattice parameter has been measured unchanged in nanowire samples after growth, and it was concluded that no zinc had diffused in the catalyst [KIR 07]. This is also consistent with our proper TEM experiments of the gold droplet. In Figure 11.18(b), (111) Au planes are measured at 2.4 Å near the particle surface.

Simon *et al.* [SIM 07] explain the absence of zinc in the gold droplet by the high equilibrium zinc partial pressure over Au-Zn compounds, which prevents from alloying the whole body of the particle. However, the authors suggest that alloying could locally occur at 900°C between Au and Zn on a thin surface layer. At this temperature, a concentration of at least 12.5% Zn would lead to a liquid state. The occurrence of this liquid surface induces a preferential path where Zn atoms efficiently diffuse to reach the growing interface. The diffusion coefficient of zinc in liquid Au-Zn is not known but estimated to be around 10–8 $m^2.s^{-1}$, four orders of magnitude higher than the calculated diffusion coefficient in solid gold, approximately 10–12 $m^2.s^{-1}$ at 900°C. This assumption around a liquid skin at the surface is supported by several observations (nonetheless, not systematical) of a thin ZnO layer (~2 nm) covering the gold droplet after growth, as shown in Figure 11.18(a). It is likely to be the result of the oxidation of zinc atoms at the catalyst surface with oxygen remaining in the chamber after the growth stopped, i.e. after shutting the precursor flows and during the cooling down of the samples.

We first infer that the catalyzed process in ZnO NWs growth is different from the VLS case of silicon since zinc will not diffuse through a fully liquid-alloyed droplet. But the most important feature is that zinc atom will not precipitate, i.e. change from liquid to solid, like silicon atoms do when reaching their solubility limit in gold. In fact, if we consider the Au-Zn phase diagram shown in Figure 11.19, it is clear that there is no eutectic point that can induce the precipitation of a pure zinc solid phase. The eutectic at 684 °C may only allow the precipitation of the β'-Au-Zn phase (then, carrying out experiments between 684 and 735°C, the growing nanowire would be an Au-Zn alloy). From our experience, as from the literature, it was shown that catalyzed growth occurs at higher temperature, in the range 800–1,000 °C. Although several papers claim "VLS growth" below 700°C, none of these works shows any gold droplet at the tip of 1D ZnO nanostructures. The catalyst stays at the substrate surface; if we can actually agree that there is an influence of the metal particles on the morphology (like there being an influence of the substrate nature), we cannot talk about VLS process in the sense given by Wagner's paper, and commonly accepted by crystal growers.

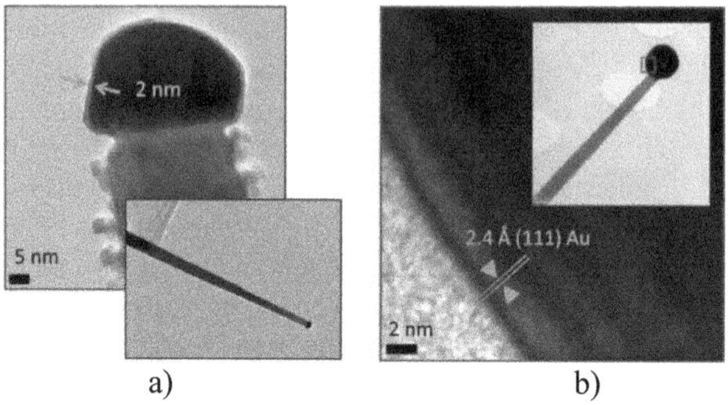

Figure 11.18. *TEM images of gold droplets after ZnO NW growth. a) A thin ZnO layer is present at the surface. b) The surface only shows (111) Au planes*

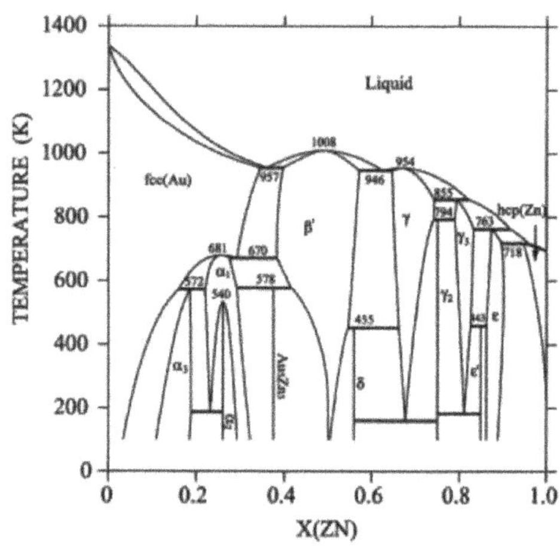

Figure 11.19. *Au-Zn phase diagram [LIU 03]. (Reprinted with permission from Elsevier, copyright 2003)*

The most relevant mechanism giving an account of the catalyzed growth of ZnO NWs is illustrated in Figure 11.20, and can be summarized as follows:

– preferential adsorption of zinc precursor molecules on the gold surface;

– fast diffusion of zinc atoms on the droplet surface or through a liquid skin at the surface;

– oxidation of zinc at the liquid–solid–vapor point, lateral diffusion of Zn and O atoms toward inner regions of the metal/ZnO interface, to finally promote the 1D growth.

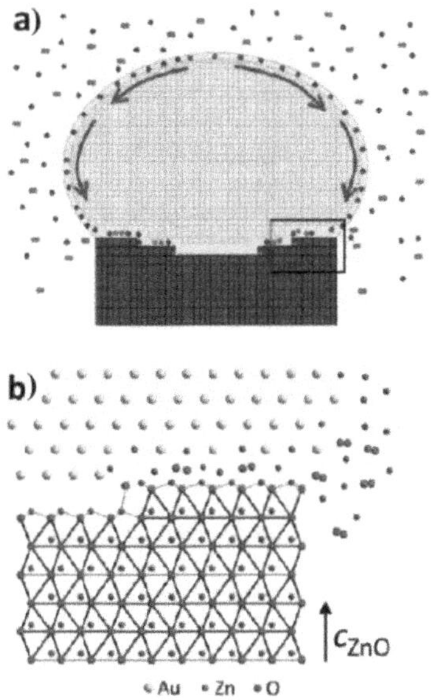

Figure 11.20. *Schematic of the catalyzed growth of ZnO NWs. (Reprinted with permission from [SIM 13]. Copyright 2013, American Chemical Society). For a color version of this figure, see www.iste.co.uk/co/consonni/nanowires1/zip*

11.6.5. *Polarity of VLS grown ZnO nanowires*

In ZnO, {0001} is a polar surface and the crystal can exhibit positively charged Zn-terminated (0001) plane (also labeled +C), or negatively charged O-terminated (000-1) plane (-C). To date, most of the results published on the crystal polarity of catalyst-free ZnO NWs have reported on their +C (Zn terminated) orientation [CHE 08, PER 12], and this has been found whatever the growth technique or the substrate employed. To assess the growth polarity of individual wires, convergent electron beam diffraction (CBED) is a useful tool, with the help of JEMS software [STA 87]. In a recent work [SAL 13] we have shown that the VLS NWs are -C oriented (O-polar) whereas the catalyst-free NWs are +C oriented (Zn-polar), independently to the nature of the substrate (sapphire or ZnO) or to the growth conditions. This is illustrated in Figure 11.21, which shows bright-field images, experimental and simulated CBED patterns for VLS and catalyst-free NWs extracted from our particular sample (see section 11.6.2 and Figure 11.14), for which both kinds of structures were found.

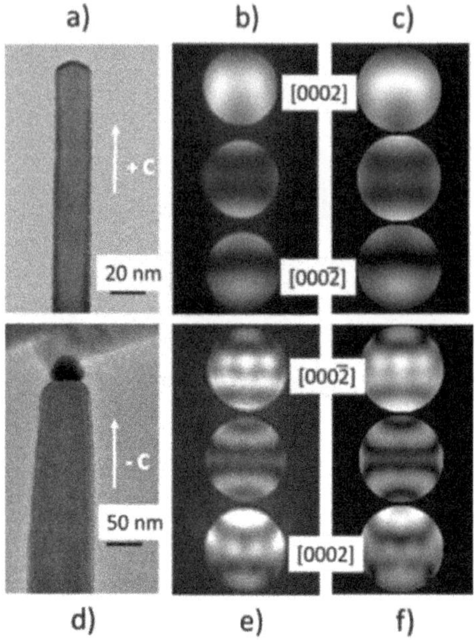

Figure 11.21. *TEM images of a) spontaneously grown and d) VLS grown ZnO NWs; b) and e) experimental and c) and f) simulated corresponding CBED pattern images. (Reprinted with permission from [SAL 13]. Copyright 2013, AIP Publishing LLC)*

Hence, opposite polarities are found between the two processes, spontaneous or catalyzed. Bearing in mind that metal impurities/aggregates could induce inverted domain boundaries, and could be at the origin of the growth of +C NWs on –C pyramids (see section 11.4.6), this result sheds more light on the particular influence of the metallic droplet/surface on the polarity of ZnO growth.

In this section, we have shown that gold-catalyzed growth of 1D ZnO nanostructures can be achieved by using the MOCVD technique. Compared with vapor-phase-transport (which, up to now, has been the most commonly used technique), it has several substantial advantages: (1) fast response of the MOCVD system to switch growth start and stop, (2) individually varying the precursors partial pressures is allowed and (3) the possibility of alternating the materials and growing sharp axial and radial heterojunctions by changing the precursor flows. Doping experiments using catalyzed growth would be also worth carrying out, as the pseudo-VLS process could efficiently incorporate column-III or -V impurities at the growing interface. Vertical arrays of ZnO NWs are produced on ZnO substrates or ZnO/sapphire templates previously covered with metal particles. Needle-shaped (with, for the largest, well-defined lateral facets) and ribbon-shaped (or flat) nanostructures were also observed, which indicated that lateral growth occurs following a non-catalyzed VS process. We show that, depending on the growth conditions (reactor pressure, VI/II mole ratio), this VS lateral growth can be enhanced.

11.7. Acknowledgements

The author wishes to acknowledge, for their contributions: Diana Nathalie Montenegro Martinez and Said Agouram (from University of Valencia, Spain), Pierre Ferret (CEA-LETI, France), as well as Corinne Sartel, ChristèleVilar, and Pierre Galtier (GEMAC, CNRS-UVSQ, France).

11.8. Bibliography

[ADA 11] ADACHI Y., OHASHI N., OHGATI T., *et al.*, "Polarity of heavily doped ZnO films grown on sapphire and SiO_2 glass substrates by pulse laser deposition", *Thin Solid Films*, vol. 519, p. 5875, 2011.

[ARN 56] ARNOLD S.M., ELOISE KOONCE S., "Filamentary growths on metals at elevated temperatures", *Journal of Applied Physics*, vol. 27, p. 964, 1956.

[ASH 11] ASHRAF S., JONES A.C., BACSA J., *et al.*, "MOCVD of vertically aligned ZnO nanowires using bidentate ether adducts of dimethylzinc", *Chemical Vapor Deposition*, vol. 17, no. 1–3, p. 45, 2011.

[BAM 72] BAMFORD C.H., TIPPER C.F.H., *Reactions of Non-metallic Inorganic Compounds*, Elsevier, Amsterdam, 1972.

[BAN 03] BANG K.H., HWANG D.K., LIM S.W., *et al.*, "Effects of growth temperature on the properties of ZnO/GaAs prepared by metalorganic chemical vapor deposition", *Journal of Crystal Growth*, vol. 250, p. 437, 2003.

[BAU 08] BAUER J., GOTTSCHALCH V., WAGNER G., "VLS growth of GaAs/(InGa)As/GaAs axial double-heterostructure nanowires by MOVPE", *Journal of Crystal Growth*, vol. 310, p. 5106, 2008.

[BEH 09] BEHRENDS A., BAKIN A., WAAG A., "Investigation of ZnO nanopillars fabrication in a new Thomas Swan close coupled showerhead MOCVD reactor", *Microelectronics Journal*, vol. 40, p. 280, 2009.

[BUR 51] BURTON W.K., CABRERA N., FRANK F.C., "The growth of crystals and the structure equilibrium of their surfaces", *Philosophical Transactions of the Royal Society of London. Series A*, vol. 243, p. 299, 1951.

[CHA 06] CHAN S.K., LIU N., CAI Y., *et al.*, "Molecular beam epitaxy-grown ZnSe nanowires", *Journal of Electronic Materials*, vol. 35, p. 1246, 2006.

[CHE 09] CHEN Y.J., HO C.H., LAI H.Y., *et al.*, "The formation of well-oriented dense ZnO film using MOCVD", *JOM*, vol. 61, no. 6, p. 23, 2009. Available at http://www.tms.org/pubs/journals/JOM/aboutJOM.aspx

[CHE 02] CHEN Y., KO H.J., HONG S.K., *et al.*, "Morphology evolution of ZnO (000-1) surface during plasma-assisted molecular-beam epitaxy", *Applied Physics Letters,* vol. 80, no. 8, p. 1358, 2002.

[CHE 05] CHEN Y.X., LEWIS M., ZHOU W.L., "ZnO nanostructures fabricated through a double-tube vapor-phase transport synthesis", *Journal of Crystal Growth,* vol. 282, p. 85, 2005.

[CHE 06] CHEN H, ZHONG J., SARAF G., *et al.*, "Interface properties onf ZnO nanotips grown on Si substrates", *Journal of Electronics Materials,* vol. 35, p. 1241, 2006.

[CHE 08] CHERNS D., SUN Y., "Defect reduction by epitaxial lateral overgrowth of nanorods in ZnO/(0001) sapphire films", *Applied Physics Letters,* vol. 92, p. 051909, 2008.

[CHI 06] CHIN S.C., CHI C.Y., LU Y.C., *et al.*, "Nanostructure study of ZnO thin films on sapphire grown with different temperature conditions", *Journal of Crystal Growth,* vol. 293, p. 344, 2006.

[DJU 10] DJURISIC A.B., NG A.M.C., CHEN X.Y., "ZnO nanostructures for optoelectronics: material properties and device applications", *Progress in Quantum Electronics,* vol. 34, p. 191, 2010.

[DUM 92] DUMONT H., MARBEUF A., BOUREE J.E., *et al.*, "Mass-spectrometric study of thermal decomposition of diethylzinc and diethyltellurium", *Journal of Materials Chemistry*, vol. 2, p. 923, 1992.

[FAN 06] FAN H.J., FUHRMANN B., SCHOLZ R., *et al.*, "Vapour-transport-deposition growth of ZnO nanostructures: switch between c-axial wires and a-axial belts by indium doping", *Nanotechnology,* vol. 17, p. S231, 2006.

[GOT 08] GOTTSCHALCH V., WAGNER G., BAUER J., *et al.*, "VLS growth of GaN nanowires on various substrates", *Journal of Crystal Growth,* vol. 310, p. 5123, 2008.

[GUZ 12] GUZIEWICZ E., GODLEWSKI M., WACHNICKI L., *et al.*, "ALD grown zinc oxide with controllable electrical properties", *Semiconductor Science and Technology*, vol. 27, p. 074011, 2012.

[HON 07] HONG Y.J., AN S.J., JUNG H.S., *et al.*, "Position-controlled selective growth of ZnO nanorods on Si substrates using facet-controlled GaN micropatterns", *Advanced Materials*, vol. 19, no. 24, p. 4416, 2007.

[ISA 13] ISAKOV I., PANFILOVA M., SOURRIBES M.J.L., *et al.*, "Growth of ZnO and ZnMgO nanowires by Au-catalysed molecular-beam epitaxy", *Physica Status Solidi C*, vol. 10, p. 1308, 2013.

[JEO 04] JEONG M.C., OH B.Y., LEE W., *et al.*, "Comparative study on the growth of ZnO nanowires and thin films by metalorganic chemical vapor deposition (MOCVD)", *Journal of Crystal Growth* , vol. 268, p. 149, 2004.

[KAB 13] KABISCH S., GLUBA M.A., KLIMM C., *et al.*, "Polarity driven morphology of zinc oxide nanostructures", *Applied Physics Letters,* vol. 103, p. 103106, 2013.

[KAP 96] KAPSTEIJN F., RODRIGUEZ-MARISOL J., MOULIJN J.A., "Heterogenous catalytic decomposition of nitrous oxide", *Applied Catalysis B: Environmental,* vol. 9, p. 25, 1996.

[KAR 10] KAR J.P., DAS S.N., CHOI J.H., *et al.*, "Study of the morphological evolution of ZnO nanostructures on various sapphire substrates", *Applied Surface Science*, vol. 256, p. 4995, 2010.

[KAR 11] KAR J.P., DAS S.N., CHOI J.H., *et al.*, "Growth, modulation and photoresponse characteristics or vertically aligned ZnO nanowires", *Applied Surface Science,* vol. 257, p. 4973, 2011.

[KHR 11] KHRANOVSKYY V., TSIAOUSSIS I., HULTMAN L., *et al.*, "Selective homoepitaxial growth and luminescent properties of ZnO nanopillars", *Nanotechnology*, vol. 22, p. 185603, 2011.

[KIR 07] KIRKHAM M., WANG X., WANG Z.L., *et al.*, "Solid Au nanoparticles as a catalyst for growing aligned ZnO nanowires: a new understanding of the vapour-liquid-solid process", *Nanotechnology*, vol. 18, p. 365304, 2007.

[KIT 08] KITAMURA K., YATSUI T., OHTSU M., *et al.*, "Fabrication of vertically aligned ultrafine ZnO nanorods using metal-organic vapor phase epitaxy with a two-temperature growth method", *Nanotechnology*, vol. 19, p. 175305, 2008.

[KLI 04] KLING R., KIRCHNER C., GRUBER T., *et al.*, "Analysis of ZnO and ZnMgO nanopillars grown by self-organization", *Nanotechnology,* vol. 15, p. 1043, 2004.

[KWA 09] KWAK C.H., KIM B.H., PARK S.H., *et al.*, "In situ and ex situ ZnO nanorod growth on ZnO homo-buffer layers", *Journal of Crystal Growth,* vol. 311, p. 4491, 2009.

[LEV 05] LEVIN I., DAVYDOV A., NIKOOBAKHT B., *et al.*, "Growth habits and defects in ZnO nanowires grown on GaN/sapphire substrates", *Applied Physics Letters*, vol. 87, p. 103110, 2005.

[LIA 10] LIANG J.H., LAI H.Y., CHEN Y.J., "Morphology transition of ZnO films with DMZn flow rate in MOCVD process", *Applied Surface Science*, vol. 256, p. 7305, 2010.

[LIU 03] LIU H.S., ISHIDA K., JIN Z.P., *et al.*, "Thermodynamic assessment of the Au-Zn binary system", *Intermetallics*, vol. 11, p. 987, 2003.

[LIU 04] LIU X., WU X., CAO H., *et al.*, "Growth mechanism and properties of ZnO nanorods synthesized by plasma-enhanced chemical vapor deposition", *Journal of Applied Physics*, vol. 95, no. 6, p. 3141, 2004.

[LIU 10] LIU J.G., GU S.L., ZHU S.M., *et al.*, "The influence of O/Zn ratio and growth temperature on carbon impurity incorporation in ZnO grown by metal-organic chemical vapor deposition", *Journal of Crystal Growth,* vol. 312, p. 2710, 2010.

[MAE 03] MAEJIMA K., UEDA M., FUJITA S., *et al.*, "Growth of ZnO nanorods on A-plane (11-20) sapphire by metal-organic vapor epitaxy", *Japanese Journal of Applied Physics,* vol. 42, Part 1, no. 5A, p. 2600, 2003.

[MAE 06] MAEJIMA K., FUJITA S., "Chemical vapor reactions of ZnO growth by metal-organic vapor phase epitaxy", *Journal of Crystal Growth,* vol. 293, p. 305, 2006.

[MAR 13] MARZOUKI A., SAYARI A., JOMARD F., *et al.*, "Carrier gas and VI/II ratio effects on carbon clusters incorporation into ZnO films grown by MOCVD", *Materials Science in Semiconductor Processing*, vol. 16, p. 1022, 2013.

[MON 12] MONTENEGRO D.N., SOUISSI A., MARTINEZ-TOMAS C., *et al.*, "Morphology transitions in ZnO nanorods grown by MOCVD", *Journal of Crystal Growth,* vol. 359, p. 122, 2012.

[MUR 04] MURPHY T.E., WALAVALKAR S., PHILLIPS D.J., "Epitaxial growth and surface modeling of ZnO on c-plane Al2O3", *Applied Physics Letters,* vol. 85, no. 26, p. 6338, 2004.

[NA 09] NA S.H., PARK C.H., "First-principles study of the surface of wurtzite ZnO and ZnS – implications for nanostructure formation", *Journal of the Korean Physical Society,* vol. 54, no. 2, p. 867, 2009.

[OGA 03] OGATA K., MAEJIMA K., FUJITA S., *et al.*, "Growth mode control of ZnO toward nanorod structures or high-quality layered structures by metal-organic vapor phase epitaxy", *Journal of Crystal Growth* , vol. 248, p. 25, 2003.

[OHS 04] OHSHIMA E., OGINO H., NIIKURA I., *et al.*, "Growth of the 2-in-size bulk ZnO single crystals by the hydrothermal method", *Journal of Crystal Growth,* vol. 260, p. 166, 2004.

[PAN 03] PANEV N., PERSSON A.I., SKOLD N., *et al.*, "Sharp exciton emission from single InAs quantum dots in GaAs nanowires", *Applied Physics Letters,* vol. 83, no. 11, p. 2238, 2003.

[PAR 02] PARK W.I., KIM D.H., JUNG S.W., *et al.*, "Metalorganic vapor-phase epitaxial growth of vertically well-aligned ZnO nanorods", *Applied Physics Letters*, vol. 80, no. 22, p. 4232, 2002.

[PAR 05a] PARK J.Y., LEE D.J., YUN Y.S., *et al.*, "Temperature-induced morphological changes of ZnO grown by metalorganic chemical vapor deposition", *Journal of Crystal Growth,* vol. 276, p. 158, 2005.

[PAR 05b] PARK J.Y., LEE J.M., JE J.H., *et al.*, "Early stage growth behavior of ZnO nanoneedle arrays on Al_2O_3 (0001) by metalorganic chemical vapor deposition", *Journal of Crystal Growth,* vol. 281, p. 446, 2005.

[PAR 06] PARK D.J., KIM D.C., LEE J.Y., *et al.*, "Synthesis and microstructural characterization of growth direction controlled ZnO nanrods using a buffer layer", *Nanotechnology,* vol. 17, p. 5238, 2006.

[PAR 07a] PARK D.J., KIM D.C., LEE J.Y., *et al.*, "Growth and characterization study of multidimensional hierarchical ZnO nanostructures", *Physica status solidi (b),* vol. 244, no. 5, p. 1567, 2007.

[PAR 07b] PARK J.S., HONG S.K., MINEGISHI T., *et al.*, "Polarity control of ZnO films on (0001) Al2O3 by Cr-compound intermediate layers", *Applied Physics Letters,* vol. 90, p. 201907, 2007.

[PAR 09] PARK D.J., LEE J.Y., KIM D.C., *et al.*, "Surface morphology and domain structure during the evolution of ZnO nanorods into films", *Journal of Physics D: Applied Physics*, vol. 42, p. 035413, 2009.

[PER 10] PERILLAT-MERCEROZ G., JOUNEAU P.H., FEUILLET G., *et al.*, "MOCVD growth mechanisms of ZnO nanorods", *Journal of Physics: Conference Series*, vol. 209, p. 012034, 2010.

[PER 12] PERILLAT-MERCEROZ G., THIERRY R., JOUNEAU P.H., *et al.*, "Compared growth mechanisms of Zn-polar ZnO nanowires on O-polar ZnO and on sapphire", *Nanotechnology*, vol. 23, p. 125702, 2012.

[RAM 10] RAMGIR N.S., SUBBANAJUI K., YANG Y., *et al.*, "Reactive VLS and the reversible switching between VS and VLS growth modes for ZnO nanowire growth", *Journal of Physical Chemistry C*, vol. 114, p. 10323, 2010.

[RAM 13] RAMDANI M.R., HARMAND J.C., GLAS F., *et al.*, "Arsenic pathways in self-catalyzed growth of GaAs nanowires", *Crystal Growth and Design*, vol. 13, p. 91, 2013.

[RAZ 89] RAZEGHI M., *The MOCVD Challenge*, Hilger, London, 1989.

[RES 07] REISER A., LADENBURGER A., PRINZ G.M., *et al.*, "Controlled catalytic growth and characterization of zinc oxide nanopillars on a-plane sapphire", *Journal of Applied Physics*, vol. 101, p. 054319, 2007.

[RIB 11] RIBEAUD A., Croissance par dépôt chimique en phase vapeur de films minces de ZnO et ZnMgO, Doctoral Thesis, Versailles Saint-Quentin-en-Yvelines University, 2011.

[ROS 09] ROSINA M., FERRET P., JOUNEAU P.H., *et al.*, "Morphology and growth mechanism of aligned ZnO nanorods grown by catalyst-free MOCVD", *Microelectronics Journal*, vol. 40, p. 242, 2009.

[RUT 64] RUTH V., HIRTH J.P., "Kinetics of diffusion-controlled whisker growth", *Journal of Chemical Physics*, vol. 41, p. 3139, 1964.

[SAL 08] SALLET V., FALYOUNI F., ZEUNER A., *et al.*, "Some aspects of the MOCVD growth of ZnO nanorods by using N2O", *Journal of the Korean Physical Society*, vol. 53, no. 5, p. 3051, 2008.

[SAL 10] SALLET V., AGOURAM S., FALYOUNI F., *et al.*, "Structural characterization of one-dimensional ZnO-based nanostructures grown by MOCVD", *Physica Status Solidi B*, vol. 247, p. 1683, 2010.

[SAL 13] SALLET V., SARTEL C., VILAR C., *et al.*, "Opposite crystal polarities observed in spontaneous and vapour-liquid-solid grown ZnO nanowires", *Applied Physics Letters*, vol. 102, p. 182103, 2013.

[SCH 67] SCHWOEBEL R.L., "A diffusion model for filamentary crystal growth", *Journal of Applied Physics*, vol. 38, no. 4, p. 1759, 1967.

[SEA 55] SEARS G.W., "A growth mechanism for mercury whiskers", *Acta Metallurgica*, vol. 3, no. 4, p. 361, 1955.

[SIM 13] SIMON H., KREKELER T., SCHAAN G., *et al.*, "Metal-seeded growth mechanism of ZnO nanowires", *Crystal Growth and Design,* vol. 13, no. 2, p. 572, 2013.

[SON 08] SONG T., CHOUNG J.W., PARK J.G., *et al.*, "Surface polarity and shape-controlled synthesis of ZnO nanostructures on GaN thin films based on catalyst-free metalorganic vapor phase epitaxy", *Advanced Materials*, vol. 20, p. 4464, 2008.

[STA 87] STADELMANN P., "EMS – a software package for electron diffraction analysis and HREM image simulation in material science", *Ultramicroscopy,* vol. 21, p. 131, 1987.

[STR 99] STRINFELLOW G.B., *Organometallic Vapor-Phase Epitaxy: Theory and Practice*, Academic Press, 1999.

[THI 09] THIANDOUME C., SALLET V., TRIBOULET R., *et al.*, "Decomposition kinetics of tertiarybutanol and diethylzinc used as precursor sources for the growth of ZnO", *Journal of Crystal Growth,* vol. 311, p. 1411, 2009.

[THI 11] THIERRY R., Croissance de nanofils de ZnO et d'hétérostructures cœur-coquilles ZnO/ZnMgO par MOVPE, Doctoral Thesis, University of Grenoble, 2011.

[WAG 64] WAGNER R.S., ELLIS W.C., "Vapor-liquid-solid mechanism of single crystal growth", *Applied Physics Letters,* vol. 4, no. 5, p. 89, 1964.

[WAN 09] WANG Z.L., "ZnO nanowire and nanobelt platform for nanotechnology", *Materials Science and Engineering R,* vol. 64, p. 33, 2009.

[YAN 98] YAN Y., PENNYCOOK S.J., DAI J., *et al.*, "Polytypoid structures in annealed In2O3-ZnO fims", *Applied Physics Letters,* vol. 73, no.18, p. 2585, 1998.

[YE 05] YE C., FANG X., HAO Y., *et al.*, "Zinc oxide nanostructures: morphology derivation and evolution", *Journal of Physical Chemistry B,* vol. 109, p. 19758, 2005.

[YE 07a] YE J.D., GU S.L., LIU W., *et al.*, "Competitive adsorption and two-site occupation effects in metal-organic chemical vapor deposition of ZnO", *Applied Physics Letters,* vol. 90, p. 174107, 2007.

[YE 07b] YE Z.Z., HUANG J.Y., XU W.Z., *et al.*, "Catalyst-free MOCVD growth of aligned ZnO nanotip arrays on silicon substrate with controlled tip shape", *Solid State Communications,* vol. 141, p. 464, 2007.

[YUA 04] YUAN H., ZHANG Y., "Preparation of well-aligned ZnO whiskers on glass substrate by atmospheric MOCVD", *Journal of Crystal Growth,* vol. 263, p. 119, 2004.

[ZHA 97] ZHANG L., PERSAUD R., MADEY T.E., "Ultrathin metal oxide surface: growth of Au on TiO_2 (110)", *Physical Review B,* vol. 56, no. 16, p. 10549, 1997.

[ZHA 04] ZHANG B.P., BINH N.T., WAKATSUKI K., *et al.*, "Pressure-dependent ZnO nanocrystal growth in a chemical vapor deposition process", *Journal of Physical Chemistry B,* vol. 108, p. 10899, 2004.

12

Pulsed-Laser Deposition of ZnO Nanowires

12.1. Introduction

Pulsed-laser deposition (PLD) is a physical vapor deposition technique and is well known for its application as a preferable growth method for thin films made of various semiconducting oxides, e.g. $BaTiO_3$, $SrTiO_3$, ZrO_2, CuO, Ga_2O_3, etc. Especially in the ZnO community, PLD is one of the most established growth methods due to not only its variety and versatility but also its simplicity. It has proven to be able to deliver samples of high structural, electrical and optical quality. Even material libraries can be made with this method [WEN 13]. In this regard, it is comprehensible, that PLD is also considered to be one of the methods of choice for the fabrication of ZnO nanostructures.

Non-directional growth methods like vapor-phase transport or solution-based processes can provide ZnO nanowires with very high aspect ratios and high crystallinity, although they are very simple to carry out (for more details, refer to Chapter 13). Nevertheless, these methods mainly lack control over the main wire properties like size (diameter and height), lateral density, position as well as reproducibility. Metal-organic vapor phase epitaxy (MOVPE) can overcome several of these disadvantages but it is more elaborated and complex (for more details, refer to Chapter 11). Further, MOVPE requires the use of large amounts of very expensive and mostly toxic metal-organic precursors. Among these methods, PLD embodies the perfect trade-off between simplicity of the exploited method and possible control over the wire properties. In the following, the basic principles of one-step and two-step PLD growth will be explained and several examples for the application of PLD-grown ZnO nanostructures will be given.

Chapter written by Christoph Peter DIETRICH and Marius GRUNDMANN.

12.2. Principles of high-pressure and hot-walled pulsed-laser deposition

PLD is a physical vapor deposition technique that is well established in thin film growth, in particular for oxides. A crucial parameter for the thin film deposition by PLD is the pressure of the oxygen background gas because it mainly determines the morphology of the deposited films, especially their surface roughness [LOR 08]. At high oxygen pressures, thin films start to have very rough surfaces. Even the formation of islands on their surfaces may occur. While this situation is highly undesirable for thin film growth, it is beneficial for the growth of nanostructures that preferentially grow at nucleation seeds like these islands. Following this simple principle, researchers from Leipzig University came up with the idea to grow ZnO nanostructures in a specially designed growth chamber at very high pressures and by using an inert gas (like argon) as a background gas [LOR 05]. In this way, the high pressure leads to a thermalization of the reacting species and to an efficient growth of ZnO nanostructures. The chosen pressure ranges between 50–200 mbar Ar (compared to 10^{-4}–1 mbar O_2 in conventional PLD).

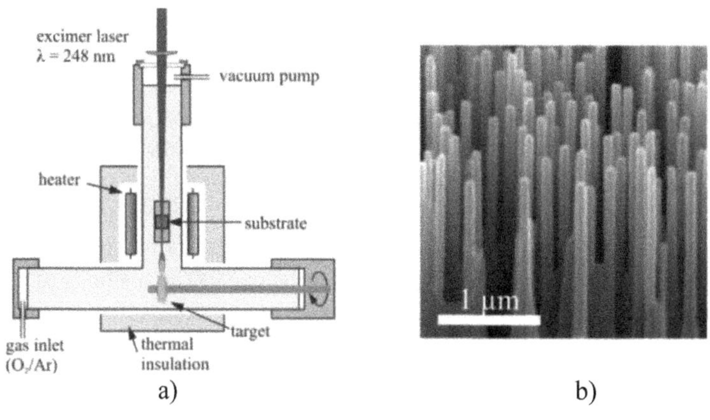

Figure 12.1. *a) Schematic view of a high-pressure pulsed-laser deposition chamber for the fabrication of ZnO nanowires; b) SEM image of ZnO nanowires grown by using high-pressure PLD (adapted from [LOR 10]). For a color version of this figure, see www.iste.co.uk/consonni/nanowires1.zip*

The schematic layout of a HP PLD chamber is shown in Figure 12.1(a). As can be seen, the laser beam of a high-pressure (HP) KrF excimer laser operating at 248 nm with ns pulses [LOR 05] (or an Nd-doped yttrium aluminium garnet (Nd: YAG) laser operating at 266 nm with fs pulses [ZHA 05]) is focused by a lens onto the target. Subsequently, the material is ablated from the target and forms a plasma plume. The position of the focusing lens hereby determines the laser fluence (around 2 J/cm^2) on the target. This energy density in relation to the ablation threshold is a crucial parameter especially in the fabrication of oxide heterostructures since it directly controls the interface abruptness [BRA 10]. During growth, the target rotates (Figure 12.1) for a homogeneous ablation and deposition. The formed plasma is directed toward the substrate (that is placed at a distance of several cm away from the target) and interacts with the background gas of the growth chamber. Subsequently, the reacting species resublime onto the substrate (that can be heated up to a temperature of 1000°C) and form nanostructures with various shapes and morphologies depending on the growth conditions. These can be optimized by adjusting several growth parameters, including the target composition, the growth temperature, the distance between target and substrate, the pressure inside the chamber and the composition of the atmosphere (oxygen/argon). Typical ZnO nanowires grown by HP PLD at 860°C and 100 mbar Ar-flow on *a*-plane sapphire can be seen in Figure 12.1(b). The shown wires have diameters ranging from 62 to 74 nm and lengths up to 3.5 μm [LOR 10].

A process that is very similar to HP PLD is the so-called hot-wall (HW) PLD that also enables the growth of ZnO nanostructures [SON 07, KIM 12a, KIM 12b]. In this process, the background pressure is considerably lower (1.6 mbar) and the complete growth takes place in a tube furnace that is heated up to a temperature of 800°C instead of only heating a part of the chamber (as in HP PLD). However, only few reports exist on this growth process, which is most probably caused by the fact that the synthesis of ZnO nanostructures by HW PLD is very sophisticated. Up until now, it was only applied to the growth of Ga-doped ZnO nanostructures and showed wires with poor morphology [KIM 12a, KIM 12b].

In general, the structural quality of PLD-grown ZnO nanowires is very high. transmission electron microscopy (TEM) images of ZnO nanowires reveal high crystallinity of the grown structures by showing no structural defects in the ZnO lattice [LOR 10]. Further, the determined nanowire lattice parameters agree well with values determined from ZnO single crystals and, thus, confirm the single crystal quality of the deposited nanowires [LIU 11]. In general, ZnO nanowires grow parallel to the substrate normal due to the lowest surface formation energy of ZnO along the [0001]-direction. That means the wire axis always equals the wurtzite crystal's *c*-axis. The nanowire cross section is usually hexagonal in this case due to the hexagonal wurtzite crystal structure. However, since PLD is an out-of-equilibrium process, it might also happen that under certain circumstances wires with a non-hexagonal shape are synthesized, e.g. a dodecagonal cross-section [NOB 07].

The growth mechanism of ZnO nanowires on sapphire or ZnO substrates has been subject to intense discussion. While it already a generally accepted fact that the atomistic growth of nanowires can be explained by a vapor–liquid–solid (VLS) mechanism [WAG 70], when using a metal template as a growth catalyst, it surprisingly turned out to be only half the truth when investigating the tips of gold-assisted grown ZnO nanowires that showed no indication for a metal-supported growth at all. Subsequently, Fan *et al.* proposed a combined VLS and vapor–solid–solid (VSS) process for the metal-assisted growth [FAN 06a]. Nevertheless, reports showed that ZnO nanowires can even be grown in the absence of any metal catalysts. This most likely indicates a pure VSS mechanism for the formation of ZnO nanowires [CAO 10]. A detailed TEM study of MOVPE-grown ZnO nanowires confirmed this assumption and demonstrated that the substrate polarity has a tremendous influence on the nanowire growth [PER 12]. Accordingly, nanowires preferentially grow Zn-polar on O-polar ZnO substrates.

12.3. Tuning the nanowire morphology

PLD is a very versatile method that allows precise control over the ZnO nanowire morphologic properties including dimension (length and diameter), position, density, orientation and shape. The length of a nanowire can easily be controlled by the number of laser pulses that are applied during the growth but it is, of course, also limited. If the nanowire is too long, it might bend or break since the nanowire post is no longer able to keep up the nanowire's weight. The longest, PLD-grown ZnO nanowires (with constant diameter), therefore, have lengths of around 10–15 μm [ZIM 10]. The diameter of a PLD-grown nanowire simply depends on the local material supply from the plasma at the position of nucleation. This means, it can be controlled by either changing the growth temperature, the substrate-target distance or the chamber pressure and ranges from several tens of nanometers to several micrometers [ZUN 07, LIU 11].

Most commonly, ZnO nanowires are grown on *a*-plane sapphire substrates that are by far less expensive than ZnO substrates. However, the growth of ZnO nanowires on bare sapphire can only be realized in a very high density ($> 10\ \mu m^{-2}$) of nanowires [GUO 08, CAO 10]. Such a high density might be beneficial in applications that make use of collective nanowire effects but are not useful for single nanowire devices. In addition, a very high nanowire density prohibits the growth of nanowire heterostructures, since the shadowing effects of neighboring nanowires prevent homogeneous coating. The high density of ZnO nanowires on sapphire is most probably caused by the lattice mismatch between ZnO and sapphire that leads to a strain-induced formation of nucleation sites in high density during the epitaxial growth of ZnO on sapphire. These shadowing effects and the high density can be circumvented by prepatterning substrates with catalytic metal nanoparticles, e.g. gold droplets [FAN 06b]. Established patterning procedures include phase-shift lithography [GRE 04], electron beam lithography [MAR 03], nanosphere lithography [RAH 07], deposition with nanoporous masks [FAN 05], etc. Unfortunately, the metal-assisted growth of ZnO nanostructures with HP PLD has a very low yield and further supports the diffusion of metal particles into the structures since the growth temperature and the metal diffusion coefficients are very high. Alternatively, the density of ZnO nanowires can be controlled in a more gentle way by growing a ZnO wetting layer by conventional low-pressure PLD on top of the sapphire substrate and prior to the ZnO nanowires. In this way, the strain induced by the epitaxial relationship between sapphire and ZnO is already drastically reduced and subsequently the density of nucleation sites for nanowire growth is lowered. In addition, a large amount of material from the plasma plume will preferentially deposit on the wetting layer instead of the nucleation sites and, therefore, lower the amount of material that is available for nanowire growth. With this procedure it is possible to locally lower the density of nanowires to 10^{-2} μm^{-2} [CAO 10].

a) b) c)

Figure 12.2. *SEM images of ZnO nanowire arrays with different lateral densities grown directly a) on bare sapphire, b) in a transition region or c) on a ZnO wetting layer. The lateral density is controlled by changing the target-substrate distance. The white scale bars represent 2 µm. (Adapted from [CAO 09])*

Nevertheless, the nanowire density also strongly depends on the distance between substrate and target as well as the lateral extension of the plasma plume. In this regard, a nanowire density gradient forms along the substrate with increasing distance from the plasma plume (Figure 12.2). The figure shows several scanning electron microscopy (SEM) images of ZnO nanowires on the same sapphire substrate (with ZnO wetting layer) for different positions on the substrate. In this way, the nanowire density varies between 10 μm^{-2} (for the largest substrate–target distance) and 10^{-2} μm^{-2} (for the smallest substrate–target distance) on a single 10×10 mm^2 substrate but is locally (2×2 mm^2) nearly constant. In this case, the high density at larger distances can be explained by the absence of a wetting layer and the high momentum of particles being accelerated away from the target and, thus, preferentially deposit at larger distances. In general, the density along the substrate strongly depends on the extension of the plasma and the relative position of the substrate to it.

As can be seen in Figures 12.1 and 12.2, ZnO nanowires usually grow with their wire axis perpendicular to the substrate surface which is a result not only of the epitaxial relationship between substrate and nanowire but also of the polarity. In this regard, ZnO nanowires always grow along their [0001]-direction. If the orientation of the underlying substrate is changed, it is possible to vary the angle between nanowire and substrate surface. By using *m*-plane sapphire substrates instead of *a*-plane sapphire, it was clearly demonstrated that nanowires grow with a tilt of 30° to the substrate normal by keeping the [0001]-direction during growth [ZUN 07]. Figure 12.3 shows two families of tilted nanowires pointing in opposite directions (due to the symmetry of the system [GRU 10]). This clearly indicates a fixed crystallographic relationship and also demonstrates that *m*-plane ZnO epitaxially grows on *m*-plane sapphire.

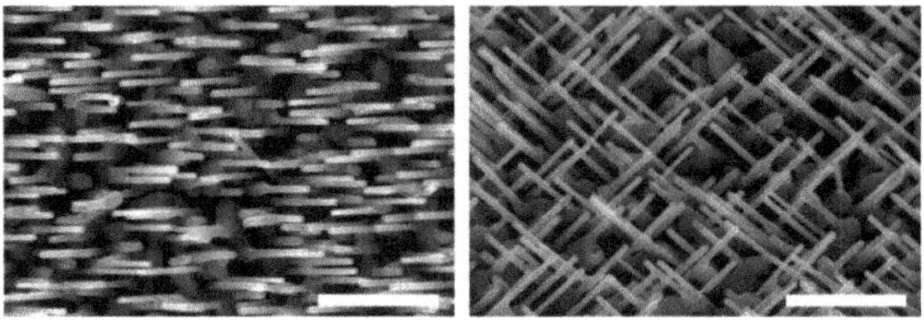

Figure 12.3. *SEM images of tilted ZnO nanowires grown on m-plane sapphire substrates (top and side view). The white bars represent 2 μm. (Adapted from [ZUN 07])*

The term nanowire usually addresses one-dimensional rods with high aspect ratios and a wire axis pointing away from the surface. However, instead of growing nanowires being fixed to the substrate by one end, it is also possible to fabricate arrays of nanowires directly lying in plane on the substrate. In this case, the wire axis lies in the surface plane. Most likely, in-plane nanowires can be synthesized on substrates with a large miscut because they provide regular terraces on their surface that are energetically highly favorable for the nucleation of adatoms. A local growth of nanowires at the steps of these terraces requires a very low growth rate which can be realized by using eclipse PLD as done by Son *et al.* [SON 08]. Eclipse PLD uses a shadow mask in front of the target in order to suppress the formation of a plasma plume and minimize the amount of ablated target material deposited on the substrate per unit time. In this way, a growth rate of only 0.04 Å per pulse can be achieved and results in the formation of in-plane nanowires with diameters in the range of 20 nm and lengths of several tens of micrometers. The very reproducible alignment of substrate terraces together with their regular lateral arrangement provides a perfect platform for the uniform growth of in-plane ZnO nanowires that can be utilized as highly efficient gas sensors. Their sensor applicability was already proven by demonstrating a strong deviation in sensitivity measurements when exposing the system to certain amounts of ethanol [SON 08].

12.4. Doped binary nanowires and ternary alloy nanowires

The versatility of applications on the nanometer scale is unquestioned and is a result of the excellent electronic properties of semiconductor nanowires. However, the implementation of a certain material in devices, or the usage of specific electrical or optical properties for distinct purposes, always requires a high degree of flexibility and adaptability. In order to even improve the already outstanding properties of ZnO nanowires, a simple way is needed to adjust its electrical, optical and magnetic characteristics. This can usually be achieved by doping or alloying. In the case of PLD-grown ZnO nanowires, the composition of the nanowires can easily be tuned by changing the composition of the target material and with that the material species available for deposition on the substrates. In this way, ZnO nanowires were already successfully doped (or alloyed) with Al, Ga, P, Co, Mn, Mg, etc. (a selection is shown in Figure 12.4) by adding the respective metal oxides to the PLD targets. However, the transfer of material from target to substrate is, thereby, not stoichiometric. The crucial parameter that determines the amount of dopants in the host lattice is the so-called transfer coefficient that changes with growth temperature and pressure but also depends on the material itself, especially on the weight and lateral dimensions of the atomic species since these quantities mainly determine the kinetics inside the growth chamber. For the incorporation of Ga, P, Co, Mn and Mg, the transfer coefficient is usually less than 1, which means that more material is necessary in the target in order to reach a certain composition inside the nanowire.

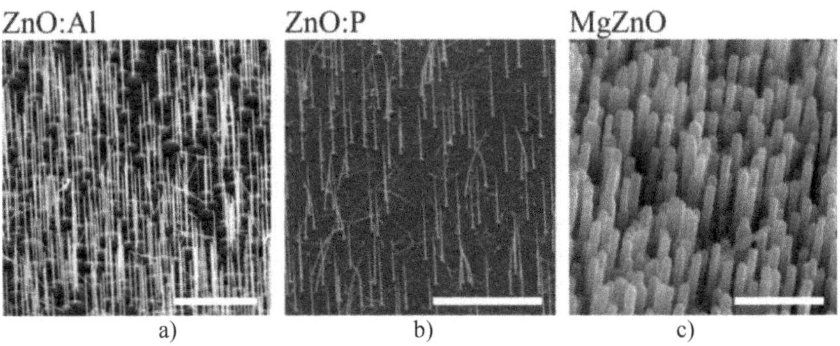

Figure 12.4. *SEM images of a) ZnO: Al nanowires (adapted from [ZIM 07]), b) ZnO:P nanowires and c)(Mg,Zn)O nanowires (adapted from [LOR 05]). The white bars represent 10 µm for ZnO:Al,P nanowires and 3 µm for (Mg,Zn)O nanowires*

For the incorporation of Al, the transfer coefficient is far above 1. This might be caused by the easy diffusion of Al into the ZnO crystal. This assumption is well supported by the fact that Al is also identified as the most prominent donor in ZnO and its traces can be found in almost all ZnO structures. ZnO:Al nanowires grown on *a*-plane sapphire are depicted in Figure 12.4(a). An energy-dispersive X-ray spectroscopy (EDX) spectrum that can be seen in Figure 12.5(a) clearly shows an Al peak. Since Al is a very active donor in ZnO, it also drastically changes its electrical properties. This is clearly evidenced by a decrease of the resistivity of ZnO:Al nanowires (containing 1% Al atoms) by four orders of magnitude compared to nominally undoped ZnO nanowires (note that the used target only contained 0.01% Al_2O_3). This means, the electrical properties of ZnO nanowires can be tuned in a very broad range by mixing only a very small amount of Al_2O_3 with the target. Similar effects are also expected for ZnO:Ga [LEE 08], thus making PLD-grown ZnO:Al and ZnO:Ga nanowires perfect candidates for the realization of high electron mobility nanowire transistors.

Since ZnO is a natural *n*-type semiconductor, some effort has to be put into the establishment of stable *p*-type conductivity. The most prominent acceptor candidates for *p*-type doping in ZnO nanostructures are the elements nitrogen [YAN 09], phosphorus [CAO 07, LIN 09], antimony [WAN 11] and arsenic [LEE 04] when substituting an oxygen atom, or lithium [LEE 11] and sodium [HE 11] when substituting a zinc atom. However, for all these dopants, reproducible *p*-type conductivity has only rarely been reported and still remains challenging. This is most likely caused by the high amount of intrinsic donor-like defects in ZnO that can easily compensate any acceptor. PLD-grown ZnO nanowires are believed to solve this drawback due to their high structural quality and crystallinity. Following this notion, P-doped ZnO nanowires were grown on *a*-plane sapphire substrates by HP PLD [CAO 07] (Figure 12.4(b)). EDX measurements confirmed the incorporation of 1% P atoms into the ZnO crystal when using a target containing 2% P_2O_5 and photoluminescence (PL) measurements revealed P-related transitions. However, final evidence for the establishment of stable p-type conductivity could not be presented. Recent electrical investigation on P-doped ZnO wires shows that P atoms in ZnO wires act more like donors than acceptors [DIE 11a] by incorporation as donor-like complexes. Besides P-doping, codoping of Ag and Al in ZnO was claimed to form acceptor-like complexes and thus *p*-type conductivity. Subsequently, Ag/Al-doped ZnO nanowires grown by HW PLD could be synthesized [KIM 12a], but final evidence for *p*-type conductivity still remains elusive.

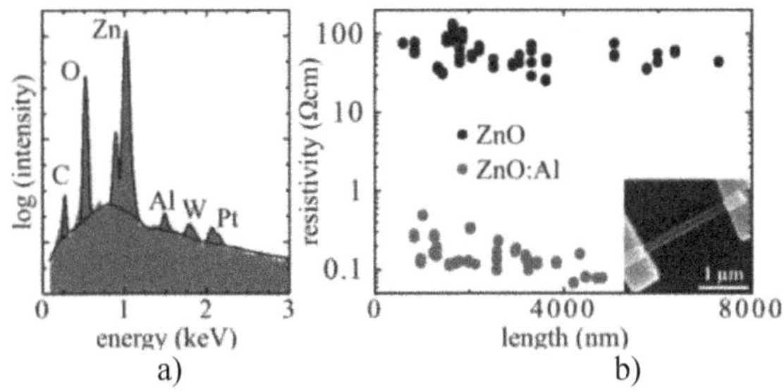

Figure 12.5. *a) EDX spectrum of a ZnO:Al nanowire showing a clear trace of aluminum; b) room-temperature resistivity for nominally undoped (black dots) and Al-doped (gray dots) ZnO nanowires versus nanowire length (inset shows a single ZnO:Al nanowire contacted for resistivity measurement). (Adapted from [ZIM 10]). For a color version of this figure, see www.iste.co.uk/consonni/nanowires1.zip*

Besides tuning the electrical properties, nanowires were also used as a platform for the realization of room-temperature ferromagnetism by doping ZnO nanowires with magnetic Mn or Co atoms. Subsequently, ZnO:Co and ZnO:Mn could successfully be synthesized by HP PLD and the incorporation of Mn and Co atoms could also be confirmed [RAH 06]. However, the formation of a ferromagnetic order in ZnO turned out to be strongly dependent on the incorporation sites of the magnetic particles. The crucial condition for ferromagnetism in these cases is the incorporation of magnetic particles on Zn lattice sites. While some PLD-grown ZnO:Co nanowires showed clear evidence for a room-temperature ferromagnetism by a hysteresis loop in the magnetization, when Co atoms are incorporated on zinc lattice sites [CHE 05], other reports on PLD-grown ZnO:Co and ZnO:Mn nanowires could not reveal the establishment of a stable ferromagnetism when distorted and transition metal enriched environments were formed [ANK 09]. The main difference in incorporation might be caused by the different growth regimes chosen. In this regard, ferromagnetism could only be confirmed in ZnO nanowires grown by using conventional low-pressure PLD, also achieving much higher contents of ferromagnetic particles (up to 5%).

As already pointed out, PLD is a highly non-equilibrium process and enables the fabrication of nanostructures with compositions beyond the chemical solution limits (that are usually only a few percent). This makes it possible to grow ZnO-based nanowire alloys, e.g. (Mg,Zn)O nanowires, with compositions covering a very broad range. The admixing of metal oxides to ZnO is thereby only limited by the occurrence of phase segregations caused by a mixture of the different crystal structures of ZnO and the metal oxide additions. However, this limit can be pushed to very high admixture ratios by choosing proper (non-equilibrium) growth conditions. The standard way of tuning the band gap of ZnO to higher energies is by the admixture of Mg atoms to the ZnO host. Mg nicely incorporates into the crystal on Zn-lattice sites and causes a local distortion of the lattice which in turn causes an increase of the material's band gap (if Mg is present in a considerable amount). Since MgO crystallizes in the rocksalt crystal structure, the formation of rocksalt inclusions in (Mg,Zn)O may occur at higher Mg contents. The highest Mg content achieved so far in PLD-grown (Mg,Zn)O nanowires in single wurtzite phase is 20% [LOR 05] and results in an increase of the band gap by 400 meV (see Figure 12.4(a) for an SEM image of (Mg,Zn)O nanowires).

Besides using (Mg,Zn)O, it was already demonstrated that the band gap of ZnO thin films can be increased up to 10.6 eV by the admixture of BeO [RYU 06, WEN 12] without phase segregations since both semiconductors crystallize in the wurtzite structure. However, compared to ZnO, BeO is highly toxic and hazardous. This circumstance prohibits its use in an HP, nanowire PLD growth chamber. The same is true for CdO, which is widely established as the only admixture material reliably reducing the ZnO bandgap [SAD 06], but is extremely toxic and carcinogenic. With respect to health issues, it is impossible to synthesize either (Be,Zn)O or (Zn,Cd)O nanowires in a HP PLD system. Nevertheless, nanowire heterostructures with single layers containing these materials are indeed producible in a conventional low-pressure PLD system, as will be discussed below.

12.5. Fabrication of nanowire heterostructures

Nanowires have proven their applicability as nanoscale building blocks for electric or photonic devices [WIL 09]. Nevertheless, certain well-established and highly functional thin film technologies make use of distinct quantum or interface effects as they are only present in semiconductor heterostructures. The successful transfer of these effects into the nanowire world, therefore, opens a whole new era of nanoscale devices, such as nanowire solar cells or nanowire light-emitting diodes.

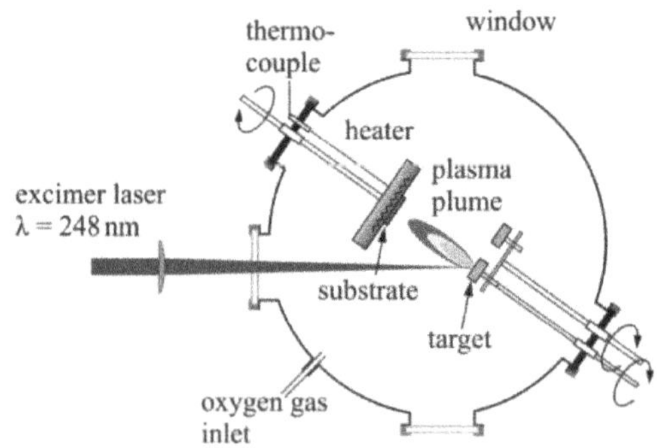

Figure 12.6. *Schematic layout of a conventional low-pressure PLD chamber. (Adapted from [LAN 12]). For a color version of this figure, see www.iste.co.uk/consonni/nanowires1.zip*

In general, nanowire heterostructures cannot be grown by HP PLD since the structural quality of the deposited layers and the interfaces is poor (Chapters 3–4 in Volume 2). Therefore, nanowire heterostructures are most commonly grown in a two-step process. In a first step, pure ZnO or (Mg,Zn)O nanowires are fabricated by, for example, HP PLD on a-plane sapphire substrates. In the second step, these nanowires are transferred to a conventional low-pressure PLD chamber for the deposition of the surrounding layers. The schematic layout of such a low-pressure PLD chamber is shown in Figure 12.6. Similar to HP PLD, the 248 nm line of an excimer laser is focused onto a rotating target and ablates material that – under the formation of a plasma plume – is deposited on a heated, rotating substrate. Several target holders enable the deposition of different heterostructure layers. The main difference in the nanowire growth is the chamber pressure that is in the range of 10^{-3}–10^{-2} mbar O_2. The grown nanowire heterostructures are often called core/shell heterostructures, with the core referring to nanowire templates grown by HP PLD and the shell referring to the surrounding layers grown by low-pressure PLD. If the dimensions of the nanowire heterostructure are of less importance for the nanowire's application, the first PLD-step can also be replaced by a simple vapor-phase transport growth. In this way, the fabrication of several two-material, single heterostructures with a variety of possible applications have already been demonstrated by coating ZnO nanowires with a single PLD-layer of a different material, including ZnO/ZnS [WAN 10] and ZnO/ZnSe nanowires [WAN 08] for solar cell applications/photovoltaics, ZnO/Er_2O_3 nanowires with Er_2O_3 as high-k dielectric for nanowire transistors [LI 07], ZnO/(La, Sr)CoO_3 for photocatalysis [JIA 09] or ZnO/ZnMnO for room-temperature ferromagnetism [KAY 09].

A more elaborated class of nanowire heterostructures are multilayered structures. Quantum heterostructures are of high interest since they provide superior quantum efficiency due to quantum confinement in quantum wells (QWs) or quantum dots (QDs). The rising complexity of the structure scales with an increasing severity of the growth process. This means, nanowire heterostructures can only be fabricated in systems with not only a high degree of freedom but also an even higher degree of controllability as it is only provided by advanced epitaxial techniques such as PLD [CAO 09], MOVPE [PAR 03, THI 12] or molecular beam epitaxy (MBE) [MOF 06]. For ZnO, quantum confinement can easily be achieved by a three-layer stack of (Mg,Zn)O barriers sandwiching a very thin, active ZnO layer. This usually results in the formation of a QW emitting energetically above the transition of ZnO A-excitons. By changing the thickness of the active layer or the composition of the barriers, it is possible to tune the QW emission in a certain energy range (several hundreds of meV).

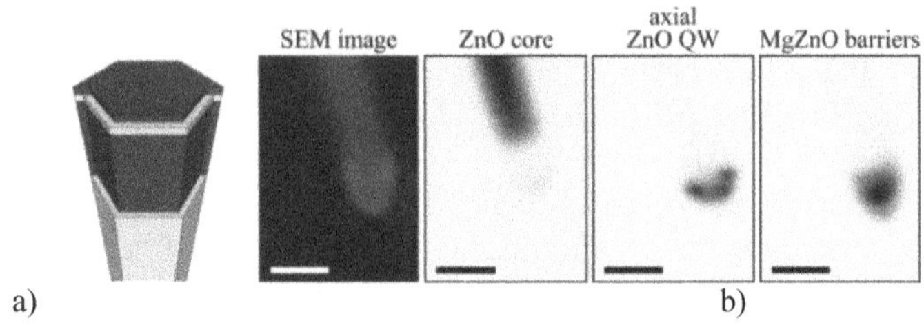

Figure 12.7. *a) Nanowire heterostructure with axial quantum well (ZnO is indicated in light gray, (Mg,Zn)O in dark gray); b) SEM image and spatial-resolved CL intensity of an axial nanowire heterostructure showing ZnO core, well and (Mg,Zn)O barrier luminescence. The scale bars represent 200 nm. (Adapted from [CZE 08]). For a color version of this figure, see www.iste.co.uk/consonni/nanowires1.zip*

By coating a pure ZnO nanowire with a (Mg,Zn)O/ZnO QW heterostructure, two different QWs can be formed: an axial QW on top of the polar tip of the nanowire (Figure 12.7(a)) or a radial QW on top of the non-polar *m*-plane side facets of the nanowire (Figure 12.8. bottom left). The difference in polarity is caused by the anisotropy of the wurtzite crystal structure and, in the case of polar QWs, results in a reduction of the QW transition energy due to the emergence of the quantum confined Stark effect.

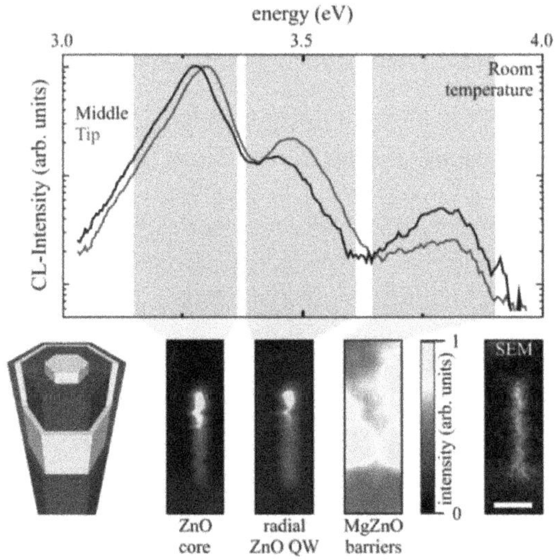

Figure 12.8. *Top: CL spectra of a core/shell nanowire heterostructure with a radial QW recorded at the tip and middle of the nanowire. Bottom: spatial-resolved CL images of the same core/shell nanowire showing core, QW and barrier luminescence. The scale bar in the SEM image represents 500 nm. (Adapted from [LAN 11b]). For a color version of this figure, see www.iste.co.uk/consonni/nanowires1.zip*

For directional growth processes like PLD, the nanowire density strongly determines the formation of radial QWs. If the lateral nanowire density is too high, shadowing effects occur and cause an inhomogeneous radial coating of the nanowire. A reduced surface diffusion even enhances this effect [CAO 09]. Nevertheless, in this density regime it is still possible to fabricate axial QW heterostructures as demonstrated in [CZE 08] by depositing a ZnO QW with 2 nm thickness inside $Mg_{0.28}Zn_{0.72}O$ barriers on top of high-density ZnO nanowires with diameters ranging between 150 nm – 2 µm (Figure 12.7). The growth parameters were the same as for QW thin film growth (670°C and 0.016 mbar). This resulted in the formation of an axial QW with a transition energy of 3.44 eV (nearly 100 meV blue shifted from the ZnO luminescence). High-resolution, spatial resolved cathodoluminescence (CL) images clearly demonstrated the formation of a homogeneous layer of ZnO inclusions into the (Mg,Zn)O surrounding (Figure 12.7). Besides that, a second QW heterostructure with the same growth parameters was grown in the same chamber that is usually used for HP PLD (*in situ*), instead of applying a two-step process in two different PLD chambers (*ex situ*). This second sample also showed transitions that were assigned to quantum-confined carriers, but, in contrast to the two-chamber nanowire heterostructures, the transitions had unusual small linewidths indicating the presence of QDs rather than of a QW. The formation of QDs would be reasonable since the deposited (Mg,Zn)O and ZnO are expected to be very inhomogeneous and rough, based on the fact that the *in situ* process is optimized for low-pressure nanowire growth and not for HP thin film growth.

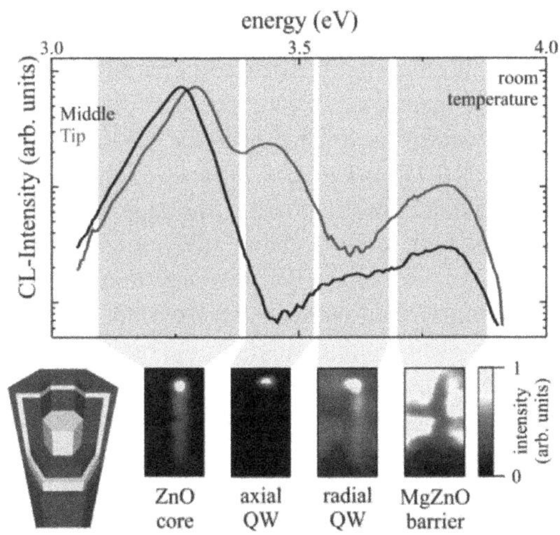

Figure 12.9. *Top: CL spectra of a core/shell nanowire heterostructure with radial and axial QW recorded at the tip and middle of the nanowire. Bottom: spatial-resolved CL images of the same nanowire showing core, radial QW, axial QW and barrier luminescence. (Adapted from [LAN 11b]). For a color version of this figure, see www.iste.co.uk/consonni/nanowires1.zip*

The quality of radial nanowire QW heterostructures depends on the lateral nanowire density. It has to be less than 1 μm^{-2} in order to obtain a homogeneous shell thickness along the whole nanowire length [CAO 09]. This reduction can be achieved by a ZnO wetting layer deposited prior to the nanowires [CAO 10] providing nanowire densities of only 10^{-2} μm^{-2}. By choosing typical thin film QW parameters for the shell growth, nanowire QWs grow in radial and axial directions of low-density ZnO nanowires (Figure 12.9). The quality of the heterointerfaces between the (Mg,Zn)O barriers and the ZnO core, as well as the ZnO active layer, can be increased by reducing the pulse repetition rate of the laser (2 Hz) and the growth rate (around 1 nm/pulse) [CAO 09]. Further, a reduction of the laser fluence (to 1.8 J/cm^2) dramatically improves the abruptness of the interfaces and with that the nanowire QW performance [BRA 10]. With these improvements, record confinement energies could be achieved in PLD-grown nanowire QW heterostructures that allow a very precise tuning of the QW transition energy in a wide energy range of up to 400 meV [LAN 11b].

The relative orientation of the nanowires (during deposition of QW shell layers) with respect to the plasma plume strongly affects the growth rate of radial and axial QWs. Usually, nanowires direct toward the PLD plasma plume [CAO 09, LAN 11b]. In this regard, thickness ratios between axial and radial QW thicknesses of up to 5 were reported [CAO 09], meaning that axial QWs are always thicker than radial QWs, by a factor of 5, in the same nanowire heterostructure. This also means that radial and axial QWs always emit at different energies (Figure 12.9). On the one hand, this is caused by the fact that fewer particles attach to the nanowire side facet because the direction of the impinging particles is parallel to these facets. On the other hand, the different nanowire facets possess different surface formation energies. The formation energy is much lower for the [0001]-oriented tip of the nanowire than for to the [10-10]-oriented side facet of the nanowire meaning that particles more easily attach to the tip than to the side facets. In order to obtain nanowire heterostructures with only a radial QW, it is, therefore, necessary to choose a QW thickness for the radial QW in a way that the active layer in axial direction exhibits no quantum confinement (Figure 12.8, left) [LAN 11b]. On the downside, this limits the possible tuning range for the radial QW energy.

Besides its spectral tunability to higher energies (by alloying it with MgO), ZnO was also a promising candidate to fill the gap of optically active semiconductors in the green region of the electromagnetic spectrum by alloying it with CdO and thus reducing its band gap. $Zn_xCd_{1-x}O$, with a composition-dependent band gap between 3.3 eV (x = 0) and 2.3 eV (x = 1) is able to bridge the gap between the green and the ultraviolet (UV) spectral range. However, during the fabrication of PLD-grown (Zn,Cd)O, researchers encountered fundamental problems in alloying ZnO with CdO. The application of well-established ZnO growth conditions always limited the incorporation of Cd atoms into ZnO to only a few percent, thus also limiting the lowest possible transition energy of (Zn,Cd)O A-excitons. These limitations are even more pronounced in ZnO/(Zn,Cd)O QW heterostructures since a small band gap difference between ZnO and (Zn,Cd)O only results in a poor quantum confinement [CHE 05]. Moreover, diffusion of Cd in ZnO leads to an intermixing of the heterointerfaces and a further reduction of the already poor quantum confinement. Besides that, carriers in (Zn,Cd)O/ZnO heterostructures most likely recombine non-radiative in ZnO than radiative in (Zn,Cd)O due to structural defects that form at the interface caused by the lattice mismatch between both semiconductors.

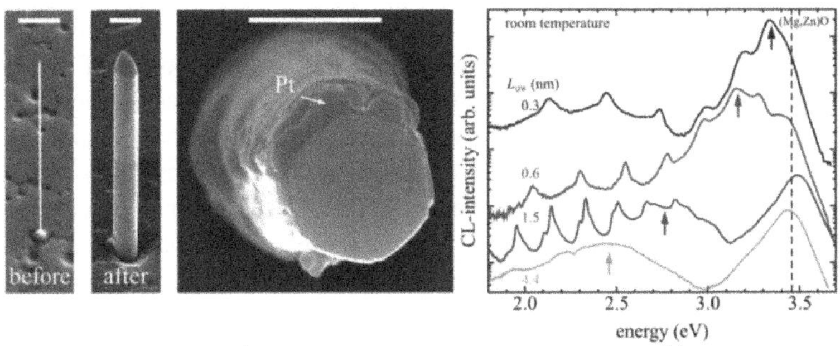

a) b)

Figure 12.10. *a) SEM images of a multiple QW nanowire heterostructure with ZnO nanowire core, (Mg,Zn)O barriers and (Zn,Cd)O quantum wells before and after coating. The white bars represent 750 nm; b) CL spectra for multiple QW nanowire heterostructures with different well thicknesses L_{QW}. The arrows mark the QW emission energies. (Adapted from [LAN 13]). For a color version of this figure, see www.iste.co.uk/consonni/nanowires1.zip*

However, these drawbacks were solved recently. On the one hand, it was found out that Cd more easily incorporates into the ZnO lattice, and diffusion effects are strongly suppressed when much lower growth temperatures (of around 300°C) are applied during the growth. This enables Cd contents of up to 25% in ZnO and A-exciton transition energies as low as 2.5 eV [LAN 11a]. On the other hand, nanowire heterostructures with active (Zn,Cd)O layers and (Mg,Zn)O barriers were fabricated (instead of using ZnO barriers) in order to reduce the lattice mismatch inside the QW. This procedure allowed the fabrication of nanowire QW heterostructures with tunable QW transition energies covering an unprecedented energetic range of 900 meV [LAN 13] with the lowest reported transition energy at 2.46 eV (exactly covering the green spectral range). These improvements of the PLD growth process even facilitate the fabrication of multiple QW heterostructures without loss of structural quality. SEM images of a multiple QW nanowire heterostructure containing four radial QWs can be seen in Figure 12.10(a). It is obvious that even after depositing four QWs, the cross-section of the nanowire heterostructure remains hexagonal. This hexagonal cross-section provides a perfect platform for the circulation of light trapped by total internal reflection, so-called whispering gallery modes (WGMs) [DIE 12]. WGMs are already known from pure hexagonal ZnO microwires [DIE 11b]. These optical modes are visible in the CL spectrum (Figure 12.10(b)) as additional sharp peaks. Since WGM photons both spectrally and spatially overlap with the nanowire QW excitons, the system opens a new way for the investigation of nanowire polaritons. For an overview over quantum and optical confinement in ZnO nanostructures, refer to Chapters 3–4 in Volume 2.

An alternative approach to achieve the strong coupling regime between photons and ZnO nanowire excitons is the coating of nanowires with Bragg mirrors and, by that, confining light inside the nanowire. Since the thickness of the Bragg mirrors influences the spectral position of the Bragg stop band as well as the cavity mode, it is necessary to have Bragg mirrors with same thickness in both radial and axial nanowire directions for perfect optical confinement. To compensate for the different growth rates in both directions, the orientation of the ZnO nanowires with respect to the plasma plume has to be changed before coating. This procedure is commonly called oblique-incidence PLD. For the homogeneous coating of ZnO nanowires, a tilt angle of around 30° proved to be optimal [SCH 10] and provided nearly constant layer thicknesses around the whole nanowire. Besides tilting, the rotation of the substrate also has to be adjusted accordingly. For the successful growth of Bragg mirrors, the material combination Al_2O_3/(yttrium stabilized ZrO_2)YSZ is beneficial, since both materials are oxides, transparent in the active region of ZnO and provide a huge difference in refractive indices. For sufficiently high reflectivities of the Bragg mirrors, several pairs of Al_2O_3/YSZ layers need to be deposited on the ZnO mirrors (see SEM image in Figure 2.11(a)). In this regard, 8.5 Al_2O_3/YSZ pairs provide reflectivities of up to 98% (see reflectivity spectrum in Figure 2.11(b)).

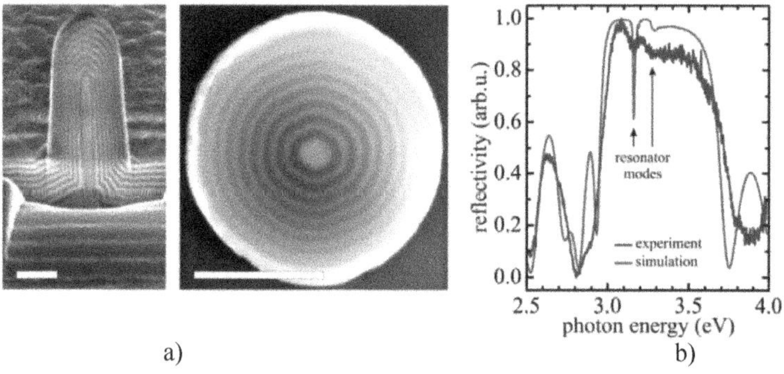

Figure 12.11. *a) SEM images of PLD-grown ZnO nanowires coated with a Bragg mirror consisting of 8.5 pairs of alternating Al_2O_3/YSZ stacks. The white scale bars represent 1 μm. b) Reflectivity of a ZnO nanowire coated with a Bragg mirror. (Adapted from [SCH 10]). For a color version of this figure, see www.iste.co.uk/consonni/nanowires1.zip*

12.6. *Summary and outlook*

HP PLD of pure ZnO nanowires has proven its high degree of controllability and is able to deliver nanowires with precise dimension, position, lateral density and orientation. Further, doping and alloying of ZnO nanowires by PLD can easily be achieved and allows an accurate tuning of the nanowire's electrical, magnetic and optical properties to pave the way for future nanoscale devices such as high electron mobility transistors, solar cells, light-emitting diodes and gas sensors.

Further, a two-step process combining high- and low-pressure PLD (and with that the advantages of thin film and nanowire growth) enables the fabrication of high quality nanowire multilayer heterostructures in radial and axial directions of the nanowire including highly efficient QWs and highly reflective Bragg mirrors. The composition of the QW barriers as well as the QW thickness and composition can be adjusted in a very broad range allowing a band gap tuning of PLD-grown ZnO-based nanowire QWs by more than 1 eV between 2.5 and 3.6 eV. The morphological quality of the fabricated heterostructures even provides efficient photonic confinement for the formation of Fabry–Pérot modes and WGMs. The presence of these optically confined photons push the PLD-grown ZnO nanowire heterostructures into the prominent area of nanocavities and facilitates the investigation of strong light-matter interaction between the optical modes and extremely robust ZnO excitons.

12.7. Bibliography

[ANK 09] ANKIEWICZ A.O., GEHLHOFF W., MARTINS J.S., *et al.*, "Magnetic and structural properties of transition metal doped zinc-oxide nanostructures", *Physica Status Solidi B.*, vol. 246, no. 4, pp. 766–770, 2009.

[BRA 10] BRANDT M., LANGE M., STÖLZEL M., *et al.*, "Control of interface abruptness of polar (Mg,Zn)O/ZnO quantum wells grown by pulsed laser deposition", *Applied Physics Letters*, vol. 97, no. 5, p. 052101, 2010.

[CAO 07] CAO B.Q., LORENZ M., RAHM A., *et al.*, "Phosphorus acceptor doped ZnO nanowires prepared by pulsed-laser deposition", *Nanotechnology*, vol. 18, no. 45, p. 455707, 2007.

[CAO 09] CAO B.Q., ZUNIGA-PEREZ J., BOUKOS N., *et al.*, "Homogeneous core/shell ZnO/ZnMgO quantum well heterostructures on vertical ZnO nanowires", *Nanotechnology*, vol. 20, no. 30, p. 305701, 2009.

[CAO 10] CAO B.Q., ZUNIGA-PEREZ J., CZEKALLA C., *et al.*, "Tuning the lateral density of ZnO nanowire arrays and its application as physical templates for radial nanowire heterostructures", *Journal of Materials Chemistry*, vol. 20, pp. 3848–3854, 2010.

[CHE 05] CHEN J.J., YU M.H., ZHOU W.L., *et al.*, "Room-temperature ferromagnetic Co-doped ZnO nanoneedle array prepared by pulsed laser deposition", *Applied Physics Letters*, vol. 87, no. 17, p. 173119, 2005.

[CHE 10] CHENG C., LIU B., SIE E.J., *et al.*, "(Zn,Cd)O/ZnO coaxial multiple quantum well nanowire heterostructures and optical properties", *Journal of Physical Chemistry C*, vol. 114, no. 9, pp. 3863–3868, 2010.

[CZE 08] CZEKALLA C., GUINARD J., HANISCH C., *et al.*, "Spatial fluctuations of optical emission from single ZnO/(Mg,Zn)O nanowire quantum wells", *Nanotechnology*, vol. 19, no. 11, p. 115202, 2008.

[DIE 11a] DIETRICH C.P., BRANDT M., LANGE M., *et al.*, "Defect properties of ZnO and ZnO:P microwires", *Journal of Applied Physics*, vol. 109, no. 1, p. 013712, 2011.

[DIE 11b] DIETRICH C.P., LANGE M., STURM C., *et al.*, "One- and two-dimensional cavity modes in ZnO microwires", *New Journal of Physics*, vol. 13, no. 10, p. 103021, 2011.

[DIE 12] DIETRICH C.P., LANGE M., STÖLZEL M., *et al.*, "Microwire (Mg,Zn)O/ZnO and (Mg,Zn)O/(Zn,Cd)O non-polar quantum well heterostructures for cavity applications", *Applied Physics Letters*, vol. 100, no. 3, p. 031110, 2012.

[FAN 05] FAN H.J., LEE W., SCHOLZ R., *et al.*, "Arrays of vertically aligned and hexagonalle arranged ZnO nanowires: a new template-directed approach", *Nanotechnology*, vol. 16, no. 6, pp. 913–917, 2005.

[FAN 06a] FAN H.F., LEE W., HAUSCHILD R., *et al.*, "Template-assisted large-scale ordered arrays of ZnO pillars for optical and piezoelectric applications", *Small*, vol. 6, no. 4, p. 561, 2006.

[FAN 06b] FAN H.F., WERNER P., ZACHARIAS M., "Semiconductor nanowires: from self-organization to growth control", *Small*, vol. 2, no. 6, pp. 700–717, 2006.

[GRE 04] GREYSON E.C., BABAYAN Y., ODON T.W., "Directed Growth of Ordered Arrays of Small-Diameter ZnO Nanowires", *Advanced Materials*, vol. 16, no. 15, pp. 1348–1352, 2004.

[GRU 10] GRUNDMANN M., BÖNTGEN T., LORENZ M., "The occurrence of rotation domains in heteroepitaxy", *Physical Review Letters*, vol. 105, no. 14, p. 146102, 2010.

[GUO 08] GUO R., NISHIMURA J., MATSUMOTO M., *et al.*, "Density-controlled growth of ZnO nanowires via nanoparticle-assisted pulsed-laser deposition and their optical properties", *Japanese Journal of Applied Physics*, vol. 47, pp. 741–745, 2008.

[HE 11] HE H., LIN S., YUAN G., *et al.*, "Single-crystalline sodium-doped p-type ZnO and ZnMgO nanowires via combination of thin-film and nano techniques", *Journal of Physical Chemistry C*, vol. 115, no. 39, pp. 19018–19022, 2011.

[JIA 09] JIAN D.L., GAO P.-X., CAI W.J., *et al.*, "Synthesis, characterization, and photocatalytic properties of ZnO/(La,Sr)CoO_3 composite nanorod arrays", *Journal of Materials Chemistry*, vol. 19, pp. 970–975, 2009.

[KAY 09] KAYDASHEV V.E., KAIDASHEV E.M., PERES M., *et al.*, "Structural and optical properties of $Zn_{0.9}Mn_{0.1}O$/ZnO core-shell nanowires designed by pulsed laser deposition", *Journal of Applied Physics*, vol. 106, no. 9, p. 093501, 2009.

[KIM 12a] KIM K., LEE D.-H., LEE S.Y., *et al.*, "Effect of Ag/Al co-doping method on optically p-type ZnO nanowires synthesized by hot-walled pulsed laser deposition", *Nanoscale Research Letters*, vol. 7, no. 1, p. 273, 2012.

[KIM 12b] KIM K., PARL D.-H., DEBNATH P.C., *et al.*, "Characterizations of Ga-doped ZnO nanowires depending on growth temperature and target-substrate distance in hot-walled pulsed laser deposition", *Journal of Nanoscience and Nanotechnology*, vol. 12, no. 4, pp. 3559–3562, 2012.

[LAN 11a] LANGE M., DIETRICH C.P., BENNDORF G., *et al.*, "Thermal stability of ZnO/(Zn,Cd)O/ZnO double heterostructures grown by pulsed laser deposition", *Journal of Crystal Growth*, vol. 328, no. 1, pp. 13–17, 2011.

[LAN 11b] LANGE M., DIETRICH C.P., ZUNIGA-PEREZ J., *et al.*, "(Mg,Zn)O/ZnO quantum well nanowire heterostructures with large confinement energies", *Journal of Vacuum Science and Technology A*, vol. 29, no. 3, p. 03A104, 2011.

[LAN 12] LANGE M., Herstellung und Charakterisierung von planaren und drahtförmigen Heterostrukturen mit ZnO- und (Zn,Cd)O-Quantengräben, Doctoral Thesis, University of Leipzig, 2012.

[LAN 13] LANGE M., DIETRICH C.P., LORENZ M., *et al.*, "Excitonic and optical confinement in microwire heterostructures with nonpolar (Zn,Cd)O/(Mg,Zn)O multiple quantum wells", *Journal of Physical Chemistry C*, vol. 117, no. 17, pp. 9020–9024, 2013.

[LEE 04] LEE W., JEONG M.-C., MYOUNG J.-M., "Optical characteristics of arsenic-doped ZnO nanowires", *Applied Physics Letters*, vol. 85, no. 25, pp. 6167–6169, 2004.

[LEE 08] LEE S.Y., SONG Y.-W., JEON K.A., "Synthesis and analysis of resistance-controlled Ga-doped ZnO nanowires", *Journal of Crystal Growth*, vol. 310, no. 20, pp. 4477–4480, 2008.

[LEE 11] LEE J.S., CHA S.N., KIM J.M., *et al.*, "p-type conduction characteristics of lithium-doped ZnO nanowires", *Advanced Materials*, vol. 23, no. 36, pp. 4183–4187, 2011.

[LI 07] LI S.Z., GAN C.L., CAI H., *et al.*, "Enhanced photoluminescence of ZnO/Er_2O_3 core-shell structure nanorods synthesized by pulsed laser deposition", *Applied Physics Letters*, vol. 90, no. 23, p. 263106, 2007.

[LIN 09] LIN S.S., HONG J.I., SONG J.H., *et al.*, "Phosphorus doped $Zn_{1-x}Mg_xO$ nanowire arrays", *NanoLetters*, vol. 9, no. 11, pp. 3877–3882, 2009.

[LIU 11] LIU W.Z., XU H.Y., WANG L., *et al.*, "Size-controlled growth of ZnO nanowires by catalyst-free high-pressure pulsed laser deposition and their optical properties", *AIP Advances*, vol. 1, no. 2, p. 022145, 2011.

[LOR 05] LORENZ M., KAIDASHEV E.M., RAHM A., *et al.*, "$Mg_xZn_{1-x}O$ ($0 \leq x < 0.2$) nanowire arrays on sapphire grown by high-pressure pulsed-laser deposition", *Applied Physics Letters*, vol. 86, no. 14, p. 143113, 2005.

[LOR 08] LORENZ M., "Pulsed Laser deposition of ZnO-based thin films" in ELLMER K., KLEIN A., RECH B. (eds.), *Transparent Conductive Zinc Oxide. Basics and Applications in Thin Film Solar Cells*, Springer Series in Materials Science, Springer, Berlin, vol. 104, pp. 303–358, 2008.

[LOR 10] LORENZ M., RAHM A., CAO B.Q., *et al.*, "Self-organized growth of ZnO-based nano- and microstructures", *Physica Status Solidi (b)*, vol. 247, no. 6, pp. 1265–1281, 2010.

[MAR 03] MÅRTENSEN T., BORGSTRÖM M., SEIFERT W., *et al.*, "Fabrication of individually seeded nanowire arrays by vapour-liquid-solid growth", *Nanotechnology*, vol. 14, no. 12, pp. 1255–1258, 2003.

[MOF 06] MOFOR A.L., EL-SHAER A., SULEIMAN M., *et al.*, "A two-step obtainment of quantum confinement in ZnO nanorods", *Nanotechnology*, vol. 17, no. 19, p. 4859, 2006.

[NOB 07] NOBIS T., RAHM A., CZEKALLA C., *et al.*, "Optical whispering gallery modes in dodecagonal zinc oxide microcrystals", *Superlattices & Microstructures*, vol. 42, nos. 1–6, pp. 333–336, 2007.

[PAR 03] PARK W.I., YI G.-C., KIM M.Y., *et al.*, "Quantum confinement observed in ZnO/ZnMgO nanorod heterostructures", *Advanced Materials*, vol. 15, no. 6, pp. 526–529, 2003.

[PER 12] PERILAT-MERCEROZ G., THIERRY R., JOUNEAU P.-H., *et al.*, "Compared growth mechanisms of Zn-polar ZnO nanowires on O-polar ZnO and on sapphire", *Nanotechnology*, vol. 23, no. 12, p. 125702, 2012.

[RAH 06] RAHM A., KAIDASHEV E.M., SCHMIDT H., *et al.*, "Growth and characterization of Mn- and Co-doped ZnO nanowires", *Microchimica Acta*, vol. 156, nos. 1–2, pp. 21–25, 2006.

[RAH 07] RAHM A., LORENZ M., NOBIS T., *et al.*, "Pulsed-laser deposition and characterization of ZnO nanowires with regular lateral arrangement", *Applied Physics A*, vol. 88, no. 1, pp. 31–34, 2007.

[RYU 06] RYU Y.R., LEE T.S., LUBGUBAN J.A., *et al.*, "Wide-band gap oxide alloy: BeZnO", *Applied Physics Letters*, vol. 88, no. 5, p. 052103, 2006.

[SAD 06] SADOFEV A., BLUMSTENGEL S., CUI J., *et al.*, "Visible band-gap (Zn,Cd)O heterostructures grown by molecular beam epitaxy", *Applied Physics Letters*, vol. 89, no. 20, p. 201907, 2006.

[SON 07] SON H.J., JEON K.A., KIM C.E., *et al.*, "Synthesis of ZnO nanowires by pulsed laser deposition in furnace", *Applied Surface Science*, vol. 253, no. 19, pp. 7848–7850, 2007.

[SON 08] SON J.Y., KIM S.J., CHO J.H., *et al.*, "Synthesis of horizontally aligned ZnO nanowires localized at terrace edges and application for high sensitivity gas sensor", *Applied Physics Letters*, vol. 93, no. 5, p. 053109, 2008.

[SCH 10] SCHMIDT-GRUND R., HILMER H., HINKEL A., *et al.*, "Two-dimensional confined photonic wire resonators – strong light–matter coupling", *Physica Status Solidi (b)*, vol. 247, no. 6, pp. 1351–1364, 2010.

[THI 12] THIERRY R., PERILLAT-MERCEROZ G., JOUNEAU P.H., *et al.*, "Core–shell multi-quantum wells in ZnO/ZnMgO nanowires with high optical efficiency at room temperature", *Nanotechnology*, vol. 23, no. 8, p. 085705, 2012.

[WAG 70] WAGNER R.G., in LEVITT A.E. (ed.), *Whisker Technology*, Wiley-Interscience, VLS Mechanism of Crystal Growth, New York, 1970.

[WAN 08] WANG K., CHEN J.J., ZHOU W.L., *et al.*, "Direct growth of highly mismatched type II ZnO/ZnSe core/shell nanowire arrays on transparent conducting oxide substrates for solar cell applications", *Advanced Materials*, vol. 20, no. 17, pp. 3248–3253, 2008.

[WAN 10] WANG K., CHEN J.J., ZENG Z.M., *et al.*, "Synthesis and photovoltaic effect of vertically aligned ZnO/ZnS core/shell nanowire arrays", *Applied Physics Letters*, vol. 96, no. 12, p. 123105, 2010.

[WAN 11] WANG G.P., CHU S., ZHAN N., *et al.*, "ZnO homojunction photodiodes based on Sb-doped p-type nanowire array and n-type film for ultraviolet detection", *Applied Physics Letters*, vol. 98, no. 4, p. 041107, 2011.

[WEN 12] VON WENCKSTERN H., SCHMIDT-GRUND R., BUNDESMANN C., *et al.*, "The (Mg,Zn)O Alloy", in FENG Z.C. (ed.), *Handbook of Zinc Oxide and Related Materials*, Materials, Taylor and Francis/CRC Press, Florida, FL, vol. 1, p. 257-320, 2012.

[WEN 13] VON WENCKSTERN H., ZHANG Z., SCHMIDT F., *et al.*, "Continuous composition spread using pulsed-laser deposition with a single, segmented target", *Cryst. Eng. Comm.*, 2013.

[WIL 09] WILLANDER M., NUR O., ZHAO Q.X., *et al.*, "Zinc oxide nanorods based photonic devices: recent progress in growth, light emitting diodes and lasers", *Nanotechnology*, vol. 20, no. 33, p. 332001, 2009.

[YAN 09] YANG X.Y., WALCOTT A., WANG G.M., *et al.*, "Nitrogen-doped ZnO nanowire arrays for photoelectrochemical water splitting", *NanoLetters*, vol. 9, no. 6, pp. 2331–2336, 2009.

[ZHA 05] ZHANG Y., RUSSO R.E., MAO S.S., "Femtosecond laser assisted growth of ZnO nanowires", *Applied Physics Letters*, vol. 87, no. 13, p. 133115, 2005.

[ZIM 10] ZIMMERMANN G., LANGE M., CAO B.Q., *et al.*, "Resistivity control of ZnO nanowires by Al doping", *Physica Status Solidi (RRL)*, vol. 4, nos. 3–4, pp. 82–84, 2010.

[ZUN 07] ZUNIGA-PEREZ J., RAHM A., CZEKALLA C., *et al.*, "Ordered growth of tilted ZnO nanowires: morphological, structural and optical characterization", *Nanotechnology*, vol. 18, no. 19, p. 195303, 2007.

13

Preparation of ZnO Nanorods and Nanowires by Wet Chemistry

13.1. Introduction

ZnO nanorods (NRs) and nanowires (NWs) can be synthesized by a large variety of techniques, including high temperature vapor phase processes and low temperature solution routes [KAW 10, DJU 10, SIN 10, XU 11a, UTA 12, CUI 12, PAN 13]. Many studies on ZnO NW/NR growth have been centered on the use of the vapor phase techniques, such as pulsed laser deposition, thermal evaporation or metal organic vapor phase epitaxy (for more details, refer to Chapter 11 and 12) [HUA 01, LI 01, WU 02]. However, since the late 1990s and early 2000s, hydrothermal/chemical and electrochemical processes have emerged as the two very important low temperature approaches for ZnO NW/NR synthesis [PAU 99, LI 00, VAY 01a, VAY 03]. Gas phase approaches usually produce high-quality ZnO NWs/NRs, but their commercial potential is constrained by the high temperature growth (above 500°C) and the high cost of the deposition system. In contrast, solution routes to ZnO NW/NR growth are usually performed at low temperature (less than 100°C) and are very promising for commercialization due to their simplicity, low cost and large-scale production. By fine control of the growth conditions and kinetics, high-quality products with controlled morphological, structural, optical and electronic properties can easily be obtained. We will first describe the preparation of ZnO NRs and NWs by hydrothermal techniques and chemical bath deposition (CBD) and then focus on the electrochemical deposition (ECD) techniques. The nanostructures can be grown as dispersed objects in the bulk solutions or arrayed onto various surfaces.

Chapter written by Thierry PAUPORTÉ.

13.2. Preparation of ZnO nanorods and nanowires by chemical bath deposition and hydrothermal techniques

13.2.1. *Principle*

The technique consists of selectively producing ZnO nanostructures as precipitates in aqueous solutions, organic solvents or in mixtures of both. Fundamentally, the reaction solution systems contain two precursors, namely Zn(II) ions as a zinc source and a precursor of hydroxide ions (OH^-). Most often, a chelating agent is also added to increase the initial solubility of zinc. Alkaline agents, such as NH_4OH or NaOH, can be used directly. Another technique consists of the use of a hydroxide precursor, typically hexamethylene tetramine (HMTA) or also urea or triethanolamine, which will increase the bath pH upon thermal decomposition.

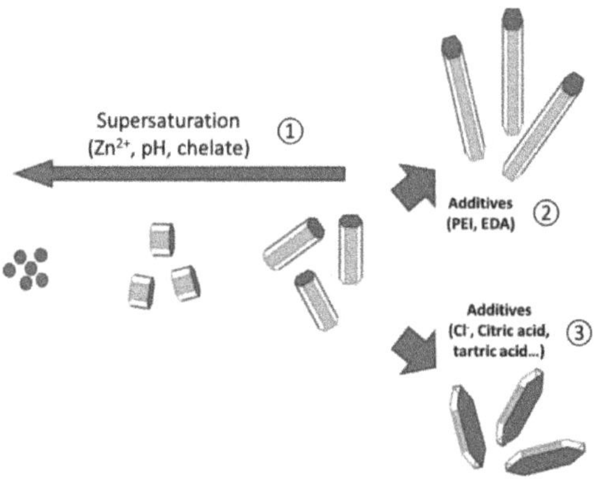

Figure 13.1. *Control of the nanostructure shape in the chemical precipitation processes in bulk solution. For a color version of this figure, see www.iste.co.uk/consonni/nanowires1.zip*

Depending on the process temperature, typically between 50°C and 100°C, the technique is called CBD for the lowest temperatures and hydrothermal growth for the highest. The driving force of crystallization is the degree of the supersaturation that is tuned by varying the solution pH and/or the Zn(II) concentration of the chemical systems. By controlling the nucleation and the growth, NRs, NWs, nanosheets or nanoplates can be selectively produced (Figure 13.1). In the presence of a selected substrate, the nanostructures can be arranged and organized on the surface, and various nanostructured layers can be produced (Figure 13.2). The orientation of the ZnO rods/wires, sheets and plates is controlled by manipulating the physicochemical specific conditions of the bath and the substrate properties. The addition of particular organic molecules as additives in solution influences the morphology of the crystals grown (Figures 13.1 and 13.2). The overall architecture of crystals is basically controlled by tuning the crystal growth, including the growth site, growth rate and growth direction. Therefore, the degree of supersaturation, the presence of seeds and the adsorption of additives are the three key parameters for the control of the growth modes and of the final structures.

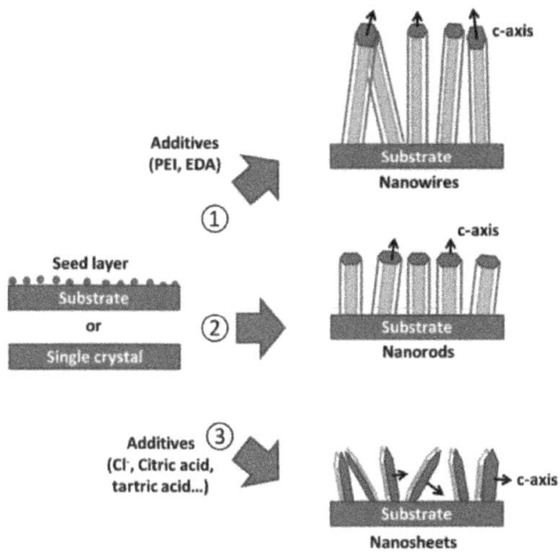

Figure 13.2. *Control of the nanostructure shape in the chemical precipitation processes on surfaces*

13.2.2. *Growth of ZnO nanostructure in bulk solution*

13.2.2.1. *Variation of the degree of supersaturation*

The solubility of ionic zinc species depends on the pH of the solution (Figure 13.3) [GOU 05, MU 09]. Thus, the degree of supersaturation can be controlled by the change in pH. Crystalline ZnO was found to be directly deposited in the range of pH from 9 to 13 [YAM 02]. The degree of supersaturation can be simply increased by adding NaOH to a zinc precursor solution [KAW 06a, KAW 08]. In addition, ammonium compounds, including NH_4OH, NH_4Cl and NH_4NO_3, are often used as complexing agents (chelating agents) because they lower the degree of supersaturation in the system [YAM 02, BOY 02, GOV 04, GOV 02, MAS 06, KIM 07a]. The shape of the resultant ZnO crystals is governed by the balance of the nucleation rate and the growth rate. The concentrations of alkali species, Zn(II) ions, and chelating agents can be delicately tuned in this procedure.

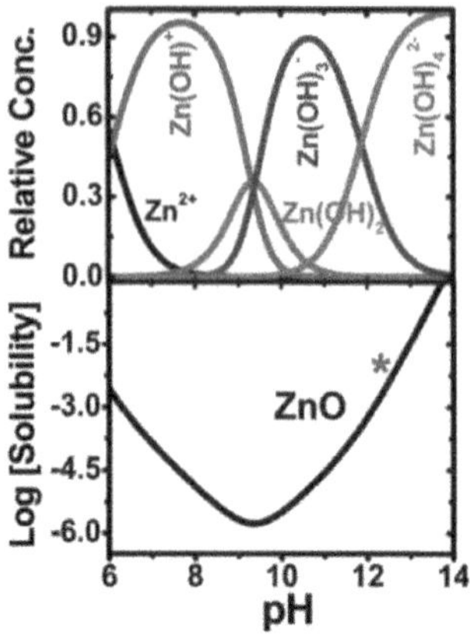

Figure 13.3. *Speciation distribution (upper) and solubility (lower) of Zn(II) vs pH in aqueous solution at 70°C. The asterisk represents the slightly supersaturated state of the deposition solution containing 6.0 mM of* $Zn(NO_3)_2$ *and 0.25 M NaOH. (Copyright [MU 09]). For a color version of this figure, see www.iste.co.uk/consonni/nanowires1.zip*

The main reactions involved in the growth are illustrated in the following equations [DEM 01, DEM 02]:

$$Zn^{2+} + 2OH^- \leftrightarrow Zn(OH)_2 \quad [13.1]$$

$$Zn(OH)_2 + 2OH^- \leftrightarrow [Zn(OH)_4]^{2-} \quad [13.2]$$

$$[Zn(OH)_4]^{2-} \leftrightarrow ZnO_2^{2-} + 2H_2O \quad [13.3]$$

$$ZnO_2^{2-} + H_2O \leftrightarrow ZnO + 2OH^- \quad [13.4]$$

$$ZnO + OH^- \leftrightarrow ZnOOH^- \quad [13.5]$$

In the above reactions, the O^{2-} in ZnO comes from the alkali, not from the solvent H_2O. Therefore, the growth of ZnO does not necessarily require the solvent to be H_2O [ZHA 02]. It can be organic solvents, such as methanol [CHE 04], ethanol [LIU 03] and butanol [CAO 06], or even ionic liquids [HOU 07, ALA 09].

Figure 13.1 describes the crystallization in bulk solution. The size and shape are influenced by the balance of the nucleation rate and the growth rate depending on the degree of the supersaturation. Fine grains are produced under a high degree of supersaturation through a high rate of homogeneous nucleation. On the other hand, NRs elongated in the *c*-direction are obtained through gradual crystal growth with a low rate of the nucleation under a low degree of supersaturation. In this case, however, the size distribution of the resultant particles is broadened due to the variation of the nucleation time. To increase the homogeneity of the prepared nanostructures, the nucleation step and the growth step can be separated. Seeds are added to the growth medium to achieve a better monodispersivity (Figure 13.4 pathway ①). They can be ZnO nanoparticles or a $Zn(OH)_2$ colloidal suspension subsequently grown and dehydrated as ZnO by increasing the temperature. Due to the anisotropic nature of the ZnO wurtzite unit cell with polar (0001) and apolar (1000) faces, the ZnO grains have a tendency to grow as rods elongated in the c-direction. If the initial seeds are aggregates of nanoparticles, NR/NW assemblies (spherulitic multipods) are produced as shown in Figure 13.4 pathway ② [PAN 05]. The ZnO rods and NWs are obtained when the growth speed along the *c*-axis is faster than that of the lateral facets. By using additives in the bath, this tendency can be favored. On the other hand, other additives can adsorb on the (001) ZnO faces and slow down the growth along the *c*-axis resulting in ZnO nanosheet formation (Figures 13.1 and 13.2).

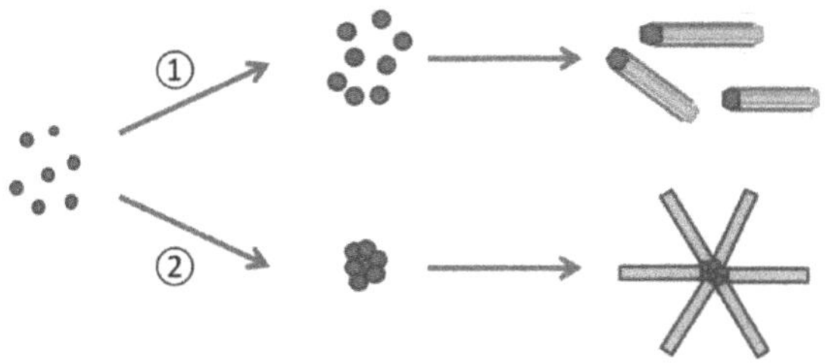

Figure 13.4. *Effect of ZnO seeds on the nanostructure shape*

Under alkaline conditions, the reactions can take place at room temperature by adjusting the ratio of Zn(II) and OH^-, giving rise to ZnO NWs with diameters even below 10 nm. ZnO NWs with various aspect ratios (ARs) have been prepared by simply adjusting OH^- concentration and reaction time [CAO 06]. The growth of polar inorganic nanocrystals is sensitive to the reaction solvents, and their morphologies can be tuned and controlled by the crystal–solvent interfacial interactions [HOU 07]. In such cases, the morphology of ZnO is largely directed by the polarity and saturated vapor pressure of the solvents [ZHA 02]. The AR of ZnO NWs, which is dictated by the relative growth rates of polar and nonpolar surfaces, can be readily tuned by varying the polarity of the solvents (Figure 13.5). Polar solvent molecules have stronger interactions with the polar surfaces of ZnO, and thus hinder the precursor molecule adsorption. The AR of the ZnO nanostructures increases going from the more polar solvent methanol to the less polar solvent 1-butanol. All the as-grown ZnO NWs showed two well-faceted basal planes along the *c*-axis [LIU 03].

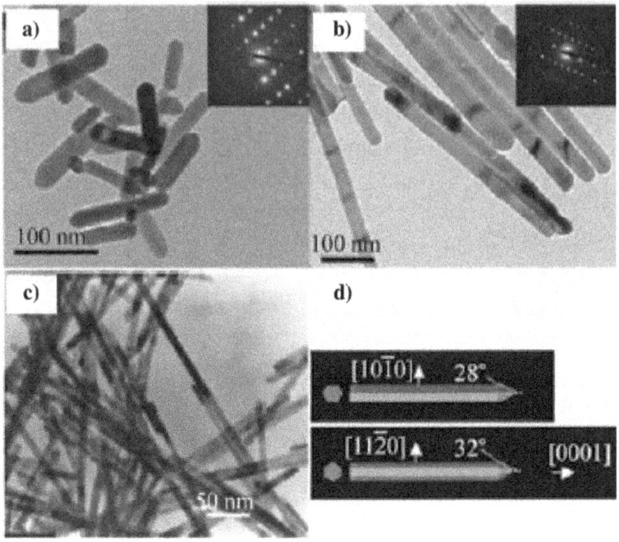

Figure 13.5. *Transmission electron microscopy (TEM) images of ZnO nanowires synthesized in solvents having different polarities a) in methanol, b) in ethanol, and c) in 1-butanol. Even though the reaction temperature and the growth time are different, we can still see the effect of the solvent polarity on the NW AR. Insets in (a) and (b) are selected area electron diffraction patterns. d) Schematic illustration of growing +c ends of ZnO with two common interplanar angles. (Copyright [CHE 04] and [LIU 03])*

When the solvent contains non-polar hexane, ultrathin ZnO NWs with diameters of 2 nm could be synthesized from a simple acetate precursor [YIN 04]. These ultrathin NWs also self-assembled into uniform stacks of NWs aligned parallel to each other with respect to the long axis [YIN 04]. Near-UV absorption and photoluminescence (PL) measurements were able to determine that quantum confinement effects were present in these ultrathin NWs, with an excitonic ground state of about 3.55 eV [YIN 04]. The ultrathin NWs were possibly grown by oriented coalescence of quantum dots. Pacholski *et al.* suggested that oriented attachment of preformed quasi-spherical ZnO nanoparticles should be a major reaction path during the formation of single crystalline NWs [PAC 02, ZHA 05a].The necks between the attached nanoparticles were later filled up, and the NW surfaces were thus smoothened by Ostwald ripening [PAC 02].

Another very popular approach to produce crystalline ZnO is to use aqueous solutions containing HMTA and zinc nitrate [VAY 01a, VAY 03, BOY 02, GOV 04, GOV 02, ZHO 09, GAR 07, PEN 06, TIA 03; GRE 03, KEN 09]. HMTA, also called methenamine or hexamine, with formulas $C_6H_{12}N_4$, is a highly water soluble, non-ionic tetradentate cyclic ternary amine (Figure 13.6).

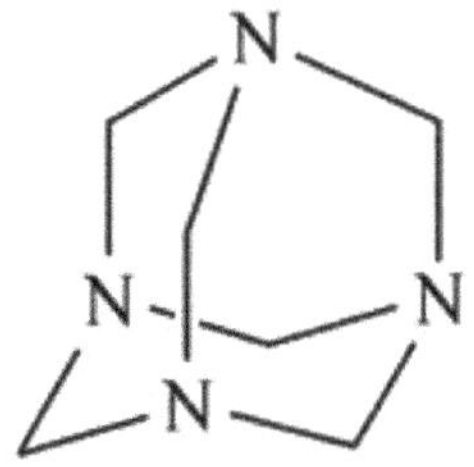

Figure 13.6. *Molecular structure of hexamethylene tetramine (HMTA)*

Classically, the synthesis is performed from a mixture of HMTA and Zn(II) salt heated at 50–95°C. The substrate is placed in a sealed bottle filled with the solution. The precursor undergoes a slow thermal decomposition, which produces NH_4OH and HCHO [XU 11a]:

$$(CH_2)_6N_4 + 6\ H_2O \leftrightarrow 4\ NH_3 + 6\ HCHO \qquad [13.6]$$

$$NH_3 + H_2O \leftrightarrow NH_3\text{-}H_2O \qquad [13.7]$$

$$NH_4OH \leftrightarrow NH_4^+ + OH^- \qquad [13.8]$$

HMTA is believed to act as a weak base, which slowly hydrolyzes in the water solution when the temperature is increased and gradually releases OH^- (reactions [13.6]–[13.8] through NH_4OH base production. The degree of supersaturation of mild basic solutions is easily kept at a low level. Nucleation and crystal growth occur in bulk solution and also on the surface of solid phases even in the absence of a seed layer (SL) [VAY 01a]. ZnO is produced by the reaction mechanism:

$$Zn^{2+} + 2\ OH^- \leftrightarrow Zn(OH)_2 \qquad [13.9]$$

$$Zn(OH)_2 \rightarrow ZnO + H_2O \qquad [13.10]$$

Zinc nitrate salt provides Zn^{2+} ions required for building-up the ZnO NWs (reactions [13.9]–[13.10]. Water molecules in the solution provide O_2^- ions.

All of the reactions [13.6]–[13.9] are in equilibrium and can be controlled by adjusting the reaction parameters, such as precursor concentration, growth temperature and growth time, to push the reaction equilibrium forward or backward. The controlled degree of supersaturation under basic conditions causes the ZnO NRs to elongate along the *c*-axis direction. The length of the rods/wires is usually on the micrometer scale because of the long-term growth at a low degree of supersaturation. In general, precursor concentration determines the NW density. The growth time and temperature control the ZnO NW morphology and aspect ratio. By this technique, the AR can be easily controlled since it was shown to increase by decreasing the equimolar zinc nitrate/HMTA mixture concentration [VAY 03].

13.2.2.2. *Control of the grain shape by use of additives in the solution*

Water soluble capping agents can be added in the solution to modify the growth habits of the ZnO nanostructures (Figure 13.7) [ZHA 05]. The capping agents that are commonly used in the literature in the hydrothermal growth of ZnO nanostructures can be separated into two categories:

1) Those that adsorb onto the side apolar ZnO surfaces and enhance the growth along the *c*-axis, such as amines like polyethylenimine (PEI) [LAW 05, WU 08, ZHO 08, XU 10] and ethylenediamine (EDA) [GOV 04, LIU 03, SHE 07].

2) Those that cap onto the basal polar plane of the ZnO nanostructures and promote lateral growth. Cl^- was described to have this effect [XU 05a]. Also organic compounds with carboxylate groups, such as citrate ions ($C_3H_5O(COO)_3^{3-}$) [KIM 06, TIA 02, YAN 08], tartric acid (HOOC-CHOH-CHOH-COOH), maleic acid (HOOC-CH-CH-COOH) or phenolphthalein, have this behavior [YAH 07].

The isoelectric point of ZnO powder is at around pH = 9.5 [ZAN 07]. The charge on the ZnO surface sites is predominately positive or negative for pH values below or above the isoelectric point, respectively [KIM 07a]. PEI is a non-polar polymer with a large amount of amino side-groups ($-NH_2$), which can be protonated over a wide range of pH values (3–11), and therefore become positively charged. The pH value of the growth solution can be adjusted to fall in the range that leads to the protonation of PEI, and therefore the linear PEI molecule with its high-positive charge density adsorbs strongly on the negatively charged surfaces due to electrostatic attraction (Figure 13.7(a)) [WU 08]. Thus, the lateral growth of the NWs is largely hindered, and wires with high AR are produced [ZHO 08]. Above their capping behavior, PEI additives also help to grow longer NWs on substrates (see section 13.2.3.2) by extending the growth time due to the coordination of PEI to Zn^{2+} ions [ZHO 08]. This is similar to the effect of adding $NH_3{\cdot}H_2O$ to increase the solubility of the nutrient precursor, and can be attributed to the decrease of the free $[Zn^{2+}]$ that usually combines with OH^- and precipitates in the form of $Zn(OH)_2$ in the bulk solution. Also after the growth, less precipitate was expected to form in the bulk solution in the presence of PEI coordination than without PEI. Thus, longer NW arrays could be produced by prolonging the growth time without refreshing the growth solution, because the Zn^{2+} depleted during the growth would be replenished through the decomposition of PEI-Zn^{2+} complexes [ZHO 08].

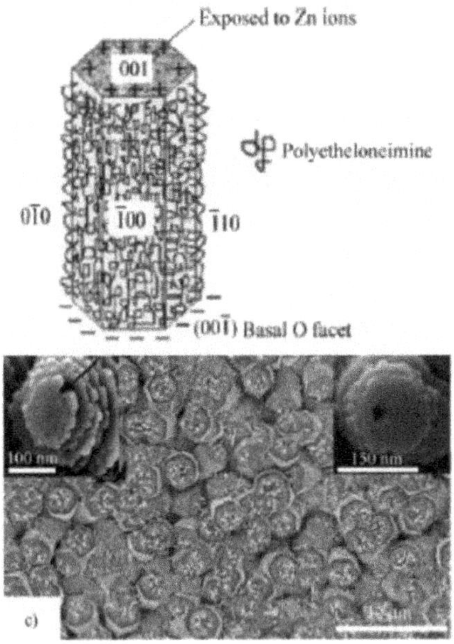

Figure 13.7. *Schematic view of the PEI capping agent interaction with ZnO planes. (Copyright [KIM 06]) b) Large arrays of well-aligned helical ZnO whiskers on top of ZnO rod base (copyright [TIA 02]). For a color version of this figure, see www.iste.co.uk/consonni/nanowires1.zip*

Citrate ions are characterized by three negative charges under the normal growth environment. Experimental results in the literature as well as theoretical calculations suggest that citrate ions preferentially adsorb to the Zn of the (0001) surface, and thus inhibit the growth along [0001] and force the growth along the $\langle 01\bar{1}0\rangle$ or $\langle 2\bar{1}\bar{1}0\rangle$ directions [KIM 06, TIA 02, HID 96]. With citrate ions, rather than high AR NWs, flat hexagonal nanoplates were produced, as shown in Figure 13.7(b) [KIM 06, TIA 02]. The formed nanostructures have high roughness factors and/or large areas of exposed polar basal planes. It was also suggested that due to the presence of the citrate ions, the surface tension of the growth solution was reduced, which lowered the energy needed to form a new phase, and thus ZnO nanostructures could nucleate at a lower supersaturation level [YAN 08]. Additive effects similar to that of citrate have been reported for poly(dialkyldimethylammonium chloride) (PDADMAC) with tertiary amine groups and sodium poly(styrene sulfonate) (PSS) with sulfonate groups [GAR 07]. They yield to significant AR reductions at low concentration due to the binding of the growth modifier on crystal surface

[GAR 07]. Yahiro *et al.* showed that NWs consisting of oriented ZnO nanograins were prepared in the presence of pyranine, a pH-sensitive fluorescent arylsulfonates dye [YAH 06]. These structures are regarded as a new type of assembled crystals named mesocrystal.

13.2.3. *Chemical growth of ZnO nanowire/nanorod arrays on substrate*

In many cases, the ZnO NWs and NRs must be grown on substrates whose chemical nature and crystallinity depend on their application. For many opto-electronic applications, it is important to grow the ZnO structures on conductive substrates with a high degree of structural, morphological and assembly control.

13.2.3.1. *Influence of the substrate on the ZnO nucleation and growth.*

The procedure for the growth of NRs and NWs on solid substrate depends on the chemical nature and crystallinity of the supporting material. Three cases can be distinguished:

1) Amorphous or polycrystalline substrates of materials different to ZnO;

2) Polycrystalline ZnO, doped-ZnO substrate and ZnO seeded surfaces;

3) Epitaxial growth.

In cases (1) and (3), the synthesis process occurs in two steps, first the heteronucleation on the substrate and then the growth of the nanostructure. We can separate the nucleation and growth stages and then better control the final structure by using a pre-deposited SL made of small seed grains (case (2)).

13.2.3.1.1. Growth on amorphous and polycrystalline substrates

The direct growth of ZnO NRs/NWs on amorphous or polycrystalline substrates is possible by CBD/hydrothermal techniques. This has been reported for instance in the $Zn(NO_3)_2$/HMTA system on glass, SiO_2/Si, on transparent conducting oxides such as F-doped SnO_2 (FTO) or ITO layers supported on glass or plastic [VAY 01a]. The formed NWs/NRs are then sparsely nucleated on the substrate. They are large and poorly oriented [YAN 09]. Xu *et al.* [XU 08a] have studied the seedless growth of ZnO NW arrays on textured gold layers in the $Zn(NO_3)_2$/HMTA equimolar system. The 50 nm thick Au(111) oriented gold layer was deposited on Si(100) and flexible kapton polymer substrates. Sparsely dispersed ZnO NWs arrays, upright-oriented and fairly uniform in height were obtained. They found that the density of wire increased with the precursor concentration up to 5 mM. When the zinc nitrate concentration was increased further, the density of NWs remained approximately steady with a slight tendency to decrease. In their experiments, 70°C was found to be the optimum temperature for obtaining high AR and well-defined hexagonal prismatic crystallites with rather high density. They explained the optimum temperature by the moderate mobility of the ions with a diffusion length within a small range in the vicinity of the substrate. The accumulation of ions at the local region results in a rather high density of NWs. At 95°C, pyramid-shaped ZnO NRs with low surface density were grown. The upright orientation observed in [XU 08a] can be explained by the structural relationship between ZnO and Au(111) as shown in Figure 13.12 (see section 13.2.3.1.3).

13.2.3.1.2. Use of seeded surfaces

In many papers of literature, a pre-deposited nanocrystalline ZnO SL is used to promote the growth of ZnO at low supersaturation. The separation of the nucleation and the crystal growth is essential for morphological control with the seeds. Spin-coating, dip-coating, spray pyrolysis or radio frequency magnetron-sputtering [SON 07, GRE 05] are the techniques usually applied to prepare this layer. They are adapted to a large variety of substrates. The presence of a SL increases markedly the NW density on the surface. For spin and dip-coating, the chemical nature of the solution is important. In the literature, ZnO quantum dots dissolved in ethanol and dip-coated on the substrate have been reported [LAW 05, GUI 12]. Chao *et al.* have reported the use of GaN seeds deposited on sapphire [CHA 10].

Zinc precursor solutions can also be employed. Seeds have been grown by the heat treatments of zinc acetyl acetonate or zinc nitrate precursor layers deposited by dip-coating that yielded unaligned NRs [GRE 05]. Most often, the precursor is a 5 mM solution of zinc acetate ($Zn(CH_3COO)_2$) in ethanol. This solution is dip-coated or spin-coated at a rate of 2000 rpm for 30 s on the substrate. The deposited layer is then heated to produce the ZnO seeds. The decomposition temperature is classically 250 or 350°C for 20–30 min to yield a ZnO island layer highly textured with its (0001) plane parallel to the substrate. Since the SL is highly textured, the NW/NR arrays are grown upright, with their *c*-axis perpendicular to the surface. The texturing of the acetate-derived seeds must stem from either the particular kinetics of zinc acetate decomposition and/or the specific adsorbates present during its conversion to ZnO islands. Below 200°C, Greene *et al.* [GRE 05] observed that $Zn(CH_3COO)_2$ is not decomposed, and dissolves in the NW/NR growth bath. Increasing the temperature of the treatment gives rise to a decrease of the island density and to an increase of the particle size (up to 70 nm at 600°C) [YAN 09].

The effect of seed size and density on the ZnO NW size and density remains controversial. Most of the papers report that the ZnO NW diameter increases, and the density decreases with the ZnO seed size enlargement [YAN 09, SON 07,

MA 07, JI 09]. However, some other papers report the inverse effect [LIU 08, GUI 12]. Yang *et al.* have investigated the effect of particle size and density on the NR density and size [YAN 09]. The degree of orientation of the NRs grown in the $Zn(NO_3)_2$/HMTA system was dominated by the seed density. Moreover, they have shown that increasing the density of seeds increases the density of rod and decreases their diameter (Figure 13.8(a)). By comparing layers with similar seed density but different mean seed size, they have also observed no change in rod diameter and concluded that this parameter mainly depends on the density of ZnO nanoparticulate seeds.

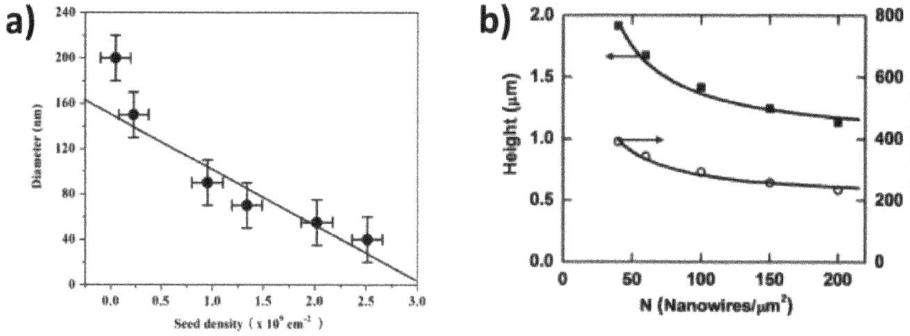

Figure 13.8. *a) Dependence of the ZnO nanorod density on ZnO seed density on the substrate. (Copyright [YAN 09]) b) Nanowire height (■) and diameter (○) as a function of nanowire density as predicted by the one-dimensional growth model. (Copyright [BOE 09])*

Boercker *et al.* have pointed out in their theoretical modeling that ZnO NW growth by CBD should be limited by the mass transport of the precursor in solution [BOE 09]. It is expected that the average diameter and the length of ZnO NWs are strongly related to their density. It has been emphasized that these characteristic dimensions (length and diameter) should be inservely proportional to their density (Figure 13.8(b)). Guillemin *et al.* [GUI 12] suggest that structural morphology of the SL can result in different nucleation sites, depending on the growth conditions and on the specific features of the empty space region in between ZnOnanoparticles (NPs) (i.e. intergranular zones). For instance, the SL porosity may be critical for the nucleation process, since it leads to the strengthening of the contribution of the grain boundary or intergranular zones. If the porosity is low (respectively high), the surface (respectively grain boundary) nucleation sites may be preferential and depend on the nature of the orientation (i.e. texture) of the ZnO seeds, irrespective of their average diameter.

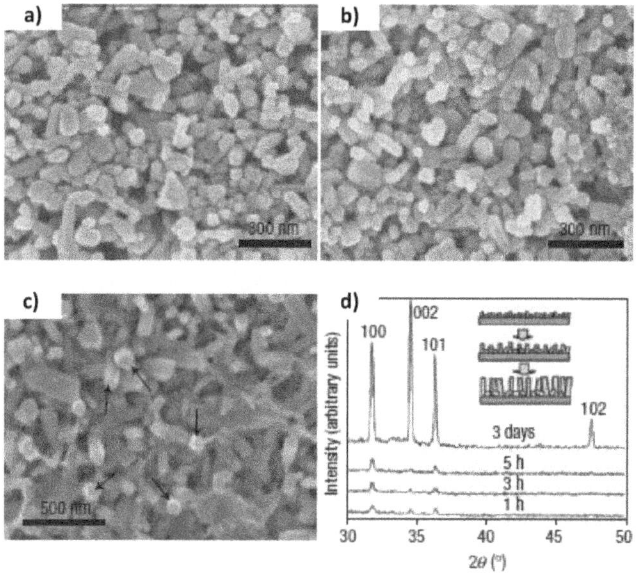

Figure 13.9. *ZnO hydrothermal growth as a function of time on a seeded substrate. a) 30 min, b) 1 h, (c) Faceted ZnO appears after 3 h of growth. The arrow point to a few nanorods that are starting to grow in the vertical direction. d) XRD patterns of the ZnO films as a function of time. The inset shows a schematic illustration of the growth process (copyright [TIA 03])*

The early stages of the layer texturation of the NW/NR arrays in the $Zn(NO_3)_2$/HMTA system were analyzed by Tian *et al.* [TIA 03]. These authors used a ZnO nanoparticle nucleation seed. In the early stages of growth, oriented structures were not formed. A SEM image of the ZnO nanoparticle seeds after 30 min of crystal growth showed little sign of crystal growth (Figure 13.9(a)). These particles consisted mostly of rounded rectangular and rod-shaped crystals with a wide size distribution. After 1 h of growth (Figure 13.9(b)), the surface roughness became visible on the ZnO seeds, indicating the initiation of crystal growth. After 3 h (Figure 13.9(c)), short and faceted hexagonal rods were observed, although most of these rods were not yet well-aligned. However, as the rod-like randomly oriented crystals grew further, they began to overlap, and their growth became physically limited as the misaligned NRs began to impinge on other neighboring crystals, giving rise to the preferred orientation of the film. The XRD patterns of the layer recorded at various growth times, and the alignment mechanism is presented in Figure 13.9(d). In the absence of the nucleation seeds, ZnO crystals were sparsely nucleated on the substrate and had freedom to grow to large sizes in all orientations.

Lee *et al.* [LEE 07] have shown that the humidity level during seeding by zinc acetate strongly influences the alignment, density and NR size of the ZnO NR arrays prepared in the $Zn(NO_3)_2$/HMTA system. At low humidity, ZnO NRs were relatively sparse and poorly aligned. In addition, SEM cross-section showed that multiple ZnO NRs appear to originate from the same point on the substrate, suggesting that a single seed may be responsible for the growth of multiple rod structures. ZnO NRs synthesized on substrates seeded between 32% relative humidity (RH), and 60% RH were densely packed and well-aligned, the great majority of ZnO NRs grew from individual points on the substrate. On the other hand, when the seeding RH was above 60%, the polydispersity in the diameter and length of the ZnO NRs increased, although good alignment was maintained.

13.2.3.1.3. Epitaxial growth of ZnO nanorods and nanowires on single-crystalline substrate

Seeding substrates with ZnO nanoparticles and preferential vertical growth of ZnO NWs/NRs are straightforward and easy means to yield textured layers. However, in this case, all the ZnO grains do not have exactly the same out of plane orientation, and they are not in-plane ordered. Moreover, the wires are not directly connected to the substrate. More organized structures can be prepared using single crystalline substrates. When a single crystalline ZnO(0001) substrate is utilized rather than a polycrystalline layer as a substrate for crystal growth, the vertical and in-plane crystallographic orientation of ZnO rods is preferentially achieved through epitaxial growth. This was for instance shown for both Zn-face and O-face of a single crystalline ZnO(001) substrate [KAW 06b, GUI 13]. The growth was carried out by Kawano *et al.* [KAW 06] in the $ZnSO_4/NH_4Cl/NaOH$ system using mirror polished single crystalline ZnO wafers. As shown in Figure 13.10, the six-fold facets (e.g. (1000) faces) of individual rods were well-oriented in the same directions on the (0001) planes of single crystalline ZnO substrates. This indicated that in-plane alignment of ZnO crystals was achieved on the both substrates. The density of the ZnO rods depended on the chemical structure and the polarity of the ZnO surface. Although a dense film consisting of oriented grains was obtained on the Zn-face substrate, hexagonal rods grew sparsely on the O-face one.

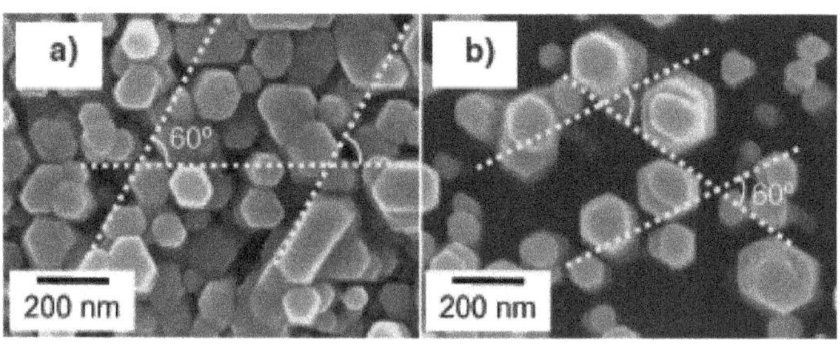

Figure 13.10. *SEM views of the top ZnO rods grown on a) a Zn-face ZnO and b) an O-face ZnO substrate. The dotted lines show that all the hexagonal rods are oriented in the same direction. (Copyright [KAW 06])*

ZnO and GaN have the same wurtzite hexagonal structure and a low lattice constant misfit (1.9% for the a-parameter), their room temperature lattice parameters being $a_{ZnO} = 0.325$ nm, $c_{ZnO} = 0.521$ nm and $a_{GaN} = 0.319$ nm, $c_{GaN} = 0.519$ nm. They also have similar thermal expansion coefficients. Consequently, the epitaxial growth of ZnO NWs/NRs can be easily achieved on single crystal GaN layers [PAU 99, CHA 10, QUA 06, KIM 07b, LE 06, LUP 12a, LUP 12b, JHA 12]. The density of NWs/NRs is controlled by the supersaturation, where it increases with the precursor concentration in the $Zn(NO_3^-)$/HMTA system [CHA 10]. Needle-shaped ZnO crystals with a pointed top and hexagonal ZnO prisms with a flat end were epitaxially grown on GaN-buffered Al_2O_3(0001) substrates in a solution system of $Zn(NO_3)_2/NH_4NO_3/NH_4OH$ and $Zn(NO_3)_2$/sodium citrate/NH_4OH, respectively [KIM 07b]. The needle-shaped wires with a pointed top are frequently observed when the growth rate was relatively high not only on a substrate but also in precipitates. Metastable high-index faces would appear in the acicular forms rather than the stable facets when relatively fast growth in c-direction occurs. On the other hand, the c-face of hexagonal prisms was stabilized with the adsorption of citrate ions on the ZnO(001) plane.

Le *et al.* [LE 06] have investigated the growth of ZnO in the $Zn(CH_3COO)_2/NH_4OH$ system. If the growth of ZnO on silicon, sapphire and glass substrates was not observed, heteroepitaxial ZnO NWs/NRs were obtained on GaN. Clearly, the excellent lattice matching condition between ZnO and GaN promotes the epitaxial nucleation of the ZnO NRs on the GaN substrate. Heterogeneous nucleation on a lattice-matched substrate has a lower free activation energy barrier than homogeneous nucleation. The wire shape, density and AR were varied with the growth temperature. Moreover, increasing the NH_4^+/Zn^{2+} molar ratio (and/or decreasing the supersaturation) increased the length and ARs and decreased the density of the NW arrays. Lupan *et al.* [LUP 12a, LUP 12b] successfully grew ZnO NR arrays on p-type Mg-doped GaN (0001) substrates by heteronucleation in the $Zn(SO_4).7H_2O/NH_4OH$ system at 95°C (Figure 13.11). The NRs were in-plan aligned with the same direction as GaN and perfectly vertically oriented. A heterojunction was formed between n-ZnO and p-GaN and UV light emission was observed when forward bias was applied. Moreover, by adding $CdCl_2$ or $CuCl_2$ at low concentration (several µM), ZnO was doped. The oxide direct bandgap of Cd:ZnO and Cu:ZnO was reduced depending on the concentration of dopant in the initial bath.

ZnO NWs grown on textured Au(111) layers are perfectly vertically up-right oriented. This can be assigned to structural relationship between the two compounds [XU 11a]. The hexagonal Zn plane of ZnO is matched to the close-packed Au plane of Au(111). The in-plane epitaxial relationship is (1x1)Au(111) $[\bar{1}10]$//(1x1)ZnO(0001)$[11\bar{2}0]$, and the mismatch is 12.7% (Figure 13.12).

Figure 13.11. *SEM images: a) side-view and b) top-view of epitaxial ZnO nanorods hydrothermally grown on p-GaN substrate at 98°C for 15 min (copyright LUP 12b)*

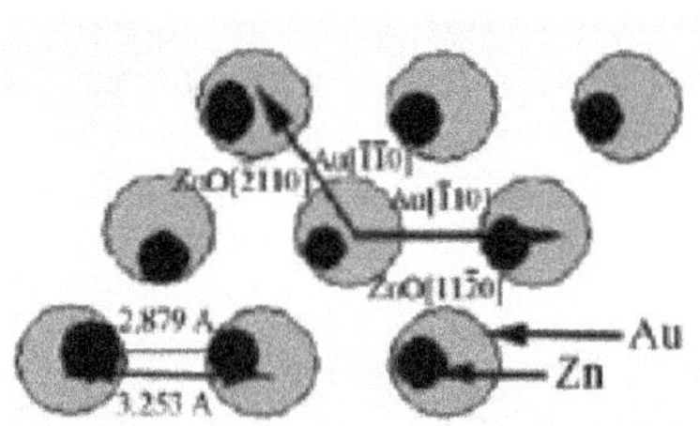

Figure 13.12. *ZnO(0001)/Au(111) in-plane structural relationship. The Au substrate atoms are light and the Zn atoms of ZnO are dark. The (1 x 1)Au-(111)[$\bar{1}$10]//(1 x 1)ZnO(0001) [11$\bar{2}$0] coincidence lattice has a mismatch of 12.7% (copyright [XU 11a])*

The epitaxial growth was not restricted to the direction perpendicular to the substrate surface. By combining the saturated solution of $Zn(NO_3)_2$ and NaOH and the degenerate p-type Si(001), a surprising tilted epitaxial deposition of ZnO nanospears was reported by Switzer *et al.* [MU 09] (Figure 13.13). The authors suggest that there may be a pH decrease at the Si/solution interface due to Si dissolution, which would lower the solubility of Zn(II) in the vicinity of the substrate and favor heteronucleation and growth. On the basis of the data of X-ray pole figure analysis, ZnO nanospears 51°-tilted relative to the surface normal were produced. The tilted nanostructure brought the (20$\bar{2}$3) planes of ZnO into in-plane coincidence with the (001) planes of Si and allowed the epitaxial growth.

13.2.3.2. *Increasing the NW length and layer roughness*

For many applications, it is required to increase the length and the AR of ZnO NWs densely arrayed on supporting substrates. We have seen in section 13.2.2.2 that the PEI molecule with its high-positive charge density adsorbs strongly on the lateral planes of ZnO due to electrostatic attraction [LAW 05]. Thus, the lateral growth of the NWs is largely hindered, and wires with high AR are produced [ZHO 08]. Moreover, due to its complexing action toward Zn^{2+}, this compound decreases the supersaturation of the growth solution, reduces the homogeneous nucleation and increases the wire length (Figure 13.14).

Figure 13.13. *SEM images of as-deposited ZnO nanospears on Si(001) substrate at a) low- (bar marker is 500 nm) and b) high-magnification (bar marker is 500 nm) (copyright [MU 09])*

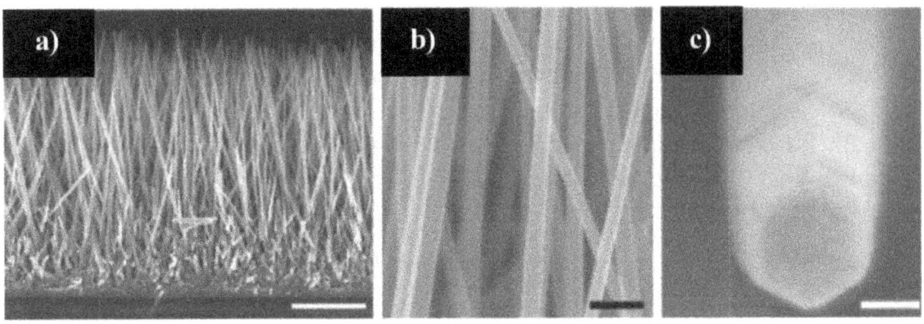

Figure 13.14. *a) Array of ZnO nanowires with high roughness, high AR and high density prepared on seeded FTO substrate in the the $Zn(NO_3)_2$/HMTA system by refreshing periodically the growth bath. The scale bar is 5 μm. b) Magnified view of the oriented wires (scale bar 500 nm). c) Magnified view of a wire tip (scale bar 50 nm). (copyright [LAW 05])*

At rather high supersaturation, most of the zinc precursor added in the bath precipitates in the bulk solution due to the homogeneous nucleation. Therefore, the length of the grown ZnO NWs/NRs is limited by the fast exhaustion of the zinc precursor. Xu *et al.* [XU 10] have described a strategy based on the $Zn(NO_3)_2$/HMTA system to prevent the formation of ZnO particles in the bulk solution and to allow the rapid growth of ZnO NWs on seeded surfaces. They have introduced ammonium, which forms complexes with the zinc ions, $Zn^{2+} + nNH_3 \leftrightarrow Zn(NH_3)_n^{2+}$, where n = 1, 2, 3 or 4. The complexes also serve as a buffer for Zn^{2+}; they continuously supply Zn^{2+} while lowering the degree of supersaturation of the reaction system. The use of ammonium hydroxide substantially suppresses ZnO homogeneous nucleation. However, by using ammonium hydroxide alone to effectively prevent formation of ZnO in the liquid phase, a high concentration of ammonium hydroxide is required, which also results in very slow growth of the wires on the seeded substrate due to excessively low degree of superaturation. To resolve this problem, PEI was added to further suppress the homogeneous nucleation process. The authors found that, in the presence of both PEI and ammonium hydroxide, the formation of ZnO in the bulk solution was effectively prevented while ZnO wires still grew on the seeded substrate at a reasonably high growth rate. A plausible explanation is that PEI preferentially adsorbs to certain crystal faces of ZnO clusters and inhibits further crystal growth along these faces. In the homogeneous nucleation case, the ZnO clusters initially formed due to thermal fluctuation are so small that the polymer chain of PEI adsorbed to certain crystal faces is long enough to cover other faces and, as a result, the steric hindrance of the PEI chain inhibited the crystal growth along all faces. Consequently, the clusters have a very small chance to grow to the critical size that thermodynamically favors the growth of crystals. On the seeded substrates, however, the preexisting seeds are large enough (5–20 nm) that despite PEI adsorption to certain crystal faces, the growing faces remain exposed to the reactants. Therefore, growth of ZnO NWs can normally occur on the seeded substrates. The significance of this discovery is that ZnO NWs grow on seeded substrates without precipitation in the bulk solution at a reasonable degree of supersaturation through coupled use of ammonium hydroxyl and PEI. Figure 13.15 compares the length growth speed of NWs for the classical technique that requires refreshment of the solution every 2.5 h described in [LAW 05] and that with the improved growth technique [XU 10].

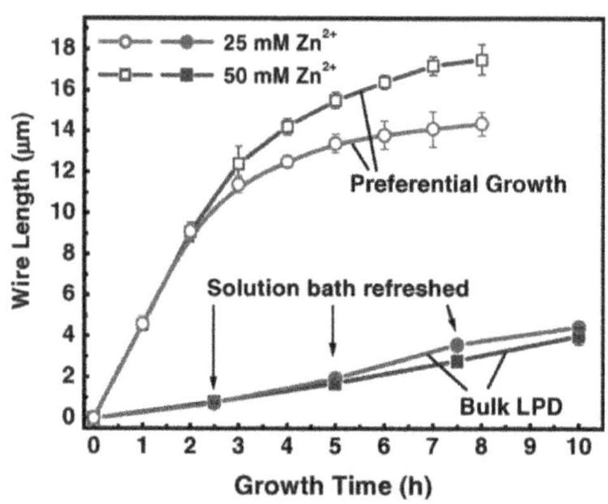

Figure 13.15. *Comparison of wire growth by using the preferential growth process developed in study [XU 10] and the bulk liquid-phase deposition method reported in the literature. The wire length is plotted as a function of growth time at different growth conditions (copyright [XU 10]). For a color version of this figure, see www.iste.co.uk/consonni/nanowires1.zip*

The same group [XU 11b] has also developed an innovating multistep technique to grow vertically aligned wire arrays with lengths up to 40–50 µm and an internal roughness of more than 500. Their multilayer assembly of NW array process is presented in Figure 13.16(a). A first-layer of ZnO NW array is grown on seeded TCO substrate (step a). Then the array is coated by a self-assembled monolayer (SAMs) of octadecyltrichlorosilane, $CH_3(CH_2)_{17}SiCl_3$ (OTS) (step b). The SAM coating at the top end of the wires is then selectively removed by exposure to UV light (step c). This treatment leaves the SAMs on the side walls of the wires. Finally, a new ZnO NW layer is grown using the top ends of the first-layer wires as seeds (step d). The aqueous solution comes into contact with only the top ends of the first-layer wires and cannot enter the gaps between the wires because of the SAMs on their side walls. Several layers can be grown as shown in Figure 13.16(b). Finally, the SAMs from the side walls of the wires are removed by calcination.

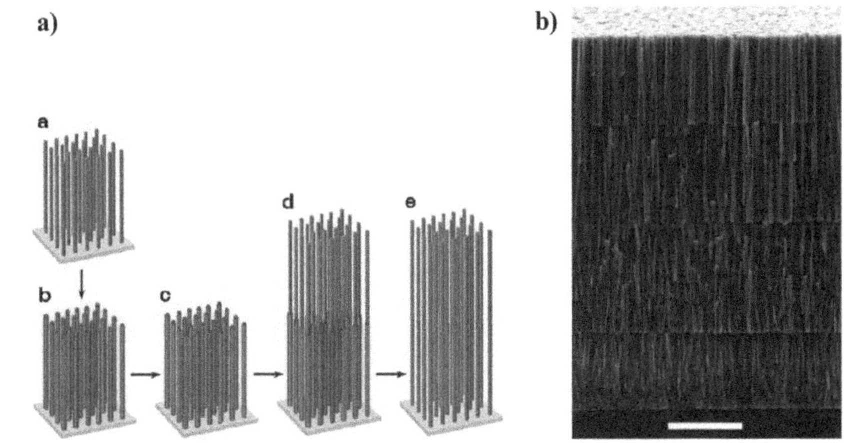

Figure 13.16. *a) Schematic process for synthesizing a two-layer assembly of ZnO nanowire arrays on TCO. The pink color represents the OTS covered ZnO NWs. b) SEM image of a four-layer assembly of ZnO nanowire arrays. The thickness of each layer was ~10 µm. Scale bars 10 µm (copyright [XU 11b]). For a color version of this figure, see www.iste.co.uk/consonni/nanowires1.zip*

Chen *et al.* [CHE 13] recently described the preparation of a ZnO NW array on ITO by a continuous flow injection (CFI) process. Peristaltic pumps were used to carry zinc acetate and HMTA aqueous precursors into the mixer (Figure 13.17(a)). The mixed solution was subsequently added into the reactor maintained at 90°C. The obtained NW arrays are presented in Figure 13.17(b). In the NH_3-assisted CFI process, ammonia was added to lower the supersaturation. The authors have shown that the structural quality was significantly better in the case of the CFI process compared to the classical multibatch process for which, the solution is changed every 4 h and not continuously renewed.

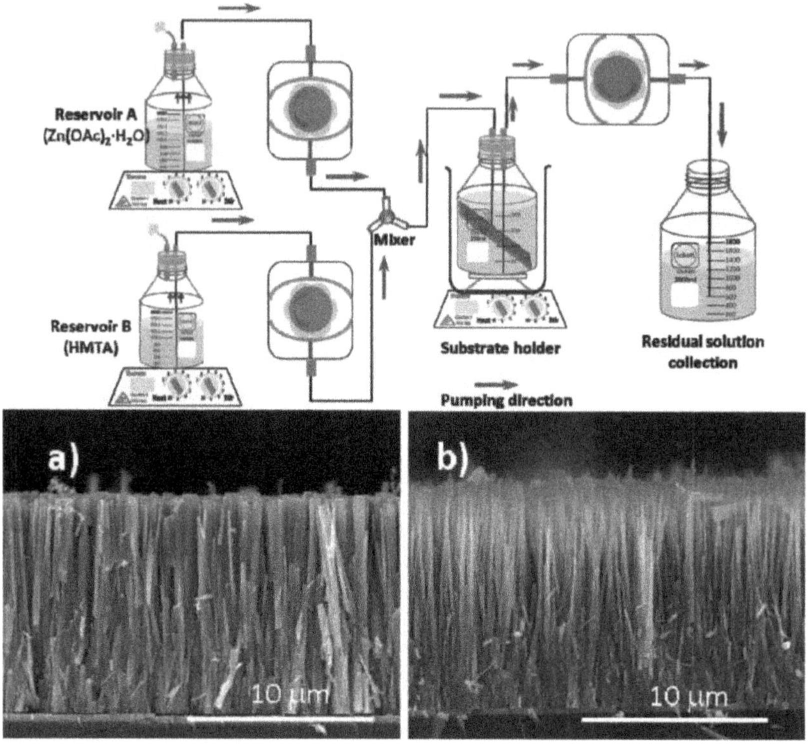

Figure 13.17. *a) Experimental set-up for long ZnO NWs growth. b) SEM images of a) ZnO grown by the CFI process, 24 h) and b) ZnO NWs (NH_3-assisted CFI, 24 h) (copyright [CHE 13]). For a color version of this figure, see www.iste.co.uk/consonni/nanowires1.zip*

13.2.3.3. *Arrays of microtubes*

Arrays of ZnO nanotubes have been prepared in the zinc nitrate/HMTA system using very long growth times [VAY 01b]. The formation mechanism is presented in Figure 13.18(a). On the first day, an array of ZnO NRs is grown at 90°C. When the sample is aged a second day, under the same condition, then there is a preferential dissolution of the less stable (0001)-Zn face of the rods. A homogeneous hollow arrayed macro/nanotube layer was produced (Figure 13.18(b)).

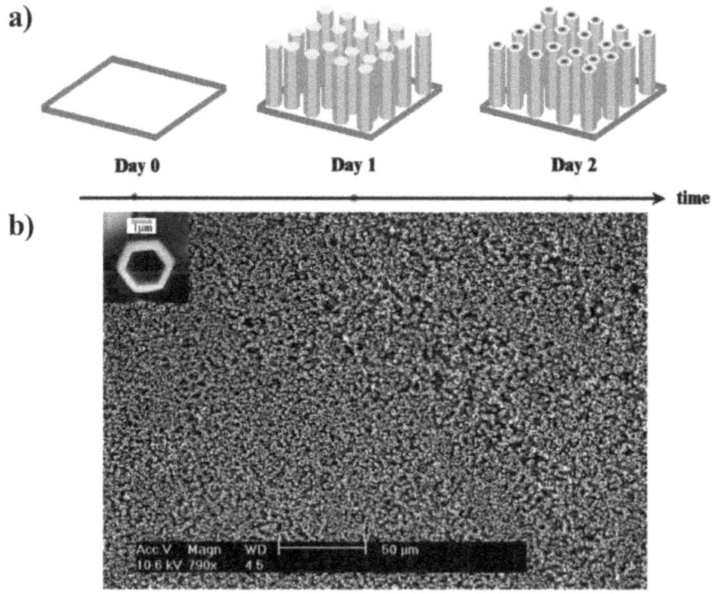

Figure 13.18. *a) Growth process; b) SEM top view of a microtube ZnO array. The inset is a zoom top view of a microtube (copyright [VAY 01b])*

13.2.3.4. Patterning of ZnO nanostructures

Application prospects of ZnO NWs/NRs largely rely on the ability to precisely localize and pattern the nanomaterials. Direct soft patterning and selective growth processes of ZnO have been explored by several groups.[GRE 04, XU 08b, KAN 11, LI 11, KOH 99, MAI 99, SAI 02, SAI 04, HSU 05, MAS 06, KIM 12]. To date, the techniques to fabricate the desired patterns of well-aligned ZnO nanostructures include photolithography [GRE 04], electron beam lithography [XU 08b], metal oxide-seed-mediated hydrothermal synthesis [KAN 11] and the patterned growth of ZnO nanostructures on catalyst pre-patterned substrates [LI 11]. In most fabrication techniques, the ZnO nanostructure etching process is involved. Overall, dry etching is favored because the etch rate is easily controlled in comparison with the wet etching process. However, the wet etching process is advantageous in that it is a simple and cost-effective way to get large-scale production of patterned ZnO nanostructures. It is well-established that ZnO is easily etched in all kinds of acid and alkali solutions. For example, Kohls *et al.* [KOH 99] and Mais *et al.* [MAI 99] reported the successful patterning techniques for planar optical amplification in which they used the mixture of HCl and H_3PO_4 as an etchant for ZnO film at an etch rate of 170 nm/s. Another approach consisted of self-assembling a monolayer (SAMs) of phenylsilane on silicon substrate, which was photopatterned [SAI 02, SAI 04]. Pd catalyst particles were selectively deposited on the surface phenyl regions of SAMs by immersing the substrate in a NaCl-based Pd/Sn colloidal solution. ZnO was subsequently deposited on the Pd catalyst particles by electroless ZnO growth using a solution containing $Zn(NO_3)_2$ and dimethylamine-borane. Cathodoluminescent images of visible light were successfully demonstrated for 1 μm wide lines. In more recent works, the use of Pd catalyst was avoided. In [HSU 05], patterned SAMs of 11-mercaptoundecanoic acid ($HSC_{10}H_{20}COOH$) were deposited on silver surfaces using polydimethalsiloxane stamps. Then ZnO was grown at 50–60°C by the HMTA decomposition technique. ZnO NR formation was observed on uncovered Ag substrate, whereas the COOH-terminated regions covered by the SAMS remained free of ZnO. The deposition of ZnO was inhibited by the formation of a COO^{-}-HMTA-H^{+} complex.

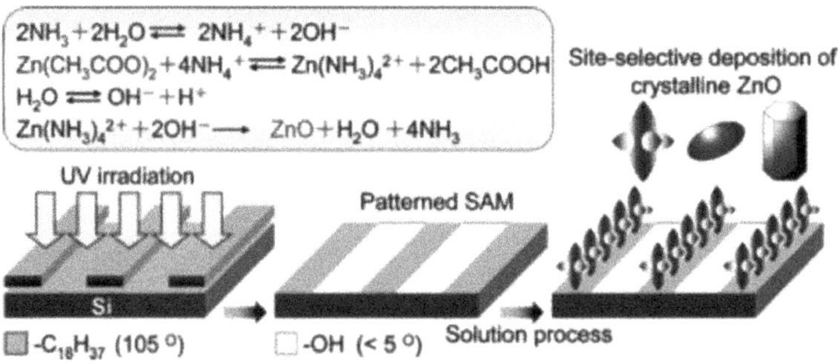

Figure 13.19. *Process for self-assembly patterning of light-emitting crystalline ZnO nanoparticles in aqueous solution (copyright [GRE 03]). For a color version of this figure, see www.iste.co.uk/consonni/nanowires1.zip*

Masuda *et al.* [MAS 06] prepared micropatterned arrays of ZnO nanocrystals by using site-selective deposition of ZnO on self-assembled silane monolayers. The various silanes with hydrophobic chains were deposited on silicon wafers. Then, the SAMs were patterned by illumination with UV light through a mask. Illuminated silanes were decomposed, and patterned SAMs on the silicon surface were obtained (Figure 13.19). The degradation of silane SAMs could be followed by contact angle measurements with a value decreasing from 105° to 5°. ZnO crystals were subsequently chemically deposited from alkaline solution in the presence of ammonia complexes on the hydrophobic parts of the substrate covered by a silane monolayer. The shape of the crystal could be changed (hexagonal rods, ellipses and multineedles) according to the ammonia/zinc acetate concentration ratio in the growth solution (Figure 13.19 inset) [MAS 06].

Recently, Kim *et al.* [KIM 12] have reported a simple protocol for the growth of patterned ZnO NWs by etching of ZnO SL in the tetramethyl ammonium hydroxyl

(TMAH) solution. First, a ZnO SL was fabricated by the sol–gel method using zinc acetate solution and patterned by using a photolithographic method. The patterned ZnO SL was etched in the TMAH solution. In the second step, ZnO NW-patterned arrays were hydrothermally grown using an equimolar $Zn(NO_3)_2$/HMTA growth bath. The remarkable point of this ZnO SL patterning was that the development of UV-exposed photoresist and etching of ZnO SL were processed in aqueous TMAH solution without interruption. ZnO NWs exhibited high purity and degree of crystallinity. Moreover, they showed a very good pattern fidelity (Figure 13.20).

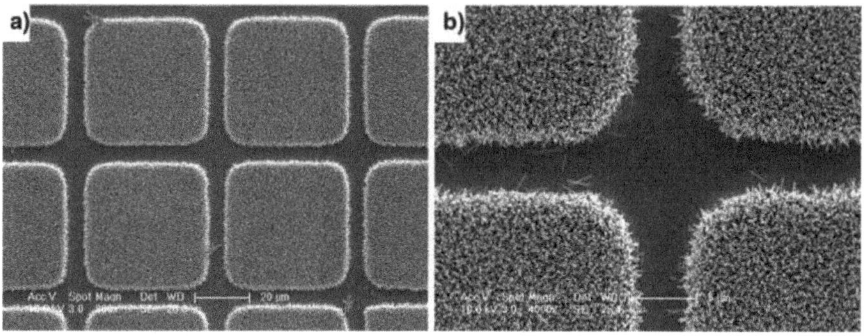

Figure 13.20. *FESEM images of a) square pattern arrays of the ZnO nanowires (50 $\times$50 μm area, 5 μm spacing) and b) a magnified view of the pattern (copyright [KIM 12])*

13.3. Preparation of ZnO nanorods and nanowires by electrodeposition

13.3.1. *Principles of ZnO electrodeposition*

Electrodeposition is a wet synthetic technique that requires the exchange of electrical charges (electron) to induce the deposition of the products on a substrate. The substrate is the electrode, and therefore it must be conductive. However, electrodeposition presents additional advantages compared to the chemical methods: the deposited layer has a good electrical continuity with the conductive substrate and the amount of deposited material is directly controlled during the growth by the measure of the electrical charge exchanged during the process.

The mechanism of ZnO film electrodeposition is based on the monitoring of the interfacial pH of an electrode by an electrochemical reaction, which produces hydroxide ions (OH^-). The flux of OH^- is controlled by the current density. Zinc ions present in solution react with hydroxides and zinc oxide precipitates. Comparison of ZnO and $Zn(OH)_2$ solubility curves calculated between 25 and 90°C reveals that

ZnO is the thermodynamically most stable phase [GOU 05]. Figure 13.21(a) compares the solubility curves of ZnO and $Zn(OH)_2$ calculated at 70°C in the presence of 0.1 M Cl^-. S' is defined as the total Zn(II) species concentration in solution. The solubility curves also show that supersaturation can only be attained if the initial pH of the solution is not too low. Experimentally, it has been observed within the range of 5 and 6.5. In solution, Zn^{2+} ions can be complexed by OH- and Cl^- ions and various Zn(II) species are found to depend on the species concentration and the solution pH. From the speciation diagrams, it can be concluded that the main reacting species at the ZnO deposition pH are Zn^{2+}, $Zn(OH)^+$ and $ZnCl^+$ in chloride solutions (Figure 13.21(b)). The chloride ion is classically used as a supporting anion in ZnO electrodeposition experiments [GOU 05, PEU 96, PEU 98, PAU 00, PAU 01a, GOU 06a]

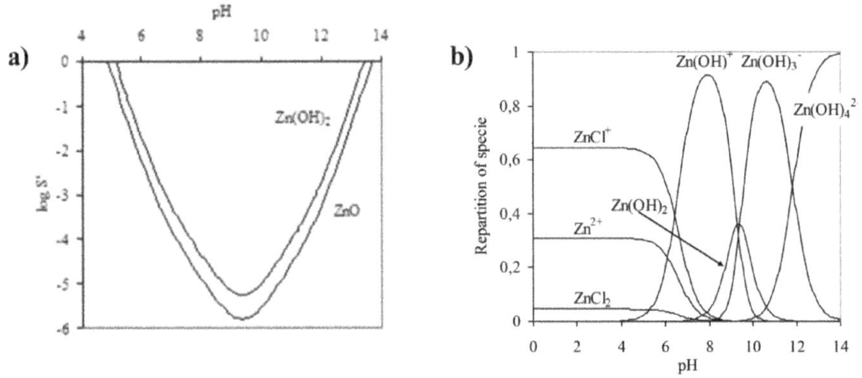

Figure 13.21. *a) Solubility curves of ZnO and $Zn(OH)_2$ at 70°C in 0.1 M Cl^-. b) Speciation diagram of Zn(II) in 0.1 M Cl^- (copyright [GOU 05])*

13.3.2. *ZnO growth mechanism*

Various precursors for generating hydroxide ions have been described in literature, in which the most significant are: O_2, H_2O_2 and NO_3^-.

13.3.2.1. *Molecular oxygen precursor*

The reduction of molecular oxygen for the preparation of ZnO was first proposed by Lincot *et al.* [PEU 96, PEU 98]. The deposition reaction mechanism is:

$$O_2 + 2H_2O + 4e^- \rightarrow 4OH^- \qquad [13.11]$$

$$Zn^{2+} + 2OH^- \rightarrow ZnO + H_2O \qquad [13.12]$$

The overall deposition reaction is:

$$Zn^{2+} + \tfrac{1}{2}O_2 \rightarrow ZnO \qquad [13.13]$$

ZnO is directly precipitated for growth temperatures ranging between 40 and 95°C [GOU 05]. The mechanism was confirmed by in-situ pH measurements with the help of a minielectrode [PEU 98]. In the absence of Zn^{2+} in solution, the pH in the vicinity of the electrode increased from 6 to about 10 due to reaction in [13.11] In the presence of zinc ions, hydroxide ions produced were consumed by the oxide precipitation reaction, and the interfacial pH was fixed at a value close to that calculated from the solubility value (6.7). In chloride medium, for a Zn(II) concentration larger than 10 mM, the stable precipitate compound was no longer ZnO but a zinc hydroxychloride [PEU 98].

The deposition rate is limited by the low solubility of molecular oxygen in aqueous solution (~0.8 mM at 70°C [GOU 05]). Pauporté *et al.* [GOU 06a] have published a kinetic study of oxygen reduction on zinc oxide electrode in conditions similar to those which were classically used in the electrodeposition process. The reduction kinetic was measured on rotating disk electrodes. Their rotation speed was varied in order to get the kinetic constants of the system by using the Koutecky-Levich approach. In the absence of Zn^{2+} in solution, the Tafel slope was 139 $mV.dec^{-1}$ for a ZnO electrode, a value close to that measured on a bare platinum electrode (133 mV dec^{-1}). It was ascribed to the limitation of the oxygen reduction reaction rate by the first electron transfer with the formation of an adsorbed intermediate. The main difference between the noble metal and the oxide electrode was a shift of the curves toward more negative potentials. In the presence of Zn^{2+} ions, the current density decreased markedly, and the Tafel slope was determined to be 282 mV dec^{-1}. The electrode was partially blocked by a reaction intermediate upon the formation of zinc oxide.

To increase the crystallographic quality of ZnO film, electrodeposition was carried out in non-aqueous dimethylsulfoxide (DMSO) solution at 150°C by Gal *et al.* [GAL 00]. The films were highly transparent and crystallographically oriented with the *c*-axis perpendicular to the substrate. The oxygen concentration in the electrolyte was varied by mixing oxygen with dry nitrogen and controlling the ratio between them. As oxygen concentration in the electrolyte was decreased, the resistivity of the ZnO layer also decreased.

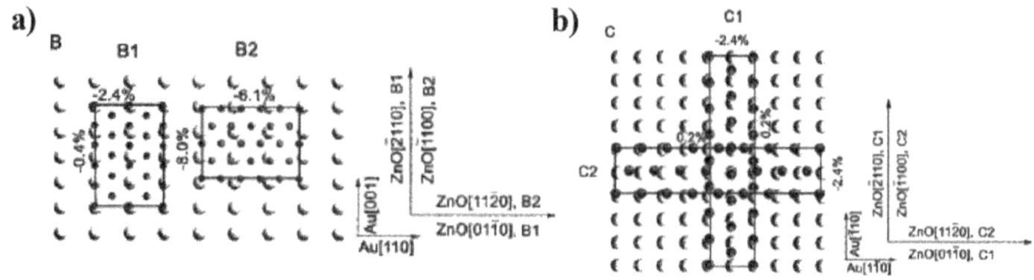

Figure 13.22. *Epitaxial relationships for growth of ZnO on a) Au(110), and b) Au(100). The Au substrate atoms are light, and the Zn atoms of ZnO are dark. The coincidence lattice (4 x 2)- Au(110)[001]//(5 x 2)ZnO(0001)[$\bar{2}$ 110] in part B1 has a mismatch of -0.4% along the Au[001] and ZnO[$\bar{2}$ 110] directions and a mismatch of -2.4% along the Au[110] and ZnO[01 $\bar{1}$ 0] directions. The coincidence lattice (3 x 3)Au(110)[110]// (2 x 5)ZnO(0001)[11 $\bar{2}$ 0] in part B2 has a mismatch of -6.1% along the Au[110] and ZnO [11 $\bar{2}$ 0] directions and a mismatch of -8.0% along the Au[001] and ZnO[$\bar{1}$ 100] directions. The coincidence lattice (9 x 2)Au(100)[-110]//(8 x 1)ZnO(0001)[$\bar{2}$ 110] in part C1 has a mismatch of 0.2% along the Au[$\bar{1}$ 10] and ZnO- [$\bar{2}$ 110] directions and -2.4% along the Au[1 $\bar{1}$ 0] and ZnO[01 $\bar{1}$ 0] directions. The coincidence lattice (2 x 9)Au(100)[1 $\bar{1}$ 0]// (1 x 8)ZnO(0001)[11 $\bar{2}$ 0] in part C2 has a mismatch of 0.2% along the Au[1 $\bar{1}$ 0] and ZnO[11 $\bar{2}$ 0] directions and a mismatch of –2.4% along the Au[$\bar{1}$ 10] and ZnO[$\bar{1}$ 100] directions (copyright [LIU 01])*

Pauporté *et al.* have prepared electrochemically high-quality ZnO by epitaxial growth on either n-type (0001)GaN or p-type (0001)GaN single crystal electrodes [PAU 99, PAU 00, PAU 06, PAU 02, PAU 03]. GaN and ZnO both have the hexagonal crystal structure with similar crystallographic parameters. The in-plane crystallographic relationship was shown to be ZnO [10 $\bar{1}$ 0]// GaN [10 $\bar{1}$ 0]. Epitaxial ZnO on n-type GaN was found to be of high quality attested by the 0.07° full width at half maximum (FWHM) of the (0002) $\theta/2\theta$ peak and 0.74° FWHM of the (10 $\bar{1}$ 1) plane in the Ψ-scan diagram. Epitaxial dense and smooth ZnO single crystalline films [PAU 99] as well as ZnO nanopillars [PAU 02] or NWs [PAU 06] could be grown on GaN single crystals. The epitaxial growth of ZnO NWs has also been achieved on p-type (0001)GaN substrate. The in-plane structural relationship was measured by glancing-angle XRD measurements. The GaN FWHM was 0.42°, and the epitaxial ZnO FWHM was 1.2°. PL measurements have shown that the epitaxial relationship markedly improves the structural quality of electrodeposited ZnO and that the resulting intrinsic defect concentration in ZnO is very low [PAU 00, LUP 10a].

ZnO nanopillars have been epitaxially electrodeposited onto Au(111), Au(110) and Au(100) single crystal substrates by Switzer and coworkers [LIU 01]. The nanopillars grew with the same [0001] out of plane orientation on all three substrates, but ZnO in-plane orientation varied with crystallographic orientation of the gold surface. The structural relationships are presented in Figures 13.12 and 13.22. The bandgap of ZnO deposited from an oxygen precursor in a chloride medium was measured at room temperature to be between 3.25 eV and 3.55 eV, depending on the deposition growth conditions, on the substrate and on the post-deposition treatment [PAU 99, PAU 00, PAU 09a, LUP 10]. A significant energy shift was observed after annealing in air at 500°C with a bandgap of 3.27 eV [PAU 00]. The PL of as-electrodeposited layers was measured at 1.8 K [PAU 99, PAU 00, PAU 09a]. A broad excitonic emission was observed at 3.4 eV for an as-deposited film. The PL intensity was stronger for layers grown in the perchlorate medium compared to chloride medium, and it was concluded that perchlorate anion yields material of higher optical quality [PAU 00]. After film annealing at 500°C for 1 h, the emission became sharper and was centered at 3.37 eV. The PL of ZnO films electrodeposited using molecular oxygen precursors has been studied in detail by Pauporté *et al.* [PAU 09a]. UV emission of electrodeposited ZnO at room temperature is strongly influenced by the growth deposition parameters such as bath temperature, the type of anion, the deposition potential and the oxygen and zinc ion concentrations. They have shown that working at a relatively high bath temperature and at low overvoltage dramatically increases the material's structural quality, since the UV excitonic emission is markedly enhanced, whereas the visible one, due to deep defects, becomes negligible. An enhancement was also observed by working at low oxygen concentrations and by replacing chloride ions by perchlorate. The UV emission markedly shifts toward a higher energy when a chloride medium is used. This was assigned to a high n-type doping of ZnO giving rise to a Burstein–Moss effect.

13.3.2.2. *Hydrogen peroxide precursor*

Pauporté *et al.* have studied the use of aqueous hydrogen peroxide solution as hydroxide precursor [PAU 01a, PAU 01b]. In the presence of zinc ions, the following two half-reactions occur:

$$H_2O_2 + 2e^- \rightarrow 2OH^- \qquad [13.14]$$

$$Zn^{2+} + 2OH^- \rightarrow ZnO + H_2O \qquad [13.15]$$

The resulting overall deposition process is then:

$$Zn^{2+} + H_2O_2 + 2e^- \rightarrow ZnO + H_2O \qquad [13.16]$$

The reduction of hydrogen peroxide on the working electrode leads to the production of hydroxide ions, and thus an increase in the local pH. When supersaturation is reached, ZnO precipitates to form a layer at the working electrode surface. Compared to O_2, hydrogen peroxide is highly soluble in water, and its use avoids the problem of gas handling. Moreover, H_2O_2 reduction, like O_2 reduction, is a "clean" reaction, since the only reduction products are hydroxide ions, which are consumed for zinc oxide formation. No undesirable by-products that may pollute the deposition bath are formed. Films of ZnO were deposited at 70 °C and –1 V *vs.* saturated calomel electrode, noted SCE (at standard conditions, the potential of the SCE is +0.244 V *vs.* the standard hydrogen electrode (SHE)), at various H_2O_2 concentrations in the presence of Zn^{2+} in excess and using perchlorate as a supporting anion to avoid the formation of basic zinc salts at large Zn^{2+} concentrations [PAU 01b]. As observed with the molecular oxygen precursor, the films were highly textured and grew along the [0002] direction. Two growth regimes were observed: (1) up to 25 mM H_2O_2, the films were dense, and the growth rate was proportional to H_2O_2 concentration in the deposition bath; (2) at higher H_2O_2 concentration, the growth rate markedly increased, and highly transparent porous ZnO was formed.

The deposition mechanism was studied in a more detailed manner in chloride mediums [PAU 01a 89]. A parametric study was carried out by varying the substrates and peroxide concentrations at constant zinc content. The film growth was under kinetic control, since the catalytic activity of ZnO surfaces for the reduction of hydrogen peroxide species was low under these conditions. The nature of the substrate (tin oxide and gold) and its treatment prior to the deposition experiments had a clear influence on the electrochemical behavior of the system. It was possible to normalize the film growth behavior by taking the mean current density as a growth variable. The dependency of film growth rate on this parameter is shown in Figure 13.23(a). For values below the limiting diffusion current of zinc ions ~0.65 $mA.cm^{-2}$, the faradaic efficiency (defined as the fraction of charge consumed for the electrochemical formation of the film) was close to 1, whereas above this value the faradaic efficiency decreased markedly. A competition mechanism was suggested in which hydroxide ions, produced in excess at the surface, diffuse toward the solution and react partly with zinc ions diffusing toward the electrode surface (Figure 13.23(b)). The reaction in solution reduces the availability of zinc ions for the heterogeneous deposition reaction at the electrode and then decreases the deposition rate. Surprisingly, a rate plateau was found at high current density (Figure 13.23(a)), which meant that a constant flux of Zn^{2+} could escape the precipitation reaction in solution and reach the surface, or that a film was formed by the aggregation of ZnO particles formed by homogeneous precipitation.

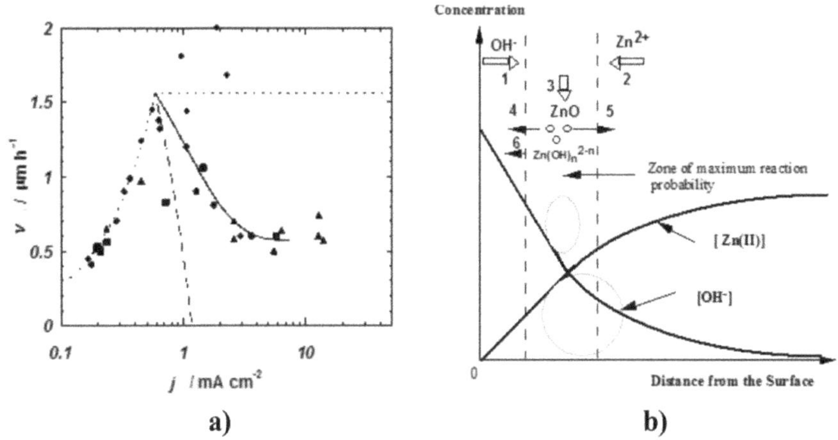

Figure 13.23. *a) Variation of the growth rate of ZnO films formed from hydrogen peroxide precursor as a function of the mean current density. The dots are experimental data, and the lines are fit curves with different models presented in Ref. [89]. b) General scheme of the reaction steps occurring in bulk solution. Step 1 is the diffusion of electrogenerated OH^- from the electrode, step 2 is the diffusion of dissolved zinc, step 3 is the formation of ZnO particles in solution, step 4 is the diffusion toward the electrode surface and step 5 is toward bulk solution. Step 6 displays a potential change in the dissolved form of zinc in the vicinity of the electrode with the pH change (copyright [PAU 01a])*

13.3.2.3. *Nitrate ion precursor*

Nitrate ions can be employed as the hydroxide precursor for ZnO electrodeposition [IZA 96a, IZA 96b, IZA 97, YOS 04, ISH 01, ISH 02, ZHE 02, IZA 03a, IZA 03b, WAN 05a, ZHA 05, CAO 07, CHE 06a, CHE 06b]. Upon the reduction of nitrate in the presence of zinc ions, the following two half-reactions occur:

$$NO_3^- + H_2O + 2e^- \rightarrow NO_2^- + 2OH^- \qquad [13.17]$$

$$Zn^{2+} + 2OH^- \rightarrow ZnO + H_2O \qquad [13.18]$$

Thus, the overall deposition process is:

$$Zn^{2+} + NO_3^- + 2e^- \rightarrow ZnO + NO_2^- \qquad [13.19]$$

Employing the nitrate ion is very convenient and simple, since it does not require, for instance, gas handling. However, the reduction reaction [13.17] produces NO_2^- by-product, which progressively accumulates in the deposition bath. Moreover, it has been observed that the deposition solution cannot be stored because of its quite rapid decomposition. ZnO thin films are classically prepared at 60–70°C and are well-crystallized with the hexagonal wurtzite structure. Electrodeposition potential has a strong effect on film crystallographic orientation [IZA 96b]. At low overvoltage, –0.7 V *vs.* Ag/AgCl (at standard conditions, the potential of the Ag/AgCl reference is +0.235 V vs. the SHE), the film surface presents the hexagonal facets of the ZnO(0002) plane, and the XRD pattern is dominated by the (0002) plane reflection. However, at lower applied potentials the film crystallographic orientation changes, and the *c*-axis is oriented parallel to the substrate. Crystallographic orientation changes were also observed with zinc nitrate concentrations [IZA 96a].

Reaction [13.17] causes an increase in the local pH at the cathode and is, to some extent, self-limiting. In the reaction mechanism, we can clearly see that reaction [13.18] consumes the OH^- reaction product and decreases the local pH. Its presence allows reaction [13.17] to proceed at a higher rate. The mechanism of ZnO film growth with a nitrate hydroxide precursor was studied by Yoshida *et al.* [YOS 04]. It was shown that the current density, and thus the film growth rate, was promoted by Zn^{2+} concentrations. The improvement of the charge transfer kinetics was interpreted as a "catalytic" role of Zn^{2+} adsorbed on the film surface to the reduction of nitrate ions. Zn^{2+} adsorption follows a Langmuir isotherm:

$$\theta = \frac{KC_{Zn^{2+}}}{1 + KC_{Zn^{2+}}} \qquad [13.20]$$

where θ stands for zinc ion coverage. K is an equilibrium constant, and $C_{Zn^{2+}}$ is the concentration of zinc ions in solution. However, the adsorbed Zn^{2+} ions cannot be catalysts in a true sense because they are spontaneously consumed to generate the ZnO film. Another specificity of nitrate precursor is that ZnO can be deposited at a potential as low as –1.5 V *vs.* SCE whereas, with the other precursors, metallic zinc is then formed below –1.2 V versus SCE in agreement with classical Pourbaix (potential–pH) diagrams [GOU 05]. Izaki and coworkers [ISH 01, ISH 02] showed that the conductivity of ZnO films deposited from zinc nitrate solution could be controlled by adjusting the concentration of dimethylamine borane (DMAB) added to the bath. Boron is included in the film and acts as an electron donor. Boron film content increased with increasing DMAB bath concentration, and the film conductivity was significantly improved.

The as-deposited ZnO films presented good structural and optical properties. Highly (0002) textured layers deposited on gold film substrates were luminescent at room temperature [IZA 03a]. However, the PL signal was observed within a narrow deposition potential range. Films deposited at –0.6 V versus Ag/AgCl had two emission peaks of similar intensity located at 3.30 eV and 2.28 eV. In the film deposited at –0.65 V versus Ag/AgCl, the defect emission was shifted to 2.8 eV. It is probable that the film quality was significantly improved by an epitaxial relationship between the gold layer and ZnO (see section 13.3.2.1.) [LIU 01]. PL measurements at 13 K confirmed that the films prepared at –0.6 V versus Ag/AgCl were of better quality than those prepared at lower potentials [LIU 01].

Low temperature electrodeposition of ZnO was reported with a nitrate precursor. Zhang *et al.* [ZHA 05] worked at a bath temperature of 0°C and a potential of –1.30 V versus SCE. The ZnO films were highly transparent and presented a bandgap measured at 3.37 eV. They were made of nanocrystallites with grain sizes between 10 and 15 nm and a preferred growth orientation along the *c*-axis.

13.3.3. *Electrochemical growth of ZnO nanorod and nanowire arrayed layers*

The deposition of ZnO NRs and NWs by electrochemical techniques has many advantages compared to other techniques, which include: (1) well-suited for scaling-up; (2) low temperature operation; (3) each NR/NW is a single crystal of high quality even without annealing; (4) homogeneous deposition on arbitrary substrate shapes, (5) tuning of morphology and size by manipulating the deposition parameters; (6) precise control of deposition positions by selective patterning of the substrate; (7) excellent electrical contact between structures and substrate and (8) minimum inter-reaction or inter-diffusion between deposit and substrate due to the low temperature processes.

Figure 13.24. *SEM views illustrating the effect of zinc concentration on film morphology (copyright [ELB 08]): a) dense ZnO film electrodeposited on activated conductive FTO glass surface from a solution containing 5 mM $ZnCl_2$ and molecular oxygen; b) ZnO nanorod array electrodeposited on the same surface from a solution containing only 0.2 mM $ZnCl_2$; c) ZnO multipod structures grown electrochemically on non-etched GaN (0001) single crystal*

ZnO 1D nanostructures elongated along the c-direction are typically found for growth in aqueous solutions. This shape is greatly favored by the polar structure of ZnO with a positively Zn^{2+}-terminated (0001) and a negatively O^{2-}-terminated

$(000\bar{1})$ polar surface, which induces a net dipole moment along the *c*-axis. Thus, the surface energy of the polar {0001} plane is higher than those of nonpolar $\{01\bar{1}0\}$ and $\{2\bar{1}\bar{1}0\}$ planes. Therefore, preferential growth along the *c*-axis is energetically favorable, and its growth is faster [CAO 07].

The concentration of Zn(II) precursor has been shown to be the key parameter that governs the deposit morphology [PAU 06, ELB 08]. At high Zn(II) concentrations, dense and smooth films are obtained (Figure 13.24(a)). Decreasing the Zn(II) concentration leads to a decrease or blocking of the lateral growth rate of the prismatic crystallites. As a consequence, the AR of ZnO crystallites increases, and arrays of hexagonal NRs are formed (Fig.13.24b). The phenomenon has been described for either oxygen [PAU 99, PAU 06, ELB 08, BAD 07a, ELI 07, PAU 10] or nitrate [CHE 06b, CAO 07, DEB 08, KHA 12, LUD 13] hydroxide ion precursors. The substrate property is also important, since it energetically governs the nucleation step, and NRs have been grown at high zinc concentration on non-activated substrates [PEU 96]. Combining high supersaturation and low density of surface nucleation site leads to the electrodeposition of spherulitic multipods of ZnO as illustrated in Figure 13.24(c) with the example of non-etched GaN(0001) single crystal substrate.

The NRs/NWs electrodeposited from oxygen precursor are elongated along their *c*-axis and are single crystals [LUP 10a] (Figure 13.25(a)). In the low-$ZnCl_2/KCl/O_2$ system, the lateral growth of the NWs is rapidly blocked due to the zinc ion depletion and pH increase (Figure 13.25(b)), whereas the crystallites continue to grow in their *c*-axis direction (Figure 13.25(c)) [ELB 08, PAU 10]. Figure 13.25(d) shows that the AR increases continuously with the deposition time [ELB 08, PAU 10]. ARs higher than 30 were achieved for a growth temperature, T_d, of 90°C. Figure 13.25(d) also illustrates that the AR is much lower for a T_d of 70°C. In [ELI 07], the growth of very long NWs was achieved by exploiting the electrodeposition growth parameters. The authors have shown that, on polycrystalline FTO substrates, the initial first wires are quite disoriented. However, for longer deposition time, only the initially best-vertically oriented wires continue to grow. This is a self-alignment process in which the wires that are more accessible to the Zn(II) precursor pursue their growth. It results in up-right oriented NW arrays.

Arrays of free standing ZnO NRs have also been grown epitaxially on GaN (0001) single crystals. Using a 0.1 mM $ZnCl_2$ solution at 80°C bubbled by a mixture of O_2 and Ar, after 90 min, NWs with an AR value of about 8 were grown epitaxially on n-type GaN (Figure 13.26(a)) [PAU 06]. In Figure 13.26(b), the top view of the layer clearly shows the hexagonal cross-section of the columns and that all the rods present the same in-plane orientation since the facets of the various crystallites are oriented in the same direction (Figure 13.26(b)).

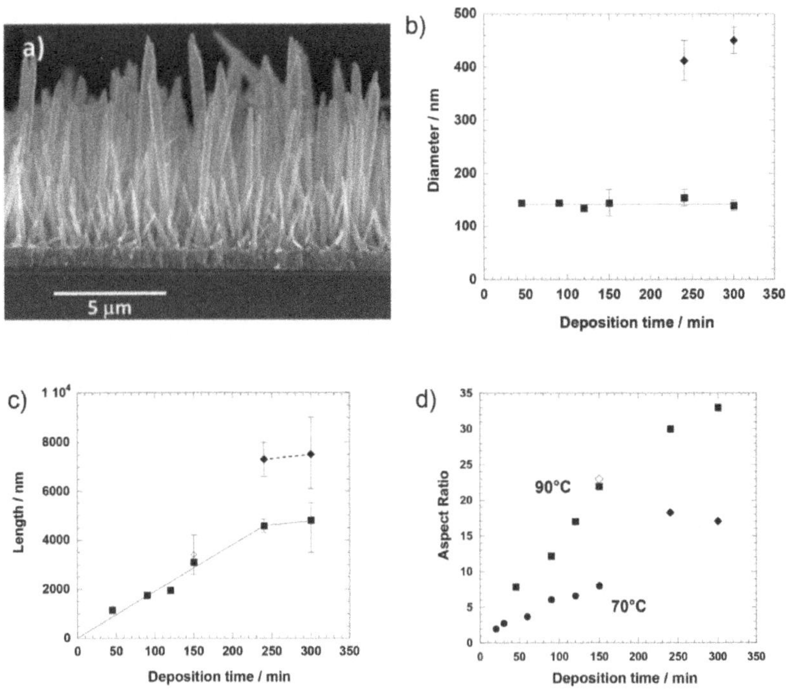

Figure 13.25. *ZnO NWs electrodeposited at 90°C in the $ZnCl_2/KCl/O_2$ with 0.2 mM $ZnCl_2$ on unseeded FTO substrate. a) SEM lateral view for 5h of deposition time; (b–d) effect of deposition time on the mean dimensional characteristics of the NWs. a) Diameter (T_d = 90°C); b) length (T_d = 90°C); c) effect of deposition time and growth temperature on the aspect ratio. After 4 h of deposition time, bigger wires appear, and the layer is less homogeneous. In b–c, the double dots at long deposition time reflect the presence of larger NWs (copyright [PAU 10])*

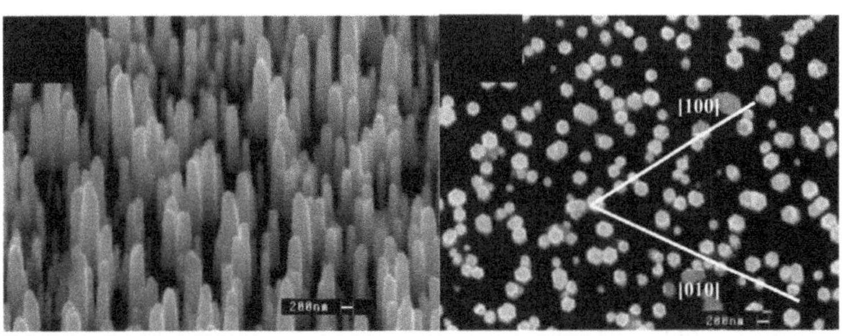

Figure 13.26. *Epitaxial ZnO NWs grown on n-type GaN(0001) single crystal layer: a) side view; b) top view with the crystallite lateral facet direction (copyright [PAU 06])*

Similar experiments have been performed on p-GaN(0001) layers (Figure 13.27). Due to the epitaxial relationship between GaN and ZnO, the rods are perfectly oriented in the vertical direction (Figure 13.27(b) inset). Increasing the $ZnCl_2$ precursor concentration leads to the increase of the NW density and to the decrease of the average distance between the columns [LUP 10].

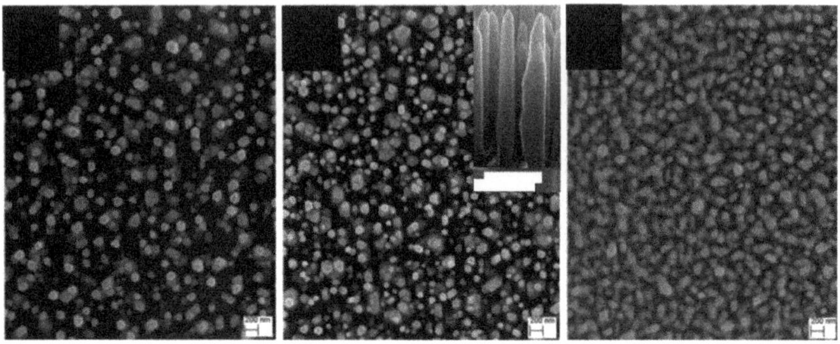

Figure 13.27. *SEM top-views of the ZnO nanowires electrodeposited on p-type Mg:GaN(0001) substrate: a) 0.07 mM $ZnCl_2$, b) 0.10 mM $ZnCl_2$ (the inset is a side-view of the wires), and c) 0.25 mM $ZnCl_2$. The scale bars are 200 nm (copyright [LUP 10a])*

The NWs electrochemical growth conditions have a marked effect on their structural quality. Pauporté *et al.* [PAU 09a] have reported an extensive study of the nanostructures by PL measurements using a reference sample to calibrate the near-band edge emission intensity. A more intense UV emission was observed from ZnO NR and NW arrays compared to ZnO dense films. The visible emission was very weak for the NWs compared to the electrodeposited films (Figure 13.28). NWs prepared at 85°C in chloride and perchlorate solutions containing 0.2 mM Zn(II) for 1 h 30 min had different aspects. The former presented a high AR of 15 and above, whereas the wires grown in perchlorate medium had a sharp conical top tip. Moreover, it was shown that the nanostructures prepared in chloride medium had a better structural quality (stronger near-band edge UV emission and a smaller visible one) compared to those grown in chloride medium [PAU 09a].

In the case of the $Zn(NO_3)_2$ precursor, the deposition current can be controlled by either $Zn(NO_3)_2$ concentration and/or $NaNO_3$ or KNO_3 supporting electrolyte concentration. At low Zn(II) concentrations, the lateral growth is hindered, whereas the growth speed along the *c*-axis could be controlled by the NO_3^- concentration or the applied current density [LUD 13].

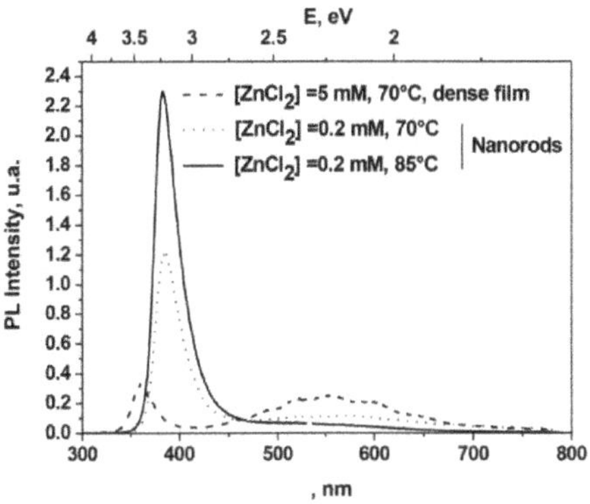

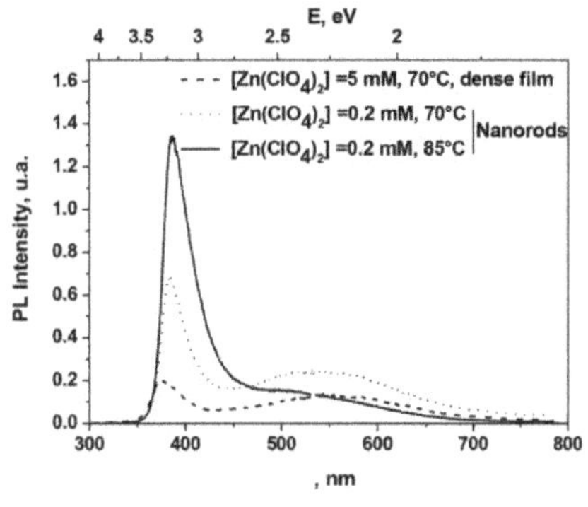

Figure 13.28. *PL emission at room temperature of electrodeposited ZnO layers. Comparison of a dense film deposited at 70°C and ZnO NW arrayed layers deposited at 70 and 85°C. a) KCl medium; b) $LiClO_4$ medium (copyright [PAU 09a])*

13.3.4. *Substrate surface treatment and seed layers*

The treatment of the substrate used as an electrode markedly influences the density of wires. It has been shown that treating the FTO surface with nitric acid yields a higher density of NWs with a lower diameter [ELI 07]. In the case of GaN, an ammonia-based wet etching treatment procedure with several steps was developed and described in detail in Ref. [LUP 10a]. It favors the ZnO nucleation and the epitaxial growth.

As observed in the case of films deposited by thermal decomposition of amino complexes (see section 13.2.3.1.2), the presence of a ZnO SL induces morphological change in electrochemically grown ZnO [WAN 05a, ZHA 05, BAD 07, ELI 07]. A systematic study of the preparation of ZnO NW arrays on an SL has been done [ELI 07]. Compact SLs, prepared by either electrodeposition at room temperature or spray pyrolysis, were composed of nanosized grains. After SL preparation, ZnO was grown at a low Zn^{2+} concentration (0.1-1mM) in a potassium chloride medium using a molecular oxygen precursor. High density free standing ZnO NW arrays with a rather high roughness were prepared by ECD. It has been shown that the electrodeposition of an SL at room temperature in the potentiostatic mode gives rise to films with a better homogeneity compared to the galvanostatic mode [PAU 09b]. However, the presence of the SL is not required with ECD for the deposition of NW arrays. Because OH^- are continuously generated at the electrode surface, the supersaturation is localized in the vicinity of the electrode, and the heterogeneous nucleation is favored.

Cao *et al.* [CAO 07] studied the electrochemical growth of zinc oxide on a (0001)-oriented SL. A zinc nitrate bath was used. In the first step, a 50 nm thick SL was deposited by sputtering on a silicon wafer of low resistivity. In the second step, zinc oxide was deposited at various constant applied currents. The authors observed a morphological transition with increasing cathodic current density. At low current density, it resulted in a flat and compact layer. At higher current densities, large-area high density well-aligned ZnO NWs arrays were obtained. The ZnO NWs were all straight, smooth and relatively vertical to the substrate with uniform diameters of about 100 nm. Their length could be easily controlled by the growth time, and they grew along the *c*-axis. At intermediate current density, the films, observed with the naked eyes, seemed to be composed of white ZnO particles. SEM imaging shows a singular morphology, since the particles were connective microcalabashes, covered by densely arrayed nanospikes. Each spike was a single crystal and grew along the *c*-axis. At least at high current density, two-dimensional ZnO nanosheets with tens of nanometers in thickness and several microns in dimension were formed by a fast non-equilibrium process. The film did not show any preferential orientation. All the nanostructured films were luminescent at room temperature with a UV near band edge (NBE) emission centered at 375 nm. The defect-related green emission was promoted by the cathodic current density of deposition and dominated in the case of the nanosheets.

13.3.5. *Electrochemical growth of ZnO nanowires on templates*

Electrodeposition in membrane templates is an alternative method to produce ZnO NWs of homogeneous size. Either anodic alumina membranes (AAMs) [ZHE 02, WAN 05b, RAM 08] or nuclear-track-etched polymer membranes [LEP 05, LAI 06] have been used as matrices for their preparation. A continuous thin film of conductive gold is deposited on one side of the membrane by sputtering or by high-vacuum evaporation to act as an electrode, and the electrodeposition is conducted starting from the bottom of the pore, at the gold electrode surface. It propagates continuously through the pores, which are progressively filled. The amount of solid deposited is theoretically proportional to the electrical charge passed. Therefore, the method should be straightforward for the preparation of NRs and NWs of perfectly defined shape and AR, since their diameter is defined by the pore size and their length by the quantity of electricity passed through the electrode during the growth process.

The hexagonal distribution of pores in AAM is presented in Figure 13.29(a). Zheng *et al.* [ZHE 02] used AAM templates and aqueous nitrate baths at 60°C. They obtained ZnO NWs with a diameter equal to that of the pores. Selected area electron diffraction (SAED) pattern showed that the ZnO NWs synthesized by the method were not single crystals. Under UV excitation, ZnO NWs presented a broad orange luminescence centered at 608 nm. ZnO NW cathodic electrodeposition in AAM template was also achieved in non-aqueous dimethyl sulfoxide (DMSO) solutions [WAN 05b] by the method reported by Gal *et al.* [GAL 00]. The solution contained chloride salts and a dissolved molecular oxygen precursor. The deposit was polycrystalline. The electrodeposition of ZnO NWs in membranes at applied constant potential can be followed by the current density, since this parameter increases with the deposition time, and the increase is more moderate once the full filling is achieved (Figure 13.29(b)) [RAM 08].

The main drawback of using AAM is the difficulty of dissolving the membrane after the synthesis and then releasing the ZnO NWs. The classical treatment consists of an etching in a concentrated NaOH solution of typically 0.5 M [ZHE 02, WAN 05b]. Unfortunately, ZnO is not stable in those conditions and partly dissolves.

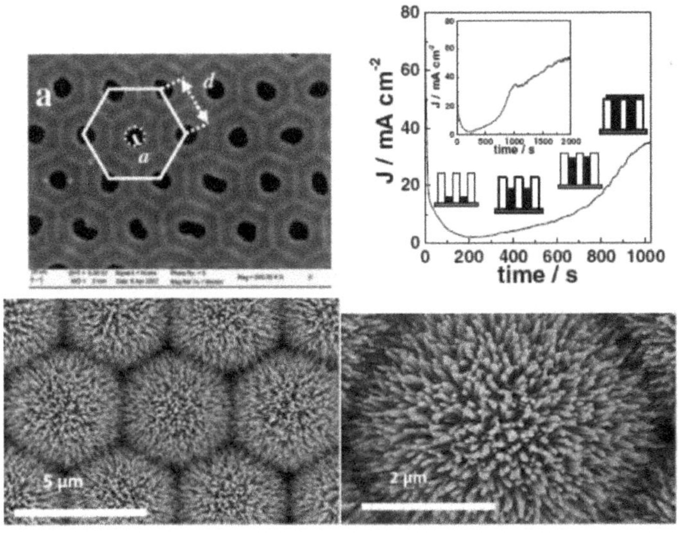

Figure 13.29. *a) SEM images of an AAM showing the interpore distance, d, the pore radius, a, and the hexagonal distribution of them. b) Current density transient recorded upon an AAM filling by electrodeposition at constant applied voltage (copyright [RAM 08]). (c,d) Organized hollow micro-urchins grown by electrodeposition (copyright [ELI 10])*

The use of polymer membranes made of polycarbonate seems more promising, since they can be easily dissolved in an organic solvent, such as dichloromethane [LEP 05, LAI 06]. However, the pores are randomly distributed in these templates, whereas they are perfectly ordered in AAM templates (Figure 13.29(a)). In [LAI 06], ZnO wires displaying characteristics of single crystals were synthesized with NO_3^- precursor at 90°C. The crystals were grown along the [10 $\bar{1}$ 1] direction, which also corresponds to the preferential orientation of the non-templated films [100]. Single crystal ZnO NWs were prepared in polycarbonate membranes of pore sizes ranging between 10 and 90 nm using an aqueous stirred solution maintained at 70°C containing 5 mM hydrogen peroxide precursor [LEP 05]. No preferential growth direction of the wires was detected. The surface roughness increased with decreasing pore size and was related to the pore quality of the commercial membrane substrates. It is noteworthy that with both precursors the NW diameter was much larger than the nominal size of the membrane pore with values at 100–120 nm and 50 nm, respectively. It was also shown that the NW length could not be directly calculated from the electrical charge exchanged during the growth and that the deposition efficiency decreased with the wire length.

Quasi-1D ZnO NWs ordered as patterned three-dimensional hollow hierarchical urchin-like structures have been prepared on transparent conductive substrates by electrodeposition (Figures 13.29(c)–(d)). [ELI 10, GUE 12a The ZnO NWs were grown on self-assembled hexagonally ordered polystyrene microspheres rendered electrically conductive. The NW length and density were tailored by playing on the electrodeposition parameters.

13.3.6. *Effect of electrochemical bath composition and use of cation additives for doping*

Xu *et al.* [XU 05b] have studied the effects of various capping agents in the $Zn(NO_3)_2$ system. They have shown that adsorption of chloride takes place preferentially onto the (0001) planes to hinder the crystal growth along the *c*-axis. It results in the formation of platelet-like crystals organized onto the substrate. The role of the supporting electrolyte in the bath was studied for the $O_2/ZnCl_2$ system [TEN 07]. More vertical NWs were found using NaCl compared to KCl salt, but the NW dimensions were not changed. On the other hand, KCl concentration had a significant effect on the NW dimension. Cl^- has been shown to lower the molecular oxygen reduction, then the deposition current and to increase the deposition faradaic efficiency. Cl^- is supposed to adsorb on the (0001) surface of ZnO and to reduce the growth speed along the *c*-axis. For KCl concentrations higher than 1mM, the lateral growth of the wires was not hindered, and the diameter increased continuously [TEN 07]. Pauporté *et al.* have compared the chlorine contained in NWs prepared in KCl and $LiClO_4$ [PAU 09a]. More Cl was observed to be included in ZnO NWs with the former salt compared to the latter.

The doping ZnO NW/NR synthesized by electrochemical techniques can be done easily by adding the cation of the dopant as an additive in the electrodeposition bath. The dopant is then integrated in the oxide material upon its deposition by precipitation. Many dopants have been investigated by Pauporté's group such as: Er, Eu, Cu, Cd, Ag, Mg and Al [GOU 06, GOU 07, PAU 07, LUP 13a; LUP 11a, PAU 12, LUP 11b, LUP 11c, PAU 13]. The concentration of dopant is controlled by the initial concentration of cations in the bath. Using the $ZnCl_2/KCl/O_2$ system, cadmium doping has been obtained by using $CdCl_2$ [LUP 11a, PAU 12], copper doping by using $CuCl_2$ [LUP 11b, LUP 11c] and silver doping with $Ag(NO_3)$ [PAU 13]. A reduction of the ZnO bandgap was measured experimentally for ZnO NWs doped by these elements.

By performing density functional theory (DFT) calculations, the bandgap of doped-ZnO has been calculated for various molar concentrations of dopants. In the case of copper, the calculations showed that the dopant gives rise to an additional acceptor level due to Cu 3d-O 2p orbitals and a donor level due to Cu 3d orbital [LUP 11b]. Their width increases with the dopant concentration, and consequently the bandgap is reduced (Figure 13.30). In the case of Cd:ZnO, it was shown that the bandgap reduction is due to an enlargement of the oxide lattice parameters. These parameters decrease continuously with the Cd content (Figure 13.30(b)) [LUP 11a]. On the contrary, a bandgap enlargement has been found experimentally for Mg doping [LUP 13a]. It is also possible to increase the conductivity of electrodeposited ZnO NWs by Al-doping.

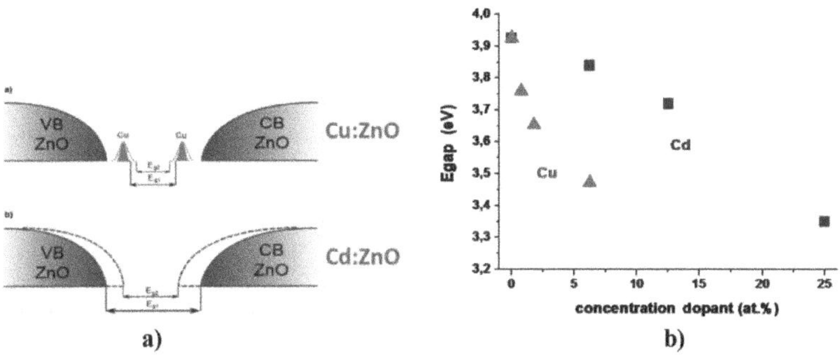

Figure 13.30. *a) Schematic of ZnO bandgap reduction by Cu and Cd doping. b) DFT computed ZnO bandgap as a function of the dopant molar content*

Eu^{3+} has been inserted in the lattice of ZnO NWs by electrodeposition. The electrodeposition of Eu-doped ZnO NWs is described in [LUP 13b]. According to EDX, SIMS and XPS studies, cationic Eu species are detected in the layers showing the incorporation of Eu into the ZnO matrix. The oxidation states of europium ions in the NWs were determined at +3 (74%) and +2 (26%). PL studies demonstrated red emission from the Eu-doped ZnO NW arrays. Due to Eu incorporated during the NW growth, a sharp PL due to 5D0 7F2 transition of Eu^{3+} was measured at around 612 nm.

13.4. Applications of ZnO nanorods/nanowires prepared by wet chemistry and by electrochemistry

During the last decade, a tremendous amount of literature has been published on the applications of ZnO NRs/NWs prepared by wet (electro-)chemistry. These applications cover various fields, an exhaustive review of which is beyond the scope of the present chapter. The light-emitting diode (LED) application will be detailed in later Chapter 9 in Vol. 2 [CON 14] and the photovoltaic application in Chapter 10 in Vol. 2. With this in mind, hereafter, we just skim through the main literature.

ZnO NWs prepared by chemical or electrochemical techniques have been used in dye-sensitized solar cells [LAW 05, XU 10, CHE 13, LUP 10b, GUE 11, GUE 12]. The main results are gathered in Table 1 of [LUP 10b]. This application requires very large internal roughness for correct sensitization that can be achieved by reducing the wire diameter and increasing both their density and length. The dye can be replaced by a thin layer of a visible/near-infrared light absorbing semiconductor. These devices, called extra-thin absorber (ETA) solar cells, require less internal surface area, since the inorganic semiconductors used are more absorbing than dyes. Lévy-Clément *et al.* have employed electrodeposited ZnO NWs for the fabrication of ETA solar cells [LEV 02, TEN 05, LEV 05]. Electrodeposited ZnO NWs have also been used as n-type semiconducteur in nanostructured p-n junction solar cells combined with electrodeposited p-type Cu_2O [MUS 11, CUI 10].

LEDs are another interesting application of ZnO NWs/NRs. In [KON 04, KON 05], ZnO NRs were connected by a 0.5–1.5 m thick film of p-type poly(3, 4-ethylene-dioxythiophene) poly(styrenesulfonate) (PEDOT/PSS) deposited from a solution by spin coating. The diode emitted a white light due to a broad defect-related emission centered at about 620 nm. In Pauporté's group, heterojonctions have been prepared by growing n-type ZnO NWs on a single crystal p-type GaN layer by either electrodeposition or hydrothermal growth [LUP 12a; LUP 12b; LUP 10b; LUP 11; PAUP 12; LUP 11a; LUP 11b; PAUP 13; LUP 10c]. The ZnO NW arrays were epitaxially grown. The heterostructures have been integrated in LED devices and emitted a strong single UV peak at low applied forward bias (less than 4 V) (Figure 13.31(a)). In the case of Ag, Cd or Cu-doped ZnO NWs, the emission peak was shifted toward longer wavelengths due to the bandgap reduction (Figure 13.31(b)). In the case of Mg dopant, the emission peak was shifted to a shorter wavelength (Figure 13.31(b)) [LUP 13a].

ZnO has received great attention as a photocatalyst for the degradation and mineralization of environmental pollutants due to its large bandgap and low fabrication cost [PAU 07b, GUO 11]. Under UV illumination, ZnO will generate electron/hole pairs able to produce hydrogen by splitting of water, synthesize H_2O_2 or reduce graphene oxides to graphene. In particular, the generated holes can be used to oxidize/decompose organic pollutants, such as rhodamine 6G, methyl orange, methylene blue and formaldehyde.

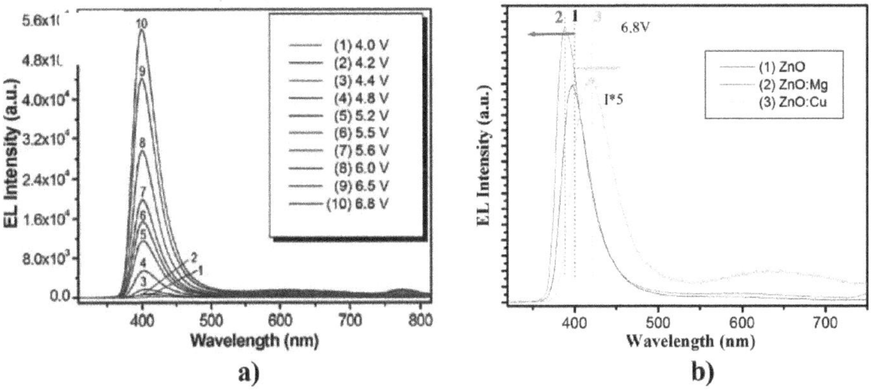

Figure 13.31. *a) Room-temperature electroluminescence spectra of a ZnO NR/p-GaN light emitting diode for various applied forward bias. The ZnO NWs were grown by a hydrothermal technique. b) Emission spectra of pure ZnO, Cu-doped ZnO and Mg-doped ZnO-based LEDs at 6.8 V (copyright [LUP 12a]). For a color version of this figure, see www.iste.co.uk/consonni/nanowires1.zip*

Field emission finds applications in photoelectric panel displays, X-ray sources and microwave devices [XU 05c]. ZnO NWs, with the advantage of allowing relatively high oxygen partial pressure during its operation, have become a good candidate for use in field emission cathodes [SEM 11].

Wettability of a solid surface is of critical importance for many industrial applications. As-grown ZnO NW arrays are superhydrophilic due to their high roughness and high surface energy [PAU 10]. When the nanostructured surfaces are covered by a SAM of amphiphilic molecules, such as stearic acid or octadecylsilane, they are converted to hydrophobic or to superhydrophobic surfaces [BAD 07a, PAU 10, BAD 08, BAD 07b]. NW arrays with high roughness amplify the wetting behavior of the surfaces in agreement with the ad-hoc models presented in [PAU 10]. In [BAD 09], the NW surface has been functionalized by a ferrocene derivative redox molecule. In that case, the wetting behavior could be controlled by applying an external voltage and was varied from hydrophobic to superhydrophilic.

ZnO NWs have also been used as active components in gas sensors [LUP 12c, WA 06], visible light detectors [SUN 10], visible-blind UV detectors [GAO 09, ZHA 13], biosensors [WEI 06], strain sensors [LIU 11] and nanogenerators [WAN 07, ZHU 10].

13.5. Conclusions

The preparation of ZnO nanostructures, especially of 1D NRs and NWs, by wet (electro-)chemistry is an exciting emerging field of research. In the present review, we have shown the possibility of developing low-cost and low-temperature routes for the wet synthesis of these structures with controlled shape, as well as structural, optical and electrical properties. They can be obtained in bulk solution or grown arrayed on many various substrates. The overall architecture of the crystals is basically controlled by tuning the crystal growth, including the growth site, growth rate and growth direction. The degree of supersaturation, the presence of seeds and the adsorption of additives are the three key parameters to control the growth modes and the final structures. In the case of electrodeposition, the supersaturation in the vicinity of the conductive substrate and ZnO precipitation are finely controlled by the electrochemical parameters. Moreover, the growth mechanism ensures an excellent electrical continuity between the deposited layer and the substrate.

The chemical and electrochemical techniques are performed at low temperature and do not necessarily require a post-deposition heat treatment to get high-quality material. The inter-reaction and inter-diffusion phenomena are drastically reduced. ZnO can be doped to tailor the bandgap of the oxide semiconductor or to provide new properties. For technological applications, various techniques of patterning have been developed. Last but not the least, a tremendous number of papers have demonstrated many potential applications of ZnO NRs/NWs prepared by wet (electro-)chemistry in various advanced technological fields.

13.6. Bibliography

[ALA 09] ALAMMAR T., MUDRING A.V., "Facile ultrasound-assisted synthesis of ZnO nanorods in an ionic liquid", *Materials Letters*, vol. 63, pp. 732–735, 2009.

[BAD 07a] BADRE C., PAUPORTÉ T., TURMINE M., *et al.*, "ZnO nanowire array film with super water repellent properties", *Nanotechnology*, vol. 18, p. 365705, 2007.

[BAD 07b] BADRE C., DUBOT P., LINCOT D., *et al.*, "Effects of nanorod structure and conformation of fatty acid self assembled layers on superhydrophobicity of zinc oxide surfaces", *Journal of Colloids & Interface Science*, vol. 316, p. 233–237, 2007.

[BAD 08] BADRE C., PAUPORTÉ T., TURMINE M., *et al.*, "Superhydrophobic properties of ZnO nanowire arrays modified by octadecylsilane", *Physica E*, vol. 40, pp. 2454–2456, 2008.

[BAD 09] BADRE C., PAUPORTÉ T., "Nanostructured ZnO-based surface with a reversible electrochemically ajustable wettability", *Advanced Materials*, vol. 21, no. 6, pp. 697–701, 2009.

[BOE 09] BOERCKER J.E., SCHMIDT J.B., AYDIL E.S., "Transport limited growth of zinc oxide nanowires", *Crystal Growth & Design*, vol. 9, pp. 2783–2783, 2009.

[BOY 02] BOYLE D.S., GOVENDER K., O'BRIEN P., "Novel low temperature solution deposition of perpendicularly oriented rods of ZnO: substrate effects and evidence of the importance of counter-ions in the control of crystallite growth", *Chemical Communications*, pp. 80–81, 2002

[CAO 06] CAO H.L., QIAN X.F., GONG Q., *et al.*, "Shape- and size- controlled synthesis of nanometer ZnO from a simple solution route at room temperature", *Nanotechnology*, vol. 17, pp. 3632–3636, 2006.

[CAO 07] CAO B., TENG X., HEO S.H., *et al.*, "Different ZnO nanostructures fabricated by seed layer assisted electrochemical route and their photoluminescence and field emission properties", *The Journal of Physical Chemistry C*, vol. 111, pp. 2470–2476, 2007.

[CHA 10] CHAO C.H., LIN W.H., CHEN C.H., *et al.*, "Low-temperature heteroepitaxy of morphology-controlled ZnO micro/nanorod arrays on GaN substrates", *Chemical Letters*, vol. 39, pp. 202–203, 2010

[CHE 04] CHENG B., SAMULSKI E.T., "Hydrothermal synthesis of one dimensional ZnO nanostructures with different aspect ratios", *Chemical Communications*, pp. 986–987, 2004

[CHE 06a] CHEN Z, TANG Y, ZHANG L, *et al.*, "Electrodeposited nanoporous ZnO films exhibiting enhanced performance in dye sensitized solar cells", *Electrochim. Acta*, vol. 51, pp. 5870–5875, 2006.

[CHE 06b] CHEN Q.P., XUE M.Z., SHENG Q.R., *et al.*, "Electrochemical growth of nanopillar zinc oxide film by applying a low concentration of zinc nitrate precursor", *Electrochem. Solid-State Lett.*, vol. 9, pp. C58–C61, 2006.

[CHE 13] CHEN L.Y., YIN Y.T., "Efficient electron transport in ZnO nanowire/ nanoparticle dye-sensitized solar cells via continuous flow injection process", *RSC Adv.*, vol. 3, pp. 8480–8488, 2013.

[CON 14] CONSONNI V., FEUILLET G., (eds.), *Wide Band Gap Semiconductor Nanowires 2: Heterostructures and Optoelectronic Devices*, ISTE, London, and John Wiley & Sons, New York, 2014.

[CUI 10] CUI J., GIBSON U.J., "A simple two-step electrodeposition of Cu_2O/ZnO nanopillar solar cells", *The Journal of Physical Chemistry C*, vol. 114, no. 14, pp. 6408–6412, 2010.

[CUI 12] CUI J., "Zinc oxide nanowires", *Materials Characterization*, vol. 64, pp. 43–52, 2012.

[DEB 08] DEBABRATA P., TONG L.K, "Controlled growth of two-dimensional and one-dimensional ZnO nanostructures on indium tin oxide coated glass by direct electrodeposition", *Langmuir*, vol. 24, no. 17, pp. 9707–9716, 2008.

[DEM 01] DEM'YANETS L.N., KOSTOMAROV D.V., "Mechanism of zinc oxide single crystal growth under hydrothermal conditions", *Ann. Chim. Sci. Mat.*, vol. 26, pp.193–198, 2001.

[DEM 02] DEM'YANETS L.N., KOSTOMAROV D.V., KUZMINA I.P., "Chemistry and kinetics of ZnO growth from alkaline hydrothermal solutions", *Inorg. Mater.*, vol. 38, pp. 124–131, 2002.

[DJU 10] DJURISIC A.B., NG A.M.C, CHEN X.Y., "ZnO nanostructures for optoelectronics: material properties and device applications", *Prog. Quantum Electronics*, vol. 34, pp. 191–259, 2010.

[ELB 08] ELBELGHITI H., PAUPORTÉ T., LINCOT D., "Mechanistic study of ZnO nanorod array electrodeposition", *Physica Status Solidi (a)*, vol. 205, pp. 2360–2364, 2008.

[ELI 07] ELIAS J., TENA-ZAERA R., LEVY-CLÉMENT C., "Electrodeposition of ZnO nanowires with a controlled dimensions for photovoltaic applications, role of buffer layer", *Thin Solid Films,* vol. 515, pp. 8553–8557, 2007.

[ELI 10] ELIAS J., LÉVY-CLÉMENT C., BECHELANY M., *et al.*, "Urchin-like ZnO thin films by electrochemical deposition", *Advanced Materials*, vol. 22, pp. 1607–1612, 2010.

[GAL 00] GAL D., HODES G., LINCOT D., *et al.*, "Electrochemical deposition of zinc oxide films from non-aqueous solution: a new buffer/window process for thin film solar cells", *Thin Solid Films*, vol. 361–362, pp.79–83, 2000.

[GAO 09] GAO P., WAN Z.Z., LIU K. H., *et al.*, "Photoconducting response on bending of individual ZnO nanowires", *Journal of Materials Chemistry*, vol. 19, pp. 1002–1005, 2009.

[GAR 07] GARCIA S.P., SEMANCIK S., "Controlling the morphology of zinc oxide nanorods crystallized from aqueous solutions: the effect of crystal growth modifiers on aspect ratio", *Chem. Mater.*, vol. 19, pp. 4016–4022, 2007.

[GON 09] GONZALEZ-VALLS I., LIRA-CANTU M., "Vertically-aligned nanostructures of ZnO for excitonic solar cells: a review", *Energy Environ. Sci.*, vol. 2, pp. 19–34, 2009.

[GOU 05] GOUX A., PAUPORTÉ T., CHIVOT J., *et al.*, "Temperature effects on ZnO electrodeposition", *Electrochim. Acta*, vol. 50, pp. 2239–2248, 2005.

[GOU 06a] GOUX A., PAUPORTÉ T., LINCOT D., "Oxygen reduction on zinc oxide electrodes in KCl aqueous solution at 70°C", *Electrochim. Acta*, vol. 51, pp. 3168–3172, 2006.

[GOU 06b] GOUX A., PAUPORTÉ T., LINCOT D., "Deposition of mixed zinc oxide/ lanthanide films by electrochemical precipitation. The ZnO/ Er system", *J. Electroanal. Chem.*, vol. 587, pp. 193–202, 2006.

[GOU 07] GOUX A., PAUPORTÉ T., LINCOT D., "Preparation of ZnO/Eu mixed films by electrochemical precipitation", *Electrochim. Acta,* vol. 53, pp. 50–58, 2007.

[GOV 02] GOVENDER K., BOYLE D.S., O'BRIEN P., *et al.*, "Room-temperature lasing observed from ZnO nanocolumns grown by aqueous solution deposition", *Advanced Materials*, vol. 14, no. 17, p. 1221, 2002.

[GOV 04] GOVENDER K., BOYLE D.S., KENWAY P.B., *et al.*, "Understanding the factors that govern the deposition and morphology of thin films of ZnO from aqueous solution", *Journal of Materials Chemistry*, vol. 14, pp. 2575–2591, 2004.

[GRE 03] GREENE L.E., LAW M., GOLDBERGER J., *et al.*, "Low-temperature wafer-scale production of ZnO nanowire arrays", *Angew. Chem., Int. Ed.*, vol. 42, pp. 3031–3034, 2003.

[GRE 04] GREYSON E.C., BABAYAN Y., ODOM T.W., "Directed growth of ordered arrays of small-diameter ZnO nanowires", *Advanced Materials*, vol. 16, pp. 1348–1352, 2004.

[GRE 05] GREENE L.E., LAW M., TAN D.H., *et al.*, "General route to vertical ZnO nanowire arrays using textured ZnO seeds", *NanoLetters*, vol. 5, no. 7, pp. 1231–1236, 2005.

[GUE 11] GUÉRIN V.M., PAUPORTÉ T., "From nanowires to hierarchical structures of template-free electrodeposited ZnO for efficient dye-sensitized solar cells", *Energy Environ. Sci.*, vol. 4, pp. 2971–2979, 2011.

[GUE 12a] GUÉRIN V.M., ELIAS J., NGUYEN T.T., *et al.*, "Ordered networks of ZnO nanowire hierarchical urchin-like structures for improved dye-sensitized solar cells", *Phys. Chem. Chem. Phys.*, vol. 14, pp. 12948–12955, 2012.

[GUE 12b] GUÉRIN V.M., RATHOUSKY J., PAUPORTÉ T., "Electrochemical design of ZnO hierarchical structures for dye-sensitized solar cell", *Sol. Energy Mater. Sol. Cells*, vol. 102, pp. 8–14, 2012.

[GUI 12] GUILLEMIN S., CONSONNI V., APPERT E., *et al.*, "Critical nucleation effects on the structural relationship between ZnO seed layer and nanowires", *The Journal of Physical Chemistry C*, vol. 116, pp. 25105–25111, 2012.

[GUI 13] GUILLEMIN S., RAPENNE L., ROUSSEL H., *et al.*, "Formation mechanisms of ZnO nanowires: the crucial role of crystal orientation and polarity", *The Journal of Physical Chemistry C*, vol. 117, pp. 20738–20745, 2013.

[GUO 11] GUO M.Y., FUNG M.K., FANG F., *et al.*, "ZnO and TiO_2 1D nanostructures for photocatalytic applications", *J. Alloys Compounds*, vol. 509, no. 4, pp. 1328–1332, 2011.

[HID 96] HIDBER P.C., GRAULE T.J., GAUCKLER L.J., "Citric acid – A dispersant for aqueous alumina suspensions", *J. Am. Ceram. Soc.*, vol. 79, pp. 1857–1867, 1996.

[HOU 07] HOU X.M., ZHOU F., SUN Y.B., *et al.*., "Ultrasound-assisted synthesis of dentritic ZnO nanostructure in ionic liquid", *Materials Letters*, vol. 61, pp. 1789–1792, 2007.

[HSU 05] HSU J.W.P., TIAN Z.R., SIMMONS N.C., *et al.*, "Directed spatial organization of zinc oxide nanorods", *NanoLetters*, vol. 5, pp. 83–86, 2005.

[HUA 01] HUANG M.H., MAO S., FEICK H., *et al.*, "Room-temperature ultraviolet nanowire nanolasers", *Science*, vol. 292, no. 5523, pp. 1897–1899, 2001.

[ISH 01] ISHIZAKI H., IZAKI M., ITO T., "Influence of $(CH_3)_2NHBH_3$ concentration on electrical properties of electrochemically grown ZnO film", *J. Electrochem. Soc.*, vol. 148, pp. C540–C543, 2001.

[ISH 02] ISHIZAKI H., IMAIZUMI M., MATSUDA S., *et al.*, "Incorporation of boron in ZnO film from an aqueous solution containing zinc nitrate and dimethylamine-borane by electrochemical reaction", *Thin Solid Films*, vol. 411, pp. 65–68, 2002.

[IZA 03a] IZAKI M., WATASE S., TAKAHASHI H., "Low-temperature electrodeposition of room temperature ultraviolet-light-emitting zinc oxide", *Advanced Materials*, vol. 15, pp. 2000–2002, 2003.

[IZA 03b] IZAKI M., WATASE S., TAKAHASHI H. "Room-temperature ultraviolet light-emitting zinc oxide micropatterns prepared by low-temperature electrodeposition and photoresist", *Applied Physics Letters*, vol. 83, pp. 4930–4932, 2003.

[IZA 96a] IZAKI M, OMI T., "Electrolyte optimization for cathodic growth of zinc oxide films", *J. Electrochem. Soc.*, vol. 143, pp. L53–L55, 1996.

[IZA 96b] IZAKI M., OMI T., "Transparent zinc oxide films prepared by electrochemical reaction", *Applied Physics Letters,* vol. 68, pp. 2439–2440, 1996.

[IZA 97] IZAKI M, OMI T "Characterization of transparent zinc oxide films prepared by electrochemical reaction", *J. Electrochem. Soc.*, vol. 144, pp. 1949–1952, 1997.

[JI 09] JI L.W., PENG S.M., WU J.S., *et al.*, "Effect of seed layer on the growth of well-aligned ZnO nanowires", *J. Phys. Chem. Solids*, vol. 70, pp. 1359–1362, 2009.

[JHA 12] JHA S.K., LUAN C., TO C.H., *et al.*, "ZnO-nanorod-array/p-GaN high-performance ultra-violet light emitting devices prepared by simple solution synthesis", *Applied Physics Letters,* vol. 101, p. 211116, 2012.

[KAN 11] KANG H.W., YEO J., HWANG J.O., *et al.*, "Simple ZnO nanowires patterned growth by microcontact printing for high performance field emission device", *The Journal of Physical Chemistry C,* vol. 115, no. 23, pp. 11435–11441, 2011.

[KAW 06a] KAWANO T., IMAI H., "Fabrication of ZnO nanoparticles with various aspect ratios through acidic and basic routes", *Cryst. Growth Des.*, vol. 6, pp. 1054–1056, 2006.

[KAW 06b] KAWANO T., YAHIRO J., MAKI H., *et al.*, "Epitaxial growth of wurtzite ZnO crystals in an aqueous solution system", *Chem. Lett.*, vol. 35, pp. 442–443, 2006

[KAW 08] KAWANO T., IMAI H., "A simple preparation technique for shape-controlled zinc oxide nanoparticles: formation of narrow size-distributed nanorods using seeds in aqueous solutions", *Colloids Surf. A*, vol. 319, pp. 130–135, 2008.

[KAW 10] KAWANO T., IMAI H., "Nanoscale morphological design of ZnO crystal grown in aqueous solutions", *J. Ceramic Soc. Jap.*, vol. 118, no. 11, pp. 969–976, 2010.

[KEN 09] KENANAKIS G., VERNARDOU D., KOUDOUMAS E., *et al.*, "Growth of c-axis oriented ZnO nanowires from aqueous solution: the decisive role of a seed layer for controlling the wires' diameter", *Journal of Crystal Growth,* vol. 311, pp. 4799–4804, 2009.

[KHA 12] KHAJAVI M.R., BLACKWOOD D.J., CABANERO G., *et al.*, "New insight into growth mechanism of ZnO nanowires electrodeposited from nitrate-based solutions", *Electrochim. Acta*, vol. 69, pp. 181–189, 2012.

[KIM 06] KIM J.H., ANDEEN D., LANGE F.F., "Hydrothermal growth of periodic, single-crystal ZnO microrods and microtunnels", *Advanced Materials*, vol. 18, pp. 2453–2457, 2006.

[KIM 07a] KIM J. H., KIM E.-M., ANDEEN D., *et al.*, "Growth of heteroepitaxial ZnO thin films on GaN-Buffered Al2O3(0001) substrates by low-temperature hydrothermal synthesis at 90 degrees C", *Adv. Funct. Mater.*, vol. 17, pp. 463–471, 2007.

[KIM 07b] KIM J.H., KIM E.M., ANDEEN D., *et al.*, "Growth of heteroepitaxial ZnO thin films on GaN-Buffered Al_2O_3(0001) substrates by low-temperature hydrothermal synthesis at 90 degrees C", *Adv. Funct. Mater.*, vol. 17, p. 463471, 2007.

[KIM 12] KIM Y.H., BAEK N.S., KIM K.H., *et al.*, "Simplifying patterning process of ZnO nanowires by one step development and etching process", *J. Sol-Gel Sci. Techn.*, vol. 64, no. 2, pp. 304–308, 2012.

[KOH 99] KOHLS M., SCHMIDT T., KATSCHOREK H., *et al.*, "A simple colloidal route to planar micropatterned Er@ZnO amplifiers", *Advanced Materials*, vol. 11, no. 4, pp. 288–292, 1999.

[KON 01] KONG Y.C., YU D.P., ZHANG B., *et al.*, "Ultraviolet-emitting ZnO nanowires synthesized by a physical vapor deposition approach", *Applied Physics Letters*, vol. 78, no. 4, pp. 407–409, 2001.

[KON 04] KÖNENKAMP R., WORD R C, SCHLEGEL C., "Vertical nanowire light-emitting diode", *Applied Physics Letters*, vol. 85, pp. 6004–6006, 2004.

[KON 05] KÖNENKAMP R., WORD R.C., GODINEZ M., "Ultraviolet electroluminescence from ZnO/polymer heterojunction light emitting diodes", *NanoLetters*, vol. 5, pp. 2005–2008, 2005.

[LAI 06] LAI M, RILEY J., "Templated electrosynthesis of zinc oxide nanorods", *Chem. Mater.*, vol. 18, pp. 2233–2237, 2006.

[LAW 05] LAW M., GREENE L., JOHNSON J.C., *et al.*, "Nanowire dye-sensitized solar cells", *Nature Mater.*, vol. 4, pp. 455–459, 2005.

[LE 06] LE H.Q., CHUA S.J., KOH Y.W., *et al.*, "Systematic studies of the epitaxial growth of single-crystal ZnO nanorods on GaN using hydrothermal synthesis", *Journal of Crystal Growth*, vol. 293, pp. 36–42, 2006.

[LEE 07] LEE Y.J., SOUNART T.L., SCRYMGEOUR D.A., *et al.*, "Control of ZnO nanorod array alignment synthesized via seeded solution growth", *Journal of Crystal Growth*, vol. 304, no. 1, pp. 80–85, 2007.

[LEP 05] LEPRINCE-WANG Y., YACOUBI-OUSLIM A., WANG G.Y., "Structure study of electrodeposited ZnO nanowires", *Microelectronics J.*, vol. 36, pp. 625–628, 2005.

[LEV 02] LÉVY-CLÉMENT C., KATTY A., BASTIDE S., *et al.*, "A new CdTe/ZnO columnar composite film for Eta-solar cells", *Physica E*, vol. 14, pp. 229–232, 2002.

[LEV 05] LÉVY-CLÉMENT C., TENA-ZAERA R., RYAN M.A., *et al.*, "CdSe-sensitized p-CuSCN/Nanowire n-ZnO heterojunction", *Advanced Materials*, vol. 17, pp. 1512–1515, 2005.

[LI 00] LI Y., MENG G.W., ZHANG L.D., *et al.*, "Ordered semiconductor ZnO nanowire arrays and their photoluminescence properties", *Applied Physics Letters,* vol. 76, no. 15, pp. 2011–2013, 2000.

[LI 01] LI J.Y., CHEn, X.L., LI, H., *et al.*, "Fabrication of zinc oxide nanorods", *Journal of Crystal Growth*, vol. 233, nos. 1–2, pp. 5–7, 2001.

[LI 11] LI H., ZHANG J., ZHOU X., *et al.*, "Aminosilane micropatterns on hydroxyl-terminated substrates: fabrication and applications", *Langmuir*, vol. 26, pp. 5603–5609, 2011.

[LIU 01] LIU R., VERTEGEL A.A., BOHANNAN E.W., *et al.*, "Epitaxial electrodeposition of zinc oxide nanopillars on single-crystal gold", *Chem. Mater.*, vol. 13, pp. 508–512, 2001.

[LIU 03] LIU B., ZENG H.C., "Hydrothermal synthesis of ZnO nanorods in the diameter regime of 50 nm", *J. Am. Chem. Soc.*, vol. 125, pp. 4430–4431, 2003.

[LIU 08] LIU J., SHE J., DENG S., *et al.*, "Ultrathin seed-layer for tuning density of ZnO nanowire arrays and their field emission characteristics", *The Journal of Physical Chemistry C*, vol. 112, pp. 11685–11690, 2008.

[LIU 11] LIU N.S., FANG G.J., ZENG W., *et al.*, "Novel ZnO nanorod flexible strain sensor and strain driving transistor with an ultrahigh 107 scale "On"-"Off" ratio fabricated by a single-step hydrothermal reaction", *The Journal of Physical Chemistry C*, vol.115, pp. 570–575, 2011.

[LUD 13] LUDWIG W., OHM W., CORREA-HOYOS J.M., *et al.*, "Electrodeposition parameters for ZnO nanorod arrays for photovoltaic", *Physica Status Solidi A*, vol. 210, no. 8, pp. 1557–1563, 2013.

[LUP 10a] LUPAN O., PAUPORTÉ T., VIANA B., *et al.*, "Epitaxial electrodeposition of ZnO nanowire arrays on p-GaN for efficient UV-light-emitting diode fabrication", *ACS Appl. Mater. Interfaces*, vol. 2, no. 7, pp. 2083–2090, 2010.

[LUP 10b] LUPAN O., GUÉRIN V.M., TIGINYANU I.M., *et al.*, "Vertically-oriented well-aligned ZnO nanowires arrays electrodeposited on ITO-coated glass and their integration in dye-sensitized solar cells", *J. Photochem. Photobiol. A*, vol. 211, pp. 65–73, 2010.

[LUP 10c] LUPAN O., PAUPORTÉ T., VIANA B., "Low-voltage UV-electroluminescence from ZnO-nanowire array/ p-GaN light emitting diodes", *Advanced Materials*, vol. 22, no. 30, pp. 3298–3302, 2010.

[LUP 11a] LUPAN O., PAUPORTÉ T., LE BAHERS T., *et al.*, "High aspect ratio ternary $Zn_{1-x}Cd_xO$ nanowires by electrodeposition for light emitting diode application", *The Journal of Physical Chemistry C*, vol. 115, no. 30, pp. 14548–14558, 2011.

[LUP 11b] LUPAN O., PAUPORTÉ T., LE BAHERS T, *et al.*, "Wavelength emission tuning of ZnO nanowires-based light emitting diodes by Cu-doping: experimental and computational insights", *Adv. Funct. Mater.*, vol. 21, no. 18, pp. 3564–3572, 2011.

[LUP 11c] LUPAN O., PAUPORTÉ T., VIANA B., *et al.*, "Electrodeposition of Cu-doped ZnO nanowire arrays and heterojunction formation with p-GaN for color tunable light emitting diode applications", *Electrochim. Acta*, vol. 56, pp. 10543–10549, 2011.

[LUP 12a] LUPAN O., PAUPORTÉ T., CHOW L., *et al.*, "Comparative study of the ZnO and $Zn_{1-x}Cd_xO$ nanorod emitters hydrothermally synthesized and electrodeposited on p-GaN", *Appl. Surf. Sci.*, vol. 259, pp. 399–405, 2012.

[LUP 12b] LUPAN O., T. PAUPORTÉ T., VIANA B., *et al.*, "UV-blue and green electroluminescence from Cu-doped ZnO nanorod emitters hydrothermally synthesized on p-GaN", *J. Nanoelectron. Optoelectron.* vol. 7, pp. 712–718, 2012.

[LUP 12c] LUPAN O., CHOW L., PAUPORTÉ T., *et al.*, "Highly sensitive and selective hydrogen single nanowire sensor operating at room temperature", *Sensors and Actuators B*, vol. 173, pp. 772–780, 2012.

[LUP 13a] LUPAN O., VIANA B., PAUPORTÉ T.H., *et al.*, "Controlled mixed violet/blue-red electroluminescence from Eu:NanoPhosphors/ZnO-nanowires/p-GaN light emitting diodes ", *The Journal of Physical Chemistry C*, vol. 117, pp. 26768–26775, 2013.

[LUP 13b] LUPAN O., PAUPORTÉ T., VIANA B., *et al.*, "Eu-doped ZnO nanowire arrays grown by electrodeposition", *Appl. Surf. Sci.*, vol. 282, pp. 782–788, 2013

[MA 07] MA T., GUO M., ZHANG M., *et al.*, "Density-controlled hydrothermal growth of well-aligned ZnO nanorod arrays", *Nanotechnology*, vol. 313, pp. 705–710, 2007.

[MAI 99] MAIS N., REITHMAIER J.P., FORCHEL A., *et al.*, "Er-doped nanocrystalline ZnO planar waveguide structures for 1.55 μm amplifier applications", *Applied Physics Letters*, vol. 75, p. 2005, 1999.

[MAS 06] MASUDA Y., KINOSHITA N., SATO F., *et al.*, "Site-selective deposition and morphology control of UV- and visible-light-emitting ZnO crystals", *Cryst. Growth Des.*, vol. 6, p. 7578, 2006.

[MU 09] MU G., GU R.V., KULP E.A., *et al.*, "Tilted epitaxial ZnO nanospears on Si(001) by chemical bath deposition", *Chem. Mater.*, vol. 21, pp. 3960–3964, 2009.

[MUS 11] MUSSELMAN K.P., MARIN A., WISNET A., *et al.*, "A novel buffering technique for aqueous processing of zinc oxide nanostructures and interfaces, and corresponding improvement of electrodeposited ZnO-Cu_2O photovoltaics", *Adv. Funct. Mater.*, vol. 21, pp. 573–582, 2011

[PAC 02] PACHOLSKI C., KORNOWSKI A., WELLER H., "Self-assembly of ZnO: from nanodots to nanorods", *Angew. Chem. Int. Ed.*, vol. 41, pp. 1188–1191, 2002.

[PAN 05] PAN A.P., YU R., XIE S., *et al.*, "ZnO flowers made up of thin nanosheets and their optical properties", *Journal of Crystal Growth*, vol. 282, pp. 165–172, 2005.

[PAN 13] PANDA D., TSENG T.Y., "One-dimensional ZnO nanostructures: fabrication, optoelectronic properties and device applications", *J. Mater Sci.*, vol.48, pp. 6849–6877, 2013.

[PAU 00] PAUPORTÉ T., LINCOT D., "Electrodeposition of semiconductors for optoelectronic devices: results on zinc oxide", *Electrochim. Acta*, vol. 45, pp. 3345–3353, 2000.

[PAU 01a] PAUPORTÉ T., LINCOT D., "Hydrogen peroxide oxygen precursor for zinc oxide electrodeposition. II- mechanistic aspects", *J. Electroanal. Chem.*, vol. 517, pp. 54–62, 2001.

[PAU 01b] PAUPORTÉ T., LINCOT D., "Hydrogen peroxide oxygen precursor for zinc oxide electrodeposition. I- deposition in perchlorate medium", *J. Electrochem. Soc.*, vol. 148, pp. C310–C314, 2001.

[PAU 02] PAUPORTÉ T., CORTÈS R., FROMENT M., *et al.* "Electrocrystallization of epitaxial zinc oxide onto gallium nitride", *Chem. Mater.*, vol. 14, pp. 4702–4708, 2002.

[PAU 03] PAUPORTÉ T., YOSHIDA T., CORTÈS R., *et al.*., "Electrochemical growth of epitaxial Eosin/ZnO hybrid films", *The Journal of Physical Chemistry B*, vol. 107, pp. 10077–10082, 2003.

[PAU 06] PAUPORTÉ T., LINCOT D., VIANA B., *et al.*, "Toward laser emission of epitaxial nanorod array of ZnO grown by electrodeposition", *Applied Physics Letters*, vol. 89, p. 233112, 2006.

[PAU 07a] PAUPORTÉ T., PELLÉ F., VIANA B., *et al.*, "Luminescence of nanostructured ZnO/Eu mixed films prepared by electrodeposition", *The Journal of Physical Chemistry C.*, vol. 111, pp. 15427–15432, 2007.

[PAU 07b] PAUPORTÉ T., RATHOUSKY J., "Electrodeposited mesoporous ZnO thin films as efficient photocatalysts for the degradation of dye pollutants", *The Journal of Physical Chemistry C*, vol. 111, pp. 7639–7644, 2007.

[PAU 09a] PAUPORTÉ T., JOUANNO E., PELLÉ F., *et al.*, "Key growth parameters for the electrodeposition of ZnO films with an intense UV-light emission at room temperature", *The Journal of Physical Chemistry C.*, vol. 113, pp. 10422–10431, 2009.

[PAU 09b] PAUPORTÉ T., JIRKA I., "Electrochemical growth of homogeneous nanocrystalline ZnO thin films at room temperature", *Electrochim. Acta,* vol. 54, pp. 7558–7564, 2009.

[PAU 99] PAUPORTÉ T., LINCOT D., "Heteroepitaxial electrodeposition of zinc oxide on gallium nitride", *Applied Physics Letters*, vol.75, pp. 3817–3820, 1999.

[PAU 10] PAUPORTÉ T., BATAILLE G., JOULAUD L., *et al.*, "Well-aligned ZnO nanowire arrays prepared by seed layer-free electrodeposition and their Cassie-Wenzel transition after hydrophobization", *The Journal of Physical Chemistry C.*, vol. 114, pp.194–202, 2010.

[PAU 12] PAUPORTÉ T., LUPAN O., VIANA B., "Tunable electroluminescence from low-threshold voltage LED structure based on $Zn_{1-x}Cd_xO$ nanorods/p-GaN heterojunction", *Physica Status Solidi (a)*, vol. 209, no. 2, pp. 359–363, 2012.

[PAU 13] PAUPORTÉ T., LUPAN O., VIANA B., "Electrodeposited ZnO nanowire-based light-emitting diodes with tunable emission from near-UV to blue", *Proceedings of the SPIE*, vol. 8626, p. 86261N-1-4, 2013.

[PEN 06] PENG Y., XU A.W., DENG B., *et al.*, "Polymer-controlled crystallization of zinc oxide hexagonal nanorings and disks", *The Journal of Physical Chemistry B*, vol. 110, pp. 2988–2993, 2006.

[PEU 96] PEULON S., LINCOT D., "Cathodic electrodeposition from aqueous solution of dense or open structured zinc oxide films", *Advanced Materials*, vol. 8, pp. 166–170, 1996.

[PEU 98] PEULON S., LINCOT D., "Mechanistic study of cathodic electrodeposition of zinc oxide and zinc hydroxychloride films from oxygenated aqueous zinc chloride solutions", *J. Electrochem. Soc.*, vol. 145, pp. 864–874, 1998.

[QUA 06] QUANG L.H., CHUA S.J., LOH K.P., *et al.*, "The effect of post-annealing treatment on photoluminescence of ZnO nanorods prepared by hydrothermal synthesis", *Journal of Crystal Growth*, vol. 287, pp. 157–161, 2006.

[RAM 08] RAMIREZ D., PAUPORTÉ T., GOMEZ H., *et al.*, "Electrochemical mechanistic study of Zn and ZnO nanowires growth inside alumina templates", *Physica Status Solidi (a)*, vol. 205, pp. 2371–2375, 2008

[SAI 02] SAITO N., HANEDA H., SEKIGUCHI H., *et al.*, "Low-temperature fabrication of light emitting zinc oxide micropatterns using self assembled monolayers", *Advanced Materials*, vol. 14, pp. 418–421, 2002.

[SAI 04] SAITO N., HANEDA H., KOUMOTO K., "Pattern-deposition of light-emitting ZnO particulate film through biomimetic process using self assembled monolayer template", *Microelectronics J.*, vol. 35, pp.349–352, 2004.

[SEM 11] SEMET V., BINH V.T., PAUPORTÉ T., *et al.*, "Field emission behavior of vertically-aligned ZnO nanowire planar cathodes", *Journal of Applied Physics*, vol. 109, p. 054301, 2011.

[SHE 07] SHEN L.M., BAO N.Z., YANAGISAWA K., *et al.*, "Direct growth of comet-like superstructures of Au-ZnO submicron rod arrays by solvothermal soft chemistry process", *J. Solid State Chem.*, vol. 180, pp. 213–220, 2007.

[SIN 10] SINGH D.P., "Synthesis and growth of ZnO nanowires", *Science of Advanced Materials*, vol. 2, pp. 245–272, 2010.

[SON 07] SONG J., LIM S., "Effect of seed layer on the growth of ZnO nanorods", *The Journal of Physical Chemistry C*, vol. 111, pp.596–600, 2007.

[SUN 10] SUN K., JING Y., PARK N., *et al.*, "Solution synthesis of large-scale, high-sensitivity ZnO/Si hierarchical nanoheterostructure photodetectors", *J. Am. Chem. Soc.*, vol. 132, pp. 15465–15467, 2010.

[TEN 05] TENA-ZAERA R., KATTY A., BASTIDE S., *et al.*, "ZnO/CdTe/CuSCN, a promising heterostructure to act as inorganic eta-solar cell", *Thin Solid Films*, vol. 483, pp. 372–377, 2005.

[TEN 07] TENA-ZAERA R., ELIAS J., WANG G., *et al.*, "Role of chloride ions on the electrochemical deposition of ZnO nanowire arrays from O_2 reduction", *The Journal of Physical Chemistry C*, vol. 111, pp. 16706–16711, 2007.

[TIA 02] TIAN Z.R., VOIGT J.A., LIU J., *et al.*, "Biomimetic arrays of oriented helical ZnO nanorods and columns", *J. Am. Chem. Soc.*, vol. 124, pp. 12954–12955, 2002.

[TIA 03] TIAN Z.R., VOIGT J.A., LIU J., *et al.*, "Complex and oriented ZnO nanostructures", *Nat. Mater.*, vol. 2, pp. 821–826, 2003.

[UTA 12] UTAMA M.I.B., ZHANG J., CHEN R., *et al.*, "Synthesis and optical properties of II-VI 1D nanostructures", *Nanoscale*, vol.4, p.1422, 2012.

[VAY 01a] VAYSSIERES L., KEIS K., LINDQUIST S.E., *et al.*, "Purpose-built anisotropic metal oxide materials: 3D highly oriented microrod array of ZnO", *The Journal of Physical Chemistry B*, vol. 105, pp.3350–3352, 2001.

[VAY 01b] VAYSSIERES L., KEIS K., HAGFELDT A., *et al.*, "Three-dimensional array of highly oriented crystalline ZnO microtubes", *Chem. Mater.*, vol. 13, no. 12, pp. 4395–4398, 2001.

[VAY 03] VAYSSIERES L., "Growth of arrayed nanorods and nanowires of ZnO from aqueous solutions", *Advanced Materials*, vol. 15, pp. 464–466, 2003.

[WA 06] WANG J.X., SUN X.W., YANG Y., *et al.*., "Hydrothermally grown oriented ZnO nanorod arrays for gas sensing applications", *Nanotechnology*, vol. 17, pp. 4995–4998, 2006.

[WAN 05a] WANG Q, WANG G, JIE J, *et al.*, "Annealing effect on optical properties of ZnO films fabricated by cathodic electrodeposition", *Thin Solid Films*, vol. 492, pp. 61–65, 2005.

[WAN 05b] WANG Q., WANG G., XU B., *et al.*, "Non-aqueous cathodic electrodeposition of large-scale uniform ZnO nanowire arrays embedded in anodic alumina membrane", *Materials Letters*, vol. 59, pp. 1378–1382, 2005.

[WAN 07] WANG Z.L., "The new field of nanopiezotronics", *Mater. Today*, vol. 10, no. 5, pp. 20–28, 2007.

[WEI 06] WEI A., SUN X.W., WANG J.X., *et al.* "Enzymatic glucose biosensor based on ZnO nanorod array grown by hydrothermal decomposition", *Applied Physics Letters*, vol. 89, p. 123902, 2006.

[WU 02] WU Y.Y., YAN H.Q., Yang P.D., "Semiconductor nanowire array: potential substrates for photocatalysis and photovoltaics", *Topics in Catalysis*, vol. 19, no. 2, pp. 197–202, 2002.

[WU 08] WU W.B., HU G.D., CUI S.G., *et al.*, "Epitaxy of vertical ZnO nanorod arrays on highly (001)-oriented ZnO seed monolayer by a hydrothermal route", *Cryst. Growth Des.*, vol. 8, pp. 4014–4020, 2008.

[XU 05a] XU, L.F., GUO Y., LIAO, Q., *et al.*, "Morphological control of ZnO nanostructures by electrodeposition", *The Journal of Physical Chemistry B*, vol. 109, pp. 13519–13522, 2005.

[XU 05b] XU L., GUO Y., LIAO Q., *et al.*, "Morphological control of ZnO nanostructures by electrodeposition", *The Journal of Physical Chemistry B*, vol. 109, pp. 13519–13522, 2005.

[XU 05c] XU N.S., HUQ S.E., "Novel cold cathode materials and applications", *Mat. Sci. Eng. R*, vol. 48, pp. 47–189, 2005.

[XU 08a] XU S., LAO C., WEINTRAUB B., *et al.*, "Density-controlled growth of aligned ZnO nanowire arrays by seedless chemical approach on smooth surfaces", *J. Mater. Res.*, vol. 23, no. 8, pp. 2072–2077, 2008.

[XU 08b] XU S., WEI Y., KIRKHAM M., *et al.*, "Patterned growth of vertically aligned ZnO nanowire arrays on inorganic substrates at low temperature without catalyst", *J. Am. Chem. Soc.*, vol. 130, no. 45, pp. 14958–14959, 2008.

[XU 10] XU C.K., SHIN P., CAO L.L., *et al.*, "Preferential growth of long ZnO nanowire array and its application in dye-sensitized solar cells", *The Journal of Physical Chemistry C*, vol. 114, pp. 125–129, 2010.

[XU 11a] XU S., WANG Z.L., "One-dimensional ZnO nanostructures: solution growth and functional properties", *Nano Res.*, vol. 4, no. 11, pp. 1013–1098, 2011.

[XU 11b] XU C., WU J., DESAI U.V., *et al.*, "Multilayer assembly of nanowire arrays for dye-sensitized solar cells", *J. Am. Chem. Soc.*, vol. 133, pp. 8122–8125, 2011.

[YAH 06] YAHIRO J., OAKI Y., IMAI H., "Biomimetic synthesis of wurtzite ZnO nanowires possessing a mosaic structure", *Small*, vol. 2, pp. 1183–1187, 2006.

[YAH 07] YAHIRO J., KAWANO T., IMAI H., "Nanometric morphological variation of zinc oxide crystals using organic molecules with carboxy and sulfonic groups", *Journal of Colloids & Interface Science*, vol. 310, p. 302311, 2007.

[YAM 02] YAMABI S., IMAI H., "Growth conditions for wurtzite zinc oxide films in aqueous solutions", *Journal of Materials Chemistry*, vol. 12, pp. 3773–3778, 2002.

[YAN 08] YANG Z., LIU Q.H., YU H.C., *et al.*, "Substrate-free growth, characterization and growth mechanism of ZnO nanorod close-packed arrays", *Nanotechnology*, vol. 19, p. 035704, 2008.

[YAN 09] YANG L.L., ZHAO Q.X., WILLANDER M., *et al.*, "Effective way to control the size of well-aligned ZnO nanorod arrays with two-step chemical bath deposition", *Journal of Crystal Growth*, vol. 311, pp.1046–1050, 2009.

[YIN 04] YIN M., GU Y., KUSKOVSKY I.L., *et al.*, "Zinc oxide quantum rods", *J. Am. Chem. Soc.*, vol. 126, pp. 6206–6207, 2004.

[YOS 04] YOSHIDA T., KOMATSU D., SHIMOKAWA N., *et al.*, "Mechanism of cathodic electrodeposition of zinc oxide thin films from aqueous zinc nitrate baths", *Thin Solid Films*, vol. 451–452, pp. 166–169, 2004.

[ZAN 07] ZANG J.F., LI C.M., CUI X.Q., *et al.*, "Tailoring zinc oxide nanowires for high performance amperometric glucose sensor", *Electroanal.*, vol.19, pp. 1008–1014, 2007.

[ZHA 02] ZHANG J., SUN L.D., YIN J.L., *et al.*, "Control of ZnO morphology via a simple solution route", *Chem. Mater.*, vol. 14, pp. 4172–4177, 2002.

[ZHA 05a] ZHANG D.F., SUN L.D., YIN J.L., *et al.*., "Attachment-driven morphology evolvement of rectangular ZnO nanowires", *The Journal of Physical Chemistry B*, vol. 109, pp. 8786–8790, 2005.

[ZHA 05b] ZHANG H., YANG D.R., LI D.S., *et al.*, "Controllable growth of ZnO microcrystals by a capping-molecule-assisted hydrothermal process", *Cryst. Growth Des.*, vol. 5, pp. 547–550, 2005.

[ZHA 05c] ZHANG L., CHEN Z., TANG Y., *et al.*, "Low temperature cathodic electrodeposition of nanocrystalline zinc oxide thin films", *Thin Solid Films*, vol. 492, pp. 24–29, 2005.

[ZHA 13] ZHANG H., BABICHEV A.V., JACOPIN G., *et al.*, "Characterization and modeling of a ZnO nanowire ultraviolet photodetector with a graphene transparent contact", *Journal of Applied Physics*, vol. 114, p. 234505, 2013.

[ZHE 02] ZHENG M.J., ZHANG L.D., LI G.H., *et al.*, "Fabrication and optical properties of large-scale uniform zinc oxide nanowire arrays by one-step electrochemical deposition technique", *Chem. Phys. Lett.*, vol. 363, pp. 123–128, 2002.

[ZHO 08] ZHOU Y., WU W.B., HU G.D., *et al.*, "Hydrothermal synthesis of ZnO nanorod arrays with the addition of polyethyleneimine", *Mater. Res. Bull.*, vol. 43, pp. 2113–2118, 2008.

[ZHO 09] ZHOU Z., DENG Y., "Kinetics study of ZnO nanorod growth in solution", *The Journal of Physical Chemistry C*, vol. 113, pp. 19853–19858, 2009.

[ZHU 10] ZHU G.A., YANG R.S., WANG S.H., *et al.*, "Flexible high-output nanogenerator based on lateral ZnO nanowire array", *NanoLetters*, vol. 10, no. 8, pp. 3151–3155, 2010.

List of Authors

Steven ALBERT
ISOM and Departamento de
Ingenieria Electronica Universidad
Politécnica de Madrid
Ciudad Universitaria
Madrid
Spain

Francesca BARBAGINI
ISOM and Departamento de
Ingenieria Electronica Universidad
Politécnica de Madrid
Ciudad Universitaria
Madrid
Spain

Ana M. BENGOECHEA-ENCABO
ISOM and Departamento de
Ingenieria Electronica Universidad
Politécnica de Madrid
Ciudad Universitaria
Madrid
Spain

Enrique CALLEJA
ISOM and Departamento de
Ingenieria Electronica Universidad
Politécnica de Madrid
Ciudad Universitaria
Madrid
Spain

Caroline CHÈZE
Paul-Drude-Institut für
Festkörperelektronik
Berlin
Germany

Vincent CONSONNI
Univ. Grenoble Alpes, LMGP
CNRS, LMGP
Grenoble
France

Le Si DANG
Univ. Grenoble Alpes, Inst Néel
CNRS, Inst Néel
Grenoble
France

Christoph Peter DIETRICH
Institut für Experimentelle Physik II
Universität Leipzig
Leipzig
Germany

Fabrice DONATINI
Univ. Grenoble Alpes, Inst Néel
CNRS, Inst Néel
Grenoble
France

Joël EYMERY
Univ. Grenoble Alpes, INAC-SP2M
CEA, INAC-SP2M
Grenoble
France

Guy FEUILLET
CEA, LETI
MINATEC Campus
Grenoble
France

Frank GLAS
CNRS
Laboratoire de Photonique et de Nanostructures
Marcoussis
France

Javier GRANDAL
Paul-Drude-Institut für Festkörperelektronik
Berlin
Germany

Marius GRUNDMANN
Institut für Experimentelle Physik II
Universität Leipzig
Leipzig
Germany

Jürgen GUTOWSKI
Institute of Solid State Physics
University of Bremen
Bremen
Germany

Bernd JENICHEN
Paul-Drude-Institut für Festkörperelektronik
Berlin
Germany

Xiang KONG
Paul-Drude-Institut für Festkörperelektronik
Berlin
Germany

Pierre LEFEBVRE
CNRS
Laboratoire Charles Coulomb
UMR 5221
Montpellier
France

Esperanza LUNA
Paul-Drude-Institut für Festkörperelektronik
Berlin
Germany

Thierry PAUPORTÉ
Institut de recherche Chimie-Paristech (IRCP)
CNRS-UMR 8247
Ecole Nationale Supérieure de Chimie de Paris
Paris
France

Julien PERNOT
Univ. Grenoble Alpes, Inst Néel
CNRS, Inst Néel
Grenoble
France

Vincent SALLET
Groupe d'étude de la matière condensée (GEMAC)
CNRS - Université de Versailles St Quentin
Versailles
France

Miguel A. SANCHEZ-GARCIA
ISOM and Departamento de Ingenieria Electronica
Universidad Politécnica de Madrid
Ciudad Universitaria
Madrid
Spain

Pierre TCHOULFIAN
Univ. Grenoble Alpes, Inst Néel
CNRS, Inst Néel
Grenoble
France

Achim TRAMPERT
Paul-Drude-Institut für Festkörperelektronik
Berlin
Germany

Tobias VOSS
Institute of Solid State Physics
University of Bremen
Bremen
Germany